Lecture Notes in Computer Science 14608

The series Lecture Notes in Computer Science (LNCS), including its subseries Lecture Notes in Artificial Intelligence (LNAI) and Lecture Notes in Bioinformatics (LNBI), has established itself as a medium for the publication of new developments in computer science and information technology research, teaching, and education.

LNCS enjoys close cooperation with the computer science R & D community, the series counts many renowned academics among its volume editors and paper authors, and collaborates with prestigious societies. Its mission is to serve this international community by providing an invaluable service, mainly focused on the publication of conference and workshop proceedings and postproceedings. LNCS commenced publication in 1973.

Nazli Goharian · Nicola Tonellotto · Yulan He ·
Aldo Lipani · Graham McDonald ·
Craig Macdonald · Iadh Ounis
Editors

Advances in Information Retrieval

46th European Conference on Information Retrieval, ECIR 2024
Glasgow, UK, March 24–28, 2024
Proceedings, Part I

Editors
Nazli Goharian
Georgetown University
Washington, WA, USA

Nicola Tonellotto
University of Pisa
Pisa, Italy

Yulan He
King's College London
London, UK

Aldo Lipani
University College London
London, UK

Graham McDonald
University of Glasgow
Glasgow, UK

Craig Macdonald
University of Glasgow
Glasgow, UK

Iadh Ounis
University of Glasgow
Glasgow, UK

ISSN 0302-9743 ISSN 1611-3349 (electronic)
Lecture Notes in Computer Science
ISBN 978-3-031-56026-2 ISBN 978-3-031-56027-9 (eBook)
https://doi.org/10.1007/978-3-031-56027-9

This Springer imprint is published by the registered company Springer Nature Switzerland AG
The registered company address is: Gewerbestrasse 11, 6330 Cham, Switzerland

Paper in this product is recyclable.

Preface

The 46th European Conference on Information Retrieval (ECIR 2024) was held in Glasgow, Scotland, UK, during March 24–28, 2024, and brought together hundreds of researchers from the UK, Europe and abroad. The conference was organised by the University of Glasgow, in cooperation with the British Computer Society's Information Retrieval Specialist Group (BCS IRSG) and with assistance from the Glasgow Convention Bureau.

These proceedings contain the papers related to the presentations, workshops, tutorials, doctoral consortium and other satellite tracks that took place during the conference. This year's ECIR program boasted a variety of novel work from contributors from all around the world. In addition, we introduced a number of novelties in this year's ECIR. First, ECIR 2024 included for the first time a new "Findings" track, which was offered to some full papers that were deemed to be solid, but which could not make the main conference track. Second, ECIR 2024 ran a new special IR4Good track that presented high-quality, high-impact, original IR-related research on societal issues (such as algorithmic bias and fairness, privacy, and transparency) at the interdisciplinary level (e.g., philosophy, law, sociology, civil society), which go beyond the purely technical perspective. Third, ECIR 2024 featured a new innovation called the "Collab-a-thon", intended to provide an opportunity for participants to foster new collaborations that could lead to exciting new research, and forge lasting relationships with like-minded researchers. Finally, ECIR 2024 introduced a new award to encourage and recognise researchers who have made significant contributions in using theory to develop the information retrieval field. The award was named after Professor Cornelis "Keith" van Rijsbergen (University of Glasgow), a pioneer in modern information retrieval, and a strong advocate of the development of models and theories in information retrieval.

The ECIR 2024 program featured a total of 578 papers from authors in 61 countries in its various tracks. The final program included 57 full papers (23% acceptance rate), an additional 18 finding papers, 36 short papers (24% acceptance rate), 26 IR4Good papers (41%), 18 demonstration papers (56% acceptance rate), 9 reproducibility papers (39% acceptance rate), 8 doctoral consortium papers (57% acceptance rate), and 15 invited CLEF papers. All submissions were peer-reviewed by at least three international Program Committee members to ensure that only submissions of the highest relevance and quality were included in the final ECIR 2024 program. The acceptance decisions were further informed by discussions among the reviewers for each submitted paper, led by a Senior Program Committee member. Each track had a final PC meeting where final recommendations were discussed and made, trying to reach a fair and equal outcome for all submissions.

The accepted papers cover the state-of-the-art in information retrieval and recommender systems: user aspects, system and foundational aspects, artificial intelligence & machine learning, applications, evaluation, new social and technical challenges, and other topics of direct or indirect relevance to search and recommendation. As in previous years, the ECIR 2024 program contained a high proportion of papers with students as first authors, as well as papers from a variety of universities, research institutes, and commercial organisations.

In addition to the papers, the program also included 4 keynotes, 7 tutorials, 10 workshops, a doctoral consortium, an IR4Good event, a Collab-a-thon and an industry day. Keynote talks were given by Charles L. A. Clarke (University of Waterloo), Josiane Mothe (Université de Toulouse), Carlos Castillo (Universitat Pompeu Fabra), and this year's Keith van Rijsbergen Award winner, Maarten de Rijke (University of Amsterdam). The tutorials covered a range of topics including explainable recommender systems, sequential recommendation, social good applications, quantum for IR, generative IR, query performance prediction and PhD advice. The workshops brought together participants to discuss narrative extraction (Text2Story), knowledge-enhanced retrieval (KEIR), online misinformation (ROMCIR), understudied users (IR4U2), graph-based IR (IRonGraphs), open web search (WOWS), technology-assisted review (ALTARS), geographic information extraction (GeoExT), bibliometrics (BIR) and search futures (SearchFutures).

The success of ECIR 2024 would not have been possible without all the help from the strong team of volunteers and reviewers. We wish to thank all the reviewers and meta-reviewers who helped to ensure the high quality of the program. We also wish to thank: the reproducibility track chairs Claudia Hauff and Hamed Zamani, the IR4Good track chairs Ludovico Boratto and Mirko Marras, the demo track chairs Giorgio Maria Di Nunzio and Chiara Renso, the industry day chairs Olivier Jeunen and Isabelle Moulinier, the doctoral consortium chairs Yashar Moshfeghi and Gabriella Pasi, the CLEF Labs chair Jake Lever, the workshop chairs Elisabeth Lex, Maria Maistro and Martin Potthast, the tutorial chairs Mohammad Aliannejadi and Johanne R. Trippas, the Collab-a-thon chair Sean MacAvaney, the best paper awards committee chair Raffaele Perego, the sponsorship chairs Dyaa Albakour and Eugene Kharitonov, the proceeding chairs Debasis Ganguly and Richard McCreadie, and the local organisation chairs Zaiqiao Meng and Hitarth Narvala. We would also like to thank all the student volunteers who worked hard to ensure an excellent and memorable experience for participants and attendees. ECIR 2024 was sponsored by a range of learned societies, research institutes and companies. We thank them all for their support. Finally, we wish to thank all of the authors and contributors to the conference.

March 2024

Nazli Goharian
Nicola Tonellotto
Yulan He
Aldo Lipani
Graham McDonald
Craig Macdonald
Iadh Ounis

Organization

General Chairs

Craig Macdonald	University of Glasgow, UK
Graham McDonald	University of Glasgow, UK
Iadh Ounis	University of Glasgow, UK

Program Chairs – Full Papers

Nazli Goharian	Georgetown University, USA
Nicola Tonellotto	University of Pisa, Italy

Program Chairs – Short Papers

Yulan He	King's College London, UK
Aldo Lipani	University College London, UK

Reproducibility Track Chairs

Claudia Hauff	Spotify & TU Delft, Netherlands
Hamed Zamani	University of Massachusetts Amherst, USA

IR4Good Chairs

Ludovico Boratto	University of Cagliari, Italy
Mirko Marras	University of Cagliari, Italy

Demo Chairs

Giorgio Maria Di Nunzio	Università degli Studi di Padova, Italy
Chiara Renso	ISTI - CNR, Italy

Industry Day Chairs

Olivier Jeunen	ShareChat, UK
Isabelle Moulinier	Thomson Reuters, USA

Doctoral Consortium Chairs

Yashar Moshfeghi	University of Strathclyde, UK
Gabriella Pasi	Università degli Studi di Milano Bicocca, Italy

CLEF Labs Chair

Jake Lever	University of Glasgow, UK

Workshop Chairs

Elisabeth Lex	Graz University of Technology, Austria
Maria Maistro	University of Copenhagen, Denmark
Martin Potthast	Leipzig University, Germany

Tutorial Chairs

Mohammad Aliannejadi	University of Amsterdam, Netherlands
Johanne R. Trippas	RMIT University, Australia

Collab-a-thon Chair

Sean MacAvaney	University of Glasgow, UK

Best Paper Awards Committee Chair

Raffaele Perego	ISTI-CNR, Italy

Sponsorship Chairs

Dyaa Albakour	Signal AI, UK
Eugene Kharitonov	Google, France

Proceeding Chairs

Debasis Ganguly	University of Glasgow, UK
Richard McCreadie	University of Glasgow, UK

Local Organisation Chairs

Zaiqiao Meng	University of Glasgow, UK
Hitarth Narvala	University of Glasgow, UK

Senior Program Committee

Mohammad Aliannejadi	University of Amsterdam, Netherlands
Omar Alonso	Amazon, USA
Giambattista Amati	Fondazione Ugo Bordoni, Italy
Ioannis Arapakis	Telefonica Research, Spain
Jaime Arguello	The University of North Carolina at Chapel Hill, USA
Javed Aslam	Northeastern University, USA
Krisztian Balog	University of Stavanger & Google Research, Norway
Patrice Bellot	Aix-Marseille Université CNRS (LSIS), France
Michael Bendersky	Google, USA
Mohand Boughanem	IRIT University Paul Sabatier Toulouse, France
Jamie Callan	Carnegie Mellon University, USA
Charles Clarke	University of Waterloo, Canada
Fabio Crestani	Università della Svizzera italiana (USI), Switzerland
Bruce Croft	University of Massachusetts Amherst, USA
Maarten de Rijke	University of Amsterdam, Netherlands
Arjen de Vries	Radboud University, Netherlands
Tommaso Di Noia	Politecnico di Bari, Italy
Carsten Eickhoff	University of Tübingen, Germany
Tamer Elsayed	Qatar University, Qatar

Liana Ermakova	HCTI/Université de Bretagne Occidentale, France
Hui Fang	University of Delaware, USA
Nicola Ferro	University of Padova, Italy
Norbert Fuhr	University of Duisburg-Essen, Germany
Debasis Ganguly	University of Glasgow, UK
Lorraine Goeuriot	Université Grenoble Alpes (CNRS), France
Marcos Goncalves	Federal University of Minas Gerais, Brazil
Julio Gonzalo	UNED, Spain
Jiafeng Guo	Institute of Computing Technology, China
Matthias Hagen	Friedrich-Schiller-Universität, Germany
Allan Hanbury	TU Wien, Austria
Donna Harman	NIST, USA
Claudia Hauff	Spotify, Netherlands
Jiyin He	Signal AI, UK
Ben He	University of Chinese Academy of Sciences, China
Dietmar Jannach	University of Klagenfurt, Germany
Adam Jatowt	University of Innsbruck, Austria
Gareth Jones	Dublin City University, Ireland
Joemon Jose	University of Glasgow, UK
Jaap Kamps	University of Amsterdam, Netherlands
Jussi Karlgren	SiloGen, Finland
Udo Kruschwitz	University of Regensburg, Germany
Jochen Leidner	Coburg University of Applied Sciences, Germany
Yiqun Liu	Tsinghua University, China
Sean MacAvaney	University of Glasgow, UK
Craig Macdonald	University of Glasgow, UK
Joao Magalhaes	Universidade NOVA de Lisboa, Portugal
Giorgio Maria Di Nunzio	University of Padua, Italy
Philipp Mayr	GESIS, Germany
Donald Metzler	Google, USA
Alistair Moffat	The University of Melbourne, Australia
Yashar Moshfeghi	University of Strathclyde, UK
Henning Müller	HES-SO, Switzerland
Julián Urbano	Delft University of Technology, Netherlands
Marc Najork	Google, USA
Jian-Yun Nie	Université de Montreal, Canada
Harrie Oosterhuis	Radboud University, Netherlands
Iadh Ounis	University of Glasgow, UK
Javier Parapar	University of A Coruña, Spain
Gabriella Pasi	University of Milano Bicocca, Italy
Raffaele Perego	ISTI-CNR, Italy

Benjamin Piwowarski	CNRS/ISIR/Sorbonne Université, France
Paolo Rosso	Universitat Politècnica de València, Spain
Mark Sanderson	RMIT University, Australia
Philipp Schaer	TH Köln (University of Applied Sciences), Germany
Ralf Schenkel	Trier University, Germany
Christin Seifert	University of Marburg, Germany
Gianmaria Silvello	University of Padua, Italy
Fabrizio Silvestri	University of Rome, Italy
Mark Smucker	University of Waterloo, Canada
Laure Soulier	Sorbonne Université-ISIR, France
Torsten Suel	New York University, USA
Hussein Suleman	University of Cape Town, South Africa
Paul Thomas	Microsoft, USA
Theodora Tsikrika	Information Technologies Institute/CERTH, Greece
Suzan Verberne	LIACS/Leiden University, Netherlands
Marcel Worring	University of Amsterdam, Netherlands
Andrew Yates	University of Amsterdam, Netherlands
Shuo Zhang	Bloomberg, UK
Min Zhang	Tsinghua University, China
Guido Zuccon	The University of Queensland, Australia

Program Committee

Amin Abolghasemi	Leiden University, Netherlands
Sharon Adar	Amazon, USA
Shilpi Agrawal	Linkedin, USA
Mohammad Aliannejadi	University of Amsterdam, Netherlands
Satya Almasian	Heidelberg University, Germany
Giuseppe Amato	ISTI-CNR, Italy
Linda Andersson	Artificial Researcher IT GmbH TU Wien, Austria
Negar Arabzadeh	University of Waterloo, Canada
Marcelo Armentano	ISISTAN (CONICET - UNCPBA), Argentina
Arian Askari	Leiden University, Netherlands
Maurizio Atzori	University of Cagliari, Italy
Sandeep Avula	Amazon, USA
Hosein Azarbonyad	Elsevier, Netherlands
Leif Azzopardi	University of Strathclyde, UK
Andrea Bacciu	Sapienza University of Rome, Italy
Mossaab Bagdouri	Walmart Global Tech, USA

Ebrahim Bagheri	Ryerson University, Canada
Seyedeh Baharan Khatami	University of California San Diego, USA
Giacomo Balloccu	Università degli Studi di Cagliari, Italy
Alberto Barrón-Cedeño	Università di Bologna, Italy
Alvaro Barreiro	University of A Coruña, Spain
Roberto Basili	University of Roma Tor Vergata, Italy
Elias Bassani	Independent Researcher, Italy
Christine Bauer	Paris Lodron University Salzburg, Austria
Alejandro Bellogin	Universidad Autonoma de Madrid, Spain
Alessandro Benedetti	Sease, UK
Klaus Berberich	Saarbruecken University of Applied Sciences, Germany
Arjun Bhalla	Bloomberg LP, USA
Sumit Bhatia	Adobe Inc., India
Veronika Bogina	Tel Aviv University, Israel
Alexander Bondarenko	Friedrich-Schiller-Universität Jena, Germany
Ludovico Boratto	University of Cagliari, Italy
Gloria Bordogna	National Research Council of Italy - CNR, France
Emanuela Boros	EPFL, Switzerland
Leonid Boytsov	Amazon, USA
Martin Braschler	Zurich University of Applied Sciences, Switzerland
Pavel Braslavski	Ural Federal University, Russia
Timo Breuer	TH Köln (University of Applied Sciences), Germany
Sebastian Bruch	Pinecone, USA
Arthur Câmara	Zeta Alpha Vector, Netherlands
Fidel Cacheda	Universidade da Coruña, Spain
Sylvie Calabretto	LIRIS, France
Cesare Campagnano	Sapienza University of Rome, Italy
Ricardo Campos	University of Beira Interior, Portugal
Iván Cantador	Universidad Autónoma de Madrid, Spain
Alberto Carlo Maria Mancino	Politecnico di Bari, Italy
Matteo Catena	Siren, Italy
Abhijnan Chakraborty	IIT Delhi, India
Khushhall Chandra Mahajan	Meta Inc., USA
Shubham Chatterjee	The University of Edinburgh, UK
Despoina Chatzakou	Information Technologies Institute, Greece
Catherine Chavula	University of Texas at Austin, USA
Catherine Chen	Brown University, USA
Jean-Pierre Chevallet	Grenoble Alpes University, France
Adrian-Gabriel Chifu	Aix Marseille University, France

Evgenia Christoforou	CYENS Centre of Excellence, Cyprus
Abu Nowshed Chy	University of Chittagong, Bangladesh
Charles Clarke	University of Waterloo, Canada
Stephane Clinchant	Naver Labs Europe, France
Fabio Crestani	Università della Svizzera Italiana (USI), Switzerland
Shane Culpepper	The University of Queensland, Australia
Hervé Déjean	Naver Labs Europe, France
Célia da Costa Pereira	Université Côte d'Azur, France
Maarten de Rijke	University of Amsterdam, Netherlands
Arjen De Vries	Radboud University, Netherlands
Amra Deli	University of Sarajevo, Bosnia and Herzegovina
Gianluca Demartini	The University of Queensland, Australia
Danilo Dess	Leibniz Institute for the Social Sciences, Germany
Emanuele Di Buccio	University of Padua, Italy
Gaël Dias	Normandie University, France
Vlastislav Dohnal	Masaryk University, Czechia
Gregor Donabauer	University of Regensburg, Germany
Zhicheng Dou	Renmin University of China, China
Carsten Eickhoff	University of Tübingen, Germany
Michael Ekstrand	Drexel University, USA
Dima El Zein	Université Côte d'Azur, France
David Elsweiler	University of Regensburg, Germany
Ralph Ewerth	Leibniz Universität Hannover, Germany
Michael Färber	Karlsruhe Institute of Technology, Germany
Guglielmo Faggioli	University of Padova, Italy
Fabrizio Falchi	ISTI-CNR, Italy
Zhen Fan	Carnegie Mellon University, USA
Anjie Fang	Amazon.com, USA
Hossein Fani	University of Windsor, UK
Henry Field	Endicott College, USA
Yue Feng	UCL, UK
Marcos Fernández Pichel	Universidade de Santiago de Compostela, Spain
Antonio Ferrara	Polytechnic University of Bari, Italy
Komal Florio	Università di Torino - Dipartimento di Informatica, Italy
Thibault Formal	Naver Labs Europe, France
Eduard Fosch Villaronga	Leiden University, Netherlands
Maik Fröbe	Friedrich-Schiller-Universität Jena, Germany
Giacomo Frisoni	University of Bologna, Italy
Xiao Fu	University College London, UK
Norbert Fuhr	University of Duisburg-Essen, Germany

Petra Galuščáková	University of Stavanger, Norway
Debasis Ganguly	University of Glasgow, UK
Eric Gaussier	LIG-UGA, France
Xuri Ge	University of Glasgow, UK
Thomas Gerald	Université Paris Saclay CNRS SATT LISN, France
Kripabandhu Ghosh	ISSER, India
Satanu Ghosh	University of New Hampshire, USA
Daniela Godoy	ISISTAN (CONICET - UNCPBA), Argentina
Carlos-Emiliano González-Gallardo	L3i, France
Michael Granitzer	University of Passau, Germany
Nina Grgic-Hlaca	Max Planck Institute for Software Systems, Germany
Adrien Guille	Université de Lyon, France
Chun Guo	Pandora Media LLC, USA
Shashank Gupta	University of Amsterdam, Netherlands
Matthias Hagen	Friedrich-Schiller-Universität Jena, Germany
Fatima Haouari	Qatar University, Qatar
Maram Hasanain	Qatar University, Qatar
Claudia Hauff	Spotify, Netherlands
Naieme Hazrati	Free University of Bozen-Bolzano, Italy
Daniel Hienert	Leibniz Institute for the Social Sciences, Germany
Frank Hopfgartner	Universität Koblenz, Germany
Gilles Hubert	IRIT, France
Oana Inel	University of Zurich, Switzerland
Bogdan Ionescu	Politehnica University of Bucharest, Romania
Thomas Jaenich	University of Glasgow, UK
Shoaib Jameel	University of Southampton, UK
Faizan Javed	Kaiser Permanente, USA
Olivier Jeunen	ShareChat, UK
Alipio Jorge	University of Porto, Portugal
Toshihiro Kamishima	AIST, Japan
Noriko Kando	National Institute of Informatics, Japan
Sarvnaz Karimi	CSIRO, Australia
Pranav Kasela	University of Milano-Bicocca, Italy
Sumanta Kashyapi	University of New Hampshire, USA
Christin Katharina Kreutz	Cologne University of Applied Sciences, Germany
Abhishek Kaushik	Dublin City University, Ireland
Mesut Kaya	Aalborg University Copenhagen, Denmark
Diane Kelly	University of Tennessee, USA

Jae Keol Choi	Seoul National University, South Korea
Roman Kern	Graz University of Technology, Austria
Pooya Khandel	University of Amsterdam, Netherlands
Johannes Kiesel	Bauhaus-Universität, Germany
Styliani Kleanthous	CYENS CoE & Open University of Cyprus, Cyprus
Anastasiia Klimashevskaia	University of Bergen, Italy
Ivica Kostric	University of Stavanger, Norway
Dominik Kowald	Know-Center & Graz University of Technology, Austria
Hermann Kroll	Technische Universität Braunschweig, Germany
Udo Kruschwitz	University of Regensburg, Germany
Hrishikesh Kulkarni	Georgetown University, USA
Wojciech Kusa	TU Wien, Austria
Mucahid Kutlu	TOBB University of Economics and Technology, Turkey
Saar Kuzi	Amazon, USA
Jochen L. Leidner	Coburg University of Applied Sciences, Germany
Kushal Lakhotia	Outreach, USA
Carlos Lassance	Naver Labs Europe, France
Aonghus Lawlor	University College Dublin, Ireland
Dawn Lawrie	Johns Hopkins University, USA
Chia-Jung Lee	Amazon, USA
Jurek Leonhardt	TU Delft, Germany
Monica Lestari Paramita	University of Sheffield, UK
Hang Li	The University of Queensland, Australia
Ming Li	University of Amsterdam, Netherlands
Qiuchi Li	University of Padua, Italy
Wei Li	University of Roehampton, UK
Minghan Li	University of Waterloo, Canada
Shangsong Liang	MBZUAI, UAE
Nut Limsopatham	Amazon, USA
Marina Litvak	Shamoon College of Engineering, Israel
Siwei Liu	MBZUAI, UAE
Haiming Liu	University of Southampton, UK
Yiqun Liu	Tsinghua University, China
Bulou Liu	Tsinghua University, China
Andreas Lommatzsch	TU Berlin, Germany
David Losada	University of Santiago de Compostela, Spain
Jesus Lovon-Melgarejo	Université Paul Sabatier IRIT, France
Alipio M. Jorge	University of Porto, Portugal
Weizhi Ma	Tsinghua University, China

Name	Affiliation
Joel Mackenzie	The University of Queensland, Australia
Daniele Malitesta	Polytechnic University of Bari, Italy
Antonio Mallia	New York University, USA
Behrooz Mansouri	University of Southern Maine, USA
Masoud Mansoury	University of Amsterdam, Netherlands
Jiaxin Mao	Renmin University of China, China
Stefano Marchesin	University of Padova, Italy
Giorgio Maria Di Nunzio	University of Padua, Italy
Franco Maria Nardini	ISTI-CNR, Italy
Mirko Marras	University of Cagliari, Italy
Monica Marrero	Europeana Foundation, Netherlands
Bruno Martins	University of Lisbon, Portugal
Flavio Martins	Instituto Superior Técnico, Lisbon
David Massimo	Free University of Bolzano, Italy
Noemi Mauro	University of Turin, Italy
Richard McCreadie	University of Glasgow, UK
Graham McDonald	University of Glasgow, UK
Giacomo Medda	University of Cagliari, Italy
Parth Mehta	IRSI, India
Ida Mele	IASI-CNR, Italy
Chuan Meng	University of Amsterdam, Netherlands
Zaiqiao Meng	University of Glasgow, UK
Tristan Miller	University of Manitoba, Canada
Alistair Moffat	The University of Melbourne, Australia
Jose Moreno	IRIT/UPS, France
Gianluca Moro	University of Bologna, Italy
Josiane Mothe	Univ. Toulouse, France
Philippe Mulhem	LIG-CNRS, France
Cataldo Musto	University of Bari, Italy
Suraj Nair	University of Maryland, USA
Hitarth Narvala	University of Glasgow, UK
Julia Neidhardt	TU Wien, Austria
Wolfgang Nejdl	L3S and University of Hannover, Germany
Thong Nguyen	University of Amsterdam, Netherlands
Diana Nurbakova	INSA Lyon, France
Hiroaki Ohshima	Graduate School of Information Science, Japan
Harrie Oosterhuis	Radboud University, Netherlands
Salvatore Orlando	Università Ca' Foscari Venezia, Italy
Panagiotis Papadakos	Information Systems Laboratory - FORTH-ICS, Greece
Andrew Parry	University of Glasgow, UK
Pavel Pecina	Charles University, Czechia

Georgios Peikos	University of Milano-Bicocca, Italy
Gustavo Penha	Spotify Research, Netherlands
Marinella Petrocchi	IIT-CNR, Italy
Aleksandr Petrov	University of Glasgow, UK
Milo Phillips-Brown	University of Edinburgh, UK
Karen Pinel-Sauvagnat	IRIT, France
Florina Piroi	Vienna University of Technology, Austria
Alessandro Piscopo	BBC, UK
Marco Polignano	Università degli Studi di Bari Aldo Moro, Italy
Claudio Pomo	Polytechnic University of Bari, Italy
Lorenzo Porcaro	Joint Research Centre European Commission, Italy
Amey Porobo Dharwadker	Meta, USA
Martin Potthast	Leipzig University, Germany
Erasmo Purificato	Otto von Guericke University Magdeburg, Germany
Xin Qian	University of Maryland, USA
Yifan Qiao	University of California, USA
Georges Quénot	Laboratoire d'Informatique de Grenoble CNRS, Germany
Alessandro Raganato	University of Milano-Bicocca, Italy
Fiana Raiber	Yahoo Research, Israel
Amifa Raj	Boise State University, USA
Thilina Rajapakse	University of Amsterdam, Netherlands
Jerome Ramos	University College London, UK
David Rau	University of Amsterdam, Netherlands
Gábor Recski	TU Wien, Austria
Navid Rekabsaz	Johannes Kepler University Linz, Austria
Zhaochun Ren	Leiden University, Netherlands
Yongli Ren	RMIT University, Australia
Weilong Ren	Shenzhen Institute of Computing Sciences, China
Chiara Renso	ISTI-CNR, Italy
Kevin Roitero	University of Udine, Italy
Tanya Roosta	Amazon, USA
Cosimo Rulli	University of Pisa, Italy
Valeria Ruscio	Sapienza University of Rome, Italy
Yuta Saito	Cornell University, USA
Tetsuya Sakai	Waseda University, Japan
Shadi Saleh	Microsoft, USA
Eric Sanjuan	Avignon Université, France
Javier Sanz-Cruzado	University of Glasgow, UK
Fabio Saracco	Centro Ricerche Enrico Fermi, Italy

Harrisen Scells	Leipzig University, Germany
Philipp Schaer	TH Köln (University of Applied Sciences), Germany
Jörg Schlötterer	University of Marburg, Germany
Ferdinand Schlatt	Friedrich-Schiller-Universität Jena, Germany
Christin Seifert	University of Marburg, Germany
Giovanni Semeraro	University of Bari, Italy
Procheta Sen	University of Liverpool, UK
Ismail Sengor Altingovde	Bilkent University, Türkiye
Vinay Setty	University of Stavanger, Norway
Mahsa Shahshahani	Accenture, Netherlands
Zhengxiang Shi	University College London, UK
Federico Siciliano	Sapienza University of Rome, Italy
Gianmaria Silvello	University of Padua, Italy
Jaspreet Singh	Amazon, USA
Sneha Singhania	Max Planck Institute for Informatics, Germany
Manel Slokom	Delft University of Technology, Netherlands
Mark Smucker	University of Waterloo, Canada
Maria Sofia Bucarelli	Sapienza University of Rome, Italy
Maria Soledad Pera	TU Delft, Germany
Nasim Sonboli	Brown University, USA
Zhihui Song	University College London, UK
Arpit Sood	Meta Inc, USA
Sajad Sotudeh	Georgetown University, USA
Laure Soulier	Sorbonne Université-ISIR, France
Marc Spaniol	Université de Caen Normandie, France
Francesca Spezzano	Boise State University, USA
Damiano Spina	RMIT University, Australia
Benno Stein	Bauhaus-Universität, Germany
Nikolaos Stylianou	Information Technologies Institute, Greece
Aixin Sun	Nanyang Technological University, Singapore
Dhanasekar Sundararaman	Duke University, UK
Reem Suwaileh	Qatar University, Qatar
Lynda Tamine	IRIT, France
Nandan Thakur	University of Waterloo, Canada
Anna Tigunova	Max Planck Institute, Germany
Nava Tintarev	University of Maastricht, Germany
Marko Tkalcic	University of Primorska, Slovenia
Gabriele Tolomei	Sapienza University of Rome, Italy
Antonela Tommasel	Aarhus University, Denmark
Helma Torkamaan	Delft University of Technology, Netherlands
Salvatore Trani	ISTI-CNR, Italy

Giovanni Trappolini	Sapienza University, Italy
Jan Trienes	University of Duisburg-Essen, Germany
Andrew Trotman	University of Otago, New Zealand
Chun-Hua Tsai	University of Omaha, USA
Radu Tudor Ionescu	University of Bucharest, Romania
Yannis Tzitzikas	University of Crete and FORTH-ICS, Greece
Venktesh V	TU Delft, Germany
Alberto Veneri	Ca' Foscari University of Venice, Italy
Manisha Verma	Amazon, USA
Federica Vezzani	University of Padua, Italy
João Vinagre	Joint Research Centre - European Commission, Italy
Vishwa Vinay	Adobe Research, India
Marco Viviani	Università degli Studi di Milano-Bicocca, Italy
Sanne Vrijenhoek	Universiteit van Amsterdam, Netherlands
Vito Walter Anelli	Politecnico di Bari, Italy
Jiexin Wang	South China University of Technology, China
Zhihong Wang	Tsinghua University, China
Xi Wang	University College London, UK
Xiao Wang	University of Glasgow, UK
Yaxiong Wu	University of Glasgow, UK
Eugene Yang	Johns Hopkins University, USA
Hao-Ren Yao	National Institutes of Health, USA
Andrew Yates	University of Amsterdam, Netherlands
Fanghua Ye	University College London, UK
Zixuan Yi	University of Glasgow, UK
Elad Yom-Tov	Microsoft, USA
Eva Zangerle	University of Innsbruck, Austria
Markus Zanker	University of Klagenfurt, Germany
Fattane Zarrinkalam	University of Guelph, Canada
Rongting Zhang	Amazon, USA
Xinyu Zhang	University of Waterloo, USA
Yang Zhang	Kyoto University, Japan
Min Zhang	Tsinghua University, China
Tianyu Zhu	Beihang University, China
Jiongli Zhu	University of California San Diego, USA
Shengyao Zhuang	The University of Queensland, Australia
Md Zia Ullah	Edinburgh Napier University, UK
Steven Zimmerman	University of Essex, UK
Lixin Zou	Wuhan University, China
Guido Zuccon	The University of Queensland, Australia

Additional Reviewers

Pablo Castells
Ophir Frieder
Claudia Hauff
Yulan He
Craig Macdonald
Graham McDonald
Iadh Ounis
Maria Soledad Pera
Fabrizio Silvestri
Nicola Tonellotto
Min Zhang

Keynotes

Human Factors and Algorithmic Fairness

Carlos Castillo

ICREA and Universitat Pompeu Fabra
chato@icrea.cat

Abstract. In this talk, we present ongoing research on human factors of decision support systems that has consequences from the perspective of algorithmic fairness. We study two different settings: a game and a high-stakes scenario. The game is an exploratory "oil drilling" game, while the high-stakes scenario is the prediction of criminal recidivism. In both cases, a decision support system helps a human make a decision. We observe that in general users of such systems must thread a fine line between algorithmic aversion (completely disregarding the algorithmic support) and automation bias (completely disregarding their own judgment). The talk presents joint work led by David Solans and Manuel Portela. [1, 2]

Keywords: Decision Support · Algorithmic Fairness

References

1. Portela, M., Castillo, C., Tolan, S., Karimi-Haghighi, M., Pueyo, A. A.: A comparative user study of human predictions in algorithm supported recidivism risk assessment. arXiv preprint arXiv:2201.11080 (2022)
2. Solans, D., Beretta, A., Portela, M., Castillo, C., Monreale, A.: Human response to an AI-based decision support system: A user study on the effects of accuracy and bias. arXiv preprint arXiv:2203.15514 (2022)

Evaluating Generative Information Retrieval Systems

Charles L. A. Clarke

School of Computer Science, University of Waterloo, Canada

Abstract. Traditionally, information retrieval systems have been evaluated with a combination of implicit feedback from user interactions, explicit user feedback, and paid assessment by human labellers. Over the past year, both academic researchers and commercial services have begun to employ large language models to replace and augment human assessment for IR evaluation. Over the same period, generative information retrieval (Gen-IR) systems and retrieval augmented generation (RAG) systems have emerged as alternatives to traditional IR systems.

The role of IR evaluation has always been to answer questions like, "Is this system better than that one?" and "Does this search result satisfy the searcher's information need?" In a world where large language models can outperform human labellers on basic relevance assessment tasks, and where the results of a search are used only as input to a generative system, how do we answer these questions, and are these even the right questions to ask? While we can now build systems we could only dream about a few years ago, we still need to know if changes to a system reflect genuine benefits to its users. In this talk, I will survey the landscape as it exists in March 2024, and talk about directions forward.

Keywords: Generative IR · Information Retrieval · Evaluation

Shaping the Future of Endangered and Low-Resource Languages: Our Role in the Age of LLMs

Josiane Mothe

INSPE, Université de Toulouse, IRIT, UMR5505 CNRS, France
josiane.mothe@irit.fr

Abstract. Isidore of Seville is credited with the adage that it is language that gives birth to a people, and not the other way around, underlining the profound role played by language in the formation of cultural and social identity. Today, of the more than 7100 languages listed, a significant number are endangered. Since the 1970s, linguists, information seekers and enthusiasts have helped develop digital resources and automatic tools to support a wide range of languages, including endangered ones. The advent of Large Language Model (LLM) technologies holds both promise and peril. They offer unprecedented possibilities for the translation and generation of content and resources, key elements in the preservation and revitalization of languages. They also present threat of homogenization, cultural oversimplification and the further marginalization of already vulnerable languages. This talk proposes an initiatory journey, exploring the potential paths and partnerships between technology and tradition, with a particular focus on the Occitan language. Occitan is a language from Southern France, parts of Spain and Italy that played a major cultural and economic role, particularly in the Middle Ages. It is now endangered according to UNESCO. This talk critically examines how human expertise and artificial intelligence can work together to offer hope for preserving the linguistic diversity that forms the foundation of our global and especially our European heritage while addressing some of the ethical and practical challenges that accompany the use of these powerful technologies.

Keywords: Endangered Languages · Large Language Models

Contents – Part I

Full Papers

Full Papers

Self Contrastive Learning for Session-Based Recommendation

Zhengxiang Shi(✉), Xi Wang, and Aldo Lipani

University College London, London, UK
{zhengxiang.shi.19,xi-wang,aldo.lipani}@ucl.ac.uk

Abstract. Session-based recommendation, which aims to predict the next item of users' interest as per an existing sequence interaction of items, has attracted growing applications of Contrastive Learning (CL) with improved user and item representations. However, these contrastive objectives: (1) serve a similar role as the cross-entropy loss while ignoring the item representation space optimisation; and (2) commonly require complicated modelling, including complex positive/negative sample constructions and extra data augmentation. In this work, we introduce Self-Contrastive Learning (SCL), which simplifies the application of CL and enhances the performance of state-of-the-art CL-based recommendation techniques. Specifically, SCL is formulated as an objective function that directly promotes a uniform distribution among item representations and efficiently replaces all the existing contrastive objective components of state-of-the-art models. Unlike previous works, SCL eliminates the need for any positive/negative sample construction or data augmentation, leading to enhanced interpretability of the item representation space and facilitating its extensibility to existing recommender systems. Through experiments on three benchmarks, we demonstrate that SCL consistently improves the performance of state-of-the-art models with statistical significance. Notably, our experiments show that SCL improves the performance of two best-performing models by 8.2% and 9.5% in P@10 (Precision) and 9.9% and 11.2% in MRR@10 (Mean Reciprocal Rank) on average across different benchmarks. Additionally, our analysis elucidates the improvement in terms of alignment and uniformity of representations, as well as the effectiveness of SCL with a low computational cost. Code is available at https://github.com/ZhengxiangShi/SelfContrastiveLearningRecSys.

Keywords: Recommendation System · Session-based Recommendation · Contrastive Learning

1 Introduction

Session-based recommendation [13,14,37,38] is a crucial aspect of modern recommender systems for various platforms such as e-commerce websites [9,12], music streaming services [3], and social media [29], with the goal of predicting a user's next interest by focusing on their current intent. Recently, Contrastive Learning (CL) [18] has been applied in session-based recommendation tasks to

N. Goharian et al. (Eds.): ECIR 2024, LNCS 14608, pp. 3–20, 2024.
https://doi.org/10.1007/978-3-031-56027-9_1

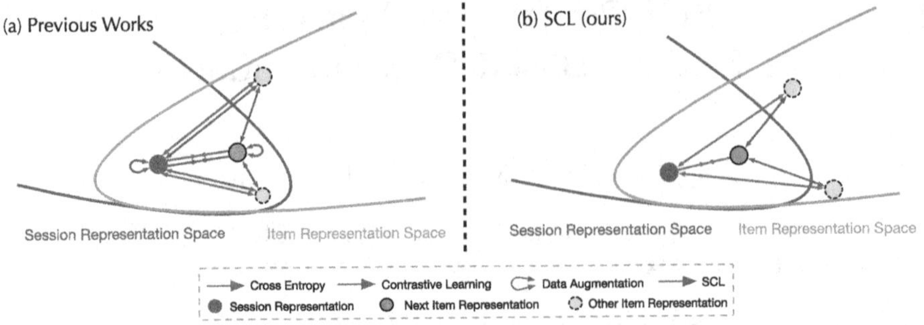

Fig. 1. The illustration of the framework of SCL. In previous works, contrastive learning (CL) objectives (depicted in green) typically involve complicated modelling, leading to a relatively lesser emphasis on optimising the item representation space. In contrast, SCL (depicted in blue) specifically addresses this issue and provides a better complement to the role of cross-entropy loss. (Color figure online)

enhance recommendation accuracy via improved representation quality, with the goal of aligning the session representation with the next item's representation, while also distinguishing it from other item representations. However, two key limitations exist within these methods.

Firstly, **the importance of optimising the item representation space *itself* by ensuring that the representations are uniformly distributed is not receiving adequate attention**. The CL objectives in previous research [17,19,35,37–39,47] serve a similar role as the cross-entropy loss while the optimisation of item representation space is not adequately addressed. As shown in Fig. 1(a), both cross-entropy loss and CL loss have the capacity to align the session representation with the representation of the next item and differentiate it from other item representations (§4). This has led to the marginalization of the role played by the uniformity of item representations. While some studies [17,19,35,39,47] have touched upon improving the uniformity of representations, these efforts generally contribute only a small fraction to the overall loss function.

Secondly, **current CL-based approaches often utilise complex techniques, including the sophisticated creation of positive and negative pairs and extra data augmentations, leading to limited adaptability across models**. Specifically, two state-of-the-art (SOTA) session-based recommendation models, S^2-DHCN [38] and COTREC [37] are the typical examples of complex CL-based applications. S^2-DHCN encompasses two encoder networks that generate varied session representations (positive) and compare them to corrupted session representations (negative) for noisy data-augmented CL. Similarly, COTREC requires two item representations to interact with the corresponding session representation in the CL objective, obtained through model-specific data augmentation techniques. These methods are heavily dependent on the model architecture and may not be compatible with various other models. Moreover, while recent studies have highlighted the importance of uniformity in user/item representations for recommendation tasks, this has simultaneously triggered a rise in the use of extra data augmentation methods, such as applying noise perturbation [47] or dropout [39,50] to augment representations, as shown in Fig. 1(a).

In this work, we argue that the importance of the uniformity of item representations has been considerably undervalued and that intricate CL objectives could be streamlined. We propose a novel approach, *Self Contrastive Learning* (SCL), which directly enforces the representation of each item distinct from those of all other items through a new loss objective and thus promotes a uniform distribution within the item representation space. SCL can be easily integrated into SOTA models to effectively replace other CL objectives, eliminating the need for creating complex positive/negative samples or engaging in any form of data augmentation. Different from previous approaches in recommendation systems that utilise the CL [19,31,35,40,47], SCL represents the first attempt to simply enforce uniformity of item representation without resorting to other CL objectives. Through our research, we aim to address the following research questions:

RQ$_1$ To what extent does SCL improve the session-based recommendation tasks? (§5.2)

RQ$_2$ How does SCL improve the model performance in terms of the alignment and uniformity of representations? (§5.3)

RQ$_3$ Are those sophisticated CL objectives still necessary in the presence of SCL? (§5.4)

RQ$_4$ Can SCL maintain SOTA performance with a low computational cost? (§5.5)

To address **RQ**$_1$, we experiment on three datasets, TMALL, NOWPLAYING, and DIGINETICA (§5.2). Our results show that SCL consistently improves the performance of SOTA models across various evaluation metrics and datasets. In particular, our experiments on TMALL show that SCL improves the performance of S^2-DHCN from 28.65% to 35.14% in P@10 and from 15.94% to 20.39% in MRR@10, and it also boosts the performance of COTREC from 30.44% to 35.03% in P@10 and from 17.28% to 20.46% in MRR@10, outperforming all existing approaches by large margins.

To understand how the model is improved (**RQ**$_2$), we investigate the transformations of the session and item representation in terms of alignment and uniformity (§5.3). Our study reveals that SCL learns item representations with a lower uniformity loss, leading to significant improvements in performance. Our findings suggest that SOTA approaches may have placed excessive emphasis on the alignment of session and item representations.

To answer **RQ**$_3$, we carry out an ablation study to evaluate the necessity of sophisticated CL objectives employed in prior works (§5.4). Our experiment reveals that SCL is capable of attaining the comparable model performance on its own, suggesting the advance of SCL and the redundant use of existing heavy and sophisticated CL objectives.

To understand the computational efficiency ($\mathbf{RQ}_4$), we further study the impact of selecting the k-nearest item representations in SCL on the model performance (§5.5). Our results show that SCL generally benefits from contrasting to more item representations. However, it can still achieve SOTA performance even just using a value of k equal to 2, indicating that SCL can be implemented with a low computational cost.

2 Related Work

Session-Based Recommendation. The session-based recommendation aims to predict the next item by utilizing user behaviours within a short time period [10,14,33,36]. Early studies on session-based recommendation focused on utilising temporal information from session data through the use of Markov chain models [5,13,21,22,42,44,51]. Then neural networks are widely applied for session-based recommendation [6,10,14,33,36]. Recurrent neural networks [11] have been applied to session-based recommendation models to capture the sequential order between items [48]. GRU4Rec [10] was the first model to use gated recurrent units (GRUs) [4] to model the sequential relations of item interactions. NARM [13] extended GRU4Rec by incorporating the attention mechanism [2,23,25,26] STAMP [14] also replaced the recurrent structure with attention layers to capture a user's general and current interests. FGNN [20] rethinks the sequence order of items to exploit users' intrinsic intents using GNNs. GCE-GNN [34] aggregates item information from both the item-level and session-level through graph convolution and self-attention mechanism. S^2-DHCN [38] utilizes hyper-graph convolutional networks to capture high-order item relations within individual sessions. COTREC [37] integrated self-supervised learning into the graph training through sophisticated positive and negative constructions.

Contrastive Learning. CL has achieved great success in various research domains, such as computer vision [8] and natural language processing [6,7,24,27, 28], with the goal of obtaining high-quality representations by pulling positive or similar instances closer in the representation space while simultaneously separating dissimilar, or negative instances. Recently, CL has recently been applied to sequential recommendation tasks, with several studies exploring its potential benefits in this area [43,45,46]. Bert4Rec [30] adapts the cloze objective from language modelling to a sequential recommendation by predicting random masked items in the sequence with surrounding contexts. S^3-Rec [49] utilizes intrinsic data correlations among attributes, items, subsequences, and sequences to generate contrastive signals and enhance data representations through pre-training. [16] proposed a seq2seq training strategy based on latent self-supervision and disentanglement of user intention behind behaviour sequences. CoSeRec [15] uses GNNs to capture more complex patterns than sequential patterns via CL objectives. CL4SRec [41] combines recommendation loss with CL loss of self-supervised tasks. DuoRec [19] retrieves the positive view of a given user sequence by identifying another user's sequence that shares the same next item through its proposed supervised CL. Unlike DuoRec, our task does not have access to user information. CL has also been applied to other recommendation paradigms, such as general recommendation [43] and social recommendation [45,46]. Previous work [1] also proposes a framework as self contrastive learning approach where every sample sharing a ground-truth label with an anchor is treated as a positive pair. In contrast, our approach is different because we define the positive sample exclusively as the sample itself, not considering other samples with the same label.

3 Preliminaries

Task Definition. In the session-based recommendation, the full set of item candidates is represented as $I = \{i_1, \ldots, i_n\}$, where n is the total number of item candidates. A session s, consisting of m items, is represented as a sequence $S = [i_1^s, \ldots, i_m^s]$ ordered by timestamps, where $i_k^s \in I$ $(1 \leq k \leq m)$ represents the k-th item that has been interacted with by a user. The objective is to predict the next item, i_{m+1}^s, from a full set of item candidates I, based on the corresponding session sequence S. For a session s, the output of the session-based recommendation model is a ranked list of item candidates $R = [r_1^s, \ldots, r_n^s]$, where r_*^s is the corresponding predicted ranking or preference score of the i-th item. Afterwards, the top-k items $(1 \leq k \leq n)$ will be selected as recommendations.

Contrastive Learning. Contrastive learning aims to pull the representation of an anchor sample and the representations of its corresponding positive sample pairs closer while simultaneously pushing the representations of the negative sample pairs away [7]. InfoNCE [18], where NCE stands for Noise Contrastive Estimation, is a type of contrastive loss function commonly used in recommendation systems [19,38,40,41]. Formally, let $\boldsymbol{a}$ denote an anchor representation and $\mathcal{X} \triangleq \{\boldsymbol{x}_1, \ldots, \boldsymbol{x}_{n-1}, \boldsymbol{x}_n\}$ denote the set of negative representations $(1 \leq k \leq n-1)$ and one positive representation $(k = n)$ with respect to $\boldsymbol{a}$, the InfoNCE loss is defined as:

$$\mathcal{L}_{\text{InfoNCE}} = -\log \frac{f(\boldsymbol{a}, \boldsymbol{x}_n)}{\sum_{j=1}^{n} f(\boldsymbol{a}, \boldsymbol{x}_j)}, \tag{1}$$

where f can be approximated by a real-valued scoring function and typically a function of the cosine similarity is used.

In the field of CL, two key properties, known as alignment and uniformity, have been proposed by [32] as measures of the quality of representations. The uniformity of the embedding distribution is measured as follows:

$$\ell_{\text{uniform}} \triangleq \log \mathop{\mathbb{E}}_{\substack{x \sim p_{\text{data}}, \\ x' \sim p_{\text{data}}}} e^{-2\|f(x)-f(x')\|^2}, \tag{2}$$

where p_{data} denotes the data distribution. ℓ_{uniform} is lower when random samples are farther from each other. Therefore, the examination of item representation uniformity ensures their semantic interpretability for a potential improvement in identifying the items of true interest. In contrast, instead of assessing the dispersion of item representations for uniformity, alignment gauges the expected distance between the embeddings of positively paired instances, assuming that representations are normalized, as expressed by the following equation:

$$\ell_{\text{align}} \triangleq \mathop{\mathbb{E}}_{\substack{x \sim p_{\text{data}}, \\ x' \sim p_{\text{pos}}(x)}} \|f(x) - f(x')\|^2, \tag{3}$$

where $p_{\text{pos}}(x)$ denotes the data distribution of samples that are positive to the instance x. ℓ_{align} is lower as all positive samples are closer to each other.

4 Methodology

In this section, we first discuss the potential limitations of SOTA session-based recommendation systems (§4). To address these issues, we then introduce a novel approach, *Self Contrastive Learning* (SCL), which aims to improve the uniformity in item representations by utilising a novel loss function (§4).

Motivation. In the field of session-based recommendation, existing works [34, 37,38], that utilise CL objectives, generally employ a framework in which the loss function is a combination of cross-entropy ($\mathcal{L}_{ce}$) and CL $\mathcal{L}_{cl}$ losses:

$$\mathcal{L} = \mathcal{L}_{ce} + \alpha \mathcal{L}_{cl}, \tag{4}$$

where $\mathcal{L}_{ce}$ aims to maximize the likelihood of selecting the correct next item, and $\mathcal{L}_{cl}$ aims to improve the learned representations, with the scalar coefficient α controlling the relative importance of these two objectives. Typically, the InfoNCE loss, as in Eq. 1, is used as $\mathcal{L}_{cl}$. However, these two objectives are similar in nature, as shown in Fig. 1. Specifically, let $\boldsymbol{s}$ denote a learned session representation and $L = \{(\boldsymbol{x}_k, y_k)\}_{k=1}^{n}$ denote a set of n learned item representations and their corresponding ground-truth labels, where y_k is 1 if the k-th item is the user's next click item and 0 otherwise. The categorical cross-entropy of classifying the next item correctly is computed as follows:

$$\mathcal{L}_{ce} = -\sum_{\boldsymbol{x}, y \in L} y \log p(\boldsymbol{x}|\boldsymbol{s}), \tag{5}$$

where p measures the probability that the item represented by $\boldsymbol{x}$ is drawn from the full set of item candidates conditioned on the session representation $\boldsymbol{s}$. The probability measure p is typically normalized using a real-valued scoring function f (e.g., cosine similarity). Thus, we can rewrite the Eq. 5 as:

$$\mathcal{L}_{ce} = -\sum_{\boldsymbol{x}, y \in L} y \log \frac{f(\boldsymbol{s}, \boldsymbol{x})}{\sum_{j=1}^{n} f(\boldsymbol{s}, \boldsymbol{x}_j)} = -\log \frac{f(\boldsymbol{s}, \boldsymbol{x}^+)}{\sum_{j=1}^{n} f(\boldsymbol{s}, \boldsymbol{x}_j)}, \tag{6}$$

where $\boldsymbol{x}^+$ is the user's next clicked item. Therefore, $\mathcal{L}_{ce}$ can be considered as an alternative expression of $\mathcal{L}_{cl}$ ($\mathcal{L}_{\text{InfoNCE}}$) when they use the same function f.

It is important to note that, while the loss functions $\mathcal{L}_{ce}$ and $\mathcal{L}_{cl}$ in Eq. 4 may have marginal variations, (*e.g.*, $\mathcal{L}_{cl}$ may use the extra temperature parameter τ in the function f in Eq. 6 and different positive and negative samples from data augmentation), their directions of optimising the representation spaces are the same: both objectives aim to push the session representation $\boldsymbol{s}$ closer to the next item representation $\boldsymbol{x}^+$ while pulling it away from other representations $\boldsymbol{x}_j$, thus improving session and item representations. Although using InfoNCE as CL objectives in conjunction with cross-entropy loss may result in a marginal improvement in performance, we argue that it is not the most effective strategy. Firstly, this may place an overemphasis on the alignment of the session and item representations, as shown in Fig. 1(a) where the green lines and red lines both have an impact on alignment. Secondly, while prior work [39] attempted to improve the uniformity of the item representation space with some auxiliary losses, the importance of blue lines (see Fig. 1(b)) appears to be diluted by other CL objectives. A more straightforward regularization approach that specifically targets the item representation distributions and effectively complements the cross-entropy loss is necessary to improve the overall recommendation performance.

Self Contrastive Learning (SCL ***).*** To address the aforementioned issues, we propose Self Contrastive Learning (SCL), a straightforward solution to improve the uniformity of the item representation space by introducing an additional loss objective, as shown in Fig. 1(b). This objective operates by directly penalizing the proximity of item representations based on our assumption that the representation of each item representation should be distant from those of all other items. Formally, given a set of n learned item representations $\mathcal{X}$, the objective of the SCL loss is calculated as follows:

$$\mathcal{L}_{\text{SCL}} = -\sum_{i=1}^{n} \log \frac{g(\boldsymbol{x}_i, \boldsymbol{x}_i)}{\sum_{j=1}^{n} g(\boldsymbol{x}_i, \boldsymbol{x}_j)}, \tag{7}$$

where the function $g(\boldsymbol{x}, \boldsymbol{x}')$ is computed by $e^{\text{sim}(\boldsymbol{x}, \boldsymbol{x}')/\tau}$, the exponential of the cosine similarity controlled by a temperature parameter τ. Using the cosine similarity, this loss pulls apart all items on the unit hypersphere. Next, we integrate $\mathcal{L}_{\text{SCL}}$ into the existing session-based recommendation models. Given the loss objective $\mathcal{L}_{model}$ from the original model (**all other CL objectives are excluded**), the overall loss function is computed as follows:

$$\mathcal{L} = \mathcal{L}_{model} + \beta \mathcal{L}_{\text{SCL}}, \tag{8}$$

where β is a hyperparameter that determines the relative importance of the two objectives. Complementary to the $\mathcal{L}_{model}$, which typical uses a $\mathcal{L}_{ce}$ to positively impact both ℓ_{align} and $\ell_{uniform}$, $\mathcal{L}_{\text{SCL}}$ has a stronger positive effect on $\ell_{uniform}$.

The advantages of SCL can be summarized in three main aspects: (1) **Improved representation spaces.** By incorporating SCL as an additional loss objective, we achieve improved uniformity in the item representation space, leading to better model performance; (2) **Simplified modelling process.** By leveraging the SCL objective, we avoid the need for complex creation of positive/negative sample pairs and data augmentation techniques, such as noise perturbation [47] or dropout [39]. In SCL, each item representation serves as the sole positive sample, and all other item representations are considered negative samples without further modifications. This greatly simplifies the construction of recommendation systems, making them more efficient and easier to implement; and (3) **Seamless integration into existing systems.** SCL can be seamlessly integrated into existing session-based recommendation systems that utilise session and item representations, without any additional modification to the architecture of the model. This high level of compatibility makes SCL widely applicable and adaptable to various settings and scenarios. Overall, these advantages make SCL a valuable solution for enhancing recommendation systems, offering improved uniformity, simplified training, and easy integration into existing models.

5 Experiments and Results

In this section, we evaluate our SCL in three benchmarks. We first describe the experimental setup, including the used datasets, baselines, evaluation metrics, and implementation details. Then we present our experimental results with respect to the four research questions introduced in Section §1.

5.1 Experimental Setup

Datasets and Baselines. **Tmall** is sourced from the IJCAI-15 competition and includes anonymized shopping logs from users on the Tmall online shopping platform, with train size 351,268, test size 25,898, items Size 40,728 and an average length 6.69. **Nowplaying** describes the music-listening behaviour of users, with train size 825,304, test size 89,824, items Size 60,417 and an average length 7.42; and **Diginetica** , from CIKM Cup 2016, comprises typical transaction data, with train size 719,470, test size 60,858, items Size 43,097 and an average length 5.12. Our proposed SCL method is compared with the following representative methods: **FPMC** [21], **GRU4REC** [10], **NARM** [13], **STAMP** [14], **SR-GNN** [36], **GCE-GNN** [34], **S^2-DHCN** [38], and **COTREC** [37].

Implementation Details. We evaluate the performance of our proposed SCL method using the metrics of P@k (Precision) and MRR@k (Mean Reciprocal Rank), where the cutoff k is set to 5, 10 or 20. We conduct experiments with the proposed SCL method using three SOTA models, GCE-GNN, S^2-DHCN, and COTREC. We first reproduce the experimental results of these models by following the settings in their original papers. Then, we apply the SCL to these three models. For the hyperparameters used in the SCL, the temperature parameter, denoted by τ, is set to 0.1, and the loss weight parameter, denoted by β, is varied within a range of 0.1 to 100. We have omitted the evaluation of COTREC on the NOWPLAYING as we were unable to replicate the results.

Table 1. Performances of all comparison methods on the dev set (RQ_1). † from the original paper. ‡ from our own reproductions. Triangles in colours indicate an improvement over our reproduced results. The highest results in each column are highlighted in bold font. SOTA performances are indicated in blue. *** indicates a p-value < 1e−20 for the improvement, ** indicates a p-value < 1e−5, and * indicates a p-value < 1e−2.

Method	Tmall				Nowplaying				Diginetica			
	P@10	MRR@10	P@20	MRR@20	P@10	MRR@10	P@20	MRR@20	P@10	MRR@10	P@20	MRR@20
FPMC	13.10	7.12	16.06	7.32	5.28	2.68	7.36	2.82	15.43	6.20	26.53	6.95
GRU4REC	9.47	5.78	10.93	5.89	6.74	4.40	7.92	4.48	17.93	7.33	29.45	8.33
NARM	19.17	10.42	23.30	10.70	13.6	6.62	18.59	6.93	35.44	15.13	49.70	16.17
STAMP	22.63	13.12	26.47	13.36	13.22	6.57	17.66	6.88	33.98	14.26	45.64	14.32
SR-GNN	23.41	13.45	27.57	13.72	14.17	7.15	18.87	7.47	36.86	15.52	50.73	17.59
GCE-GNN†	28.01	15.08	33.42	15.42	16.94	8.03	22.37	8.40	41.16	18.15	54.22	19.04
GCE-GNN‡	27.48	14.85	32.52	15.20	17.19	8.09	22.42	8.45	40.98	18.12	54.23	19.04
w/ SCL	28.67**	15.20*	33.65**	15.55*	17.44*	**8.10**	22.81*	**8.47**	**41.93****	**18.45***	**54.93***	**19.38***
Δ (%)	(▲4.3%)	(▲2.4%)	(▲3.5%)	(▲2.3%)	(▲1.5%)	(▲0.1%)	(▲1.7%)	(▲0.2%)	(▲2.3%)	(▲1.8%)	(▲1.3%)	(▲1.8%)
COTREC†	30.62	17.65	36.35	18.04	-	-	-	-	41.88	18.16	54.18	19.07
COTREC‡	30.44	17.28	36.09	17.67	-	-	-	-	40.26	17.75	53.75	18.69
w/ SCL	35.03***	**20.46*****	**39.29*****	**20.76*****	-	-	-	-	40.78*	18.00*	53.78	18.90*
Δ (%)	(▲15.1%)	(▲18.4%)	(▲8%)	(▲17.5%)	-	-	-	-	(▲1.3%)	(▲1.4%)	(▲0.1%)	(▲1.1%)
S^2-DHCN†	26.22	14.60	31.42	15.05	17.35	7.87	23.50	8.18	39.87	17.53	53.18	18.44
S^2-DHCN‡	28.65	15.94	34.54	16.35	17.23	7.70	23.00	8.10	39.54	17.31	52.76	18.22
w/ SCL	**35.14*****	20.39***	39.13***	20.67***	**17.61***	7.92*	**23.74****	8.32*	40.91**	17.79**	53.91*	18.69*
Δ (%)	(▲22.7%)	(▲27.9%)	(▲13.3%)	(▲26.4%)	(▲2.2%)	(▲2.9%)	(▲3.2%)	(▲1.7%)	(▲3.5%)	(▲2.8%)	(▲2.2%)	(▲2.6%)

5.2 Main Results (RQ1)

Table 1 presents the performance of all comparison methods, where our SCL is applied to three SOTA models, GCE-GNN, COTREC, and S^2-DHCN. Results show that SCL consistently improves the model performance in terms of P@k and MRR@k across three datasets, TMALL, NOWPLAYING, and DIGINETICA, achieving the new SOTA performance (highlighted in blue). The significance tests further corroborate the effectiveness of SCL. Particularly remarkable is that SCL achieves a notable improvement compared to the SOTA models on the TMALL dataset. COTREC model with the proposed SCL method also shows significant improvement, with an increase of 18.4% and 17.5% in terms of MRR@10 and MRR@20, respectively. Additionally, S^2-DHCN + SCL achieves a new SOTA performance on the TMALL dataset. It records a 27.9% increase of the MRR@10 from 15.94% to 20.39% and a 26.4% increase of MRR@20 from 16.35% to 20.67%. Similar improvement in performance can also be observed on the NOWPLAYING and DIGINETICA datasets.

5.3 Alignment and Uniformity (RQ2)

The substantial improvement in performance achieved by the SCL raises the research question of where these improvements come from (**RQ_2**). Figure 2 depicts the impact of the proposed SCL method on the alignment and uniformity on TMALL and DIGINETICA datasets. In general, we find that (1) SCL has improved the uniformity of item representations, leading to an improvement in

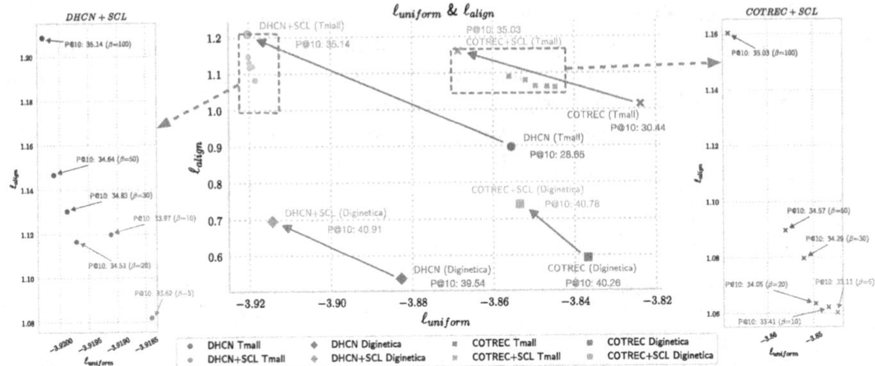

Fig. 2. The analysis of alignment loss ℓ_{align} and uniformity loss ℓ_{uniform}, where P@10 is reported as model performance. While it is generally acknowledged that a decrease in the alignment or uniformity loss leads to improved model performance, an excessive emphasis on alignment and insufficient attention to uniformity can result in sub-optimal performance.

model performance; and (2) a higher loss in alignment ℓ_{align} does not necessarily result in worse performance if the uniformity loss ℓ_{uniform} is improved.

Better Uniformity of Item Representations Brings Substantial Improvement in Performance. The sub-figure in the centre of Fig. 2 illustrates how SCL improves the uniformity of item representations of S^2-DHCN and COTREC on TMALL and DIGINETICA. This is indicated by a lower uniformity loss when SCL is applied. The uniformity loss measures the dissimilarity between the item representations themselves and a lower uniformity loss indicates that the item representations are becoming more discriminative and less correlated with each other. Specifically, the use of SCL results in a reduction of the uniformity loss of S^2-DHCN from −3.86 to −3.92 on the TMALL dataset, and this improvement is accompanied by an increase in P@10 from 28.65% to 35.14%.

The Trade-Off Between Alignment and Uniformity. We also observe that the proposed SCL method leads to an increase in the alignment loss. This indicates that the next item representations are not only becoming more discriminative to other item representations but also less correlated with the session representations. We conduct additional studies on the TMALL dataset by adjusting the alignment and uniformity loss through controlling the SCL loss weight β. The results are depicted in the two sub-figures of Fig. 2. Specifically, for the S^2-DHCN model, when the uniformity loss gradually decreases from −3.86 to −3.92 and the alignment loss increases from 1.08 to around 1.20, the model performance in P@10 is generally improved from 33.62% to 35.14%. Similar results are observed in the experiments using the COTREC model. This suggests that an excessive focus on alignment and inadequate attention to uniformity could result in sub-optimal model performance.

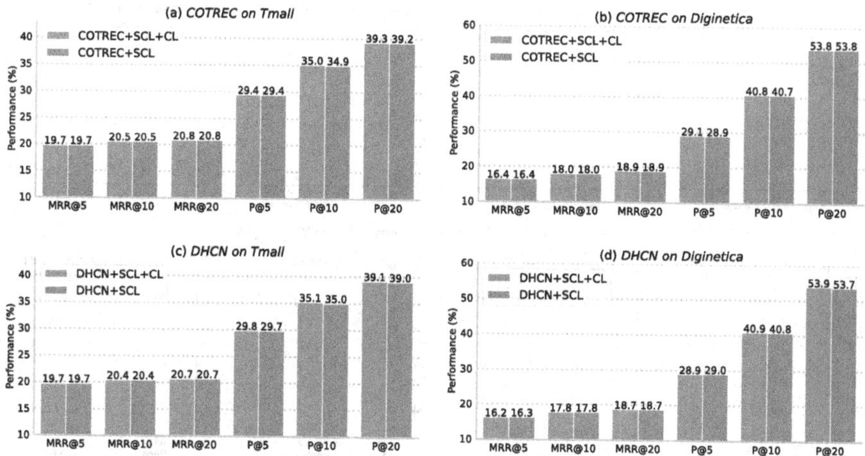

Fig. 3. Ablation studies ($\mathbf{RQ}_3$) on other CL objectives. Blue indicates the model performance using SCL and other CL objectives. Red represents the model performance using SCL only.

5.4 Sophisticated CL Objectives Are Unnecessary (RQ3)

Given the complexity of CL objectives used in the SOTA models, we investigate the necessity of these complex contrastive objectives with our proposed SCL approach ($\mathbf{RQ}_3$). We conduct experiments for COTREC and S^2-DHCN on two datasets, TMALL and DIGINETICA with two different settings as follows: (1) **Model + SCL + CL** refers to the model performance with our SCL and all CL objectives in the original model; (1) **Model + SCL** refers to the model performance with our SCL only.

Results. Figure 3 depicts the performance of the models. It can be observed that MODEL + SCL + CL and MODEL + SCL achieve very similar performance results, which implies that the utilization of other sophisticated CL objectives may not be necessary and that the proposed SCL is able to effectively improve the model performance on its own. Below we delve deeper into these findings and discuss them in more detail. Specially, for the TMALL dataset, when using the COTREC model as the backbone, the MODEL + SCL + CL and the MODEL + SCL achieve P@10 scores of 35.0% and 34.9%, respectively, and the same M@10 scores of 20.5%. Similar performance can be observed for the S^2-DHCN model and DIGINETICA dataset, implying that the complexity of these objectives may not be necessary when using SCL.

5.5 Computational Cost (RQ4)

The size of negative samples plays a critical role in the model performance for CL. However, it also has some potential drawbacks, including increased computational resources and model complexity, which may make the proposed method

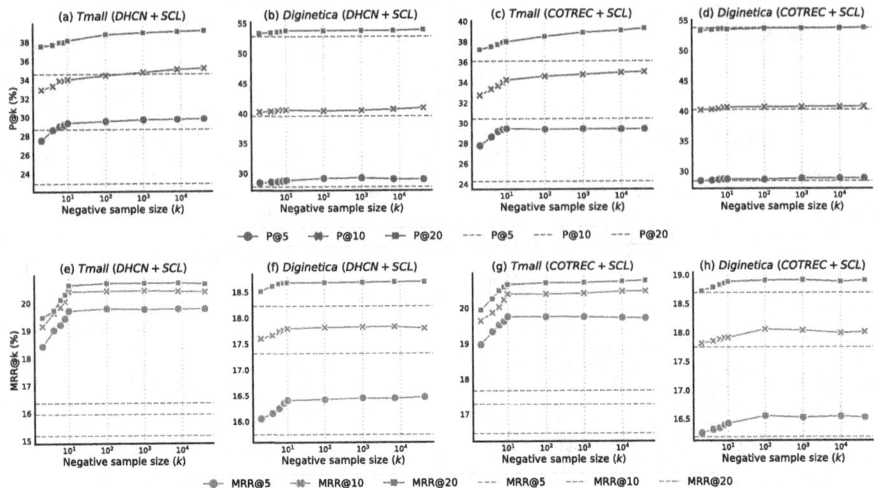

Fig. 4. The impact of negative sample sizes with S^2-DHCN + SCL and COTREC + SCL on the TMALL and DIGINETICA datasets (**RQ**$_4$). The original model performance without using SCL is represented by the corresponding dash line with the same colour. SCL could achieve SOTA performance even when the negative sample size k is equal to 2.

impractical for certain applications or settings. The proposed SCL method has a time complexity of $O(n^2 * d)$, where n is the number of item representations used in Eq. 7 and d is the dimension of item representations. To reduce the computational cost, we simplify the objective function of SCL by encompassing a k-Nearest Neighbour (kNN) with a fast dense embedding retrieval method, which reduces the time complexity from $O(n^2 * d)$ to $O(n * k * d)$, where $k \ll n$. The updated objective is as follows:

$$\mathcal{L}_{\mathrm{SCL}}^{\mathrm{knn}} = -\sum_{i=1}^{n} \log \frac{f(\boldsymbol{x}_i, \boldsymbol{x}_i)}{\sum_{\boldsymbol{x}' \in \mathcal{K}_i} f(\boldsymbol{x}_i, \boldsymbol{x}')}, \tag{9}$$

where $\mathcal{K}_i$ is a set of k nearest item representations in the distance measured by the cosine similarity for the i-th item representation, including its own representation. We conduct experiments to evaluate the impact of negative sample size k, with the values of 2, 4, 6, 8, 10, 100, 1 000, 10 000 and the full set.

Results. Figure 4 presents the model performance in P@k and MRR@k with respect to various values of the negative sample size k on the TMALL and DIGINETICA datasets, where S^2-DHCN + SCL and COTREC + SCL are evaluated. Overall, our experimental results indicate SCL could improve the performance of SOTA models even when the negative sample size k is equal to 2, and that the performance of the models generally improves as the size of negative samples increases. As the negative sample size continues to increase, the improvement

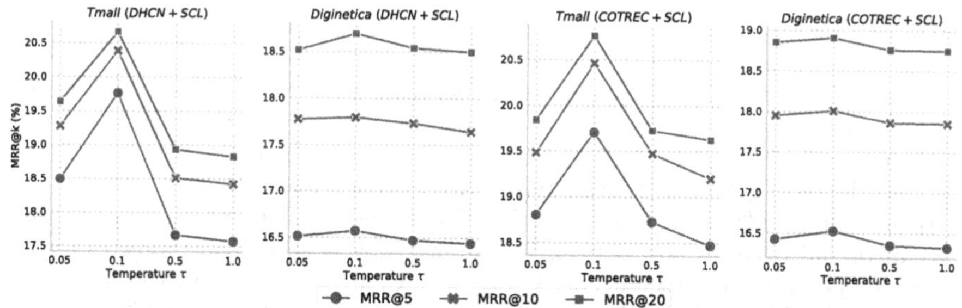

Fig. 5. The effect of the temperature τ using the S^2-DHCN + SCL and COTREC + SCL model on the TMALL and DIGINETICA datasets, where MRR@k is reported as the representative of model performance.

of the model tends to level off and become less noticeable, using a small value for k can produce comparable results to using values greater than 10 000. For example, the performance of P@k and MRR@k for the S^2-DHCN + SCL model tends to become stable once the negative sample size reaches 10 on the TMALL dataset, as shown in sub-figure (a) and (e) of Fig. 4.

5.6 Hyperparameter Sensitivity

We conduct an additional study to investigate the effect of varying the hyperparameter temperature τ on model performance. In the experiment, 4 distinct values of τ (namely 0.05, 0.1, 0.5, and 1.0) are evaluated with the S^2-DHCN + SCL and COTREC + SCL on the TMALL and DIGINETICA datasets. The experimental results are presented in Fig. 5, indicating that the model achieves optimal performance when the temperature τ is set to 0.1.

6 Conclusion

In this work, we propose *Self Contrastive Learning* (SCL), which improves the performance of SOTA models with statistical significance across three datasets. SCL targets the optimization of item representation uniformity in SOTA session-based recommendation systems. SCL complements the use of cross-entropy loss, eliminating the need for sophisticated CL objectives. This simplicity makes SCL highly adaptable across a variety of models. Moreover, we shed light on how SCL enhances representation spaces from the alignment and uniformity viewpoints, thus emphasizing the importance of uniformity in item representations. Our analysis also points out that achieving an optimal balance between alignment and uniformity loss is a crucial aspect of designing recommendation systems. Lastly, we demonstrate that the implementation of SCL is efficient and entails low computational costs.

References

1. Bae, S., Kim, S., Ko, J., Lee, G., Noh, S., Yun, S.Y.: Self-contrastive learning: single-viewed supervised contrastive framework using sub-network. Proc. AAAI Conf. Artif. Intell. **37**(1), 197–205 (2023). https://doi.org/10.1609/aaai.v37i1.25091. https://ojs.aaai.org/index.php/AAAI/article/view/25091
2. Bahdanau, D., Cho, K., Bengio, Y.: Neural machine translation by jointly learning to align and translate. In: Bengio, Y., LeCun, Y. (eds.) 3rd International Conference on Learning Representations, ICLR 2015, Conference Track Proceedings, San Diego, CA, USA, 7–9 May 2015 (2015). http://arxiv.org/abs/1409.0473
3. Brost, B., Mehrotra, R., Jehan, T.: The music streaming sessions dataset. In: Liu, L., et al. (eds.) The World Wide Web Conference, WWW 2019, San Francisco, CA, USA, 13–17 May 2019, pp. 2594–2600. ACM, USA (2019). https://doi.org/10.1145/3308558.3313641
4. Chung, J., Gülçehre, Ç., Cho, K., Bengio, Y.: Empirical evaluation of gated recurrent neural networks on sequence modeling. CoRR abs/1412.3555 (2014). http://arxiv.org/abs/1412.3555
5. Fu, X., Lipani, A.: Priming and actions: an analysis in conversational search systems. In: Association for Computing Machinery, SIGIR 2023, July 2023. https://doi.org/10.1145/3539618.3592041
6. Fu, X., Yilmaz, E., Lipani, A.: Evaluating the Cranfield paradigm for conversational search systems. In: Proceedings of the 2022 ACM SIGIR International Conference on Theory of Information Retrieval, ICTIR 2022, pp. 275–280. Association for Computing Machinery, New York (2022). https://doi.org/10.1145/3539813.3545126
7. Gao, T., Yao, X., Chen, D.: SimCSE: simple contrastive learning of sentence embeddings. In: Proceedings of the 2021 Conference on Empirical Methods in Natural Language Processing, Online and Punta Cana, Dominican Republic, pp. 6894–6910. Association for Computational Linguistics (2021). https://doi.org/10.18653/v1/2021.emnlp-main.552. https://aclanthology.org/2021.emnlp-main.552
8. He, K., Fan, H., Wu, Y., Xie, S., Girshick, R.: Momentum contrast for unsupervised visual representation learning. In: 2020 IEEE/CVF Conference on Computer Vision and Pattern Recognition (CVPR), June 2020, pp. 9726–9735 (2020). https://doi.org/10.1109/CVPR42600.2020.00975
9. Hendriksen, M., Kuiper, E., Nauts, P., Schelter, S., de Rijke, M.: Analyzing and predicting purchase intent in e-commerce: anonymous vs. identified customers. arXiv preprint arXiv:2012.08777 (2020)
10. Hidasi, B., Karatzoglou, A., Baltrunas, L., Tikk, D.: Session-based recommendations with recurrent neural networks. In: Bengio, Y., LeCun, Y. (eds.) 4th International Conference on Learning Representations, ICLR 2016, San Juan, Puerto Rico, Conference Track Proceedings, 2–4 May 2016 (2016). http://arxiv.org/abs/1511.06939
11. Hochreiter, S., Schmidhuber, J.: Long short-term memory. Neural Comput. **9**(8), 1735–1780 (1997)
12. Jannach, D., Ludewig, M., Lerche, L.: Session-based item recommendation in e-commerce: on short-term intents, reminders, trends and discounts. User Model. User-Adap. Inter. **27**(3), 351–392 (2017)
13. Li, J., Ren, P., Chen, Z., Ren, Z., Lian, T., Ma, J.: Neural attentive session-based recommendation. In: Lim, E., et al. (eds.) Proceedings of the 2017 ACM on Conference on Information and Knowledge Management, CIKM 2017, Singapore, 06–10

November 2017, pp. 1419–1428. ACM, Singapore (2017). https://doi.org/10.1145/3132847.3132926

14. Liu, Q., Zeng, Y., Mokhosi, R., Zhang, H.: STAMP: short-term attention/memory priority model for session-based recommendation. In: Guo, Y., Farooq, F. (eds.) Proceedings of the 24th ACM SIGKDD International Conference on Knowledge Discovery & Data Mining, KDD 2018, London, UK, 19–23 August 2018, pp. 1831–1839. ACM, UK (2018). https://doi.org/10.1145/3219819.3219950
15. Liu, Z., Chen, Y., Li, J., Yu, P.S., McAuley, J., Xiong, C.: Contrastive self-supervised sequential recommendation with robust augmentation. arXiv preprint arXiv:2108.06479 (2021)
16. Ma, J., Zhou, C., Yang, H., Cui, P., Wang, X., Zhu, W.: Disentangled self-supervision in sequential recommenders. In: Gupta, R., Liu, Y., Tang, J., Prakash, B.A. (eds.) The 26th ACM SIGKDD Conference on Knowledge Discovery and Data Mining, KDD 2020, Virtual Event, CA, USA, 23–27 August 2020, pp. 483–491. ACM (2020). https://dl.acm.org/doi/10.1145/3394486.3403091
17. Nie, P., et al.: MIC: model-agnostic integrated cross-channel recommender. In: Proceedings of the 31st ACM International Conference on Information and Knowledge Management, CIKM 2022, pp. 3400–3409. Association for Computing Machinery, New York (2022). https://doi.org/10.1145/3511808.3557081
18. van den Oord, A., Li, Y., Vinyals, O.: Representation learning with contrastive predictive coding. arXiv preprint arXiv:1807.03748 (2018)
19. Qiu, R., Huang, Z., Yin, H., Wang, Z.: Contrastive learning for representation degeneration problem in sequential recommendation. In: Proceedings of the Fifteenth ACM International Conference on Web Search and Data Mining, WSDM 2022, pp. 813–823. Association for Computing Machinery, New York (2022). https://doi.org/10.1145/3488560.3498433
20. Qiu, R., Li, J., Huang, Z., Yin, H.: Rethinking the item order in session-based recommendation with graph neural networks. In: Zhu, W., et al. (eds.) Proceedings of the 28th ACM International Conference on Information and Knowledge Management, CIKM 2019, Beijing, China, 3–7 November 2019, pp. 579–588. ACM, Beijing (2019). https://doi.org/10.1145/3357384.3358010
21. Rendle, S., Freudenthaler, C., Schmidt-Thieme, L.: Factorizing personalized Markov chains for next-basket recommendation. In: Rappa, M., Jones, P., Freire, J., Chakrabarti, S. (eds.) Proceedings of the 19th International Conference on World Wide Web, WWW 2010, Raleigh, North Carolina, USA, 26–30 April 2010, pp. 811–820. ACM, USA (2010). https://doi.org/10.1145/1772690.1772773
22. Shani, G., Heckerman, D., Brafman, R.I.: An MDP-based recommender system. J. Mach. Learn. Res. **6**, 1265–1295 (2005)
23. Shi, Z., Feng, Y., Lipani, A.: Learning to execute actions or ask clarification questions. In: Findings of the Association for Computational Linguistics, NAACL 2022, Seattle, United States, pp. 2060–2070. Association for Computational Linguistics (2022). https://doi.org/10.18653/v1/2022.findings-naacl.158. https://aclanthology.org/2022.findings-naacl.158
24. Shi, Z., Lipani, A.: Don't stop pretraining? Make prompt-based fine-tuning powerful learner. In: Thirty-seventh Conference on Neural Information Processing Systems. NeurIPS (2023). https://openreview.net/forum?id=s7xWeJQACI
25. Shi, Z., Ni, P., Wang, M., Kim, T.E., Lipani, A.: Attention-based ingredient parser. In: European Symposium on Artificial Neural Networks, Computational Intelligence and Machine Learning (ESANN), Bruges, Belgium (2022). https://doi.org/10.14428/esann/2022.ES2022-10

26. Shi, Z., Ramos, J., Kim, T.E., Wang, X., Rahmani, H.A., Lipani, A.: When and what to ask through world states and text instructions: IGLU NLP challenge solution. In: Advances in Neural Information Processing Systems (NeurIPS), IGLU Workshop (2023). https://nips.cc/virtual/2022/66405
27. Shi, Z., Tonolini, F., Aletras, N., Yilmaz, E., Kazai, G., Jiao, Y.: Rethinking semi-supervised learning with language models. In: Rogers, A., Boyd-Graber, J., Okazaki, N. (eds.) Findings of the Association for Computational Linguistics, ACL 2023, Toronto, Canada, July 2023, pp. 5614–5634. Association for Computational Linguistics (2023). https://doi.org/10.18653/v1/2023.findings-acl.347. https://aclanthology.org/2023.findings-acl.347
28. Shi, Z., Zhang, Q., Lipani, A.: StepGame: a new benchmark for robust multi-hop spatial reasoning in texts. In: Proceedings of the AAAI Conference on Artificial Intelligence, June 2022, vol. 36, pp. 11321–11329 (2022). https://doi.org/10.1609/aaai.v36i10.21383. https://ojs.aaai.org/index.php/AAAI/article/view/21383
29. Song, W., Xiao, Z., Wang, Y., Charlin, L., Zhang, M., Tang, J.: Session-based social recommendation via dynamic graph attention networks. In: Culpepper, J.S., Moffat, A., Bennett, P.N., Lerman, K. (eds.) Proceedings of the Twelfth ACM International Conference on Web Search and Data Mining, WSDM 2019, Melbourne, VIC, Australia, 11–15 February 2019, pp. 555–563. ACM (2019). https://doi.org/10.1145/3289600.3290989
30. Sun, F., et al.: BERT4Rec: sequential recommendation with bidirectional encoder representations from transformer. In: Zhu, W., et al. (eds.) Proceedings of the 28th ACM International Conference on Information and Knowledge Management, CIKM 2019, Beijing, China, 3–7 November 2019, pp. 1441–1450. ACM, Beijing (2019). https://doi.org/10.1145/3357384.3357895
31. Wang, L., Lim, E.P., Liu, Z., Zhao, T.: Explanation guided contrastive learning for sequential recommendation. In: Proceedings of the 31st ACM International Conference on Information and Knowledge Management, CIKM 2022, pp. 2017–2027. Association for Computing Machinery, New York (2022). https://doi.org/10.1145/3511808.3557317
32. Wang, T., Isola, P.: Understanding contrastive representation learning through alignment and uniformity on the hypersphere. In: Proceedings of the 37th International Conference on Machine Learning, ICML 2020, Virtual Event, 13–18 July 2020, vol. 119, pp. 9929–9939. Proceedings of Machine Learning Research (PMLR) (2020). http://proceedings.mlr.press/v119/wang20k.html
33. Wang, W., et al.: Beyond clicks: modeling multi-relational item graph for session-based target behavior prediction. In: Huang, Y., King, I., Liu, T., van Steen, M. (eds.) The Web Conference 2020, WWW 2020, Taipei, Taiwan, 20–24 April 2020, pp. 3056–3062. ACM/IW3C2, Taiwan (2020). https://doi.org/10.1145/3366423.3380077
34. Wang, Z., Wei, W., Cong, G., Li, X., Mao, X., Qiu, M.: Global context enhanced graph neural networks for session-based recommendation. In: Huang, J., et al. (eds.) Proceedings of the 43rd International ACM SIGIR Conference on Research and Development in Information Retrieval, SIGIR 2020, Virtual Event, China, 25–30 July 2020, pp. 169–178. ACM (2020). https://doi.org/10.1145/3397271.3401142
35. Wei, Y., et al.: Contrastive learning for cold-start recommendation. In: Proceedings of the 29th ACM International Conference on Multimedia, MM 2021, pp. 5382–5390. Association for Computing Machinery, New York (2021). https://doi.org/10.1145/3474085.3475665

36. Wu, S., Tang, Y., Zhu, Y., Wang, L., Xie, X., Tan, T.: Session-based recommendation with graph neural networks. In: The Thirty-Third AAAI Conference on Artificial Intelligence, AAAI 2019, The Thirty-First Innovative Applications of Artificial Intelligence Conference, IAAI 2019, The Ninth AAAI Symposium on Educational Advances in Artificial Intelligence, EAAI 2019, Honolulu, Hawaii, USA, 27 January–1 February 2019, pp. 346–353. AAAI Press (2019). https://doi.org/10.1609/aaai.v33i01.3301346
37. Xia, X., Yin, H., Yu, J., Shao, Y., Cui, L.: Self-supervised graph co-training for session-based recommendation. In: Proceedings of the 30th ACM International Conference on Information and Knowledge Management, Virtual Event, Queensland, Australia, pp. 2180–2190. ACM (2021). https://doi.org/10.1145/3459637.3482388. https://dl.acm.org/doi/10.1145/3459637.3482388
38. Xia, X., Yin, H., Yu, J., Wang, Q., Cui, L., Zhang, X.: Self-supervised hypergraph convolutional networks for session-based recommendation. In: Proceedings of the AAAI Conference on Artificial Intelligence, Virtual, vol. 35, pp. 4503–4511. AAAI (2021). https://doi.org/10.1609/aaai.v35i5.16578. https://ojs.aaai.org/index.php/AAAI/article/view/16578
39. Xie, R., Qiu, Z., Zhang, B., Lin, L.: Multi-granularity item-based contrastive recommendation. arXiv preprint arXiv:2207.01387 (2022)
40. Xie, X., Sun, F., Liu, Z., Gao, J., Ding, B., Cui, B.: Contrastive pre-training for sequential recommendation. arXiv preprint arXiv:2010.14395 (2020)
41. Xie, X., et al.: Contrastive learning for sequential recommendation. In: 2022 IEEE 38th International Conference on Data Engineering (ICDE), virtual, pp. 1259–1273. IEEE (2022). https://ieeexplore.ieee.org/abstract/document/9835621
42. Xu, C., et al.: Graph contextualized self-attention network for session-based recommendation. In: Kraus, S. (ed.) Proceedings of the Twenty-Eighth International Joint Conference on Artificial Intelligence, IJCAI 2019, Macao, China, 10–16 August 2019, pp. 3940–3946. ijcai.org (2019). https://doi.org/10.24963/ijcai.2019/547
43. Yao, T., et al.: Self-supervised learning for deep models in recommendations. arXiv e-prints pp. arXiv-2007 (2020)
44. Yin, H., Cui, B.: Spatio-Temporal Recommendation in Social Media. SCS, Springer, Singapore (2016). https://doi.org/10.1007/978-981-10-0748-4
45. Yu, J., Yin, H., Gao, M., Xia, X., Zhang, X., Hung, N.Q.V.: Socially-aware self-supervised tri-training for recommendation. arXiv preprint arXiv:2106.03569 (2021)
46. Yu, J., Yin, H., Li, J., Wang, Q., Hung, N.Q.V., Zhang, X.: Self-supervised multi-channel hypergraph convolutional network for social recommendation. arXiv preprint arXiv:2101.06448 (2021)
47. Yu, J., et al.: Are graph augmentations necessary? Simple graph contrastive learning for recommendation. In: Proceedings of the 45th International ACM SIGIR Conference on Research and Development in Information Retrieval, SIGIR 2022, pp. 1294–1303. Association for Computing Machinery, New York (2022). https://doi.org/10.1145/3477495.3531937
48. Zhang, Y., et al.: Sequential click prediction for sponsored search with recurrent neural networks. In: Brodley, C.E., Stone, P. (eds.) Proceedings of the Twenty-Eighth AAAI Conference on Artificial Intelligence, 27–31 July 2014, Québec City, Québec, Canada, pp. 1369–1375. AAAI Press (2014). http://www.aaai.org/ocs/index.php/AAAI/AAAI14/paper/view/8529

49. Zhou, K., et al.: S3-Rec: self-supervised learning for sequential recommendation with mutual information maximization. In: d'Aquin, M., Dietze, S., Hauff, C., Curry, E., Cudré-Mauroux, P. (eds.) The 29th ACM International Conference on Information and Knowledge Management, CIKM 2020, Virtual Event, Ireland, 19–23 October 2020, pp. 1893–1902. ACM (2020). https://doi.org/10.1145/3340531.3411954
50. Zhou, X., Sun, A., Liu, Y., Zhang, J., Miao, C.: SelfCF: a simple framework for self-supervised collaborative filtering. ACM Trans. Recomm. Syst. **1**, 1–25 (2023). https://doi.org/10.1145/3591469
51. Zimdars, A., Chickering, D.M., Meek, C.: Using temporal data for making recommendations. arXiv preprint arXiv:1301.2320 (2013)

MOReGIn: Multi-Objective Recommendation at the Global and Individual Levels

Elizabeth Gómez[1], David Contreras[2], Ludovico Boratto[3](✉), and Maria Salamó[1]

[1] Facultat de Matemàtiques i Informàtica, Universitat de Barcelona, Barcelona, Spain
egomezye13@alumnes.ub.edu, maria.salamo@ub.edu

[2] Facultad de Ingeniería y arquitectura, Universidad Arturo Prat, Iquique, Chile
david.contreras@unap.cl

[3] Department of Mathematics and Computer Science, University of Cagliari, Cagliari, Italy
ludovico.boratto@acm.org

Abstract. Multi-Objective Recommender Systems (MORSs) emerged as a paradigm to guarantee multiple (often conflicting) goals. Besides accuracy, a MORS can operate at the *global* level, where additional beyond-accuracy goals are met for the system as a whole, or at the *individual* level, meaning that the recommendations are tailored to the needs of each user. The state-of-the-art MORSs either operate at the global or individual level, without assuming the co-existence of the two perspectives. In this study, we show that when global and individual objectives co-exist, MORSs are not able to meet both types of goals. To overcome this issue, we present an approach that regulates the recommendation lists so as to guarantee both global and individual perspectives, while preserving its effectiveness. Specifically, as individual perspective, we tackle genre calibration and, as global perspective, provider fairness. We validate our approach on two real-world datasets, publicly released with this paper (https://tinyurl.com/yc6nnx5v).

Keywords: Multi-Objective Recommendation · Calibration · Provider Fairness

1 Introduction

Motivation. Since the goal of recommender systems is to provide relevant suggestions for the users, the main focus has been the effectiveness of the results [37]. Nevertheless, users might be interested in properties of the items besides their effectiveness, and there are other stakeholders who can benefit from how recommendations are produced (e.g., content providers). Hence, beyond-accuracy perspectives are central to the generation and evaluation of recommendations.

Multi-Objective Recommender Systems (MORSs) support the provision of perspectives that go beyond item relevance, such as, e.g., diversity, calibration, and fairness [49]. The optimization for these objectives can happen at the *global*

N. Goharian et al. (Eds.): ECIR 2024, LNCS 14608, pp. 21–38, 2024.
https://doi.org/10.1007/978-3-031-56027-9_2

(*aggregate*) level, thus ensuring that the system as a whole can guarantee certain properties (e.g., all providers receive a certain exposure in the recommendation lists). In alternative, a MORS can operate at the *individual* (*local*) level, and shape results that are consider the prominence of individual users towards the different goals (e.g., each user can receive a different level of diversity or the recommended genres can be calibrated to the preferences in the training set) [21].

When analyzing the current literature, a MORS either operates at the global level [15,26,28,32,44] or at the local level [7,8,34].

Open Issues. There might be scenarios in which both global and individual objectives co-exist. Indeed, a platform might decide that, as a whole, the recommendations should offer certain properties (e.g., be fair to providers of different demographic groups, or enable a certain level of novelty). Moreover, specific goals might be set for the individual users (e.g., the calibration of the genres or the diversity of the recommended items might need to follow what is observed in the training set of each user). As we show in Sect. 6, when a MORS tackles *only* global or individual perspectives, the other perspective trivially remains under-considered and cannot be guaranteed by the system.

Our Contributions. To overcome the aforementioned challenges, in this paper, we present a MORS that produces recommendations with both global and individual objectives. As a use case, we consider, as a global objective, *provider fairness* and, as an individual one, *calibrated recommendations.* This aligns our study with the rest of the MORS literature, where two beyond-accuracy objectives are considered. For the sake of clarity, we will talk about *provider-fair and calibrated recommendations* but, as we discuss in Sect. 3.2, **our approach can be generalized to any global or individual objective**.

Besides accounting for beyond-accuracy perspectives involving both global and individual objectives, the problem of providing provider-fair and calibrated recommendations becomes interesting also from a practical point of view. As we will show in Sect. 4, users tend to rate items of certain genres and that are produced in certain geographic areas, suggesting that we can account for both perspectives at the same time when generating the recommendations. Hence, at the technical level, we would need a unique solution that (i) produces effective results for the users, (ii) can provide fairness for providers belonging to different groups at the *global* level, i.e., by distributing, over the entire user base, the recommendation of items belonging to different provider groups in equitable ways, and (iii) can calibrate the recommendation lists of each *individual* user.

Our approach involves a post-processing strategy. To enable a form of provider fairness that can consider demographic groups that are not necessarily characterized by a binary group (e.g., males and females), we consider, as a sensitive attribute, the geographic provenance of the providers and have the different continents as the granularity with which we split the groups; this is aligned with recent literature on provider fairness [17,18]. As in classic calibrated fashion, we distribute the recommendations according to the item genre. Based on this characterization of the data, we present an approach that makes use of buckets to associate the continents in which the items are produced and the genre of the items. We use these buckets to post-process the recommendation lists (we will later discuss that this is the best way to regulate both aggregate- and individual-level properties) and regulate how the recommendations are distributed across the users. Thanks to the fact that each bucket contains (i) the continent in which the item is produced, to regulate provider fairness, and (ii) the genre of the item, to regulate calibration, both global and individual perspectives are captured at once by our approach. To validate our proposal, we apply it to the recommendations produced by five algorithms, and study the effectiveness of our approach on two datasets (including a novel one, released with this study), and against state-of-the-art approaches for calibrated recommendation and provider fairness.

Concretely, our contributions can be summarized as follows:

- After the identification of the research gaps (Sect. 2) and characterization of our setting (Sect. 3), we provide the foundations to our use case, by showing that calibration and provider fairness are related problems, since the genres of the items and their country of production are connected (Sect. 4);
- We present an approach to post-process the recommendation lists to meet both global and individual goals. We calibrate the results for the individual users in terms of genres, and are fair towards providers (Sect. 5);
- We face the limitation of evaluating this problem, due to the scarcity of data offering both the category of the items and the sensitive attributes of the providers, so we i) extend the MoviLens-1M dataset, to integrate the continent of production of each item, and ii) we collect and present (in Sect. 3) a novel dataset. Both resources are publicly available here[1];
- We perform experiments (Sect. 6) to validate our proposal when applied to the recommendation produced by five algorithms, covering both memory- and model-based approaches, and point-wise and pair-wise approaches. To evaluate its effectiveness in different domains, we consider movie and song recommendation as application scenarios. Based on our outcomes, we highlight possible research paths that might emerge from it (Sect. 7).

2 Related Work

MORSs. Recent literature has studied how to account for multi-objective goals from different angles. The user perspective was tackled by Li *et al* [26], which balance recommendation accuracy for users with different levels of activities. From an item perspective, Ge *et al.* [15] proposed an approach to balance item relevance and exposure. Considering both the user and item perspectives, Naghiaei *et al.* [32] propose a re-ranking approach to account for consumer and provider fairness. Other studies blend the multiple objectives into a single function, in order to obtain a Pareto-optimal solution [28,44]. Recent advances have also proposed MORSs in sequential settings, by optimizing the results for accuracy, diversity, and novelty [41]. MORS that operate at the individual level have optimized the

[1] https://tinyurl.com/yc6nnx5v.

recommendation process mainly via online interactions, such as conversational approaches [25] or via critiquing [12,43], but approaches aiming at learning individual propensities from past interactions also exist, e.g., [7,8,22,34].

Calibrated Recommendation. *Calibration* is a well-studied technique commonly used to solve the problem of unfair output [33,42,46] in recommender systems. Seymen *et al.* [40] address the problem of providing calibration in the recommendations from a constrained optimization perspective. Abdollahpouri *et al.* [1] study the connection between popularity bias, calibration, and consumer fairness in recommendation. Recently, Rojas *et al.* [38] analyze how the calibration method in [42] deals with the bias in different recommendation models. Other studies focus on analyzing user profiles to mitigate miscalibrated recommendations [27] or to mitigate popularity bias from the user's perspective [6]. Existing metrics have some limitations when applying a user-centered approach to evaluate popularity bias and calibrated recommendations. To address these limitations, Abdollahpouri *et al.* [2] present a new metric.

Provider Fairness. *Provider fairness* [9] has been studied in many common scenarios, e.g., [10,11,14,16,18,29,30]. It is usually assessed by considering metrics such as the visibility and the exposure that respectively assess the amount of times an item is present in the rankings [13,47] and *where* an item is ranked [5,48], for users to whom each provider's items are recommended. Other approaches, such as that by Karakolis *et al.* [23], consider diversity and coverage for users. Raj *et al.* [35] present a comparative analysis among several fairness metrics recently introduced to measure fair ranking. Wu *et al.* [45] formalize a family of exposure fairness metrics that model the problem of fairness jointly from the perspective of both types of stakeholders.

Contextualizing Our Work. No MORS can address both calibrated recommendation lists for the users and provider fairness. Our algorithm's aims are to provide i) each user with calibrated recommendations, ii) fair recommendations for the providers, iii) aiming at a minimum loss in effectiveness.

3 Preliminaries

3.1 Recommendation Scenario

Let $U = \{u_1, u_2, ..., u_n\}$ be a set of users, $I = \{i_1, i_2, ..., i_j\}$ be a set of items, and V be a totally ordered set of values that can be used to express a preference together with a special symbol $\perp$. The set of ratings results from a map $r : U \times I \rightarrow V$, where V is the rating domain. If $r(u,i) = \perp$, then we say that u did not rate i. To easy notation, we denote $r(u,i)$ by r_{ui}. We define the set of ratings as $R = \{(u, i, r_{ui}) : u \in U, i \in I, r_{ui} \neq \perp\}$ and they can directly feed an algorithm in the form of triplets (point-wise approaches) or shape user-item observations (pair-wise approaches). We denote with R_u the ratings associated with a user $u \in U$. We consider a temporal split of the data, where a fixed percentage of the ratings of the users (ordered by timestamp) goes to the training and the rest goes to the test set [4]. The goal is to learn a function f that estimates the relevance ($\hat{r}_{ui}$) of the user-item pairs that do not appear in the training data (i.e., $r_{ui} = \perp$). We denote as $\hat{R}$ the set of recommendations.

Let C denote the set of geographic continents in which items are organized. We consider a geographic continent as the provenance of an item provider. We denote as C_i the set of geographic continents associated with an item i. Note that, since an item could be produced by more than one provider, it might be associated with several geographic continents, and thus, $|C_i| \geq 1$ and $C_i \subseteq C$. In case two providers belong to the same geographic continent, that continent appears only once; indeed, we are dealing with group fairness so, when a group of providers is associated with an item (once or multiple times), we account for its presence. We use the geographic continents to shape demographic groups, which can be defined to group the ratings of the items produced in a continent (we denote the items in I produced in a continent $c \in C$ as I_c, where $I_c \subseteq I$).

Let G denote the set of genres in which items are organized. We denote as G_i the set of genres associated with an item i. Note that, an item can be of one or more genres, and thus, $|G_i| \geq 1$ and $G_i \subseteq G$. We denote the items in I that have a genre $g \in G$ as I_g, where $I_g \subseteq I$.

3.2 Metrics

Provider-Group Representation. In order to enable provider fairness, we should understand the attention received by a provider group in the training data. For this reason, we compute the representation of a demographic group in the data as the number of ratings for items associated with that group in the data. We define with $\mathcal{R}$ the *representation* of a group $c \in C$ as follows:

$$\mathcal{R}_c = \frac{|\{r_{ui} : u \in U, i \in I_c\}|}{|R|} \quad (1)$$

Equation (1) accounts for the proportion of ratings given to the items of a demographic group associated with a continent. This metric ranges between 0 and 1. We compute the representation of a group only considering the training set. Trivially, the sum of the representations of all groups is equal to 1.

User-Based Genre Propensity. In order to calibrate the results for the users, we need to understand how the preferences for the different item genres are distributed. For this reason, we define with $\mathcal{P}$ the *propensity* of a user of $u \in U$ to rate items of a genre $g \in G$, as follows:

$$\mathcal{P}_{ug} = \frac{|\{r_{ui} : g \in G_i\}|}{|R_u|} \quad (2)$$

Equation (2) accounts for the proportion of ratings associated with a genre for a given user. This metric ranges between 0 and 1. Trivially, the sum of the propensities of all genres for a user is equal to 1. This metric is equivalent to the distribution $p(g|u)$ [42].

Disparate Impact. We assess unfairness with the notion of *disparate impact* generated by a recommender system. Specifically, we assess disparate visibility.

Definition 1 (Disparate visibility). *Given a group* $c \in C$, *the* disparate visibility *returned by a recommender system for that group is measured as the difference between the share of recommendations for items of that group and the representation of that group in the input data:*

$$\Delta\mathcal{V}_c = \left(\frac{1}{|U|} \sum_{u \in U} \frac{|\{\hat{r}_{ui} : i \in I_c\}|}{|\hat{R}|} \right) - \mathcal{R}_c \tag{3}$$

The range of values for this score is $[-\mathcal{R}_c, 1 - \mathcal{R}_c]$; specifically, it is 0 when the recommender system has no disparate visibility, while negative/positive values indicate that the group received a share of recommendations that is lower/higher than its representation. This metric is based on that defined by Fabbri *et al.* [13].

Miscalibration. We assess the tendency of a system to recommend a user items whose genres are distributed differently from those they prefer via *miscalibration.*

Definition 2 (Miscalibration). *Given a user* $u \in U$ *and a genre* $g \in G$, *the miscalibration returned by a recommender system for that user is measured as the difference between the share of recommendations for items of that genre and the propensity of the user for that genre in the training data:*

$$\Delta\mathcal{M}_{ug} = \frac{|\{\hat{r}_{ui} : i \in I_g\}|}{|\hat{R}_u|} - \mathcal{P}_{ug} \tag{4}$$

Generalizability. The rest of our paper will consider disparate visibility ($\Delta\mathcal{V}_c$) as the *global* perspective and miscalibration ($\Delta\mathcal{M}_{ug}$) as the *individual* perspective our MORS considers. Nevertheless, our approach can be generalized to *any* metric that assesses the difference between (i) the distribution of the recommendations and (ii) what can be observed in the training set or an objective set by the platform via a policy (e.g., a given amount of content novelty or diversity).

4 Matching Item Providers and Genre Propensity

4.1 Real-World Datasets

First, we extended the MovieLens-1M dataset, so as to integrate the continent of production of each movie. Second, a domain that fits our problem is song recommendation. However, existing music datasets, such as LastFM-2B [31], do not contain song genres and sensitive attributes of the artists, so they do not fit our problem. Thus, we collected a dataset from an online music platform.

In particular, the **MovieLens-1M (Movies)** dataset comprises 1M ratings (range 1–5), from 6,040 users for 3,600 movies across 18 genres. The dataset provides its IMDB ID, which allowed us to associate it to its continent of production, thanks to the OMDB APIs (http://www.omdbapi.com/). Keep in mind that *a movie may be produced on more than one continent.* On the other hand, **BeyondSongs (Songs)** contains 1,777,981 ratings (range 1–5), provided by 30,759 users, to 16,380 songs. For each song, we collected the continent of provenance of the artist, and 14 music genres. Both resources are available online (See footnote 1).

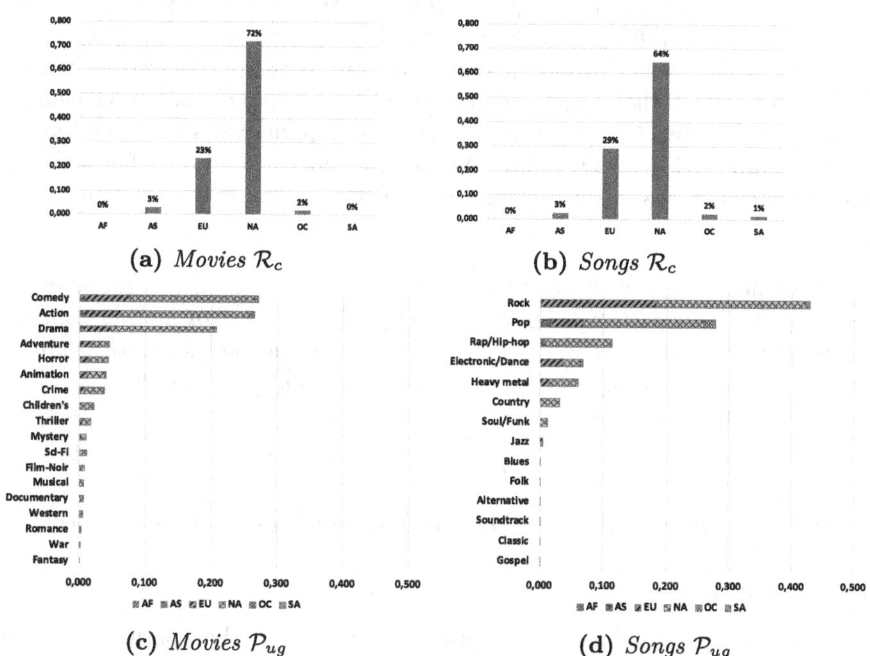

Fig. 1. Group representation (a and b) and genre propensity (c and d) in the Movies and Songs data. Acronyms stand for AF: Africa, AS: Asia, EU: Europe, NA: North America, OC: Oceania, SA: South America.

4.2 Characterizing Group Representation and Genre Propensity

We consider the temporal split of the data, where 80% of the ratings are considered for the training set and have been used to measure $\mathcal{R}_c$ and $\mathcal{P}_{ug}$. Note that, while the representation of a demographic group covers the entire training set, the propensity is measured at the user level. Hence, to characterize the link between the two phenomena we aggregate the propensity of all the users for a given genre by summing their values.

Figures 1a and 1b show the $\mathcal{R}_c$ for Movies and Songs, respectively. Both datasets depict a similar representation by continents, where the highest representation is of items from NA providers (72% in movies and 64% in songs) and the second place is for EU providers (23% and 29%). In the rest of the continents, for both datasets, it is less than 10%. Figures 1d and 1d show the $\mathcal{P}_{ug}$ in both datasets; three genres attract most of the ratings by users.

We can also observe that ratings seem to be clustered between certain genre-continent pairs. In other words, different genres are distributed differently across continents. In the Movies data (Fig. 1c), Comedy movies are largely preferred when produced by EU producers, just as Action attracted the majority of ratings for movies by NA producers. In the Songs data (Fig. 1d), the Electronic/Dance genre was consumed much more heavily when produced by EU artists than by those in the rest of the world, and Heavy metal songs are mostly consumed when they come from NA. In both datasets, users' preferences for the minority provider groups (AF, AS, OC, and SA) are also concentrated on a few selected genres, confirming this rating aggregation in certain genre-continent pairs.

Observation 1. *Users have the propensity to rate items of certain genres and that are produced by certain geographic groups (i.e., in certain continents). Calibration and provider fairness are related problems so, when producing recommendation lists, both perspectives should be accounted at the same time, in MORS fashion.*

5 Individually Calibrated and P-Fair Recommendation

5.1 Algorithm

MOREGIN adjusts the recommendations according to the continent of the providers and the representation of each demographic group and seeks to make a calibration at the individual level, following the propensity of each user to rate items of a given genre. Formally, MOREGIN (see Algorithm 1) works following four main steps. **Steps 1 and 2** are devoted to compute $\mathcal{R}_c$ and $\mathcal{P}_{ug}$, considering the ratings in the training set. **Step 3** computes the items that were predicted as relevant for a user by the recommender system and creates a bucket list, *joinBucket*, considering each continent-genre pair, which will store the predicted items. Each bucket comes with two attributes: $\mathcal{R}_c$ and $\mathcal{P}_{ug}$. Specifically, the recommender system returns a list of top-n recommendations (where n is much larger than the cut-off value k, so as to be able to perform a re-ranking). Our starting point to fill a bucket is the relevance predicted for a user u and an item i, $\hat{r}_{ui}$. That item will be stored in the buckets associated with each genre $g \in G_i$ and each continent $c \in C_i$ (even though an item may appear in more than one bucket, it can only be recommended only once). Each element in the bucket is a record that contains the item ID and the $\hat{r}_{ui}$. We sort each bucket considering three values. We sort out $\mathcal{R}_c$ and $\mathcal{P}_{ug}$, in ascending order to ensure the inclusion in the recommendation lists of items from genres and continents that are less represented in the dataset, and we sort in descending order by rating to enhance those products that are relevant to the user. Finally, **Step 4**

```
Input: recList: ranked list (records contain user, item, rating, position, genre,
       continent), which arrives sorted by user and rating and contains topn
       recommendations to the user.
       trainList: list with the training set (records contain user, item, rating,
       genre, continent), which is sorted by user and rating.
       topk: top k recommendations, we set up k = 10.
       topn: top n recommendations, we set up n= 1000.
Output: reRankedList: ranked list with Individually Calibrated and P-Fair
        Recommendation.
define MOReGIn (recList, trainList, topk, topn)
begin
    // Step 1. Compute R_c
    recBucketRep ← computeRepresentation(topk, recList, trainList);
    // Step 2. Compute P_ug
    recBucketUserProp ← computePropensity(topk, recList, trainList);
    // Step 3. Create a bucket list
    joinBucket ← recList + recBucketRep + recBucketUserProp;
    joinBucket ← sort(joinBucket);
    // Step 4. Perform selection of items with three phases
    userCounts, userGenCounts , contCounts ← ∅;
    joinBucket ← selectWithHardConstraints(joinBucket, recBucketRep,
      recBucketUserProp, userCounts, userGenCounts, contCounts) ;
      // Phase 1
    joinBucket ← selectWithSoftConstraints(joinBucket, recBucketRep,
      recBucketUserProp, 2, userCounts, userGenCounts, contCounts) ;
      // Phase 2
    joinBucket ← selectWithSoftConstraints(joinBucket, recBucketRep,
      recBucketUserProp, 3, userCounts, userGenCounts, contCounts) ;
      // Phase 3
    reRankedList ← chooseSelectedItems(joinBucket);
    reRankedList ← sort(reRankedList) ; // sort by user and rating
    return reRankedList;
end
```

Algorithm 1: Pseudocode of MOReGIn algorithm

performs a three-phase re-ranking based on the generated bucket lists. Phase 1 is where we begin, and subsequent phases occur until the top-k is complete. In detail, **Phase 1** selects items starting from the least represented continents to the most represented ones in their corresponding buckets. The algorithm selects items with these conditions: (1) the percentage of items in the recommendation list for a continent is lower or equal to the representation of the continent ($\mathcal{R}_c$); (2) the percentage of items of a given genre in the top-k is lower or equal than $\mathcal{P}_{ug} \cdot k$; and (3) the number of recommended items so far is lower than k. **Phase 2** relaxes the restrictions of phase 1 and here condition 2 is not applied. **Phase 3** selects the items that have the greater relevance for the user, until we complete the top-k. That is, conditions 1 and 2 are not considered.

```
define selectWithHardConstraints(joinBucket, recBucketRep,
    recBucketUserProp, userCounts, uGenCounts, contCounts) begin
    expectedRecordsCont ← getExpectedRecordsCont(recBucketRep);
    expectedRecUserGen ← getRecordsUserGen(recBucketUserProp);
    foreach rec ∈ joinBucket do // for each record
        userGen ← rec.user + " − " + rec.genre;
        if userGen ∈ expectedRecUserGen and
          rec.cont ∈ expectedRecordsCont then
            userCounts[rec.user] ← userCounts[rec.user] + 1;
            uGenCounts[rec.userGen] ← uGenCounts[rec.userGen] + 1;
            contCounts[rec.cont] ← contCounts[rec.cont] + 1;
            if expectedRecUserGen[rec.userGen] ≥ userGenCounts and
              expectedRecordsCont[rec.cont] ≥ contCounts and topk ≥
              userCounts[rec.user] then
                rec.phase ← 1; // selects element in phase 1
                joinBucket.update(rec); // updates the element
            end
        end
    end
    return joinBucket
end
define selectWithSoftConstraints(joinBucket, recBucketRep,
    recBucketUserProp, phaseMOReGIn, userCounts, userGenCounts,
    contCounts) begin
    expectedRecordsCont ← getExpectedRecordsCont(recBucketRep);
    foreach rec ∈ joinBucket do // for each record
        if rec.cont ∈ expectedRecordsCont then
            if phaseMOReGIn == 2 then
                if expectedRecordsCont[rec.cont] ≥ contCounts and topk ≥
                  userCounts[rec.user] then
                    userCounts[rec.user] ← userCounts[rec.user] + 1;
                    contCounts[rec.cont] ← contCounts[rec.cont] + 1;
                    rec.phase ← 2; // selects element in phase 2
                    joinBucket.update(rec); // updates the element
                end
            end
            if phaseMOReGIn == 3 then
                if topk ≥ userCounts[rec.user] then
                    contCounts[rec.cont] ← contCounts[rec.cont] + 1;
                    rec.phase ← 3; // selects element in phase 3
                    joinBucket.update(rec); // updates the element
                end
            end
        end
    end
    return joinBucket
end
```

Algorithm 2: Selection methods for the MOReGIn algorithm

6 Experimental Evaluation

6.1 Experimental Methodology

In this work, we focus on well-known state-of-the-art Collaborative Filtering algorithms: **ItemKNN** [39], **UserKNN** [20], **BPRMF** [36], **SVDpp** [24], and **NeuMF** [19]). We will report the results of the original recommendation algorithm (denoted as **OR**). We also consider two comparison baselines: (i) a greedy calibration algorithm [42] (denoted as **CL**) with a λ value of 0.99 (setup defined in [42]), which post-processes the recommendation lists generated by traditional recommender systems; and (ii) a provider fairness algorithm [17] (denoted as **PF**) that considers the providers' continent provenance as a sensitive attribute, with a re-ranking approach that regulates the share of recommendations given to the items produced in a continent (visibility) and the positions in which items are ranked in the recommendation list (exposure).

To run the recommendation models, we used the *Elliot* framework [3], which generated the recommendations for each user that fed the input of MOReGIn. As noted in Sect. 4.2, the dataset was divided into two sets, one for training (80%) and the other for testing with the most recent ratings of each user (20%).

For each user, we generated the top-1000 recommendations (denoted in the paper as the top-n; we remind the reader that these $n = 1000$ results are not shown to the users, they are only used internally by our algorithm) to then re-rank the top-k (set up to 10) through the proposed MOReGIn algorithm. We performed a grid search of the hyper-parameters for each model in the two datasets. For ItemKNN and UserKNN, in both datasets, we use 50 *neighbors*, a cosine *similarity*, and the classical *implementation*. For BPRMF, SVDpp, and NeuMf we defined 10 *epochs* and 10 *factors* on each dataset, except NeuMF in Movies that uses 12 *factors*. The *batch size* is 512 for SVDpp and NeuMF and is 1 for BPRMF on both datasets. Moreover, for BPRMF in Movies~Songs, *learning rate* = 0.1~1.346, *bias regularization* = 0~1.236, *user regularization* = 0.01~1.575, *positive item regularization* = 0.01~1.376, and *negative item regularization* = 0.01~1.624; for SVDpp in Movies~Songs, *learning rate* = 0.01~0.001, *factors regularization* = 0.1~0.001, and *bias regularization* = 0.001 in both datasets; NeuMF in Movies~Songs, the *multi-layer perceptron* = 10 in both, *learning rate* = 0.0025 in both, and *factors regularization* = 0.1~0.001.

6.2 Assessment of Disparities and Mitigation

Table 1 compares MOReGIn with the baselines in terms of the overall disparate visibility, $\Delta Total$, for each continent. It is computed as $\forall c \in C$, $\Delta Total = \sum \Delta \mathcal{V}_c$. MOReGIn almost entirely reduces the disparities in both movies and song datasets, where most results are $\Delta Total = 0.0000$. Although there is a little difference in the $\Delta Total$ between some approaches, these differences are more explicit, considering the provider provenance. For example, in the movie domain with the BPRMF algorithm, the $\Delta Total$ value in the OR approach is similar to that of PF. However, in a more detailed analysis of more representative continents such as NA and EU, there are notorious differences between the two approaches (i.e., 0.0075 for OR in contrast to −0.0066 for PF in the NA continent, see the example shown in Fig. 2a). It is important to highlight that in both domains, our proposal mitigates the disparity regardless of the provenance of the provider, in contrast to the other algorithms that show a clear dependence on the data (i.e., the continent attribute).

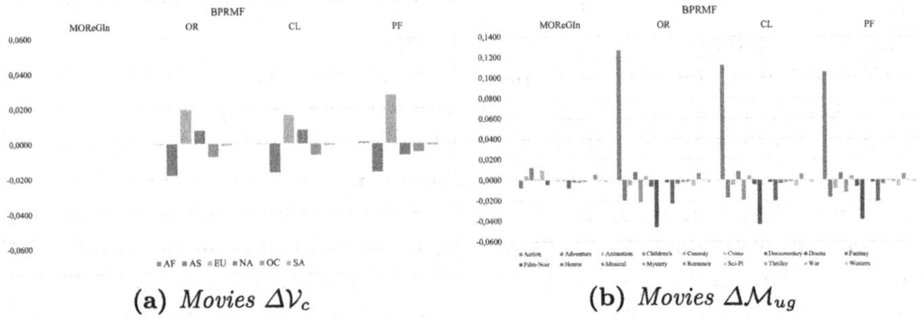

(a) *Movies* $\Delta\mathcal{V}_c$ **(b)** *Movies* $\Delta\mathcal{M}_{ug}$

Fig. 2. Disparity mitigation per continent (a) and miscalibration per genre (b) in BPRMF.

Table 1. Results of disparity mitigation of continents in the Movies and Songs datasets. Each value represents the sum of disparities, $\Delta Total$.

	MOVIES				SONGS			
	OR	CL	PF	MOReGIn	OR	CL	PF	MOReGIn
BPRMF	0.0539	0.0485	0.0576	**0.0000**	0.2637	0.0840	0.2628	**0.0000**
SVDpp	0.1154	0.1085	0.1059	**0.0000**	0.2678	0.1063	0.2445	**0.0000**
NeuMF	0.0395	0.0421	0.0638	**0.0000**	0.4434	0.4516	0.3990	**0.0000**
UserKNN	0.0345	0.0327	0.0328	**0.0000**	0.0361	0.0575	0.0370	**0.0000**
ItemKNN	0.0431	0.0418	0.0412	**0.0000**	0.0392	0.0583	0.0420	**0.0000**

Regarding the item genres, Table 2 compares MOReGIn with the baselines in terms of the overall miscalibration, $\Delta Genre$, for each continent. It is computed as $\forall g \in G$ and each user $u \in U$, $\Delta Genre = \sum \Delta\mathcal{M}_{ug}$. For both datasets, MOReGIn obtained the best $\Delta Genre$ (i.e., lowest miscalibration) in all the recommendation models. An analysis of how the algorithms behave with the different genres is shown in Fig. 2b. Although miscalibration never reaches values of $\Delta Genre$ equal to zero, our proposal always calibrates better than the baselines.

Observation 2. MOREGIN, *by taking action on the distribution of the items per genre at the user level and provider provenance at the same time, can both calibrate and be fair to the providers. This joint effort allows us to improve the capability to calibrate the results and to be fair to providers with respect to baselines devoted solely to these purposes.*

6.3 Impact on the Quality of Recommendations

We evaluate the accuracy for the different approaches via the NDCG metric.

Table 3 shows its values for MOREGIN and the rest of the baselines, in all the recommendation algorithms, for Movies and Songs. MOREGIN obtained a better NDCG than the PF model, except for BPRMF and ItemKNN in the Movies dataset, and UserKNN and ItemKNN in the Songs dataset. Similar results are obtained with the CL method. Comparing MOREGIN to a non-fair approach, MOReGIn outperforms OR models, with the exception of BPRMF and ItemKNN in the Movie domain. Except for UserKNN and ItemKNN, MOREGIN also outperforms the OR model in the Songs domain.

All recommendation quality results show that the need for fairer and calibrated recommendations impacts the recommendation quality. However, beyond-

Table 2. Results of miscalibration of genres in the Movies and Songs datasets. Each value represents the sum of miscalibrations, $\Delta Genre$.

	MOVIES				SONGS			
	OR	CL	PF	MOReGIn	OR	CL	PF	MOReGIn
BPRMF	0.2892	0.2606	0.2454	**0.0634**	5.5107	0.0772	0.4610	**0.0289**
SVDpp	0.5792	0.5026	0.5694	**0.1184**	0.5773	0.1031	0.5029	**0.0256**
NeuMF	0.4596	0.3962	0.3735	**0.2901**	1.2886	0.7494	1.2202	**0.0787**
UserKNN	0.0743	0.0862	0.0580	**0.0392**	0.0298	0.0989	0.0291	**0.0208**
ItemKNN	0.2102	0.1966	0.1954	**0.0559**	0.0890	0.0601	0.0879	**0.0205**

Table 3. NDCG for each approach and recommendation algorithm.

	MOVIES				SONGS			
	OR	CL	PF	MOReGIn	OR	CL	PF	MOReGIn
BPRMF	**0.3204**	0.3144	0.3195	0.3057	0.0034	**0.0067**	0.0031	0.0055
SVDpp	0.0830	0.0888	0.0812	**0.1024**	0.0050	0.0103	0.0051	**0.0138**
NeuMF	0.1963	0.1931	0.1956	**0.2050**	0.0183	0.0098	0.0179	**0.0314**
UserKNN	0.3051	0.2954	0.3030	**0.3053**	**0.3760**	0.1925	0.3759	0.2648
ItemKNN	**0.3229**	0.3145	0.3211	0.3131	**0.3860**	0.1668	0.3857	0.2864

accuracy perspectives, such as those offered by MOReGIn allows for compensating for the minimal loss in quality with more unbiased recommendations.

7 Conclusions and Future Work

Global and individual objectives in MORs have never been studied jointly. To study this problem, we provided data, by i) extending the MovieLens-1M dataset and ii) collecting a new dataset for song recommendation. The analysis of this data showed that when users rate items of a given genre, the geographic provenance of that item matters. Based on these insights, we proposed a new post-processing approach, named MOReGIn, that aggregates the recommended items into buckets, pairing item genres and their continent of production. Results show that MOReGIn outperforms the existing approaches at producing effective, calibrated, and provider-fair recommendations. Future work will explore different strategies to generate recommendation lists given the generated buckets. Moreover, we will consider consumer fairness as a global perspective.

Acknowledgments. D. Contreras research was partially funded by postdoctoral project (grant No. 74200094) from ANID-Chile and by the supercomputing infrastructure of the NLHPC (ECM-02). M. Salamó was supported by the FairTransNLP-Language Project (MCIN-AEI-10.13039-501100011033-FEDER) and by the Generalitat de Catalunya (2021 SGR 00313). Maria also belongs to the Associated unit to CSIC by IIIA.

References

1. Abdollahpouri, H., Mansoury, M., Burke, R., Mobasher, B.: The connection between popularity bias, calibration, and fairness in recommendation. In: Fourteenth ACM Conference on Recommender Systems, pp. 726–731 (2020)
2. Abdollahpouri, H., Mansoury, M., Burke, R., Mobasher, B., Malthouse, E.: User-centered evaluation of popularity bias in recommender systems. In: Proceedings of the 29th ACM Conference on User Modeling, Adaptation and Personalization, pp. 119–129 (2021)
3. Anelli, V.W., et al.: Elliot: a comprehensive and rigorous framework for reproducible recommender systems evaluation, pp. 2405–2414. Association for Computing Machinery, New York (2021)
4. Bellogín, A., Castells, P., Cantador, I.: Statistical biases in information retrieval metrics for recommender systems. Inf. Retr. J. **20**(6), 606–634 (2017). https://doi.org/10.1007/s10791-017-9312-z
5. Biega, A.J., Gummadi, K.P., Weikum, G.: Equity of attention: amortizing individual fairness in rankings. In: The 41st International ACM SIGIR Conference on Research & Development in Information Retrieval, SIGIR 2018, pp. 405–414. ACM, New York (2018). https://doi.org/10.1145/3209978.3210063
6. Chen, J., Wu, W., Shi, L., Zheng, W., He, L.: Long-tail session-based recommendation from calibration. Appl. Intell. **53**(4), 4685–4702 (2023). https://doi.org/10.1007/s10489-022-03718-7

7. Dokoupil, P., Peska, L., Boratto, L.: Looks can be deceiving: Linking user-item interactions and user's propensity towards multi-objective recommendations. CoRR abs/2307.00654 (2023). arXiv arXiv:2307.00654. https://doi.org/10.48550/arXiv.2307.00654
8. Dokoupil, P., Peska, L., Boratto, L.: Rows or columns? Minimizing presentation bias when comparing multiple recommender systems. In: Chen, H., Duh, W.E., Huang, H., Kato, M.P., Mothe, J., Poblete, B. (eds.) Proceedings of the 46th International ACM SIGIR Conference on Research and Development in Information Retrieval, SIGIR 2023, Taipei, Taiwan, 23–27 July 2023, pp. 2354–2358. ACM (2023). https://doi.org/10.1145/3539618.3592056
9. Ekstrand, M.D., Das, A., Burke, R., Diaz, F.: Fairness in recommender systems. In: Ricci, F., Rokach, L., Shapira, B. (eds.) Recommender Systems Handbook, pp. pp 679–707. Springer, New York (2022). https://doi.org/10.1007/978-1-0716-2197-4_18
10. Ekstrand, M.D., Kluver, D.: Exploring author gender in book rating and recommendation. User Model. User-Adap. Inter. **31**(3), 377–420 (2021). https://doi.org/10.1007/s11257-020-09284-2
11. Ekstrand, M.D., Tian, M., Kazi, M.R.I., Mehrpouyan, H., Kluver, D.: Exploring author gender in book rating and recommendation. In: Proceedings of the 12th ACM Conference on Recommender Systems, RecSys 2018, pp. 242–250. ACM (2018). https://doi.org/10.1145/3240323.3240373
12. Elahi, M., Ge, M., Ricci, F., Massimo, D., Berkovsky, S.: Interactive food recommendation for groups. In: 8th ACM Conference on Recommender Systems, RecSys 2014. CEUR-WS (2014)
13. Fabbri, F., Bonchi, F., Boratto, L., Castillo, C.: The effect of homophily on disparate visibility of minorities in people recommender systems. In: Proceedings of the Fourteenth International AAAI Conference on Web and Social Media, ICWSM 2020, pp. 165–175. AAAI Press, USA (2020). https://ojs.aaai.org/index.php/ICWSM/issue/view/262
14. Ferraro, A., Serra, X., Bauer, C.: What is fair? Exploring the artists' perspective on the fairness of music streaming platforms. In: Ardito, C., et al. (eds.) INTERACT 2021, Part II. LNCS, vol. 12933, pp. 562–584. Springer, Cham (2021). https://doi.org/10.1007/978-3-030-85616-8_33
15. Ge, Y., et al.: Toward pareto efficient fairness-utility trade-off in recommendation through reinforcement learning. In: Candan, K.S., Liu, H., Akoglu, L., Dong, X.L., Tang, J. (eds.) The Fifteenth ACM International Conference on Web Search and Data Mining, WSDM 2022, Virtual Event/Tempe, AZ, USA, 21–25 February 2022, pp. 316–324. ACM (2022). https://doi.org/10.1145/3488560.3498487
16. Gharahighehi, A., Vens, C., Pliakos, K.: Fair multi-stakeholder news recommender system with hypergraph ranking. Inf. Process. Manage. **58**(5), 102663 (2021). https://doi.org/10.1016/j.ipm.2021.102663. https://www.sciencedirect.com/science/article/pii/S0306457321001515
17. Gómez, E., Boratto, L., Salamó, M.: Provider fairness across continents in collaborative recommender systems. Inf. Process. Manag. **59**(1), 102719 (2022). https://doi.org/10.1016/j.ipm.2021.102719
18. Gómez, E., Zhang, C.S., Boratto, L., Salamó, M., Ramos, G.: Enabling cross-continent provider fairness in educational recommender systems. Fut. Gener. Comput. Syst. **127**, 435–447 (2022). https://doi.org/10.1016/j.future.2021.08.025
19. He, X., Liao, L., Zhang, H., Nie, L., Hu, X., Chua, T.S.: Neural collaborative filtering. In: Proceedings of the 26th International Conference on World Wide Web, pp. 173–182 (2017)

20. Herlocker, J.L., Konstan, J.A., Riedl, J.: An empirical analysis of design choices in neighborhood-based collaborative filtering algorithms. Inf. Retr. **5**(4), 287–310 (2002). https://doi.org/10.1023/A:1020443909834
21. Jannach, D.: Multi-objective recommendation: overview and challenges. In: Abdollahpouri, H., et al. (eds.) Proceedings of the 2nd Workshop on Multi-objective Recommender Systems co-located with 16th ACM Conference on Recommender Systems, RecSys 2022, CEUR Workshop Proceedings, Seattle, WA, USA, 18th–23rd September 2022, vol. 3268. CEUR-WS.org (2022). https://ceur-ws.org/Vol-3268/paper1.pdf
22. Jugovac, M., Jannach, D., Lerche, L.: Efficient optimization of multiple recommendation quality factors according to individual user tendencies. Exp. Syst. Appl. **81**, 321–331 (2017). https://doi.org/10.1016/j.eswa.2017.03.055. https://www.sciencedirect.com/science/article/pii/S0957417417302075
23. Karakolis, E., Kokkinakos, P., Askounis, D.: Provider fairness for diversity and coverage in multi-stakeholder recommender systems. Appl. Sci. **12**(10) (2022). https://doi.org/10.3390/app12104984. https://www.mdpi.com/2076-3417/12/10/4984
24. Koren, Y.: Factorization meets the neighborhood: a multifaceted collaborative filtering model. In: Proceedings of the 14th ACM SIGKDD International Conference on Knowledge Discovery and Data Mining, pp. 426–434. ACM (2008). https://doi.org/10.1145/1401890.1401944
25. Li, R., Kahou, S., Schulz, H., Michalski, V., Charlin, L., Pal, C.: Towards deep conversational recommendations. In: Proceedings of the 32nd International Conference on Neural Information Processing Systems, NIPS 2018, , Red Hook, NY, USA, pp. 9748–9758. Curran Associates Inc. (2018)
26. Li, Y., Chen, H., Fu, Z., Ge, Y., Zhang, Y.: User-oriented fairness in recommendation. In: Leskovec, J., Grobelnik, M., Najork, M., Tang, J., Zia, L. (eds.) The Web Conference 2021, WWW 2021, Virtual Event/Ljubljana, Slovenia, 19–23 April 2021, pp. 624–632. ACM/IW3C2 (2021). https://doi.org/10.1145/3442381.3449866
27. Lin, K., Sonboli, N., Mobasher, B., Burke, R.: Calibration in collaborative filtering recommender systems: a user-centered analysis. In: Proceedings of the 31st ACM Conference on Hypertext and Social Media, pp. 197–206 (2020)
28. Lin, X., et al.: A pareto-efficient algorithm for multiple objective optimization in e-commerce recommendation. In: Bogers, T., Said, A., Brusilovsky, P., Tikk, D. (eds.) Proceedings of the 13th ACM Conference on Recommender Systems, RecSys 2019, Copenhagen, Denmark, 16–20 September 2019, pp. 20–28. ACM (2019). https://doi.org/10.1145/3298689.3346998
29. Marras, M., Boratto, L., Ramos, G., Fenu, G.: Equality of learning opportunity via individual fairness in personalized recommendations. Int. J. Artif. Intell. Educ. **32**, 636–684 (2021). https://doi.org/10.1007/s40593-021-00271-1
30. Mehrotra, R., McInerney, J., Bouchard, H., Lalmas, M., Diaz, F.: Towards a fair marketplace: counterfactual evaluation of the trade-off between relevance, fairness & satisfaction in recommendation systems. In: Proceedings of the 27th ACM International Conference on Information and Knowledge Management, CIKM 2018, pp. 2243–2251. ACM, New York (2018). https://doi.org/10.1145/3269206.3272027
31. Melchiorre, A.B., Rekabsaz, N., Parada-Cabaleiro, E., Brandl, S., Lesota, O., Schedl, M.: Investigating gender fairness of recommendation algorithms in the music domain. Inf. Process. Manage. **58**(5), 102666 (2021)
32. Naghiaei, M., Rahmani, H.A., Deldjoo, Y.: CPFair: personalized consumer and producer fairness re-ranking for recommender systems. In: The 45th International ACM SIGIR Conference on Research and Development in Information Retrieval, SIGIR 2022, pp. 770–779. ACM (2022). https://doi.org/10.1145/3477495.3531959

33. Nixon, J., Dusenberry, M.W., Zhang, L., Jerfel, G., Tran, D.: Measuring calibration in deep learning. In: IEEE Conference on Computer Vision and Pattern Recognition Workshops, CVPR Workshops 2019, Long Beach, CA, USA, 16–20 June 2019, pp. 38–41. Computer Vision Foundation/IEEE (2019). http://openaccess.thecvf.com/content_CVPRW_2019/html/Uncertainty_and_Robustness_in_Deep_Visual_Learning/Nixon_Measuring_Calibration_in_Deep_Learning_CVPRW_2019_paper.html
34. Peska, L., Dokoupil, P.: Towards results-level proportionality for multi-objective recommender systems. In: Amigó, E., Castells, P., Gonzalo, J., Carterette, B., Culpepper, J.S., Kazai, G. (eds.) The 45th International ACM SIGIR Conference on Research and Development in Information Retrieval, SIGIR 2022, Madrid, Spain, 11–15 July 2022, pp. 1963–1968. ACM (2022). https://doi.org/10.1145/3477495.3531787
35. Raj, A., Ekstrand, M.D.: Measuring fairness in ranked results: an analytical and empirical comparison. In: The 45th International ACM SIGIR Conference on Research and Development in Information Retrieval, SIGIR 2022, pp. 726–736. ACM, New York (2022). https://doi.org/10.1145/3477495.3532018
36. Rendle, S., Freudenthaler, C., Gantner, Z., Schmidt-Thieme, L.: BPR: bayesian personalized ranking from implicit feedback. In: Proceedings of the Twenty-Fifth Conference on Uncertainty in Artificial Intelligence, UAI 2009, pp. 452–461. AUAI Press (2009)
37. Ricci, F., Rokach, L., Shapira, B.: Recommender systems: techniques, applications, and challenges. In: Ricci, F., Rokach, L., Shapira, B. (eds.) Recommender Systems Handbook, pp. 1–35. Springer, New York (2022). https://doi.org/10.1007/978-1-0716-2197-4_1
38. Rojas, C., Contreras, D., Salamó, M.: Analysis of biases in calibrated recommendations. In: Boratto, L., Faralli, S., Marras, M., Stilo, G. (eds.) Advances in Bias and Fairness in Information Retrieval, BIAS 2022. Communications in Computer and Information Science, vol. 1610, pp. 91–103. Springer, Cham (2022). https://doi.org/10.1007/978-3-031-09316-6_9
39. Sarwar, B.M., Karypis, G., Konstan, J.A., Riedl, J.: Item-based collaborative filtering recommendation algorithms. In: Proceedings of the Tenth International World Wide Web Conference, WWW 10, pp. 285–295. ACM (2001). https://doi.org/10.1145/371920.372071
40. Seymen, S., Abdollahpouri, H., Malthouse, E.C.: A constrained optimization approach for calibrated recommendations. In: Fifteenth ACM Conference on Recommender Systems, pp. 607–612 (2021)
41. Stamenkovic, D., Karatzoglou, A., Arapakis, I., Xin, X., Katevas, K.: Choosing the best of both worlds: diverse and novel recommendations through multi-objective reinforcement learning. In: Candan, K.S., Liu, H., Akoglu, L., Dong, X.L., Tang, J. (eds.) The Fifteenth ACM International Conference on Web Search and Data Mining, WSDM 2022, Virtual Event/Tempe, AZ, USA, 21–25 February 2022, pp. 957–965. ACM (2022). https://doi.org/10.1145/3488560.3498471
42. Steck, H.: Calibrated recommendations. In: Proceedings of the 12th ACM Conference on Recommender Systems, RecSys 2018, pp. 154–162. Association for Computing Machinery, New York (2018). https://doi.org/10.1145/3240323.3240372
43. Wang, Z., Meng, C., Ji, S., Li, T., Zheng, Y.: Food package suggestion system based on multi-objective optimization: a case study on a real-world restaurant. Appl. Soft Comput. **93**, 106369 (2020). https://doi.org/10.1016/j.asoc.2020.106369. https://www.sciencedirect.com/science/article/pii/S1568494620303094

44. Wu, H., Ma, C., Mitra, B., Diaz, F., Liu, X.: Multi-FR: a multi-objective optimization framework for multi-stakeholder fairness-aware recommendation. Trans. Inf. Syst. (TOIS) **41**, 1–29 (2022)
45. Wu, H., Mitra, B., Ma, C., Diaz, F., Liu, X.: Joint multisided exposure fairness for recommendation. In: Proceedings of the 45th International ACM SIGIR Conference on Research and Development in Information Retrieval, SIGIR 2022, pp. 703–714. Association for Computing Machinery, New York (2022). https://doi.org/10.1145/3477495.3532007
46. Zadrozny, B., Elkan, C.: Obtaining calibrated probability estimates from decision trees and naive Bayesian classifiers. In: ICML, vol. 1, pp. 609–616. Citeseer (2001)
47. Zehlike, M., Bonchi, F., Castillo, C., Hajian, S., Megahed, M., Baeza-Yates, R.: FA*IR: a fair top-k ranking algorithm. In: Proceedings of the 2017 ACM on Conference on Information and Knowledge Management, CIKM 2017, pp. 1569–1578. ACM, New York (2017). https://doi.org/10.1145/3132847.3132938
48. Zehlike, M., Castillo, C.: Reducing disparate exposure in ranking: a learning to rank approach. In: The Web Conference 2020, WWW 2020, pp. 2849–2855. ACM/IW3C2, New York (2020). https://doi.org/10.1145/3366424.3380048
49. Zheng, Y., Wang, D.X.: A survey of recommender systems with multi-objective optimization. Neurocomputing **474**, 141–153 (2022). https://doi.org/10.1016/j.neucom.2021.11.041

RIGHT: Retrieval-Augmented Generation for Mainstream Hashtag Recommendation

Run-Ze Fan[1,2], Yixing Fan[1,2], Jiangui Chen[1,2], Jiafeng Guo[1,2](✉), Ruqing Zhang[1,2], and Xueqi Cheng[1,2]

[1] ICT, CAS, CAS Key Lab of Network Data Science and Technology, Beijing, China
{fanrunze21s,fanyixing,chenjiangui18z,guojiafeng,zhangruqing,cxq}@ict.ac.cn
[2] University of Chinese Academy of Sciences, Beijing, China

Abstract. Automatic mainstream hashtag recommendation aims to accurately provide users with concise and popular topical hashtags before publication. Generally, mainstream hashtag recommendation faces challenges in the comprehensive difficulty of newly posted tweets in response to new topics, and the accurate identification of mainstream hashtags beyond semantic correctness. However, previous retrieval-based methods based on a fixed predefined mainstream hashtag list excel in producing mainstream hashtags, but fail to understand the constant flow of up-to-date information. Conversely, generation-based methods demonstrate a superior ability to comprehend newly posted tweets, but their capacity is constrained to identifying mainstream hashtags without additional features. Inspired by the recent success of the retrieval-augmented technique, in this work, we attempt to adopt this framework to combine the advantages of both approaches. Meantime, with the help of the generator component, we could rethink how to further improve the quality of the retriever component at a low cost. Therefore, we propose ***RetrI****eval-augmented* ***G****enerative Mainstream* ***H****ash****T****ag Recommender* (**RIGHT**), which consists of three components: (i) a retriever seeks relevant hashtags from the entire tweet-hashtags set; (ii) a selector enhances mainstream identification by introducing global signals; and (iii) a generator incorporates input tweets and selected hashtags to directly generate the desired hashtags. The experimental results show that our method achieves significant improvements over state-of-the-art baselines. Moreover, RIGHT can be easily integrated into large language models, improving the performance of ChatGPT by more than 10%. Code will be released at: https://github.com/ict-bigdatalab/RIGHT.

Keywords: Hashtag recommendation · Retrieval-augmented generation · Social media

1 Introduction

Millions of user-generated microblogs flood Twitter daily, surpassing users' comprehension. To facilitate rapid and easy understanding, hashtags (e.g., *#ChatGPT*)

N. Goharian et al. (Eds.): ECIR 2024, LNCS 14608, pp. 39–55, 2024.
https://doi.org/10.1007/978-3-031-56027-9_3

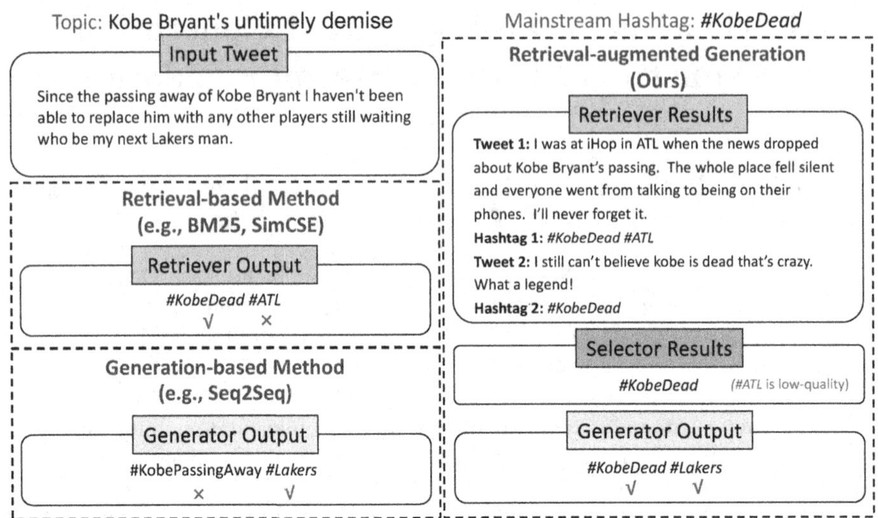

Fig. 1. Illustration of evaluating hashtag recommendation with different methods.

are extensively used to convey central ideas and topics, which also enhance content visibility to reach a broader audience [24]. Such hashtags are commonly referred to as mainstream hashtags, denoting their status as not only the most prevalent hashtags but also semantically accurate. For instance, in the context of Kobe Bryant's untimely demise, both *#KobeDead* and *#KobeDeath* can be utilized, with both possessing accurate semantic meanings. However, the former holds more widespread usage and has achieved the distinction of being a mainstream hashtag.

To provide mainstream hashtags, two main challenges need to be addressed. First, comprehending a new tweet presents challenges primarily attributable to the absence of real-time information [42,50]. This is a direct consequence of the continuous emergence of numerous new tweets in response to new topics and events. Second, accurately identifying mainstream hashtags beyond semantic correctness remains a challenging task. The reason is that numerous hashtags could be used to describe a topic, but only a few are mainstream.

To address the above challenges, a considerable amount of work has been proposed, which could be divided into two research lines [10,13,15,42,43]. Retrieval-based methods retrieve hashtags from a fixed predefined mainstream hashtag list [15,43], which could alleviate the second problem. However, their ability to fully grasp the meaning of a newly posted tweet in response to emerging topics and events is constrained. Moreover, it is a considerable cost to maintain the predefined list [42]. In contrast, generation-based methods [30,42,50] demonstrate remarkable proficiency in comprehending new tweets and generating semantically accurate hashtags, owing to their substantial pretraining knowledge. Nevertheless, they may encounter difficulties when it comes to identifying mainstream hashtags without enough mainstream information. As a result, the tweet might fail to be indexed by a mainstream hashtag on microblog services due to the tags' unpopularity, weakening the recall rate of microblog searches. Inspired by the recent success of retrieval-augmented generation technique [1,12,21,37,48], therefore, we try to adapt this method to mainstream hashtags recommendation, utilizing the advantages of both retrieval and generation approaches.

Typically, retrieval-augmented techniques incorporate the results of the retriever, whether explicitly or implicitly, into the generator to enhance the quality of generation. Utilizing this framework, the introduction of a generator endowed with strong comprehensive capabilities might mitigate the dependency on the quality of the retriever [31]. Thus, we could rethink the trade-off between the quality and the cost of the retriever. Traditional retrieval-based methods rely on a predefined list of mainstream hashtags, which can ensure the quality of the retrieved information, but maintains such a list at a significant cost. To reduce the maintenance burden, we transform the small predefined list into a larger aggregation of existing tweet-hashtags pairs, which can be automatically collected and updated without manual cost. However, this approach carries the risk of introducing numerous low-quality hashtags due to the informal characteristics of social media content. Such hashtags have the potential to mislead the generator. As illustrated in Fig. 1, both *#KobeDead* and *#ATL* are results of the retriever. Nonetheless, it is noteworthy that *#ATL* is law-quality, even though tweet 1 exhibits the highest degree of similarity with the input tweet. Consequently, it becomes imperative to further improve the quality of retrieved information without increasing the cost.

Therefore, in this study, we propose a ***R****etr**I**eval-augmented **G**enerative Mainstream **H**ash**T**ag Recommender* (**RIGHT**), which combines the retriever and the generator by the retrieval-augmented technique with inserting a selector. Specifically, our method involves three components: 1) **Retriever** is utilized to acquire relevant hashtags. We retrieve the tweets most similar to the input from the tweet-hashtags corpus and obtain the corresponding hashtags set. 2) **Selector** is used to improve the capability of identifying mainstream hashtags. We incorporate three features, the similarity between the input tweet and the retrieved tweet and its hashtags, as well as the frequency of the hashtags, to enhance the mainstream information. 3) **Generator** is leveraged to provide strong semantic comprehension and the ability of hashtag generation. We concatenate the selected hashtags with the input tweet and feed it into the generator to obtain the desired hashtags. In this way, we can utilize not only the retriever and the selector to seek the mainstream hashtags but also the generator to produce the desired hashtags flexibly.

We conduct experiments on two large-scale datasets (i.e., English Twitter (THG) and Chinese Weibo (WHG)). Experimental results show that our method achieves significant improvements over state-of-the-art baselines. Moreover, as it can be easily incorporated in black-box language models, we also apply our framework to ChatGPT by zero-shot instruction learning, bringing a 12.7% boost for THG and 18.3% for WHG in F1@1. Finally, to deeply understand this method, we present a detailed analysis.

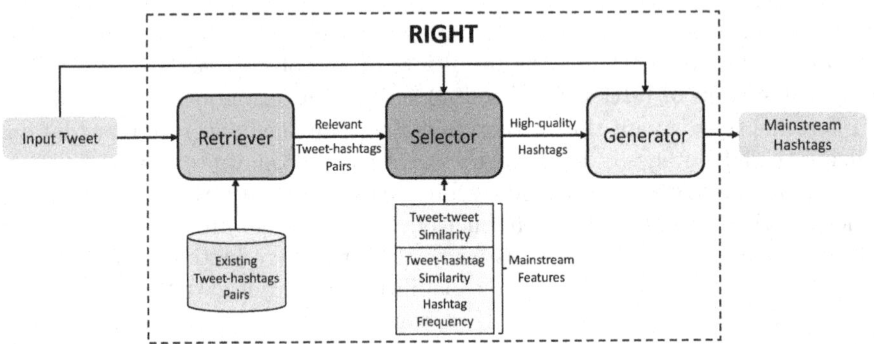

Fig. 2. Our RIGHT framework consists of a retriever, a selector, and a generator.

2 Methodology

We propose ***R****etr****I****eval-augmented* ***G****enerative Mainstream* ***H****ash****T****ag Recommender* (**RIGHT**), a simple yet effective framework for mainstream hashtag recommendation, which includes three components: retriever, selector, and generator.

Overall, we first utilize the retriever to retrieve the tweets most similar to the input from the existing tweet-hashtags corpus and obtain the corresponding hashtags set. Then, we adopt a selector to select the hashtags that are most probable mainstream from the retrieved labels using three signals. Finally, We concatenate the selected hashtags with the input tweet and feed it into the generator to obtain the desired hashtags. An overview of our method is shown in Fig. 2.

2.1 Retriever

The goal of the retriever is to retrieve the top-N tweet-hashtags pairs on the same topic with the input tweet from the existing corpus, aiming to find relevant hashtags on the same topic.

Inspired by Wang et al. [41], we view the labeled training data as our corpus and index these as input-label pairs, i.e., $\mathcal{C} = \{(\tilde{t}_i, \tilde{H}_i)\}$. Then, given the input tweet t, the retrieval model $\mathcal{R}$ matches it with all tweets in the corpus and returns the top-N most similar tweet-hashtags pairs together with their scores:

$$\{(\tilde{t}_1, \tilde{H}_1, \tilde{s}_1), \ldots, (\tilde{t}_N, \tilde{H}_N, \tilde{s}_N)\} = \mathcal{R}(t|\mathcal{C}),$$

where we denote $\tilde{s}_i$ as the similarity between t and the i-th retrieved tweet $\tilde{t}_i$. Each $\tilde{H}_i$ consists of hashtags $\{\tilde{h}_1^i, \ldots, \tilde{h}_{|\tilde{H}_i|}^i\}$. We report the results with sparse retrieval (e.g., BM25 [38]) and dense retrieval (e.g., SimCSE [11]) in experiments.

2.2 Selector

The goal of the selector is to filter the low-quality and non-mainstream hashtags existing in the results of the retriever (see Fig. 1). We consider three mainstream features: the similarity between the input tweet and the retrieved tweet and its hashtags, as well as the frequency of the hashtags. Thus, we train a selector to compute the similarity between the tweet and the hashtags and propose a simple algorithm for hashtag ranking.

Training. The training data consists of positive samples and hard negative samples. Each hashtag labeled in a tweet can be viewed as a positive sample t^+. However, a significant challenge lies in constructing hard negative samples (t^-) to facilitate the efficient selection of mainstream hashtags on the same topic by the selector. Inspired by BERT [9], we propose to create a hard negative sample by disturbing the labeled hashtag without changing the semantic meaning. Specifically, we randomly select a word to: (i) replace with its synonym 70% of the time; (ii) delete 10% of the time; (iii) swap with the adjacent word 10% of the time; (iv) insert a synonym after it 10% of the time. Thus, we obtain a training dataset $\{(t_i, t_i^+, t_i^-)|i = 1, \dots, N\}$. Finally, we utilize contrastive learning to train our selector by minimizing the following loss:

$$\mathcal{L}_{\mathcal{S}} = -\log \frac{e^{\mathrm{sim}(\mathbf{h}_{t_i}, \mathbf{h}_{t_i}^+)/\tau}}{\sum_{j=1}^{L} \left(e^{\mathrm{sim}(\mathbf{h}_{t_i}, \mathbf{h}_{t_j}^+)/\tau} + e^{\mathrm{sim}(\mathbf{h}_{t_i}, \mathbf{h}_{t_j}^-)/\tau} \right)}$$

where sim presents similarity, $\mathbf{h}_t$ indicates the representation of t, L is mini-batch size, and τ is a temperature hyperparameter.

Inference. In the inference stage, we propose a simple algorithm for hashtag ranking. Given an input tweet t and the result of the retriever $\{(\tilde{t}_1, \tilde{H}_1, \tilde{s}_1), \dots, (\tilde{t}_N, \tilde{H}_N, \tilde{s}_N)\}$, we put retrieved hashtags into a set $\{\tilde{h}_1, \dots, \tilde{h}_M\}$ where we denote M as the number of different hashtags, and record the number of occurrences $\{f_1, \dots, f_M\}$ and the corresponding score of each retrieved hashtag $\{\tilde{s}_{i,1} \dots, \tilde{s}_{i,f_i}\}$. Then, we match the input tweet t with all hashtags using the selector $\mathcal{S}$ to obtain the similarity score between the tweet and all hashtags $\{\ddot{s}_1, \dots, \ddot{s}_M\}$:

$$\ddot{s}_m = \mathcal{S}(t, \tilde{h}_m). \tag{1}$$

Finally, we average the tweet-to-tweet similarity for each hashtag and add the similarity between the tweet and the hashtag. Since hashtags that occur more frequently are more likely to be mainstream, we magnify the sum of the similarity score and ranking score with a downscaled frequency:

$$s_i = ((\frac{1}{f_i}\sum_{j=1}^{f_i} \tilde{s}_{i,j}) + \ddot{s}_i) \times (1 + ((f_i - 1)/10)),$$

and sort the hashtags by the final score from largest to smallest to select the top-k hashtags $\{\tilde{h}_1, \dots, \tilde{h}_k\}$.

Table 1. Data statistics for the English Twitter hashtag generation (THG) dataset and the Chinese Weibo hashtag generation (WHG) dataset. # T-H pairs denotes the number of tweet-hashtags pairs. #AvgHashtags denotes the average number of hashtags in each tweet-hashtags pair. AvgTweetLen denotes the avenge length (token level) of all input tweets. AvgHashtagLen denotes the average length (token level) of all hashtags.

Dataset	THG Dataset			WHG Dataset		
	Train	Validation	Test	Train	Validation	Test
# T-H pairs	201444	11325	11328	307401	2000	2000
# AvgHashtags	4.1	4.1	4.1	1.0	1.0	1.0
AvgTweetLen	39.7	39.6	39.6	87.1	86.8	87.8
AvgHashtagLen	3.1	3.0	3.0	6.6	6.5	6.5

2.3 Generator

The goal of the generator is to generate the desired hashtags, given an input tweet t and the selected hashtags $\{\tilde{h}_1, \ldots, \tilde{h}_k\}$. We concatenate the input tweet with the retrieved hashtags and separate each hashtag by a special token:

$$I = < t, \texttt{SEP1}, \tilde{h}_1, \texttt{SEP1}, \tilde{h}_2, \ldots, \texttt{SEP1}, \tilde{h}_k >,$$

and feed it into the generator $\mathcal{G}$, which will output a concatenated sequence O that includes the hashtags, with each hashtag separated by another token:

$$O = < h_1, \texttt{SEP2}, h_2, \ldots, \texttt{SEP2}, h_{|H|} > .$$

By easily splitting by the special token, we would obtain the hashtag list $H = \{h_1, h_2, \ldots, h_{|H|}\}$.

The generative model could be Transformer-based encoder-decoder architecture (e.g., T5 [35], BART [25]) or decoder-only architecture (e.g., a series of GPT [2,33,34]). Thus, the training stage focuses on the finetuning of generative models by minimizing the cross-entropy loss:

$$\mathcal{L}_{\mathcal{G}} = \sum_{(I,O)\in\mathcal{D}} -\log p(O|I;\theta_{\mathcal{G}}),$$

where I is the input sequence consisting of the input tweet and the selected hashtags, O is the output sequence consisting of the desired hashtags, and $\theta_{\mathcal{G}}$ is the parameters of the generator.

3 Experiments

3.1 Experimental Setup

Datasets. Our experiments are conducted on two large-scale datasets, which were crawled from official media and influencers of social media [30]. The details are shown in Table 1.

- **THG:** The English Twitter hashtag generation (THG) dataset has been crawled from official Twitter sources, encompassing organizations, media outlets, and other authenticated users, with the primary objective of acquiring tweets of superior quality.
- **WHG:** The Chinese Weibo hashtag generation (WHG) dataset has been acquired through the systematic extraction of microblogs from Weibo, encompassing notable sources including *People's Daily*, *People.cn*, *Economic Observe press*, *Xinlang Sports*, and various other accounts boasting over 5 million followers. These accounts span diverse domains, encompassing politics, economics, military affairs, sports, and more.

We use the training datasets as our retrieval corpus.

Evaluation Metric. Following previous work [30,42], we utilize ROUGE metrics and F1 scores at K as our evaluation metric. The average ROUGE score measures the overlap between the generated sequence of hashtags (excluding special tokens) and the reference sequence, including ROUGE-1, ROUGE-2, and ROUGE-L. For F1 scores at K, different K values result in a similar trend, so only F1@1 and F1@5 are reported. We report results on the test dataset. Noticeably, for the WHG dataset, where input posts have only one hashtag, F1@1 and F1@5 are identical, so we only report F1@1 for this dataset.

Implementation Details. Our implementation details of the retriever, selector, and generator are the following:

- **For Retriever**, we utilize BM25 [38] and SimCSE [11] (i.e., RoBERTa-Large [29] for THG and Bert-Base-Chinese [9] for WHG) as our retrievers. Following Gao et al. [11], we train our model for 3 epochs with a learning rate of 1e−5. The hyperparameter of N is set to 10, and the batch size is 6 per device.
- **For Selector**, we use the training datasets from THG and WHG to construct our hard negative samples, which are subsequently employed for training our selectors in both English and Chinese independently. We utilize RoBERTa-Large for THG and Bert-Base-Chinese for WHG. The temperature τ is 0.05 and other hyperparameters are the same as the retriever.
- **For Generator**, we fine-tune a T5-base [35] for THG and a mT5-small [44] for WHG and use Adam [22] as an optimizer. We set the weight decay and batch size as 1e−5 and 16 and grid-search the learning rate, training epochs, and the number of concatenated hashtags k from {3e−4, 1e−4, 5e−5}, {5, 10}, and {1, 3, 5, 7, 9} respectively. The maximum length is 180 for T5-base and 256 for mT5-small. The special token `SEP1` and `SEP2` are $<$ `extra_id_0` $>$ and $<$ `extra_id_1` $>$ respectively.

All models are trained on four NVIDIA Tesla K80.

Table 2. The prompt used for ChatGPT. The Chinese version is its translation.

Baseline	Instruction
ChatGPT	I want you to act as a hashtag annotator. I will provide you a tweet and your role is to annotate the relevant hashtag. You should use the related knowledge and find the topic. I want you only reply the hashtags segmented by "#" and nothing else, do not write explanations. I want you segment the word in a hashtag by space. My first tweet is {`Input Tweet`}.
RIGHT$^{\mathbf{ChatGPT}}$	I want you to act as a hashtag annotator. I will provide you with a tweet, and your role is to annotate the relevant hashtag. Using your related knowledge, you should identify the topic and reply with only the hashtags segmented by "#", without any explanations. Make sure to capitalize the first letter of the word. Make sure to split every word in a hashtag by a space. There are some potential hashtags:[{`Retrieved Top-`k `Hashtags`}]. You can decide whether use the part of them or not. My first tweet is {`Input Tweet`}.

Baselines. Our baselines consist of retrieval-based methods, generation-based methods, and retrieval-augmented generative methods:

- **Retrieval-based methods:** Following Mao et al. [30], we construct the predefined hashtags list from all hashtags in the training datasets and select top-4 hashtags for THG and top-1 for WHG according to the average number of hashtags in each data item. We apply BM25 and SimCSE to the hashtag recommendation: (i) **BM25** [38] is a traditional strong sparse retrieval based on term matching. (ii) **SimCSE** [11] is a representative dense retriever, which applies a simple contrastive learning framework to present sentence embeddings on semantic textual similarity tasks. We fine-tuned SimCSE by constructing positive samples and hard negative samples from BM25.
- **Generation-based methods:** We consider three predominant generative methods: (i) **ChatGPT** is a powerful large language model to execute various NLP tasks [23]. Specifically, we adopt `gpt-3.5-turbo` and instruction zero-shot learning to evaluate our task (Prompts are shown in Table 2). (ii) **SEGTRM Soft** [30] is the previous SOTA on our datasets, an end-to-end generative method segments selection-based deep transformer. (iii) **Seq2Seq** [42] is the first generation-based method for hashtag recommendation. Due to the unreality to assume the existence of conversations before publishing the tweet [50] and the lack of conversation contexts, we reimplement the Seq2Seq model on the pretrained language model (T5-base for THG and mT5-small for WHG) to formulate this task to a seq-to-seq paradigm.

- **Retrieval-augmented Generative Methods (Ours):** We apply our retrieval-augmented framework to ChatGPT by incorporating the retrieval results into the instruction to prompt the model to generate mainstream hashtags, denoted as $\mathbf{RIGHT}^{\mathbf{ChatGPT}}$ (Prompts are shown in Table 2). We only use the best retriever on the datasets (i.e., SimCSE for THG and BM25 for WHG), due to the high cost of ChatGPT. Moreover, we use BM25 and SimCSE as our retriever of RIGHT, denoted them as $\mathbf{RIGHT}_{\mathbf{BM25}}$ and $\mathbf{RIGHT}_{\mathbf{SimCSE}}$.

Table 3. Main results (%) on the THG and WHG datasets. Bold and underline indicate the best and second method respectively. We donate ROUGE as RG. $*$ indicates statistically significant improvements over all baselines (p-value < 0.05).

Model	THG					WHG			
	RG-1	RG-2	RG-L	F1@1	F1@5	RG-1	RG-2	RG-L	F1@1
Retrieval-based Methods									
BM25	16.23	4.17	15.11	5.92	9.84	61.98	58.76	61.81	48.20
SimCSE	28.43	10.34	26.38	12.40	15.15	59.71	55.81	59.54	47.65
Generation-based Methods									
ChatGPT	44.60	27.67	39.29	9.72	26.08	32.27	24.54	31.80	7.9
SEGTRM Soft	51.18	37.15	47.05	27.17	29.02	55.51	51.28	54.30	30.72
Seq2Seq	59.90	41.39	59.15	29.75	41.71	66.64	61.71	66.39	48.60
Retrieval-augmented Generative Methods (Ours)									
$\text{RIGHT}^{\text{ChatGPT}}$	47.54	25.63	44.47	22.39	31.09	48.17	41.51	47.75	26.15
$\text{RIGHT}_{\text{BM25}}$	<u>61.60</u>	<u>43.77</u>	<u>60.85</u>	<u>30.27</u>	<u>42.98</u>	**70.62***	**66.12***	**70.35***	**53.85***
$\text{RIGHT}_{\text{SimCSE}}$	**62.11***	**43.86***	**61.39***	**30.58***	**43.23***	<u>68.84</u>	<u>64.19</u>	<u>68.56</u>	<u>51.50</u>

3.2 Main Results

As shown in Table 3, we can observe that:

1. Among the retrieval-based methods, the performance of SimCSE outperforms BM25 in THG, while BM25 demonstrates superior performance in WHG. This difference may be attributed to that English hashtags tend to be concise summaries, while Chinese hashtags often comprise small sentences extracted directly from the input text. Consequently, dense retrieval approaches utilizing semantic matching may be more suitable for English datasets, while Chinese datasets may benefit more from sparse retrieval techniques based on term matching.
2. Among the generation-based methods, Seq2Seq performs well on both datasets, potentially attributable to the utilization of mainstream hashtag knowledge from the training dataset during fine-tuning. However, ChatGPT lags behind other generation methods, suggesting a deficiency in mainstream hashtag knowledge despite its vast repository of general knowledge.

Table 4. Ablation study results on the THG datasets. Bold indicates the best method.

Model	ROUGE-1	ROUGE-2	ROUGE-L	F1@1	F1@5
RIGHT	**62.11**	**43.86**	**61.39**	**30.58**	**43.23**
w/o Retriever	59.91	41.63	59.23	29.66	41.70
w/o Selector	60.49	42.06	59.76	30.22	41.95
w/o Generator	36.24	16.02	32.86	24.61	26.73

3. Among the retrieval-augmented generative methods, retrieval augmentation brings the performance of baselines to a new level, demonstrating the effectiveness of our method. For ChatGPT, retrieval augmentation boosts F1@1 performance by 12.67% for THG and 18.25% for WHG, indicating the substantial value of mainstream hashtag knowledge. For RIGHT, both sparse and dense retrievers show the potential to enhance performance compared with Seq2Seq. Specifically, SimCSE is particularly effective for THG, while BM25 performs better for WHG. The reason could be attributed to the superiority in the performance of retrieval-based methods is directly proportional to the enhancement of the retrieval augmentation. Moreover, different retrievers excel in different scenarios, emphasizing the importance of the careful selection of retrievers based on specific use cases. The performance of the retrieval-based approach serves as a preliminary guide for informed decision-making.

Overall, our method shows robustness across various scenarios, whether applied with a fine-tuned generation model or a large black box language model. Regardless of the retrieval approach used, our method consistently improves performance.

3.3 Analysis

Ablation Study. We conduct an ablation study to explore the impact of each component in RIGHT on THG: 1) **w/o Retriever**: We remove the retriever and randomly concatenate k hashtags from the training dataset with the input tweet. 2) **w/o Selector**: We remove the selector and directly use the top-k hashtags from the retriever's results by the similarity between the input tweet and the retrieved tweet. 3) **w/o Generator**: We remove the generator and output the top-4 hashtags produced by the selector. Table 4 presents the results, indicating that:

1. The performance improvement is considerable through the integration of the retriever, confirming that the incorporation of mainstream hashtag knowledge indeed facilitates accurate hashtag selection.
2. Without the selector, the performance gains are limited, indicating that simply being on the same topic is insufficient. It is crucial to identify and incorporate mainstream hashtags.

RIGHT

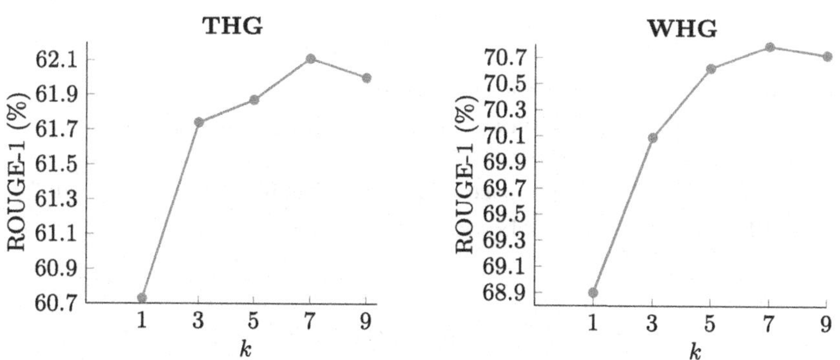

Fig. 3. Our Rouge-1 results in the different number of augmented hashtags (i.e., $k = 1, 3, 5, 7, 9$).

3. The generator is crucial in RIGHT, emphasizing the significant impact of semantic comprehension on performance. In contrast to the retrieval-based approaches, it is more powerful to directly output the hashtags in the tweet-hashtags pair that are most similar to the input.

Impact of the Number of Augmented Hashtags. To explore the impact of the number of concatenated hashtags with the input tweet, we conduct a series of experiments. Specifically, we concatenate various top-k ($k = 1, 3, 5, 7, 9$) with the input tweet for the THG and the WHG datasets. Figure 3 demonstrates the Rouge-1 results (other metrics show the same trends), showing that:

1. Retrieval augmentation aids in improving the performance when a sufficient number is considered, suggesting that augmenting more hashtags increases the probability of covering mainstream hashtags and makes the generator more robust in the presence of mismatches from certain hashtags.
2. Upon reaching a certain threshold of the number of augmented hashtags (i.e., $k = 7$), the performance converges, suggesting that the majority of mainstream hashtags might have already been augmented.

Case Study. To validate the successful recall of mainstream hashtags and the potential for further improvement, we analyze the successful and unsuccessful cases and present a representative case study in Table 5. We conclude that:

1. Some retrieved tweets share the same topic as the input tweets but have subpar labeled hashtags (e.g., "fx logix" in Table 5). Fortunately, our generator demonstrates the capability to disregard these irrelevant hashtags.
2. Some retrieved tweets are partially relevant to the input tweet. Although the retrieved hashtags align well with the topic of the retrieved tweet, it is not

Table 5. An example from the THG test set. Correct results are marked bold.

Input:	Geeks guide to microsoft teams optimization with windows virtual desktop citrix.
Label:	**microsoft**; **windows**; **citrix**; **wvd**
Retriever & Selector:	**citrix**; **wvd**; fs logix; v mware; azure; aws; **microsoft**
Seq2Seq:	**microsoft**; windows virtual desktop; **citrix**; vdi
RIGHT:	**citrix**; **wvd**

highly pertinent to the primary topic of the input tweet (e.g., "azure" in the retrieved hashtags). Nonetheless, our generator can filter out these irrelevant hashtags.

3. The generation model produces a semantically accurate but non-mainstream hashtag "windows virtual desktop" by directly copying the original word from the input tweet due to its limited knowledge of mainstream hashtags. However, our retriever and selector effectively identify the corresponding mainstream hashtag in its abbreviated form "wvd". Our RIGHT successfully replaced the original hashtag with the mainstream hashtag, indicating the effectiveness of retrieval augmentation.
4. Seq2Seq generates certain hashtags that are also retrieved by the retriever, while RIGHT fails to generate them (e.g., "microsoft"). These cases constitute less than 1% of the total. We speculate that this discrepancy may be due to the selector placing the correct hashtags toward the end of the list, leading to reduced confidence and subsequent non-adoption by the generator.

4 Related Works

Our work mainly builds on two streams of previous work: hashtag recommendation and retrieval-augmented generation.

4.1 Mainstream Hashtag Recommendation

Mainstream hashtag recommendation aims to provide users with short topical and popular tags representing the main ideas of their tweets before publication. Three primary methods have been proposed for this task [24]:

1) **Keyphrase extraction method** formulates this task as keyphrases extraction from source posts [14,47,49], which fails to produce hashtags that do not appear in the microblog posts while large freedom is allowed for users to write whatever hashtags they like. The performance of this method is much lower than other methods. 2) **Retrieval-based method** aims to retrieve from a predefined hashtag list [15,19,43,46], which is limited to generating only the hashtags that are included in the list. In reality, a wide range of hashtags can

be created every day, resulting impossibility to be covered by a fixed list and the difficulty to maintain the list. 3) **Generation-based method** was proposed [30,32,42,50] to overcome the aforementioned challenges, which formulates the task as a sequence-to-sequence generation paradigm, allowing for the creation of a wider range of hashtags that better capture the main ideas of the microblog post. However, previous studies pay limited attention to mainstream hashtags. Consequently, even though it produces semantically correct tags, the tweet might fail to be indexed by a mainstream hashtag on microblog services due to tags' unpopularity, thus weakening the recall rate of microblog searches.

To the best of our knowledge, we are the first to alleviate this issue by combining retrieval and generation methods. Meanwhile, we improve the quality of the retriever at a low cost.

4.2 Retrieval-Augmented Generation

The retrieval-augmented generation represents a novel paradigm that merges pre-trained generative models with information retrieval techniques [1,26]. Previous research in this field primarily has focused on introducing external knowledge to address knowledge-intensive tasks [4,6–8,17,18,27,39,40,45] and utilizing similar data to enhance the model performance across various natural language processing (NLP) tasks, including image captioning [37], keyphrase generation [12,21], named entity recognition [5,48], and others. Recently, this technique has also been used in large language models to alleviate issues like factual hallucination [3,20,36], knowledge out-dating [16], and the lack of domain-specific expertise [28].

Notably, we adopt this framework to the mainstream hashtag recommendation task, by introducing a selector combining global signals to improve mainstream identification.

5 Conclusion

In this study, we have proposed a simple yet effective retrieval-augmented generative recommender, designed to utilize the advantage of retrieval and generation methods for mainstream hashtag recommendation. To improve the quality of the retriever's results at a low cost, we have integrated a selector module into the conventional retrieval-augmented framework. Specifically, the retriever's role is to find relevant hashtags on the same topic, the selector is employed to enhance the identification of mainstream hashtags, and the generator is responsible for combining input tweets and selected hashtags to generate desired hashtags. We have conducted extension experiments using two extensive datasets to validate the effectiveness of our approach.

In future work, it is valuable to explore optimal strategies for combining the retrieval-based method with the generation-based method, as well as developing a co-training approach that jointly refines the three components.

Acknowledgements. This work was funded by the National Natural Science Foundation of China (NSFC) under Grants No. 62372431, and 62006218, the Youth Innovation Promotion Association CAS under Grants No. 2021100, the project under Grants No. 2023YFA1011602, JCKY2022130C039 and 2021QY1701, and the Lenovo-CAS Joint Lab Youth Scientist Project. All content represents the opinion of the authors, which is not necessarily shared or endorsed by their respective employers and/or sponsors.

References

1. Asai, A., Min, S., Zhong, Z., Chen, D.: Retrieval-based language models and applications. In: ACL, Toronto, Canada, July 2023, pp. 41–46 (2023). https://doi.org/10.18653/v1/2023.acl-tutorials.6. https://aclanthology.org/2023.acl-tutorials.6
2. Brown, T.B., et al.: Language models are few-shot learners. In: Larochelle, H., Ranzato, M., Hadsell, R., Balcan, M., Lin, H. (eds.) NeurIPS (2020). https://proceedings.neurips.cc/paper/2020/hash/1457c0d6bfcb4967418bfb8ac142f64a-Abstract.html
3. Cao, M., Dong, Y., Wu, J., Cheung, J.C.K.: Factual error correction for abstractive summarization models. In: EMNLP, November 2020 (2020). https://doi.org/10.18653/v1/2020.emnlp-main.506. https://aclanthology.org/2020.emnlp-main.506
4. Chen, D., Fisch, A., Weston, J., Bordes, A.: Reading Wikipedia to answer open-domain questions. In: ACL, July 2017 (2017). https://doi.org/10.18653/v1/P17-1171. https://aclanthology.org/P17-1171
5. Chen, J., Zhang, R., Guo, J., Fan, Y., Cheng, X.: GERE: generative evidence retrieval for fact verification. In: SIGIR 2022, pp. 2184–2189. ACM (2022). https://doi.org/10.1145/3477495.3531827
6. Chen, J., Zhang, R., Guo, J., Liu, Y., Fan, Y., Cheng, X.: CorpusBrain: pre-train a generative retrieval model for knowledge-intensive language tasks. In: CIKM 2022, pp. 191–200. ACM (2022). https://doi.org/10.1145/3511808.3557271
7. Chen, J., et al.: Continual learning for generative retrieval over dynamic corpora. In: CIKM 2023, pp. 306–315. ACM (2023). https://doi.org/10.1145/3583780.3614821
8. Chen, J., et al.: A unified generative retriever for knowledge-intensive language tasks via prompt learning. In: SIGIR 2023, pp. 1448–1457. ACM (2023). https://doi.org/10.1145/3539618.3591631
9. Devlin, J., Chang, M.W., Lee, K., Toutanova, K.: BERT: pre-training of deep bidirectional transformers for language understanding. In: NAACL, June 2019 (2019). https://doi.org/10.18653/v1/N19-1423. https://aclanthology.org/N19-1423
10. Ding, Z., Zhang, Q., Huang, X.: Automatic hashtag recommendation for microblogs using topic-specific translation model. In: COLING, December 2012 (2012). https://aclanthology.org/C12-2027
11. Gao, T., Yao, X., Chen, D.: SimCSE: simple contrastive learning of sentence embeddings. In: EMNLP, November 2021 (2021). https://doi.org/10.18653/v1/2021.emnlp-main.552. https://aclanthology.org/2021.emnlp-main.552
12. Gao, Y., et al.: Retrieval-augmented multilingual keyphrase generation with retriever-generator iterative training. In: NAACL (2022). https://doi.org/10.18653/v1/2022.findings-naacl.92. https://aclanthology.org/2022.findings-naacl.92
13. Gong, Y., Zhang, Q., Huang, X.: Hashtag recommendation using Dirichlet process mixture models incorporating types of hashtags. In: EMNLP (2015). https://doi.org/10.18653/v1/D15-1046. https://aclanthology.org/D15-1046

14. Gong, Y., Zhang, Q., Huang, X.: Hashtag recommendation using dirichlet process mixture models incorporating types of hashtags. In: Màrquez, L., Callison-Burch, C., Su, J., Pighin, D., Marton, Y. (eds.) EMNLP (2015). https://doi.org/10.18653/v1/d15-1046. https://doi.org/10.18653/v1/d15-1046
15. Gong, Y., Zhang, Q.: Hashtag recommendation using attention-based convolutional neural network. In: Kambhampati, S. (ed.) IJCAI (2016). http://www.ijcai.org/Abstract/16/395
16. He, H., Zhang, H., Roth, D.: Rethinking with retrieval: Faithful large language model inference. arXiv preprint (2022). https://arxiv.org/pdf/2301.00303.pdf
17. He, S., Fan, R.Z., Ding, L., Shen, L., Zhou, T., Tao, D.: MerA: merging pretrained adapters for few-shot learning. arXiv preprint arXiv:2308.15982 (2023)
18. He, S., Fan, R.Z., Ding, L., Shen, L., Zhou, T., Tao, D.: Merging experts into one: improving computational efficiency of mixture of experts. In: Bouamor, H., Pino, J., Bali, K. (eds.) Proceedings of the 2023 Conference on Empirical Methods in Natural Language Processing, Singapore, December 2023, pp. 14685–14691. Association for Computational Linguistics (2023). https://aclanthology.org/2023.emnlp-main.907
19. Huang, H., Zhang, Q., Gong, Y., Huang, X.: Hashtag recommendation using end-to-end memory networks with hierarchical attention. In: COLING, December 2016 (2016). https://aclanthology.org/C16-1090
20. Ji, Z., et al.: Survey of hallucination in natural language generation. ACM Comput. Surv. (2023). https://doi.org/10.1145/3571730
21. Kim, J., Jeong, M., Choi, S., Hwang, S.: Structure-augmented keyphrase generation. In: EMNLP (2021). https://doi.org/10.18653/v1/2021.emnlp-main.209. https://aclanthology.org/2021.emnlp-main.209
22. Kingma, D.P., Ba, J.: Adam: a method for stochastic optimization. In: Bengio, Y., LeCun, Y. (eds.) ICLR (2015). http://arxiv.org/abs/1412.6980
23. Kocoń, J., et al.: ChatGPT: jack of all trades, master of none. arXiv preprint (2023). https://arxiv.org/pdf/2302.10724.pdf
24. Kwak, H., Lee, C., Park, H., Moon, S.: What is Twitter, a social network or a news media? In: WWW (2010). https://ink.library.smu.edu.sg/cgi/viewcontent.cgi?article=7104&context=sis_research
25. Lewis, M., et al.: BART: denoising sequence-to-sequence pre-training for natural language generation, translation, and comprehension. In: Jurafsky, D., Chai, J., Schluter, N., Tetreault, J.R. (eds.) ACL (2020). https://doi.org/10.18653/v1/2020.acl-main.703
26. Lewis, P.S.H., et al.: Retrieval-augmented generation for knowledge-intensive NLP tasks. In: Larochelle, H., Ranzato, M., Hadsell, R., Balcan, M., Lin, H. (eds.) NeurIPS (2020). https://proceedings.neurips.cc/paper/2020/hash/6b493230205f780e1bc26945df7481e5-Abstract.html
27. Li, J., Sun, S., Yuan, W., Fan, R.Z., Zhao, H., Liu, P.: Generative judge for evaluating alignment. arXiv preprint arXiv:2310.05470 (2023)
28. Li, X., Zhu, X., Ma, Z., Liu, X., Shah, S.: Are ChatGPT and GPT-4 general-purpose solvers for financial text analytics? An examination on several typical tasks. arXiv preprint (2023). https://arxiv.org/pdf/2305.05862.pdf
29. Liu, Y., et al.: RoBERTa: a robustly optimized bert pretraining approach. arXiv preprint (2019). https://arxiv.org/pdf/1907.11692.pdf
30. Mao, Q., et al.: Attend and select: a segment selective transformer for microblog hashtag generation. Knowl. Based Syst. **254**, 109581 (2022). https://doi.org/10.1016/j.knosys.2022.109581. https://www.sciencedirect.com/science/article/pii/S0950705122007973

31. Mialon, G., et al.: Augmented language models: a survey. arXiv preprint (2023). https://arxiv.org/pdf/2302.07842.pdf
32. Ni, S., Bi, K., Guo, J., Cheng, X.: A comparative study of training objectives for clarification facet generation. In: Annual International ACM SIGIR Conference on Research and Development in Information Retrieval in the Asia Pacific Region, pp. 1–10 (2023)
33. Radford, A., Narasimhan, K., Salimans, T., Sutskever, I., et al.: Improving language understanding by generative pre-training. preprint (2018). https://cdn.openai.com/research-covers/language-unsupervised/language_understanding_paper.pdf
34. Radford, A., et al.: Language models are unsupervised multitask learners. preprint (2019). https://cdn.openai.com/better-language-models/language_models_are_unsupervised_multitask_learners.pdf
35. Raffel, C., et al.: Exploring the limits of transfer learning with a unified text-to-text transformer. J. Mach. Learn. Res. **21**, 1–67 (2020). http://jmlr.org/papers/v21/20-074.html
36. Raunak, V., Menezes, A., Junczys-Dowmunt, M.: The curious case of hallucinations in neural machine translation. In: ACL, June 2021 (2021). https://doi.org/10.18653/v1/2021.naacl-main.92. https://aclanthology.org/2021.naacl-main.92
37. Ramos, R., Elliott, D., Martins, B.: Retrieval-augmented image captioning. In: EACL (2023). https://arxiv.org/abs/2302.08268
38. Robertson, S.E., Walker, S.: Some simple effective approximations to the 2-Poisson model for probabilistic weighted retrieval. In: Croft, W.B., van Rijsbergen, C.J. (eds.) SIGIR 1994 (1994). https://doi.org/10.1007/978-1-4471-2099-5_24
39. Wang, S., Gan, T., Liu, Y., Wu, J., Cheng, Y., Nie, L.: Micro-influencer recommendation by multi-perspective account representation learning. IEEE Trans. Multimedia **25**, 2749–2760 (2022)
40. Wang, S., Gan, T., Liu, Y., Zhang, L., Wu, J., Nie, L.: Discover micro-influencers for brands via better understanding. IEEE Trans. Multimedia **24**, 2595–2605 (2021)
41. Wang, S., et al.: Training data is more valuable than you think: a simple and effective method by retrieving from training data. In: ACL, May 2022 (2022). https://doi.org/10.18653/v1/2022.acl-long.226. https://aclanthology.org/2022.acl-long.226
42. Wang, Y., Li, J., King, I., Lyu, M.R., Shi, S.: Microblog hashtag generation via encoding conversation contexts. In: NAACL, June 2019 (2019). https://doi.org/10.18653/v1/N19-1164. https://aclanthology.org/N19-1164
43. Weston, J., Chopra, S., Adams, K.: #TagSpace: semantic embeddings from hashtags. In: EMNLP, October 2014 (2014). https://doi.org/10.3115/v1/D14-1194. https://aclanthology.org/D14-1194
44. Xue, L., et al.: mT5: a massively multilingual pre-trained text-to-text transformer. In: Toutanova, K., et al. (eds.) NAACL (2021). https://doi.org/10.18653/v1/2021.naacl-main.41
45. Zhang, H., Zhang, R., Guo, J., de Rijke, M., Fan, Y., Cheng, X.: From relevance to utility: evidence retrieval with feedback for fact verification. In: Bouamor, H., Pino, J., Bali, K. (eds.) Findings of the Association for Computational Linguistics, EMNLP 2023, Singapore December 2023, pp. 6373–6384. Association for Computational Linguistics (2023). https://aclanthology.org/2023.findings-emnlp.422
46. Zhang, Q., Wang, J., Huang, H., Huang, X., Gong, Y.: Hashtag recommendation for multimodal microblog using co-attention network. In: Sierra, C. (ed.) IJCAI (2017). https://doi.org/10.24963/ijcai.2017/478

47. Zhang, Q., Wang, Y., Gong, Y., Huang, X.: Keyphrase extraction using deep recurrent neural networks on Twitter. In: EMNLP, November 2016 (2016). https://doi.org/10.18653/v1/D16-1080. https://aclanthology.org/D16-1080
48. Zhang, X., et al.: Domain-specific NER via retrieving correlated samples. In: COLING (2022). https://aclanthology.org/2022.coling-1.211
49. Zhang, Y., Li, J., Song, Y., Zhang, C.: Encoding conversation context for neural keyphrase extraction from microblog posts. In: NAACL, June 2018 (2018). https://doi.org/10.18653/v1/N18-1151. https://aclanthology.org/N18-1151
50. Zheng, X., Mekala, D., Gupta, A., Shang, J.: News meets microblog: hashtag annotation via retriever-generator. arXiv preprint (2021). https://arxiv.org/abs/2104.08723

VEMO: A Versatile Elastic Multi-modal Model for Search-Oriented Multi-task Learning

Nanyi Fei[1], Hao Jiang[2], Haoyu Lu[3], Jinqiang Long[3], Yanqi Dai[3], Tuo Fan[2], Zhao Cao[2], and Zhiwu Lu[3(✉)]

[1] School of Information, Renmin University of China, Beijing, China
feinanyi@ruc.edu.cn

[2] Huawei Poisson Lab, Hangzhou, Zhejiang, China
{jianghao66,fantuo1,caozhao1}@huawei.com

[3] Gaoling School of Artificial Intelligence, Renmin University of China, Beijing, China
{lhy1998,longjinqiang,yanqidai,luzhiwu}@ruc.edu.cn

Abstract. Cross-modal search is one fundamental task in multi-modal learning, but there is hardly any work that aims to solve multiple cross-modal search tasks at once. In this work, we propose a novel **V**ersatile **E**lastic **M**ulti-m**O**dal (VEMO) model for search-oriented multi-task learning. VEMO is versatile because we integrate cross-modal semantic search, named entity recognition, and scene text spotting into a unified framework, where the latter two can be further adapted to entity- and character-based image search tasks. VEMO is also elastic because we can freely assemble sub-modules of our flexible network architecture for corresponding tasks. Moreover, to give more choices on the effect-efficiency trade-off when performing cross-modal semantic search, we place multiple encoder exits. Experimental results show the effectiveness of our VEMO with only 37.6% network parameters compared to those needed for uni-task training. Further evaluations on entity- and character-based image search tasks also validate the superiority of search-oriented multi-task learning.

Keywords: multi-modal model · multi-task learning · cross-modal search

1 Introduction

Cross-modal search [19,30,34,41,47] is fundamental in multi-modal learning. Humans obviously possess cross-modal search ability. We can not only find images with proper descriptions (*i.e.*, cross-modal semantic search), but also match images with given entities (*i.e.*, entity-based image search), as well as find images having the given text in them (*i.e.*, character-based image search). These different types of cross-modal search tasks are certainly in need in real scenarios.

N. Fei and H. Jiang—Contribute equally.

N. Goharian et al. (Eds.): ECIR 2024, LNCS 14608, pp. 56–72, 2024.
https://doi.org/10.1007/978-3-031-56027-9_4

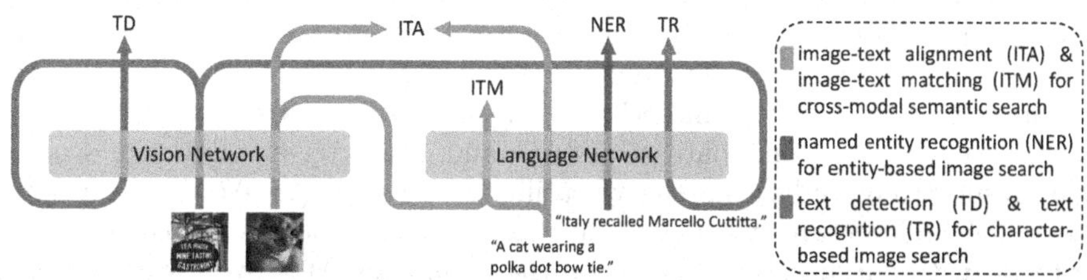

Fig. 1. Overview of our VEMO. It integrates cross-modal semantic search (including ITA & ITM), named entity recognition, and scene text spotting (including TD & TR) for search-oriented multi-task learning.

However, researches on multi-modal search models in the literature mostly focus on only one type of search task. In recent years, techniques towards cross-modal semantic search (also known as cross-modal retrieval) [11,15,16,27,33,43] have achieved great success. Significant progress has also been made in the fields of named entity recognition [17,44,46,51] and scene text spotting/optical character recognition [8,28,42,49], which are closely related to entity- and character-based image search tasks, respectively. Unfortunately, there is hardly any work that aims to solve multiple cross-modal search tasks at once.

To fill the void on search-oriented multi-task learning, we propose a **V**ersatile **E**lastic **M**ulti-m**O**dal model (VEMO), which integrates cross-modal semantic search (CMSS), named entity recognition (NER), and scene text spotting (STS) in a unified framework. The overview of our VEMO is presented in Fig. 1. Once trained, besides these three explicitly learned tasks, VEMO can go beyond and carry out entity- and character-based image search tasks with the help of NER and STS, respectively. We devise a flexible model architecture that allows the simultaneous training of these tasks. That is, we modify ViT [7] and BERT [6] as our two main networks, from which we can take out sub-modules (encoders and decoders) for corresponding tasks. Specifically, for CMSS, we need an image encoder and a text encoder for instance-level image-text alignment, and we also need a multi-modal encoder for finer-grained token-level image-text matching. For NER, we simply need a text encoder for token classification. For STS, we need an image encoder and two decoders: a location decoder for text detection and a text decoder for text recognition. All these encoders and decoders come from our modified ViT and BERT, with parameter re-use not only among tasks but also within one task. Our design keeps VEMO in a relatively small scale and also makes VEMO very flexible because we can choose modules as we need.

For the inference of CMSS, the instance-level image-text alignment is efficient because we only need to extract the query embedding and compute similarities with all pre-extracted candidate embeddings at the instance level. On the other hand, although the token-level image-text matching is very time-consuming, it is more effective because of the finer-grained modeling. To give more choices between the two extremes, we place multiple exits at different layers of the multi-modal encoder when performing image-text matching. In this way, we can freely adjust the effect-efficiency trade-off according to the usage scenarios.

Our contributions are summarized here: (1) We propose a novel **V**ersatile **E**lastic **M**ulti-m**O**dal (VEMO) model for search-oriented multi-task learning. We integrate CMSS, NER, and STS into a unified framework, where the latter two can further contribute to entity- and character-based image search tasks, respectively, indicating the versatility of VEMO. (2) We design a flexible model architecture, where we can use different modules for corresponding tasks, with parameter re-use among these modules. Moreover, we introduce selectable image-text matching exits, making VEMO more elastic. (3) Experiments show that VEMO saves a lot of parameters while achieving comparative results with independently trained uni-task models. Further evaluations on entity- and character-based image search tasks also demonstrate the effectiveness of our search-oriented model.

2 Related Work

Multi-Modal Multi-Task Search has drawn very little attention in the literature. The only related work we find is Multi-task Unified Model (MUM) [30] from Google, which is devised for their new search function: multisearch [47]. Multisearch allows the input of image and text together as search query, and aims to understand them in more natural ways to form a composed query. It is different from our VEMO in that MUM focuses on performing end-to-end search via one model to finally replace all the steps of search engines, while we simply design a multi-modal multi-task model for different types of search tasks.

Multi-Task Learning (MTL) [3,40,50] aims to leverage useful knowledge in multiple related tasks to improve the model generalization ability. Most existing MTL studies in computer vision [13,22,23,29] and natural language processing [3,35,39] focus on tasks that can be summarized into a uniform format, making the design of multi-task model neat and orderly (*e.g.*, one shared backbone with multiple task-specific heads). However, we consider tasks that differ in formalizations, thus making our model architecture much more complex.

Cross-Modal Semantic Search is also known as cross-modal retrieval. Recent works can be generally grouped into three types. Dual-stream methods [1,27,38,45] adopt separate vision and language encoders for global instance-level image-text alignment, which are efficient during evaluation. Single-stream ones [4,10,16,48] resort to one multi-modal encoder, allowing finer-grained modality interaction, which are thus more effective but time-consuming. Methods adopting hybrid architectures [15,43] integrate dual-stream and single-stream, offering both choices. Inspired by BLIP [15], we also adopt the hybrid architecture and extend it to more tasks. Notably, with multiple exits of the multi-modal encoder for token-level image-text matching, our VEMO is even more elastic.

Named Entity Recognition [17,44,46,51] aims to identify named entities in text and classify them into pre-defined categories (*e.g.*, person and location). **Scene Text Spotting** [8,28,42,49] aims to detect and recognize text in natural scenes. To our best knowledge, we are the first to combine these two tasks with cross-modal semantic search for search-oriented multi-task learning.

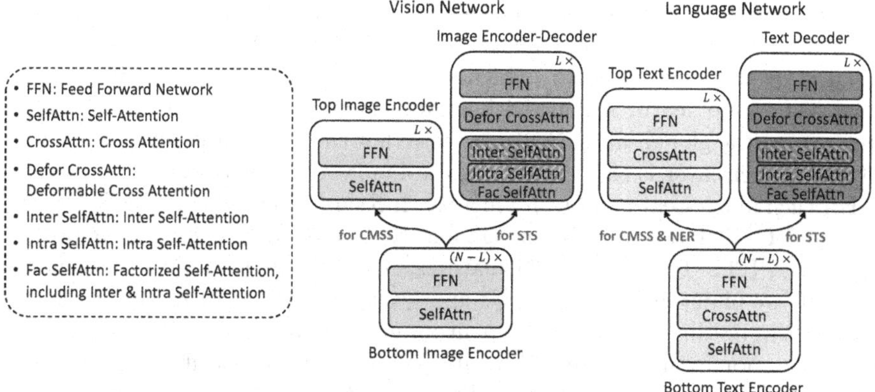

Fig. 2. The overall architecture of VEMO. Parameter re-use among different search-oriented tasks makes VEMO versatile, elastic, and meanwhile storage efficient.

3 Methodology

3.1 Framework Overview

The proposed VEMO is an end-to-end framework that handles cross-modal semantic search (CMSS), named entity recognition (NER), and scene text spotting (STS) in a unified manner. As shown in Fig. 2, VEMO contains two main networks of N layers: a vision network modified from ViT [7] and a language network modified from BERT [6]. Each network can be divided into three parts: bottom $(N - L)$ layers only work as encoding layers, while top L layers have two branches. In Sect. 3.2, we describe the CMSS task, where the bottom and top image encoders as well as the bottom and top text encoders are used. In Sect. 3.3, we introduce NER, where only part of the bottom and top text encoders are used. Then we describe STS in Sect. 3.4, which contains text detection and recognition as two sub-tasks. The bottom image encoder and part of the image encoder-decoder are used for encoding input images, followed by text detection using the image encoder-decoder and text recognition using the text decoder.

3.2 Cross-Modal Semantic Search (CMSS)

Given a query from one modality, CMSS aims to search samples from another modality that semantically match the query. As illustrated in Fig. 3(a), we consider two training losses for modality interactions at two levels: a cross-modal momentum contrastive loss for instance-level image-text alignment, and an token-level image-text matching loss for finer-grained modality interaction.

Cross-Modal Momentum Contrastive (CMMC) Loss. The CMMC loss is adopted to project samples from different modalities into a unified embedding space for image-text alignment at the global instance level. Here, the total N layers of the top and bottom image (or text) encoders are used as the uni-modal image (or text) encoder, which encodes image patches (or word tokens) into a sequence of features along with an additional `[CLS-I]` (or `[CLS-T]`) token representing the global sample feature. Note that the cross attention (CrossAttn) module in each layer is skipped for the uni-modal text encoder.

For each image-text pair (I_i, T_i) in a batch $\mathcal{B}$ sampled from the CMSS dataset, let $\mathbf{f}_i^I$ and $\mathbf{f}_i^T$ denote the global image and text embeddings after linear projection on the `[CLS-I]` and `[CLS-T]` features, respectively. Inspired by uni-modal MoCo [9], we maintain two queues $\mathcal{Q}^I$ and $\mathcal{Q}^T$ to keep the most recent N_q image-text features obtained from the momentum uni-modal encoders. For each I_i, we regard the momentum feature $\hat{\mathbf{f}}_i^T$ of its paired T_i as the positive sample, and take all samples in $\mathcal{Q}^T$ as negatives. The image-to-text contrastive loss is:

$$\mathcal{L}_{\text{i2t}} = -\frac{1}{|\mathcal{B}|}\sum\nolimits_{(I_i,T_i)\in\mathcal{B}} \log \frac{\text{pos}(\mathbf{f}_i^I, \hat{\mathbf{f}}_i^T, \tau)}{\text{pos}(\mathbf{f}_i^I, \hat{\mathbf{f}}_i^T, \tau) + \text{neg}(\mathbf{f}_i^I, \mathcal{Q}^T, \tau)}, \tag{1}$$

where τ is the temperature parameter, and

$$\text{pos}(\mathbf{f}_i^I, \hat{\mathbf{f}}_i^T, \tau) = \exp(\mathbf{f}_i^I \cdot \hat{\mathbf{f}}_i^T/\tau), \quad \text{neg}(\mathbf{f}_i^I, \mathcal{Q}^T, \tau) = \sum\nolimits_{\mathbf{q}_j^T\in\mathcal{Q}^T} \exp(\mathbf{f}_i^I \cdot \mathbf{q}_j^T/\tau). \tag{2}$$

Similarly, the text-to-image contrastive loss is defined as:

$$\mathcal{L}_{\text{t2i}} = -\frac{1}{|\mathcal{B}|}\sum\nolimits_{(I_i,T_i)\in\mathcal{B}} \log \frac{\text{pos}(\mathbf{f}_i^T, \hat{\mathbf{f}}_i^I, \tau)}{\text{pos}(\mathbf{f}_i^T, \hat{\mathbf{f}}_i^I, \tau) + \text{neg}(\mathbf{f}_i^T, \mathcal{Q}^I, \tau)}, \tag{3}$$

where $\hat{\mathbf{f}}_i^I$ is the momentum feature of I_i. The total CMMC loss is simply as:

$$\mathcal{L}_{\text{cmmc}} = \mathcal{L}_{\text{i2t}} + \mathcal{L}_{\text{t2i}}. \tag{4}$$

After loss calculation, the momentum features of paired images and texts from the current batch are then pushed into the corresponding momentum queues, meanwhile the earliest $|\mathcal{B}|$ samples in both queues are popped out.

Image-Text Matching (ITM) Loss. The ITM loss is adopted to model finer modality interactions at the token level. Concretely, we use all N layers of the top and bottom text encoders as the multi-modal encoder. Note that the self-attention (SelfAttn) and the feed forward network (FFN) in each layer share weights with those in the same layer of the uni-modal text encoder, which largely reduces the number of network parameters. The multi-modal encoder takes image-text pairs as input, and performs binary classification to predict whether the input pair is matched. For the raw input text, we append an `[ITM]` token to represent the multi-modal feature, where image information is encoded by inputting the final sequence of patch features from the uni-modal image encoder as the "keys" and "values" for the CrossAttn in each layer.

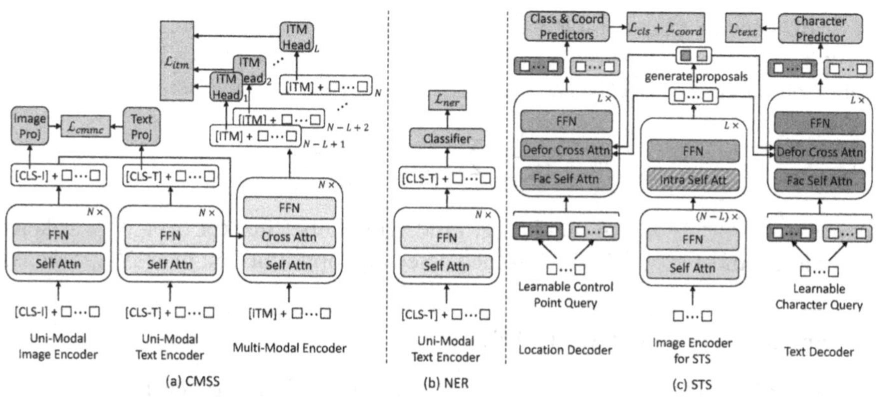

Fig. 3. Schematic illustration of three search-oriented training tasks in VEMO. Note that sub-modules with the same name and color are shared across different tasks and encoders/decoders. Multiple exits for ITM make VEMO more elastic.

Furthermore, to freely adjust the effect-efficiency trade-off, we allow the multi-modal encoder to exit at each of the top L layers. Specifically, for each input pair (I, T), let $f^{\text{ITM}}_{N-L+l}(I, T)$ denote the `[ITM]` feature at the $(N-L+l)$-th layer $(l = 1, \cdots, L)$. A two-class linear classifier h^{ITM}_l (*i.e.*, the ITM head) is adopted, which is followed by a softmax function σ, resulting in two probabilities $\sigma(h^{\text{ITM}}_l(f^{\text{ITM}}_{N-L+l}(I, T))) \in \mathbb{R}^2$ $(l = 1, \cdots, L)$. Without loss of generality, we regard its first element as the matching score of (I, T), denoted as $\text{ITM}_l(I, T)$.

To calculate the ITM loss, inspired by BLIP [15], for each image I_i (or each text T_i) in batch $\mathcal{B}$, we first sample a hard negative sample $\text{HardNeg}(I_i) \in \{T_j | j = 1, \cdots, |\mathcal{B}|, j \neq i\}$ (or $\text{HardNeg}(T_i) \in \{I_j | j = 1, \cdots, |\mathcal{B}|, j \neq i\}$). In this way, we have one positive pair (I_i, T_i) and two negative pairs $(I_i, \text{HardNeg}(I_i))$ and $(\text{HardNeg}(T_i), T_i)$ for each $(I_i, T_i) \in \mathcal{B}$. In each training iteration, we average the ITM losses calculated over all top L layers of the multi-modal encoder:

$$\begin{aligned}\mathcal{L}_{\text{itm}} =& \frac{1}{3L|\mathcal{B}|} \sum\nolimits_{(I_i,T_i)\in\mathcal{B}} \sum\nolimits_{l\in\{1,\cdots,L\}} [\text{BCE}(1, \text{ITM}_l(I_i, T_i)) \\ &+ \text{BCE}(0, \text{ITM}_l(I_i, \text{HardNeg}(I_i))) + \text{BCE}(0, \text{ITM}_l(\text{HardNeg}(T_i), T_i))],\end{aligned} \tag{5}$$

where $\text{BCE}(y, \hat{y}) = -y \log(\hat{y}) - (1 - y) \log(1 - \hat{y})$ is the binary cross-entropy.

Finally, the total loss for cross-modal semantic search is:

$$\mathcal{L}_{\text{cmss}} = \mathcal{L}_{\text{cmmc}} + \mathcal{L}_{\text{itm}}. \tag{6}$$

3.3 Named Entity Recognition (NER)

As illustrated in Fig. 3(b), for NER, we use the same uni-modal text encoder as in calculating CMMC loss. We simply adopt a two-layer classifier on top of the text encoder and compute the token classification loss. Specifically, for each text T in a batch $\mathcal{B}_{\text{ner}}$ sampled from the NER dataset, we can obtain a score matrix $\hat{\mathbf{S}} \in \mathbb{R}^{N_{\text{seq}} \times N_{\text{c}}}$, where N_{seq} and N_{c} are the token sequence length and the number of classes, respectively. The NER loss is then defined as:

$$\mathcal{L}_{\text{ner}} = \frac{1}{|\mathcal{B}_{\text{ner}}|} \sum\nolimits_{T \in \mathcal{B}_{\text{ner}}} \text{CE}(\mathbf{S}, \hat{\mathbf{S}}), \tag{7}$$

where $\mathbf{S} \in \mathbb{R}^{N_{\text{seq}} \times N_{\text{c}}}$ is the one-hot ground-truth label matrix over N_{seq} tokens of sample T, and $\text{CE}(\cdot, \cdot)$ is the batched cross-entropy function over N_{seq} tokens.

3.4 Scene Text Spotting (STS)

Inspired by TESTR [49], we design an end-to-end STS approach involving one image encoder and two decoders. As illustrated in Fig. 3(c), the architectures of both decoders are the same, *i.e.*, each decoder layer contains a factorized SelfAttn (including an intra SelfAttn and an inter SelfAttn), a deformable CrossAttn [52], and an FFN. As For the image encoder, to minimize the possible learning conflict of fully sharing it with the uni-modal image encoder used in CMSS, we only reuse the bottom $(N - L)$ layers. The top L layers of the image encoder for STS come from the location decoder (that is why we also call it the image encoder-decoder), where only the intra SelfAttn and FFN are used in each layer.

For each image I in an STS batch $\mathcal{B}_{\text{sts}}$, the encoder first extracts the sequence of image patch features. Then for each patch, we predict a coarse bounding box (*i.e.*, proposal) and a probability of having text within the box. Only proposals with top-P probability values are selected for further location and text decoding.

Location Decoder Loss. To localize text in arbitrary shapes, we expect N_{ctrl} control points to enclose a polygon. Specifically, we adopt N_{ctrl} learnable control point query tokens $\mathbf{C} \in \mathbb{R}^{N_{\text{ctrl}} \times d_{\text{ctrl}}}$, where d_{ctrl} is the dimension of each control point token embedding. To help the location decoding process, we embed the proposal information into $\mathbf{C}$ by first making P copies of $\mathbf{C}$ and then adding the transformed p-th proposal into the p-th copy. In this way, we obtain the input of the location decoder as $\mathcal{C} = \{\mathbf{C}^{(p)} \in \mathbb{R}^{N_{\text{ctrl}} \times d_{\text{ctrl}}}\}_{p=1}^{P}$, where each group focuses on one region in the image. In each layer, an intra SelfAttn is first performed over N_{ctrl} query tokens for each of the P groups independently, then an inter SelfAttn is performed over P tokens for the same j-th $(j = 1, \cdots, N_{\text{ctrl}})$ control point. In the following deformable CrossAttn, P proposals are naturally used as reference points to sample "keys" from the image patch features (see details of the deformable attention in [52]). After obtaining the output query embeddings from the location decoder, for each control point group, we devise a binary classifier to predict whether this region has text in it and a coordinate predictor to predict the coordinates of N_{ctrl} control points. Since not all P groups' corresponding image regions contain text, an injective function $\phi : \{1, \cdots, N_{\text{gt}}\} \rightarrow \{1, \cdots, P\}$ is needed, where N_{gt} is the number of ground-truth text annotations in an image. This bipartite matching problem can be efficiently solved by the Hungarian algorithm [14]. Let $s^{(p)}$ denote the classification probability of the p-th control point group and $\Phi = \{\phi(g)\}_{g=1}^{N_{\text{gt}}} \subseteq \{1, \cdots, P\}$ denote the index set of groups containing text. The classification loss is then defined as a focal loss [20]:

$$\mathcal{L}_{\text{cls}}^{\text{dec}} = -\sum\nolimits_{p \in \Phi} \alpha(1 - s^{(p)})^{\gamma} \log s^{(p)} - \sum\nolimits_{p \notin \Phi} (1 - \alpha)(s^{(p)})^{\gamma} \log(1 - s^{(p)}), \quad (8)$$

where α and γ are two hyper-parameters. Let $\mathbf{z}^{(g)} \in \mathbb{R}^{N_{\text{ctrl}} \times 2}$ $(g = 1, \cdots, N_{\text{gt}})$ denote the g-th ground-truth annotation in an image and $\hat{\mathbf{z}}^{(p)} \in \mathbb{R}^{N_{\text{ctrl}} \times 2}$ $(p = 1, \cdots, P)$ denote the predicted control point coordinates of the p-th query group. We define the coordinate regression loss as:

$$\mathcal{L}_{\text{coord}} = \sum\nolimits_{g \in \{1, \cdots, N_{\text{gt}}\}} \|\mathbf{z}^{(g)} - \hat{\mathbf{z}}^{(\phi(g))}\|. \quad (9)$$

Text Decoder Loss. Like the learnable control point query tokens for location decoding, we adopt N_{char} learnable character query tokens for the text decoder. Similarly, the query is duplicated into P copies before being input into the text decoder. By adopting a character classifier, we can finally obtain the predicted classification scores $\hat{\mathbf{t}}^{(p)} \in \mathbb{R}^{N_{\text{char}} \times (N_{\text{voc}}+1)}$ for p-th query group, where N_{voc} is the number of characters in the vocabulary and an additional null class is needed because N_{char} is larger than the length of ground-truth text in most cases. The text recognition loss is then defined as the character classification loss:

$$\mathcal{L}_{\text{text}} = \sum\nolimits_{g \in \{1, \cdots, N_{\text{gt}}\}} \text{CE}(\mathbf{t}^{(g)}, \hat{\mathbf{t}}^{(\phi(g))}), \quad (10)$$

where $\mathbf{t}^{(g)} \in \mathbb{R}^{N_{\text{char}} \times (N_{\text{voc}}+1)}$ denotes the one-hot labels over N_{char} characters in the g-th ground-truth text in an image.

Encoder Loss. Since the two decoders both rely on the output of the image encoder (*i.e.*, image patch features and the generated top-P proposals), we introduce extra constraints for the encoder. As we have mentioned, for each image patch feature, we predict a coarse bounding box and a probability of having text within the box. We thus adopt similar binary classification loss $\mathcal{L}_{\text{cls}}^{\text{enc}}$ and bounding box coordinate regression loss $\mathcal{L}_{\text{bbox}}$ like those for the location decoder. An extra generalized IoU loss [36] for bounding box regression $\mathcal{L}_{\text{giou}}$ is also used.

Overall, the final scene text spotting loss is:

$$\begin{aligned}\mathcal{L}_{\text{sts}} = \frac{1}{|\mathcal{B}_{\text{sts}}|} \sum\nolimits_{I \in \mathcal{B}_{\text{sts}}} (&\lambda_{\text{cls}} \mathcal{L}_{\text{cls}}^{\text{dec}} + \lambda_{\text{coord}} \mathcal{L}_{\text{coord}} + \lambda_{\text{text}} \mathcal{L}_{\text{text}} \\ &+ \lambda_{\text{cls}} \mathcal{L}_{\text{cls}}^{\text{enc}} + \lambda_{\text{coord}} \mathcal{L}_{\text{bbox}} + \lambda_{\text{giou}} \mathcal{L}_{\text{giou}}),\end{aligned} \quad (11)$$

where λ_{cls}, λ_{coord}, λ_{text}, and λ_{giou} are all hyper-parameters.

4 Experiments

4.1 Datasets

Our VEMO is trained on five datasets simultaneously: two for cross-modal semantic search (CMSS), one for named entity recognition (NER), and two for scene text spotting (STS).

We select two widely-used datasets for CMSS: (1) **MSCOCO** [21] is an image-text dataset of 123,287 images, with each image annotated by 5 captions. We follow [15,16,27] and split the dataset into 113,287 training, 5,000 validation, and 5,000 test images. (2) **Flickr30K** [32] is smaller, with 31,000 images and 158,915 captions in total. As in [15,16,27], we split the dataset into 29,000 training, 1,000 validation, and 1,000 test images. For performance evaluation, Recall@k (R@k, $k = 1, 5, 10$) and Recall@Mean (R@Mean) are reported, where R@Mean is the average of R@1, R@5, and R@10.

For NER, we adopt the classic **CoNLL-2003** [37], which has 203,621 training, 51,362 validation, and 46,435 testing tokens. It has four types of entities: persons, organizations, locations, and miscellaneous names. We report precision (P), recall (R), and the F1 score as the evaluation results.

For STS, two popular datasets are selected: (1) **Total-Text** [5] has 1,255 training images and 300 test images, with each image containing curved texts. (2) **ICDAR 2015** [12] contains 1,000 training images and 500 test images. It is more difficult because the images are from hand-held cameras in the wild. The standard evaluation protocols used for these datasets are followed.

4.2 Implementation Details

The vision and language networks of our VEMO are modified from ViT-B/16 [7] and BERT-Base [6], respectively. Thus the number of layers N for each network is 12. And the number of location/text decoder layers L is set to 6.

For CMSS, the input image size is 384 × 384, and the maximum length of input text is 35. The batch size $|\mathcal{B}| = 240$. The negative queue size $N_q = 57,600$. And τ in Eqs. (1)–(3) is learnable with the initialization of 0.07. For NER, we set $|\mathcal{B}_{\text{ner}}| = 128$, $N_{\text{seq}} = 128$, and $N_{\text{c}} = 12$. For STS, the input image size is $1,024 \times 1,024$. We set $|\mathcal{B}_{\text{sts}}| = 8$, $N_{\text{char}} = 25$, and $N_{\text{ctrl}} = 16$. Top-100 (*i.e.*, $P = 100$) proposals are selected for further location and text decoding. In Eq. (8), $\alpha = 0.25$ and $\gamma = 2.0$. In Eq. (11), $\lambda_{\text{cls}} = 2$, $\lambda_{\text{coord}} = 5$, $\lambda_{\text{text}} = 4$, and $\lambda_{\text{giou}} = 2$.

We employ the AdamW [26] optimizer with the initial learning rate of 1e−5. And we adopt the cosine learning rate scheduler for training a total of 5 epochs.

Before multi-task training, we load the "BLIP w/ ViT-B (14M)" model [15], fix all its parameters, and only train the image encoder-decoder and the text decoder for STS (*i.e.*, STS pre-training). Following TESTR [49], three STS pre-training datasets are used: SynthText 150K [24], MLT 2017 [31], and Total-Text [5] (see details in [49]). The initial pre-training learning rate is 1e−4. We also adopt the cosine learning rate scheduler for training 400K iterations.

Table 1. Main results (%) on 5 datasets across 3 tasks. Notations: #Params – the number of model parameters; R@Mean – the average of image-to-text and text-to-image R@Mean; Δm – the overall performance gain of each VEMO variant w.r.t. the uni-task method, defined as the difference of $\frac{1}{3}(\frac{1}{2}(\text{R@Mean(MSCOCO)}+\text{R@Mean(Flickr30K)})+\text{F1(CoNLL-2003)}+\frac{1}{2}(\text{F1(Total-Text)}+\text{F1(ICDAR 2015)}))$. Best results in each column are highlighted in bold, and second-best ones are underlined.

Method	#Params	CMSS		NER			STS						Δm
		MSCOCO	Flickr30K	CoNLL-2003			Total-Text			ICDAR 2015			
		R@Mean	R@Mean	P	R	F1	P	R	F1	P	R	F1	
Uni-Task	964.22M	<u>85.16</u>	**96.48**	75.45	<u>83.71</u>	79.37	62.94	**51.26**	56.50	38.31	**26.19**	**31.11**	-
VEMO-sum	362.50M	85.04	96.07	**79.86**	83.39	**81.59**	<u>63.72</u>	50.60	56.41	38.91	24.07	29.74	**0.41**
VEMO-log_sum	362.50M	84.87	96.06	75.04	**84.21**	79.36	**64.95**	**51.26**	**57.30**	**40.91**	24.27	30.46	−0.10
VEMO-DWA	362.50M	84.98	96.14	<u>79.37</u>	83.02	<u>81.15</u>	63.30	50.55	56.21	39.07	23.93	29.68	<u>0.22</u>
VEMO-iterative	362.50M	**85.48**	<u>96.38</u>	75.10	83.45	79.05	63.46	<u>50.99</u>	<u>56.54</u>	<u>40.42</u>	<u>25.28</u>	<u>31.10</u>	−0.07

4.3 Main Results

We first present the results by our VEMO using different multi-task training strategies, as well as the uni-task training results in Table 1. Specifically, four strategies are used to train our VEMO: (1) "sum" – three task losses are simply added together, *i.e.*, $\mathcal{L}_{\text{sum}} = \mathcal{L}_{\text{cmss}} + \mathcal{L}_{\text{ner}} + \mathcal{L}_{\text{sts}}$. (2) "log_sum" – the logarithm of three losses are added, *i.e.*, $\mathcal{L}_{\text{log\_sum}} = \log \mathcal{L}_{\text{cmss}} + \log \mathcal{L}_{\text{ner}} + \log \mathcal{L}_{\text{sts}}$. (3) "DWA" [23] – an adaptive weight is assigned to each loss according to the change rate of the loss value. Generally, if the decrease rate of one loss becomes small, then the assigned weight also becomes small. The final loss $\mathcal{L}_{\text{DWA}} = w_1\mathcal{L}_{\text{cmss}} + w_2\mathcal{L}_{\text{ner}} + w_3\mathcal{L}_{\text{sts}}$, where $w_1 + w_2 + w_3 = 3$. (4) "iterative" – we first calculate loss for one task, and then perform back propagation immediately before calculating other tasks' losses. This process is conducted iteratively among 3 tasks. As for uni-task training, we train one model for each dataset independently, using the same hyper-parameter settings as those in multi-task training.

Besides results on each dataset, we report Δm, the overall performance gain of VEMO models w.r.t. uni-task results. We also give the number of network parameters needed for each method, where for uni-task training, we report the total parameter amount of all uni-task models separately trained on five datasets.

We can observe from Table 1 that although VEMO only needs about 37.6% network parameters of those for uni-task training, VEMO variants achieve comparative results with uni-task training. This clearly validates the effectiveness of VEMO with such great reduction on the number of model parameters. Among four VEMO variants, different multi-task training strategies seem to place emphasis on different tasks, and the simplest "sum" strategy has the best overall performance. It is not surprising because unlike those strongly related tasks in conventional multi-task learning, our three chosen tasks diverse in modalities and formalizations, which leads to a more complicated multi-task balancing scenario.

We further present detailed results on MSCOCO and Flickr30K in Table 2. Note that our VEMO variants are trained in a multi-task manner and they also have multiple image-text matching exits, which clearly make the training more difficult. Despite that, Table 2 shows that VEMO variants generally achieve comparative results with current best, even beating ALIGN with 18B pre-training data. This indicates that search-oriented multi-task learning and placing multiple image-text matching exits do not harm the performance on CMSS.

Table 2. Detailed results (%) on MSCOCO and Flickr30K for CMSS. Notations: # PT Images – the number of images in pre-training data; MTL – multi-task learning; I2T/T2I – image-to-text/text-to-image. Best results in each group are in bold.

Method	# PT Images	MTL	MSCOCO						Flickr30K					
			I2T Search			T2I Search			I2T Search			T2I Search		
			R@1	R@5	R@10	R@1	R@5	R@10	R@1	R@5	R@10	R@1	R@5	R@10
UNITER-Base [4]	4M	no	64.4	87.4	93.1	50.3	78.5	87.2	85.9	97.1	98.8	72.5	92.4	96.1
OSCAR-Base [18]	4M	no	70.0	91.1	95.5	54.0	80.8	88.5	-	-	-	-	-	-
ALIGN [11]	1.8B	no	77.0	93.5	96.9	59.9	83.3	89.8	95.3	**99.8**	**100.0**	84.9	97.4	98.6
ALBEF [16]	14M	no	77.6	94.3	97.2	60.7	84.3	90.5	95.9	**99.8**	**100.0**	85.6	**97.5**	**98.9**
COTS [27]	14M	no	69.0	90.4	94.9	52.4	79.0	86.9	90.6	98.7	99.7	76.5	93.9	96.6
BLIP-Base [15]	14M	no	**80.6**	**95.2**	**97.6**	**63.1**	**85.3**	**91.1**	**96.6**	**99.8**	**100.0**	**87.2**	**97.5**	98.8
Uni-Task	14M	no	80.60	94.20	96.84	63.53	85.19	90.58	**96.00**	99.70	99.90	**87.08**	**97.48**	**98.70**
VEMO-sum	14M	yes	80.54	94.32	96.88	63.17	84.99	90.34	95.40	99.70	**100.00**	85.66	97.14	98.52
VEMO-log_sum	14M	yes	80.46	94.14	96.64	62.96	84.69	90.32	95.50	99.50	99.90	86.08	97.06	98.34
VEMO-DWA	14M	yes	80.58	94.12	96.72	63.27	84.94	90.27	95.70	99.70	**100.00**	85.86	97.06	98.52
VEMO-iterative	14M	yes	**81.08**	**94.70**	**97.12**	**63.90**	**85.27**	**90.79**	95.90	**99.80**	99.90	86.58	97.42	98.68

4.4 Selectable ITM Exits

To freely adjust the effect-efficiency trade-off for CMSS, we place multiple ITM exits at top 6 layers of the multi-modal encoder (see Fig. 3(a)). Below we give a detailed analysis of why we choose the top 6 layers.

In subfigures (a)–(d) of Fig. 4, we show results by uni-task CMSS training of our VEMO on MSCOCO. The blue dash line shows the instance-level image-text alignment result by only using separate uni-modal encoders (denoted as dual-encoder result). The orange line shows the result by using the multi-modal encoder only (denoted as single-encoder result). And the red line result is the ensemble of dual-encoder and single-encoder.

In Fig. 4(a), the model is trained in two stages. We first train the model with only one ITM head at the last layer. Then we place 11 ITM heads at every previous layer and only train these 11 heads by freezing the trained model at the first stage. We can see that single-encoder results (orange line) generally get better as the layer is closer to the top, but only top 4 layers produce better results than simple dual-encoder. It is also observed that the ensemble performance (red line) first drops and then rises, because as the multi-modal encoder uses more and more layers, the contribution of the dual-encoder decreases. In Fig. 4(b), we train the model with all 12 ITM heads at once, where the performance of 2nd–8th layers is improved a lot. This can be explained as: in Fig. 4(a), all 1st to 11th layers are only trained to provide information for the subsequent layer, while in Fig. 4(b), these layers are also trained to classify. But still, only the top 5 layers deliver better performance than simple dual-encoder. Another interesting

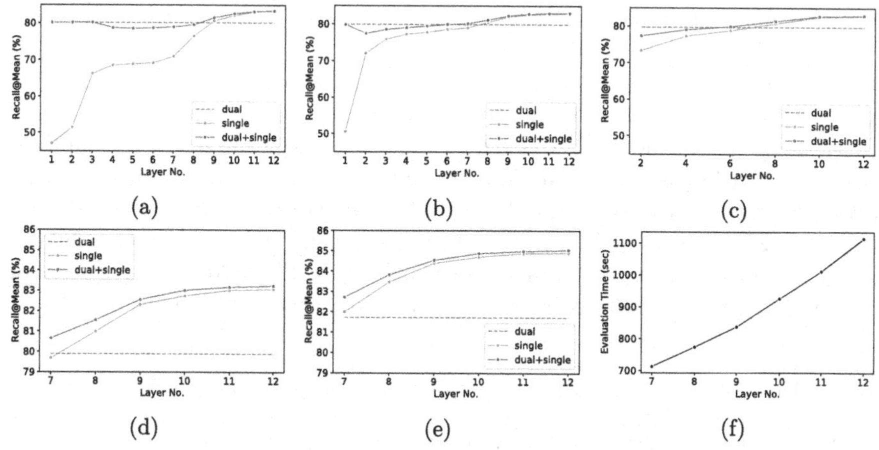

Fig. 4. (a)–(d) Results on MSCOCO for uni-task CMSS training with multiple ITM exits at different multi-modal encoder layers: (a) two-stage training (training with the last layer ITM head and then linear probing the other 11 heads); (b) training with 12 ITM heads; (c) training with 6 ITM heads at every other layer; (d) training with 6 ITM heads at top 6 layers. (e) Results on MSCOCO of multi-task trained VEMO-sum with ITM exits at top 6 layers. (f) Evaluation time for exiting at each of the top 6 layers of VEMO-sum on MSCOCO 5K test set, with one NVIDIA A100 GPU. Note that the image size is 224×224 in (a)–(d), while 384×384 in (e)–(f). (Color figure online)

phenomenon is that the 1st layer result is still incredibly low, maybe because it is very vital for the 1st layer to focus on passing information so that later layers can learn well. After these two experiments, we want to know what causes the bad performance of bottom layers. Is the simultaneous training of 12 exits too burdening, or the bottom layers themselves cannot master both passing information and classification. So in Fig. 4(c), we place only 6 exits at every other layer. It shows that the 2nd, 4th, and 6th layers perform similarly to those in Fig. 4(b), indicating their own capability limitations. Therefore, we choose to place 6 exits at top 6 layers of the multi-modal encoder. Results in Fig. 4(d) are more closer to ideal.

In Fig. 4(e) and (f), we present the results and evaluation time on MSCOCO of multi-task trained VEMO-sum, respectively. We can see that all 6 layers of the multi-modal encoder perform better than dual-encoder. As expected, using more layers gets higher results but costs more time. Selectable ITM exits blur the boundary between recall and ranking, making the model more elastic.

4.5 Further Evaluation

To demonstrate the ability of VEMO on different search tasks other than semantic search, we resort to two types of text-to-image search: (1) entity-based image search (EIS) – given a named entity, EIS aims to find images containing this entity; (2) character-based image search (CIS) – given a piece of text, CIS aims to find images that literally have this text in them. For EIS, we randomly sample 5,000 images with their captions from GoodNews [2], which is originally an image-text dataset collected from New York Times. For each image, we employ a strong news NER model [44] to extract named entities from its paired caption, which are regarded as the "ground-truth" entities for this image. We name this processed dataset as GoodNews-5K. For CIS, we directly use an STS dataset CTW1500 [25], which consists of 1,500 images (each image has several curved text annotations). As a result, GoodNews-5K and CTW1500 are essentially of the same format, *i.e.*, each image has several short text annotations.

Table 3. Results for EIS on GoodNews-5K and CIS on CTW1500.

Method	# PT Images	Search Type	GoodNews-5K				CTW1500			
			R@1	R@5	R@10	R@Mean	R@1	R@5	R@10	R@Mean
BLIP-Base	14M	CMSS	1.10	3.42	5.00	3.18	1.19	3.65	5.53	3.46
BLIP-Base	129M	CMSS	1.88	4.51	6.21	4.20	11.92	21.35	26.73	20.00
BLIP-Large	129M	CMSS	2.26	6.05	8.40	5.57	15.24	26.98	32.24	24.82
VEMO-sum	14M	CMSS	1.72	3.56	4.73	3.34	2.61	5.27	6.53	4.80
VEMO-sum	14M	EIS/CIS	**28.67**	**38.15**	**40.48**	**35.77**	**30.24**	**42.46**	**46.33**	**39.68**

We adopt three BLIP models [15] as compared methods for EIS and CIS by directly conducting CMSS. For our VEMO-sum, besides directly conducting CMSS, we can employ the NER/STS ability to address EIS/CIS more gracefully. Specifically, for EIS, VEMO-sum first extracts named entities from all captions as candidates. Then for each entity query, we use the uni-modal text encoder of VEMO-sum to calculate text-to-text similarities between it and all candidates. Finally, we assign the similarity scores to each entity candidate's corresponding image and get the search results. Similarly, for CIS, VEMO-sum first recognize texts in all images as candidates. Then we calculate text-to-text similarities, assign them to corresponding images, and finally obtain the evaluation results.

We report R@1, R@5, R@10, and R@Mean in Table 3, which clearly shows that semantic search is not suitable for either EIS or CIS. On the contrary, VEMO-sum with NER/STS ability outperforms strong semantic search methods (even with large models and 129M training data) by huge margins, validating the general applicability of our VEMO framework on different search tasks.

5 Conclusion

In this work, we investigate how to deal with multiple types of search tasks simultaneously with a single multi-modal model. Specifically, we propose a **V**ersatile **E**lastic **M**ulti-m**O**dal model termed VEMO for search-oriented multi-task learning. VEMO integrates cross-modal semantic search, named entity recognition, and scene text spotting in a unified framework, where the latter two can be further adapted to entity-based and character-based image search tasks, respectively. Furthermore, we place multiple image-text matching exits to offer more choices on the effect-efficiency trade-off for cross-modal semantic search. Extensive experiments validate the effectiveness of our VEMO with significantly fewer network parameters. We believe that search-oriented multi-task learning is meaningful, especially for devices with limited resources.

Acknowledgement. This work was supported by National Natural Science Foundation of China (62376274).

References

1. Bain, M., Nagrani, A., Varol, G., Zisserman, A.: Frozen in time: a joint video and image encoder for end-to-end retrieval. In: International Conference on Computer Vision (ICCV), pp. 1728–1738 (2021)
2. Biten, A.F., Gómez, L., Rusiñol, M., Karatzas, D.: Good news, everyone! Context driven entity-aware captioning for news images. In: IEEE/CVF Conference on Computer Vision and Pattern Recognition (CVPR), pp. 12466–12475 (2019)
3. Chen, S., Zhang, Y., Yang, Q.: Multi-task learning in natural language processing: an overview. arXiv preprint arXiv:2109.09138 (2021)
4. Chen, Y.-C., et al.: UNITER: UNiversal Image-TExt Representation learning. In: Vedaldi, A., Bischof, H., Brox, T., Frahm, J.-M. (eds.) ECCV 2020. LNCS, vol. 12375, pp. 104–120. Springer, Cham (2020). https://doi.org/10.1007/978-3-030-58577-8_7
5. Ch'ng, C., Chan, C.S., Liu, C.: Total-text: toward orientation robustness in scene text detection. Int. J. Doc. Anal. Recogn. **23**(1), 31–52 (2020)
6. Devlin, J., Chang, M.W., Lee, K., Toutanova, K.: BERT: pre-training of deep bidirectional transformers for language understanding. In: Conference of the North American Chapter of the Association for Computational Linguistics - Human Language Technologies (NAACL-HLT), pp. 4171–4186 (2018)
7. Dosovitskiy, A., et al.: An image is worth 16 $\times$ 16 words: transformers for image recognition at scale. In: International Conference on Learning Representations (ICLR) (2021)
8. Feng, W., He, W., Yin, F., Zhang, X.Y., Liu, C.L.: TextDragon: an end-to-end framework for arbitrary shaped text spotting. In: International Conference on Computer Vision (ICCV), pp. 9076–9085 (2019)
9. He, K., Fan, H., Wu, Y., Xie, S., Girshick, R.: Momentum contrast for unsupervised visual representation learning. In: IEEE/CVF Conference on Computer Vision and Pattern Recognition (CVPR), pp. 9729–9738 (2020)
10. Huang, Z., Zeng, Z., Liu, B., Fu, D., Fu, J.: Pixel-BERT: aligning image pixels with text by deep multi-modal transformers. arXiv preprint arXiv:2004.00849 (2020)
11. Jia, C., et al.: Scaling up visual and vision-language representation learning with noisy text supervision. In: International Conference on Machine Learning (ICML), pp. 4904–4916 (2021)
12. Karatzas, D., et al.: ICDAR 2015 competition on robust reading. In: International Conference on Document Analysis and Recognition (ICDAR), pp. 1156–1160 (2015)

13. Kendall, A., Gal, Y., Cipolla, R.: Multi-task learning using uncertainty to weigh losses for scene geometry and semantics. In: IEEE/CVF Conference on Computer Vision and Pattern Recognition (CVPR), pp. 7482–7491 (2018)
14. Kuhn, H.W.: The Hungarian method for the assignment problem. Nav. Res. Logist. Q. **2**(1–2), 83–97 (1955)
15. Li, J., Li, D., Xiong, C., Hoi, S.C.H.: BLIP: bootstrapping language-image pre-training for unified vision-language understanding and generation. In: International Conference on Machine Learning (ICML), pp. 12888–12900 (2022)
16. Li, J., Selvaraju, R.R., Gotmare, A., Joty, S.R., Xiong, C., Hoi, S.C.: Align before fuse: vision and language representation learning with momentum distillation. In: Annual Conference on Neural Information Processing Systems (NeurIPS), pp. 9694–9705 (2021)
17. Li, X., Feng, J., Meng, Y., Han, Q., Wu, F., Li, J.: A unified MRC framework for named entity recognition. In: Annual Meeting of the Association for Computational Linguistics (ACL), pp. 5849–5859 (2020)
18. Li, X., et al.: OSCAR: object-semantics aligned pre-training for vision-language tasks. In: Vedaldi, A., Bischof, H., Brox, T., Frahm, J.-M. (eds.) ECCV 2020. LNCS, vol. 12375, pp. 121–137. Springer, Cham (2020). https://doi.org/10.1007/978-3-030-58577-8_8
19. Liao, L., He, X., Zhao, B., Ngo, C.W., Chua, T.S.: Interpretable multimodal retrieval for fashion products. In: ACM International Conference on Multimedia (ACM-MM), pp. 1571–1579 (2018)
20. Lin, T., Goyal, P., Girshick, R.B., He, K., Dollár, P.: Focal loss for dense object detection. In: International Conference on Computer Vision (ICCV), pp. 2999–3007 (2017)
21. Lin, T.Y., et al.: Microsoft COCO: common objects in context. In: Fleet, D., Pajdla, T., Schiele, B., Tuytelaars, T. (eds.) ECCV 2014. LNCS, vol. 8693, pp. 740–755. Springer, Cham (2014). https://doi.org/10.1007/978-3-319-10602-1_48
22. Liu, B., Liu, X., Jin, X., Stone, P., Liu, Q.: Conflict-averse gradient descent for multi-task learning. In: Annual Conference on Neural Information Processing Systems (NeurIPS), pp. 18878–18890 (2021)
23. Liu, S., Johns, E., Davison, A.J.: End-to-end multi-task learning with attention. In: IEEE/CVF Conference on Computer Vision and Pattern Recognition (CVPR), pp. 1871–1880 (2019)
24. Liu, Y., Chen, H., Shen, C., He, T., Jin, L., Wang, L.: ABCNet: real-time scene text spotting with Adaptive Bezier-Curve Network. In: IEEE/CVF Conference on Computer Vision and Pattern Recognition (CVPR), pp. 9806–9815 (2020)
25. Liu, Y., Jin, L., Zhang, S., Luo, C., Zhang, S.: Curved scene text detection via transverse and longitudinal sequence connection. Pattern Recogn. **90**, 337–345 (2019)
26. Loshchilov, I., Hutter, F.: Decoupled weight decay regularization. In: International Conference on Learning Representations (ICLR) (2019)
27. Lu, H., Fei, N., Huo, Y., Gao, Y., Lu, Z., Wen, J.: COTS: collaborative two-stream vision-language pre-training model for cross-modal retrieval. In: IEEE/CVF Conference on Computer Vision and Pattern Recognition (CVPR), pp. 15671–15680 (2022)
28. Lyu, P., Liao, M., Yao, C., Wu, W., Bai, X.: Mask TextSpotter: an end-to-end trainable neural network for spotting text with arbitrary shapes. In: Ferrari, V., Hebert, M., Sminchisescu, C., Weiss, Y. (eds.) Computer Vision – ECCV 2018. LNCS, vol. 11218, pp. 71–88. Springer, Cham (2018). https://doi.org/10.1007/978-3-030-01264-9_5

29. Navon, A., ET AL.: Multi-task learning as a bargaining game. In: International Conference on Machine Learning (ICML), pp. 16428–16446 (2022)
30. Nayak, P.: MUM: a new AI milestone for understanding information. Google Blog (2021). https://www.blog.google/products/search/introducing-MUM/
31. Nayef, N., et al.: ICDAR2017 robust reading challenge on multi-lingual scene text detection and script identification - RRC-MLT. In: International Conference on Document Analysis and Recognition (ICDAR), pp. 1454–1459 (2017)
32. Plummer, B.A., Wang, L., Cervantes, C.M., Caicedo, J.C., Hockenmaier, J., Lazebnik, S.: Flickr30k entities: collecting region-to-phrase correspondences for richer image-to-sentence models. In: International Conference on Computer Vision (ICCV), pp. 2641–2649 (2015)
33. Radford, A., et al.: Learning transferable visual models from natural language supervision. In: International Conference on Machine Learning (ICML), pp. 8748–8763 (2021)
34. Rafailidis, D., Manolopoulou, S., Daras, P.: A unified framework for multimodal retrieval. Pattern Recogn. **46**(12), 3358–3370 (2013)
35. Raffel, C., et al.: Exploring the limits of transfer learning with a unified text-to-text transformer. J. Mach. Learn. Res. **21**, 140:1–140:67 (2020)
36. Rezatofighi, H., Tsoi, N., Gwak, J., Sadeghian, A., Reid, I.D., Savarese, S.: Generalized intersection over union: a metric and a loss for bounding box regression. In: IEEE/CVF Conference on Computer Vision and Pattern Recognition (CVPR), pp. 658–666 (2019)
37. Sang, E.F.T.K., Meulder, F.D.: Introduction to the CoNLL-2003 shared task: language-independent named entity recognition. In: Conference on Natural Language Learning (CoNLL), pp. 142–147 (2003)
38. Sun, S., Chen, Y.C., Li, L., Wang, S., Fang, Y., Liu, J.: LightningDOT: pre-training visual-semantic embeddings for real-time image-text retrieval. In: Conference of the North American Chapter of the Association for Computational Linguistics - Human Language Technologies (NAACL-HLT), pp. 982–997 (2021)
39. Tay, Y., Zhao, Z., Bahri, D., Metzler, D., Juan, D.: HyperGrid transformers: towards a single model for multiple tasks. In: International Conference on Learning Representations (ICLR) (2021)
40. Vandenhende, S., Georgoulis, S., Van Gansbeke, W., Proesmans, M., Dai, D., Van Gool, L.: Multi-task learning for dense prediction tasks: a survey. IEEE Trans. Pattern Anal. Mach. Intell. **44**(7), 3614–3633 (2022)
41. Wang, M., Li, H., Tao, D., Lu, K., Wu, X.: Multimodal graph-based reranking for web image search. IEEE Trans. Image Process. **21**(11), 4649–4661 (2012)
42. Wang, P., et al.: PGNET: real-time arbitrarily-shaped text spotting with point gathering network. In: AAAI Conference on Artificial Intelligence (AAAI), pp. 2782–2790 (2021)
43. Wang, W., Bao, H., Dong, L., Wei, F.: VLMo: unified vision-language pre-training with mixture-of-modality-experts. arXiv preprint arXiv:2111.02358 (2021)
44. Wang, X., et al.: Improving named entity recognition by external context retrieving and cooperative learning. In: Joint Conference of Annual Meeting of the Association for Computational Linguistics and International Joint Conference on Natural Language Processing (ACL-IJCNLP), pp. 1800–1812 (2021)
45. Wen, K., Xia, J., Huang, Y., Li, L., Xu, J., Shao, J.: COOKIE: contrastive cross-modal knowledge sharing pre-training for vision-language representation. In: International Conference on Computer Vision (ICCV), pp. 2208–2217 (2021)

46. Yamada, I., Asai, A., Shindo, H., Takeda, H., Matsumoto, Y.: LUKE: deep contextualized entity representations with entity-aware self-attention. In: Conference on Empirical Methods in Natural Language Processing (EMNLP), pp. 6442–6454 (2020)
47. Zeng, B.: Go beyond the search box: Introducing multisearch. Google Blog (2022). https://blog.google/products/search/multisearch/
48. Zhang, P., et al.: VinVL: revisiting visual representations in vision-language models. In: IEEE/CVF Conference on Computer Vision and Pattern Recognition (CVPR), pp. 5579–5588 (2021)
49. Zhang, X., Su, Y., Tripathi, S., Tu, Z.: Text spotting transformers. In: IEEE/CVF Conference on Computer Vision and Pattern Recognition (CVPR), pp. 9509–9518 (2022)
50. Zhang, Y., Yang, Q.: A survey on multi-task learning. IEEE Trans. Knowl. Data Eng. **34**(12), 5586–5609 (2022)
51. Zhu, E., Li, J.: Boundary smoothing for named entity recognition. In: Annual Meeting of the Association for Computational Linguistics (ACL), pp. 7096–7108 (2022)
52. Zhu, X., Su, W., Lu, L., Li, B., Wang, X., Dai, J.: Deformable DETR: deformable transformers for end-to-end object detection. In: International Conference on Learning Representations (ICLR) (2021)

Align MacridVAE: Multimodal Alignment for Disentangled Recommendations

Ignacio Avas(✉), Liesbeth Allein, Katrien Laenen, and Marie-Francine Moens

Department of Computer Science, KU Leuven, Leuven, Belgium
contact@ignacioavas.com,
{liesbeth.allein,katrien.laenen,sien.moens}@kuleuven.be

Abstract. Explaining why items are recommended to users is challenging, especially when these items are described by multimodal data. Most recommendation systems fail to leverage more than one modality, preferring textual or tabular data. In this work, a new model, **Align MacridVAE**, that considers the complementarity of visual and textual item descriptions for item recommendation is proposed. This model projects both modalities onto a shared latent space, and a dedicated loss function aligns the text and image of the same item. The aspects of the item are then jointly disentangled for both modalities at a macro level to learn interpretable categorical information about items and at a micro level to model user preferences on each of those categories. Experiments are conducted on six item recommendation datasets, and recommendation performance is compared against multiple baseline methods. The results demonstrate that our model increases recommendation accuracy by 18% in terms of NCDG on average in the studied datasets and allows us to visualise user preference by item aspect across modalities and the learned concept allocation (The code implementation is available at https://github.com/igui/Align-MacridVAE).

Keywords: Multimodal Recommender System · Disentangled Representation Learning · Contrastive Learning

1 Introduction

Users interact with recommendation systems (RS) to find goods to purchase, movies to watch or news articles to read. RSs identify user-relevant content by heeding the behaviour and historical interests of the user. Since online content is often presented by visual and textual signals that are expected to be complementary, forming recommendations by considering the relationship between the two modalities is possible. This is especially beneficial in contexts where data are sparse or one modality is missing. Although multimodal RSs have gained popularity over the years [24,39], most of them primarily consider visual and textual representations independently to each other and project them onto modality-specific latent spaces [5,24]. To exploit the complementarity of modalities, a

N. Goharian et al. (Eds.): ECIR 2024, LNCS 14608, pp. 73–89, 2024.
https://doi.org/10.1007/978-3-031-56027-9_5

recommendation method where multimodal content is projected onto a shared latent space and alignment between vision and language is explicitly enforced is presented in this paper.

Variational autoencoders (VAEs) [19] have been widely employed in various areas of machine learning, including RSs, to alleviate data sparsity. Recent works [23,29] leveraged VAEs to learn disentangled representations of items and user preferences. Those models are designed to handle unimodal information; extending them to handle various modalities by performing early fusion [46], e.g., concatenating feature vectors, is possible. However, this can harm the recommendation accuracy by neglecting the modality complementarity.

To that end, we proposed a new model called Align MacridVAE (**Align**ed **Macro** m**I**cro **D**isentangled **V**ariational **A**uto **E**ncoder). The model includes a time-efficient pretraining step to leverage the multimodal item representation in a joint latent space. We show that enforcing complementarity among item modalities enables this method to reveal more relevant items than other state-of-the-art methods. This alignment phase enforces a contrastive loss to handle data from different modality encoders; for example, if the textual modality is encoded with an encoder such as BERT [8], the item visual data are processed with another method such as ViT [9].

The other objective of this study is to build a RS that exhibits interpretability and transparency, which may provide the implementers and designers with insights into how the model operates and how the recommendations are generated. We leverage VAEs to achieve two levels of disentanglement. Disentanglement at the macro level helps visualising how the model interprets macro concepts related to buying preferences and item representations, such as the different item categories (e.g., dresses and shoes). Disentanglement at the micro level considers how features are related to user behaviours in each macro concept, such as shoe size or dress colour. The proposed model can also measure each modality's contribution to the overall recommendation score, providing another tool for analysing the generated recommendations.

The main contribution of this work is to introduce a novel and efficient alignment phase for use with VAEs, which allows handling sparse datasets, where the RS can suggest items that are relevant in either modality. The second contribution is the possibility of measuring the user interest by each modality, along with the macro disentangling phase, which allows user interests and item representations to be visualised.

2 Related Work

2.1 Multimodal Recommender Systems

Multimodality refers to using more than one mode of communication [20]. In linguistics, each mode corresponds to a well-defined resource for making meaning. Images, writing, speech, and videos are examples of different modes of communication. Although the available information is often multimodal, RSs usually focus on textual or tabular data from item features [7,11]. Using multiple modalities is attractive due to the complementary information of these modalities [44]. In some areas, such as e-commerce and social networks, the visual modality has as much substantial information value as the textual modality [3]. Approaches such as VBPR [12] integrate visual information into the user interaction matrix. Early fusion can be employed to integrate information from different modalities and different encoders into a single input vector. Other methods handle each information channel separately [25,26] or employ intermediate fusion through attention mechanisms [21]. More recently, approaches using graph convolution networks have been explored for multimedia recommendation [37,41,42]. However, those methods can learn entangled item and user representations, making them difficult to interpret [1]. Similar works, such as SEM MacridVAE [38], use principal component analysis [18] on the concatenation of the modality-specific representations. This can eliminate some valuable information and fail to capture complementary information if the domain of values does not overlap. We address this shortcoming with a contrastive loss, as was suggested in other multimodal RS models [11], but in contrast to these works we disentangle the obtained representations into interpretable categorical information in an unsupervised way.

2.2 Disentangled Representation Learning for Recommendation

Disentangled representation learning seeks to develop a model that detects and separates latent aspects of variation detectable in observable data in representations [4]. This technique benefits recommendation tasks because it can yield more explainable models. Representations in the latent space are mapped according to a factor of independence, allowing for the influence of certain variables on items in the space to be analysed. One common approach to learning disentangled representations is to employ VAEs [22] for cross-modal information retrieval or unimodal RS such as MacridVAE [29]. Recent research suggests that it is possible to examine items with shared values in one of the latent dimensions and identify their commonalities [11,39,45]. The model works by separating underlying factors of variation into variables with semantic meaning. This benefits learning explainable representations of data, which imitates the meaningful understanding process of humans when observing objects or relations. Although some models [24] achieve disentanglement, they fail to capture the complementarity of both modalities, and the recommendation performance thus suffers. Our work aims to align modalities for the same item to capture common parts, helping to identify relevant items that do not necessarily match completely in both modalities but rather in one of them. Other methods, such as SEM MacridVAE [38], leverage categorical information; however, this information is not always available. Instead, our proposed method learns clusters of related items, identifying semantic concepts in an unsupervised way.

3 Proposed Method

3.1 Task Definition

In the item recommendation task, the system aims to predict the preference of a user $u \in \mathcal{U}$ towards an item $i \in \mathcal{I}$, given the implicit feedback [35], presented as item reviews, from which we obtain training data for each user as r_u: which denotes a sparse vector with ones corresponding to the items that are highly rated by the user. The model also takes text and images about the items as input to predict scores $\hat{r}_u$. The item's text and images are preprocessed by modality-specific encoders like BERT [8] or ViT [9] to obtain an embedding for each modality. Those encoder weights are kept frozen during training.

Additionally, the model aims to learn disentangled representations of users and items. The disentanglement is done at two levels. At the macro level, the model identifies a small number of K clusters from the item and interaction data. At the micro level, the model tries to learn disentangled characteristics of user preferences in each of those clusters. K should be small to facilitate the inspection and transparency of the recommendation, when it is communicated to the user.

3.2 Overview of the Model

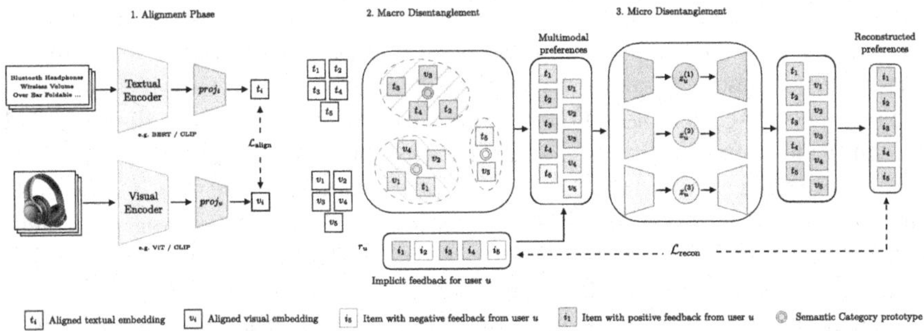

Fig. 1. Architecture of Align MacridVAE. The alignment phase trained with alignment loss $\mathcal{L}_{\text{align}}$ for a single item is shown on the left. Macro disentanglement with five items and $K = 3$ categories is shown in the middle, and micro disentanglement via VAEs is shown on the right.

In this work, a new approach to handling multiple modalities, called *Align MacridVAE*, is proposed. This model, based on MacridVAE [38] and serving as an alternative to SEM MacridVAE [38], attempts to align preferences for different modalities and learn representations in the same semantic space. The model assumes that if two items have similar textual representations, their visual representations are also similar. When a user shows interest in an item, then they may like similar items, where similarity is based on any of the visual or textual representations. The alignment phase is added for two reasons: First, it allows the original MacridVAE model to remain almost unchanged in the inference, permitting a fair comparison between SEM MacridVAE and Align MacridVAE. Second, the alignment phase is a fast alternative to fine-tuning the features in feature extractors. It is possible to incorporate joint training of the encoding of the modalities and recommendation task. However, those models require a much larger number of parameters, which might slow the fine-tuning process.

The model training is a three-step process as illustrated in Fig. 1. It begins with an alignment phase where the model learns the initial values of the textual and visual representations of items t_i and v_i, initially obtained fromt the textual and visual encoders, aiming to align those two values in the latent space. Then, the model proceeds to learn the K initial semantic clusters, represented by prototypes o_k, to form the first phase of the macro disentanglement. The cluster prototypes $\{o_k\}_{k=1}^{K}$ are learned by performing K-means clustering [30] on t_i, v_i and the affinity of item modalities to those K clusters represented by a matrix C with elements $c_{i,k}$. The model spends most of its time in the main training phase where it performs the micro disentanglement. This process refines the representations t_i, v_i and o_k, and the model learns the values $z_u^{(k)}$, which denotes the micro disentangled representation of user preferences in each of the K clusters.

3.3 Learning Text-Image Alignments

One of the goals of the method is to learn disentangled representations of the items on the textual and visual modality t_i and v_i, respectively. Those vectors belong to $\mathbb{R}^d$ so that each dimension represents an aspect of each item, like the colour, shape or type of item. For example, items in the same category (e.g., movie genre) should have similar values in each dimension.

First, visual and textual representations are aligned in a joint semantic space as shown in Fig. 1. Unlike previous approaches, this model learns an embedding for each item t_i and v_i using two linear projections, while the modality-specific encoders weights are kept frozen to accelerate the alignment process.

This process aims to align the embeddings so that t_i is close to v_i in the joint semantic space. To achieve this goal, a contrastive loss function is used in this phase. This function takes the cosine similarity of vectors and a learnable temperature scalar η to improve the training stability by making the similarity score multinomial distribution less peaky [43].

$$\mathcal{L}_{\text{align}} = \frac{1}{|\mathcal{I}|^2 - |\mathcal{I}|} \sum_{i \neq j} \ln s_{i,j} - \frac{1}{|\mathcal{I}|} \sum_{i=1}^{|\mathcal{I}|} \ln s_{i,i} \tag{1}$$

$$s_{i,j} = \text{softmax}_j\left(\hat{s}_{i,j}\right), \quad \hat{s}_{i,j} = e^{\eta}\, \text{cosine}(t_i, v_j) \tag{2}$$

Equation 1 expresses the loss function $\mathcal{L}_{\text{align}}$, a variation of the cross entropy loss that aims to bring the visual and textual modalities from the same item closer to each other and farther from the other items. The values $\hat{s}_{i,j}$ store the cosine similarities between the visual and textual modality predictions, where cosine similarity is computed as $\text{cosine}(a, b) = \frac{a^T b}{\|a\|\|b\|}$. The values $s_{i,j}$ make these similarities a probability distribution using softmax [17].

$\mathcal{L}_{\text{align}}$ can be formulated in other ways, such as by using a margin loss or another formulation that achieves alignment by placing matching modalities close together in the embedding space. The most commonly used distance functions for scalars and vectors are instantiations of the Euclidean distance and the cosine distance [33]. These metrics serve as a means of organising objects, forming concepts, and making generalisations. In this work, cosine similarity is utilised due to its strong properties and insensitivity to vector norms [28]. Using the unnormalized dot product can cause a large variance in the distance values between two representations and make the model sensitive to the input distribution.

3.4 Macro Disentanglement

After alignment, we perform disentanglement at a macro level; this disentanglement is refined in the main training phase. We initialize item clusters using K-means clustering into K macro concepts on the items using the representations t_i and v_i in the joint semantic space, with the Euclidean distance as distance metric. In this way, we can independently model preferences for visual or textual modalities.

In the original formulation of MacridVAE, clusters are initialized randomly, which can lead to poor convergence and explainability since allocations might not reflect the actual item semantics, such as categories of item types. Incorporating modality-specific information enhances cluster detection, improving training stability since two signals are provided for each item instead of one. Additionally, the model is more resilient to missing data in one modality, such as missing item images, which are common in inspected datasets and real-world scenarios [15]. From the macro disentanglement, we obtain cluster prototypes o_k, which are represented by the centres of each cluster.

3.5 Micro Disentanglement

The model adopts a similar training phase as MacridVAE, using a β-VAE [13] to learn disentangled representations of the user interest in each micro concept. In the proposed model, we can distinguish preferences by modality by considering each item as a pair of textual and visual-specific representations. In contrast, in MacridVAE, textual and visual representations are fused; that model handles $|\mathcal{I}|$ items, whereas we consider $2|\mathcal{I}|$ items (textual and visual). The goal is to learn parameters θ so that the model learns a generative distribution of the user preference towards each item modality:

$$p(r_u) = \mathbb{E}_{p_\theta(C)} \left(\int p_\theta(r_u | z_u, C) p_\theta(z_u) dz_u \right) \tag{3}$$

where $z_u = \left[z_u^{(1)}, \ldots, z_u^{(K)}\right]$, $z_u^{(k)}$ is the micro disentangled user representation for the kth cluster, and C represents the learned cluster allocation described in §3.1. The cluster assignment is a probability distribution over the K categories for the textual and visual modality. The following equation provides the values for the allocation in the textual modality:

$$c_{k,i}^{(t)} = \text{softmax}_{k \in \{1 \ldots K\}} \left(\frac{1}{\tau} \text{cosine}\left(t_i, o_k\right)\right) \tag{4}$$

where t_i is the learned textual embedding for the item. Hyperparameter τ helps training stability, similar to η introduced in §3.3, to make the softmax function less peaky. The visual modality cluster allocation is equivalent to Eq. 4 substituting t_i with v_i, yielding $c_{k,i}^{(v)}$. Finally, the cluster affinity forms a matrix $C = \left[C^{(t)}; C^{(v)}\right] \in \mathbb{R}^{K \times 2|\mathcal{I}|}$ combining the affinity of each modal-specific representation to a cluster, where each row represents the distribution of visual and textual representations in that cluster. This reduces the need for training data with categorical information data from items as employed by SEM MacridVAE, which employs one-hot encoded vector columns for building C from categorical information present in the training data. Our proposed model avoids this encoding, as items may have multiple categories or category information may not be available.

3.6 Encoder

The model assumes that when the user gives positive feedback on an item, the positive feedback is on both the visual and textual parts. Therefore, the model takes a vector of preferences $x_u \in \mathbb{R}^{2\mathcal{I}}$ based on r_u, which has ones in positions i and $i+|\mathcal{I}|$ if the user likes the ith item. The encoder follows a common VAE architecture. The values $z_u^{(k)}$ are approximated by a normal distribution $q_\theta(z_u^{(k)}|x_u, C) = \mathcal{N}\left(z_u^{(k)}; \mu_u, \text{diag}\left(\exp(\sigma_u)\right)\right)$, where the values are generated by a two-layer neural network defined by:

$$\left[\mu_u^{(k)}; \sigma_u^{(k)}\right] = W_2^{(\text{enc})} \tanh\left(W_1^{(\text{enc})}\left(C_k \odot x_u\right) + b_1^{(\text{enc})}\right) + b_2^{(\text{enc})} \tag{5}$$

where $\odot$ represents the Hadamard product [2]. Then, $z_u^{(k)}$ are drawn from the normal distribution q_θ.

3.7 Decoder

After obtaining the values of $z_u^{(k)}$, we reconstruct the user preferences using the decoder and obtain $w_u = (w_{u,1}^{(t)}, \ldots, w_{u,|\mathcal{I}|}^{(t)}, w_{u,1}^{(v)}, \ldots, w_{u,|\mathcal{I}|}^{(v)})$ with:

$$w_{u,i}^{(t)} = \ln\left(\sum_k c_{k,i}^{(t)} \cdot \exp\left(\frac{1}{\tau} \cos(z_u^{(k)}, t_i)\right)\right) \tag{6}$$

The training loss follows the VAE paradigm and is defined by the evidence lower bound (ELBO) $\max_\theta \ln(p_\theta(x_u))$:

$$\begin{aligned} \ln \hat{r}_u \geq \mathcal{L}_{\text{recon}} + \beta \mathcal{L}_{\text{KL}} &= \mathcal{L}_{\text{train}} \\ \text{with} \quad \mathcal{L}_{\text{recon}} &= \mathbb{E}_{p_\theta(C)} \left[\mathbb{E}_{q_\theta(z_u|r_u,C)} \left(\ln p_\theta(r_u|z_u, C) \right) \right] \\ \mathcal{L}_{\text{KL}} &= -D_{\text{KL}} \left[q_\theta \left(z_u | r_u, C \right) \| p_\theta \left(z_u \right) \right] \end{aligned} \tag{7}$$

3.8 Inference

The item score is a pooling function over the modality preference score:

$$\hat{r}_{u,i} = \frac{\hat{r}^{(t)}_{u,i} + \hat{r}^{(v)}_{u,i}}{2} \quad \text{with} \quad \hat{r}^{(t)}_{u,i} = \text{softmax}_i(w^{(t)}_{u,i}),\ \hat{r}^{(v)}_{u,i} = \text{softmax}_i(w^{(v)}_{u,i}) \tag{8}$$

Like other methods [24], the scoring function averages the preference for each modality. At inference time, we use the constant value $\mu_u^{(k)}$ instead of sampling the normal distribution q_θ to obtain more consistent results for the same user.

4 Experiments

Table 1. The table on top displays the dataset statistics. Density is calculated as the ratio of the number of reviews over the possible number of user and item combinations. The table at the bottom shows the model hyperparameters used in MacridVAE, Align MacridVAE, and SEM MacridVAE.

Dataset	# Users	# Items	# Ratings	Density
Movies & TV	21974	23958	216110	0.041%
Musical Instruments	14429	29040	93923	0.022%
Home & Kitchen	5968	57645	135839	0.040%
Clothing, Shoes & Jewelry	23318	38493	178944	0.020%
Movielens 25M	162541	59047	25000095	0.260%
Book-crossing	14790	33962	519613	0.103%

Table a: Dataset statistics.

Parameter	Value	Parameter	Value
Number of latent factors K	7	Epochs	50
Embedding dimension for D	200	Learning rate	0.001
Disentanglement $\mathcal{L}_{KL}$ weight β	0.2	Image Encoder	CLIP
Temperature for cosine similarity (τ)	0.1	Text Encoder	CLIP
Batch size	100	Epochs for Alignment	10

Table b: Hyperparameter settings.

To validate our model, we evaluated it on six public datasets. We first describe the experimental setup and then discuss the obtained results. The experiments are conducted to measure the recommendation quality, analyse the complementarity and alignment of the multimodal recommendation, measure disentanglement, and explore post hoc ways to provide sentence explanations.

We experimented with the VIT [9], BERT [8], and CLIP [34] feature extractors to validate whether our method successfully combines encoded information from diverse feature spaces. They use different techniques to encode the information in the latent space, affecting recommendation performance.

4.1 Datasets

This method was tested on various datasets used in the literature for multimodal recommendations. Four datasets were based on the Amazon reviews 2018 dataset [32], one on the Book-crossing dataset [47], and one on the MovieLens dataset [10] 25 M version. Table 1a shows the dataset statistics. The datasets' data formats and textual modalities are diverse, and they therefore require different techniques for preprocessing. The textual part is the concatenation of the item title and description. The item titles and synopsis for the MovieLens database were extracted from the IMDb Non-Commercial dataset [16] and the IMDb website. For the Book-crossing dataset, because many items did not have available visual modalities, the items were filtered to retrieve only those with available images to have both modalities available. Following similar works [23], implicit feedback is binarised, with only reviews of 4 or 5 stars (or the equivalent in other scales) considered as positive and the rest as negative.[1] The hyperparameters utilised in the models were found after a grid search and are the same as those utilised in the MacridVAE and SEM MacridVAE. We used the same hyperparameters to compare the three models fairly, allowing us to evaluate how much the new architecture contributes to the recommendation quality changes.

4.2 Recommendation Performance

The recommendation performance results measured in terms of normalized discounted cumulative gain (NDCG) [40] and recall [36] at the first five positions and mean average precision (MAP) [6] to emphasise the importance of the first results. Choosing a larger number of positions would capture items in the tail of the recommendations, which are less relevant. We also performed an ablation test to evaluate the impact of changing encoders. For the textual modality, we implemented BERT [8] for fair comparison of the results with the baselines [24,38], and also the CLIP model [34], which leverages alignment between multiple modalities. The visual modality is modeled using ViT, as done in [24] and the CLIP model.

Figure 2 shows the performance on the different datasets depending on the encoders used, averaged to all datasets. Similar trends can be seen when analyzing the results of the different datasets. In Table 2, we compare the proposed

---

[1] The code for processing the original datasets is available at https://github.com/igui/Align-MacridVAE-data.

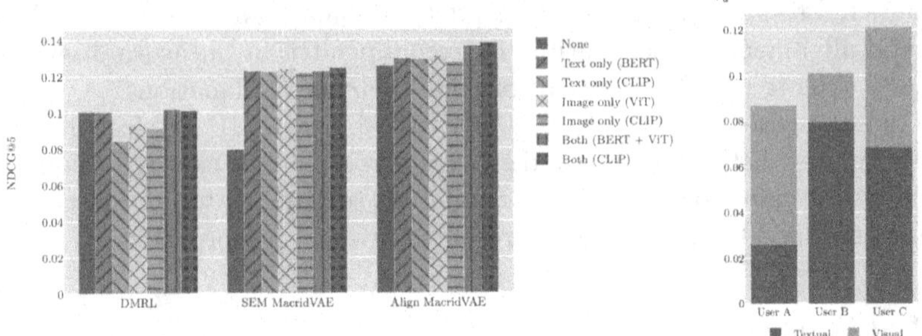

Fig. 2. On the left, a comparison of the performance based on the provided modality information and the utilised encoder (ViT, BERT or CLIP) averaged over all the datasets. On the right, the prediction for three different users measures the preference in each modality. Depending on the user and item, the weight of each item modality is different.

Table 2. Comparison of model performances on different datasets in terms of NDCG@5, Recall@5 and MAP. The best value is bolded, and the second best is underlined.

Dataset	Model	NDCG@5	Recall@5	MAP	Dataset	Model	NDCG@5	Recall@5	MAP
Clothing, Shoes & Jewelry	MultiDAE	0.18685	0.22728	0.17300	Musical Instruments	MultiDAE	0.06932	0.08623	0.06440
	MultiVAE	0.19452	0.23226	0.18107		MultiVAE	0.07734	0.09403	0.07200
	MacridVAE	0.21029	0.24141	0.19754		MacridVAE	0.06904	0.07920	0.06597
	DMRL	0.20381	0.23975	0.19121		DMRL	0.07067	0.09673	0.07486
	SEM MacridVAE	0.21271	0.24433	0.19914		SEM MacridVAE	0.07441	0.08841	0.07067
	Align MacridVAE	**0.23400**	**0.27118**	**0.21988**		Align MacridVAE	**0.08541**	**0.10138**	**0.08077**
Home & Kitchen	MultiDAE	0.03107	0.04027	0.02845	Bookcrossing	MultiDAE	0.02500	0.02894	0.02117
	MultiVAE	0.03821	0.04637	0.03590		MultiVAE	0.02365	0.02763	0.02047
	MacridVAE	0.04228	0.05170	0.03918		MacridVAE	0.03316	0.03741	0.02254
	DMRL	0.04324	0.05214	0.04072		DMRL	0.03384	0.04010	0.02736
	SEM MacridVAE	0.04674	0.05588	0.04353		SEM MacridVAE	0.03924	0.04353	0.02744
	Align MacridVAE	**0.05878**	**0.06753**	**0.05592**		Align MacridVAE	**0.04653**	**0.05347**	**0.03484**
Movies & TV	MultiDAE	0.07414	0.09198	0.07169	Movielens 25M	MultiDAE	0.24834	0.24225	0.18494
	MultiVAE	0.07936	0.09572	0.07698		MultiVAE	0.26308	0.25667	0.19240
	MacridVAE	0.08352	0.09698	0.07842		MacridVAE	0.28501	0.27671	0.19264
	DMRL	0.06627	0.08472	0.06401		DMRL	0.18709	0.16276	0.11583
	SEM MacridVAE	0.08518	0.09661	0.08180		SEM MacridVAE	**0.28857**	**0.27999**	**0.19405**
	Align MacridVAE	**0.12231**	**0.14461**	**0.11409**		Align MacridVAE	0.28091	0.27242	0.18940

model with several baselines, including MultiVAE and MultiDAE [23], DMRL [24], MacridVAE [38], and SEM MacridVAE [38]. The compared baselines do not optimally perform when presented with both the visual and textual modalities, indicating challenges in leveraging the complementary information in the different channels. These results are aligned with recent studies [46]. The results in Table 2 reveal that our model outperforms the examined baselines in almost all datasets, suggesting that the alignment phase improves recommendation performance. To analyse the value of each modality, we show the values for the modality-specific preferences $\hat{r}_{u,i}^{(t)}$ and $\hat{r}_{u,i}^{(v)}$ for the same item for different users in Fig. 2. The model infers that each user gives more priority to either text or image. While User A prioritises the visual aspect, User B prioritises the item text.

We note that each epoch of the alignment phase takes some seconds, depending on the dataset size, which is a small fraction of the time it would take to fine-tune a DNN model on the dataset. For example, an epoch of fine-tuning the CLIP model in the MovieLens 25M dataset takes approximately 40 min while an alignment iteration takes 30 s in our model.

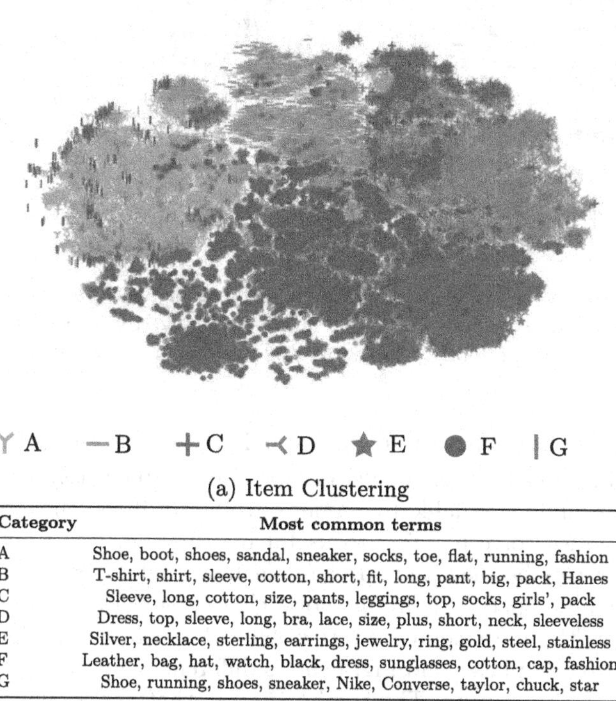

(a) Item Clustering

Category	Most common terms
A	Shoe, boot, shoes, sandal, sneaker, socks, toe, flat, running, fashion
B	T-shirt, shirt, sleeve, cotton, short, fit, long, pant, big, pack, Hanes
C	Sleeve, long, cotton, size, pants, leggings, top, socks, girls', pack
D	Dress, top, sleeve, long, bra, lace, size, plus, short, neck, sleeveless
E	Silver, necklace, sterling, earrings, jewelry, ring, gold, steel, stainless
F	Leather, bag, hat, watch, black, dress, sunglasses, cotton, cap, fashion
G	Shoe, running, shoes, sneaker, Nike, Converse, taylor, chuck, star

(b) Common terms by textual modality

Fig. 3. Item clustering common terms by textual modality. On the left is a two-dimensional visualization created with t-SNE [14] of the learned item categories after the alignment phase from the d-dimensional space of the Amazon Clothing, Shoes & Jewelry dataset. On the right are the most common words among items in each learned category. Terms are ordered by the number of occurrences among the items in the category. The list omits terms present in most categories, such as "for" and "and".

4.3 Disentanglement and Interpretability

The macro disentanglement can be judged by how well the collected categories align with macro semantic concepts. To this end, we project modality-specific item representations to a 2D space using t-SNE [14], allowing us to analyse the learned representations in a qualitative manner. Figure 3a shows the distribution of clusters. Each modality-specific representation is coloured with the

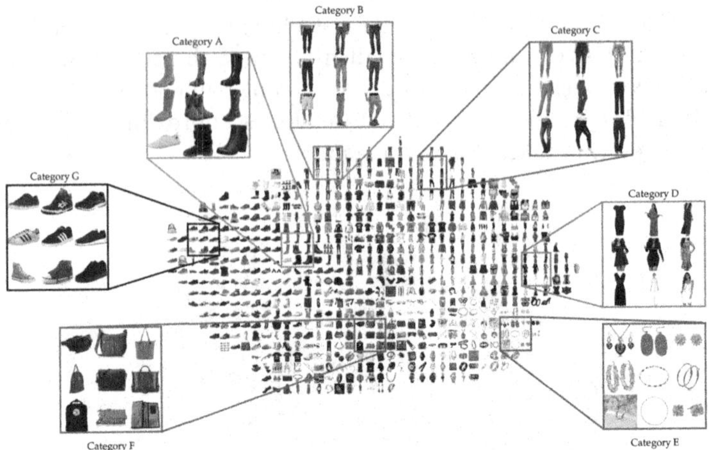

Fig. 4. Item images of selected items using the same coordinates as in Fig. 3a, for the Amazon clothing Dataset. For clarity, the space is arranged into a grid and for each cell, one item image is randomly selected among overlapping instances.

colour of the category with the highest affinity, which is the largest value of $c_{i,k}^{(\cdot)}$. The results show that the model groups items into clusters with meaningful information, understandable by humans. To verify if the clusters determine semantic categories corresponding to useful information, the individual items in each cluster can be observed to examine how well items are clustered in the latent semantic space. Figure 3a shows clusters generated by our method, Figure 3b shows the most common terms found in each of the identified categories for the textual phase, and Fig. 4 shows different item images, highlighting some items in each category. The results suggest that the model places items with distinct characteristics in separate clusters. For example, cluster E covers the concept of jewellery, while cluster B encompasses concepts such as men's clothing. Choosing the right number of concepts is important for avoiding categories that convey unrelated or latent information unrelated to understandable human buying behaviour. The model presents similar results on other datasets, such as Home & Kitchen or Musical Instruments, consistently forming semantic categories analysable by humans. These qualitative analyses illustrate the ability of Align MacridVAE to build meaningful clusters of semantically related items.

4.4 Micro Disentanglement

The quality and interpretability of the disentanglement can also be measured quantitatively [31]. A possible way to do so is to study the effect of the regularization parameter β and the recommendation quality. The amount of disentanglement can be measured by calculating the independence of each dimension for the user and item modality representations. To this end, we calculate the correlation measured as $\text{Corr} = 1 - \frac{1}{d(d-1)} \sum_{1 \le i < j \le d} \frac{\text{cov}_{ij}}{\sqrt{\text{cov}_{ii}\,\text{cov}_{jj}}}$ where cov is the

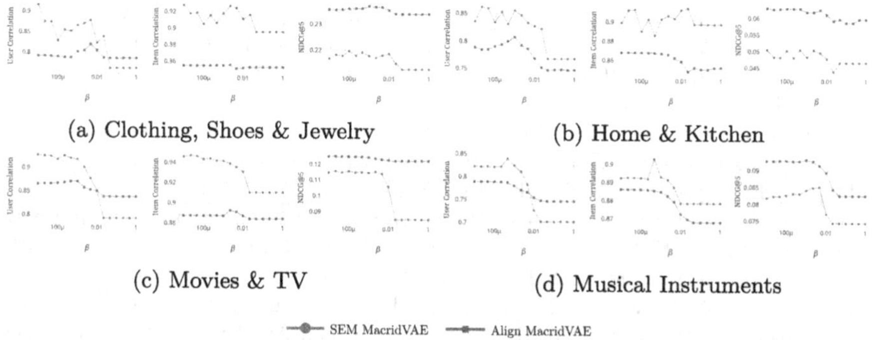

Fig. 5. Micro disentanglement effect in different datasets for Align MacridVAE and SEM MacridVAE. On the horizontal axis different values of the regularization parameter β are shown, and on the vertical axis, the values of the item correlation, user correlation, and NDCG@5 results correspond to the recommendation accuracy.

covariance matrix. Figure 5 illustrates the effect of β that regulates the importance of the $\mathcal{L}_{KL}$ term of Eq. 7. A higher value of β emphasises disentanglement at the cost of recommendation accuracy.

The results underline that Align MacridVAE attains a lower correlation in both user and item representations evidencing disentanglement, while maintaining a strong recommendation accuracy. Unfortunately, as noted by Locatello [27], disentanglement does not ensure that representations are semantic.

5 Conclusion

In this paper, we studied the problem of interpretable multimodal recommendation. We presented a novel recommendation model called Align MacridVAE, which exploits the multimodal (visual and textual) information in a transparent way. We showed the effectiveness of the alignment of visual and textual modalities in a joint semantic space and their benefits to the recommendation performance. The results show that the model can learn semantic clusters without relying on data annotated with item categories, which are interpretable. In addition, user preferences for a particular modality are detected to aid implementers and designers in understanding the inner workings of the algorithm. A preference score for each modality helps to understand the preference that users attain for each part of the items.

References

1. Anand, A., Lyu, L., Idahl, M., Wang, Y., Wallat, J., Zhang, Z.: Explainable information retrieval: a survey (2022). https://doi.org/10.48550/ARXIV.2211.02405. https://arxiv.org/abs/2211.02405

2. Ando, T.: Majorization relations for Hadamard products. Linear Algebra Appl. **223–224**, 57–64 (1995). Honoring Miroslav Fiedler and Vlastimil Ptak. https://doi.org/10.1016/0024-3795(94)00014-5. https://www.sciencedirect.com/science/article/pii/0024379594000145
3. Baltrušaitis, T., Ahuja, C., Morency, L.P.: Multimodal machine learning: a survey and taxonomy. IEEE Trans. Pattern Anal. Mach. Intell. **41**, 423–443 (2019). https://doi.org/10.1109/TPAMI.2018.2798607
4. Bengio, Y., Courville, A., Vincent, P.: Representation learning: a review and new perspectives. IEEE Trans. Pattern Anal. Mach. Intell. **35**(8), 1798–1828 (2013). https://doi.org/10.1109/TPAMI.2013.50
5. Cai, Z., Cai, Z.: PEVAE: a hierarchical VAE for personalized explainable recommendation. In: Proceedings of the 45th International ACM SIGIR Conference on Research and Development in Information Retrieval, SIGIR 2022, pp. 692–702. Association for Computing Machinery, New York (2022). https://doi.org/10.1145/3477495.3532039
6. Carterette, B., Voorhees, E.M.: Overview of information retrieval evaluation. In: Lupu, M., Mayer, K., Tait, J., Trippe, A. (eds.) Current Challenges in Patent Information Retrieval. The Information Retrieval Series, vol. 29, pp. 69–85. Springer, Heidelberg (2011). https://doi.org/10.1007/978-3-642-19231-9_3
7. Chen, H., et al.: Curriculum disentangled recommendation with noisy multi-feedback. In: Ranzato, M., Beygelzimer, A., Dauphin, Y., Liang, P., Vaughan, J.W. (eds.) Advances in Neural Information Processing Systems, vol. 34, pp. 26924–26936. Curran Associates, Inc. (2021)
8. Devlin, J., Chang, M.W., Lee, K., Toutanova, K.: BERT: pre-training of deep bidirectional transformers for language understanding. In: Proceedings of the 2019 Conference of the North American Chapter of the Association for Computational Linguistics: Human Language Technologies, Volume 1 (Long and Short Papers), Minneapolis, Minnesota, June 2019, pp. 4171–4186. Association for Computational Linguistics (2019). https://doi.org/10.18653/v1/N19-1423. https://aclanthology.org/N19-1423
9. Dosovitskiy, A., et al.: An image is worth 16 × 16 words: transformers for image recognition at scale. In: 9th International Conference on Learning Representations, ICLR 2021, Virtual Event, Austria, 3–7 May 2021 (2021)
10. GroupLens Research: Movielens Dataset (2023). https://grouplens.org/datasets/movielens/
11. Han, T., Wang, P., Niu, S., Li, C.: Modality matches modality: pretraining modality-disentangled item representations for recommendation. In: Proceedings of the ACM Web Conference 2022, WWW 2022, pp. 2058–2066. Association for Computing Machinery, New York (2022). https://doi.org/10.1145/3485447.3512079
12. He, R., McAuley, J.: VBPR: visual Bayesian personalized ranking from implicit feedback. In: Proceedings of the Thirtieth AAAI Conference on Artificial Intelligence, AAAI 2016, pp. 144–150. AAAI Press (2016)
13. Higgins, I., et al.: beta-VAE: learning basic visual concepts with a constrained variational framework. In: International Conference on Learning Representations (2016)
14. Hinton, G.E., Roweis, S.: Stochastic neighbor embedding. In: Advances in Neural Information Processing Systems, vol. 15 (2002)
15. Idrissi, N., Zellou, A.: A systematic literature review of sparsity issues in recommender systems. Soc. Netw. Anal. Min. **10**(1), 15 (2020). https://doi.org/10.1007/s13278-020-0626-2

16. IMDb.com: IMDb non-commercial datasets (2023). https://help.imdb.com/contact/developer/. Accessed 22 Jul 2023
17. Jang, E., Gu, S., Poole, B.: Categorical reparameterization with Gumbel-Softmax. In: 5th International Conference on Learning Representations, ICLR 2017, Conference Track Proceedings, Toulon, France, 24–26 April 2017. OpenReview.net (2017). https://openreview.net/forum?id=rkE3y85ee
18. Jolliffe, I.T., Cadima, J.: Principal component analysis: a review and recent developments. Philos. Trans. R. Soc. A Math. Phys. Eng. Sci. **374**(2065), 20150202 (2016). https://doi.org/10.1098/rsta.2015.0202
19. Kingma, D.P., Welling, M.: Auto-encoding variational Bayes. In: Bengio, Y., LeCun, Y. (eds.) 2nd International Conference on Learning Representations, ICLR 2014, Conference Track Proceedings, Banff, AB, Canada, 14–16 April 2014 (2014)
20. Kress, G.: Multimodality: A Social Semiotic Approach to Contemporary Communication. Routledge, London (2009). https://doi.org/10.4324/9780203970034
21. Laenen, K., Moens, M.F.: A comparative study of outfit recommendation methods with a focus on attention-based fusion. Inf. Process. Manage. **57**(6), 102316 (2020). https://doi.org/10.1016/j.ipm.2020.102316. https://www.sciencedirect.com/science/article/pii/S0306457320308116
22. Laenen, K., Moens, M.F.: Learning explainable disentangled representations of e-commerce data by aligning their visual and textual attributes. Computers **11**(12) (2022). https://doi.org/10.3390/computers11120182. https://www.mdpi.com/2073-431X/11/12/182
23. Liang, D., Krishnan, R.G., Hoffman, M.D., Jebara, T.: Variational autoencoders for collaborative filtering. In: Proceedings of the 2018 World Wide Web Conference, WWW 2018, pp. 689–698. International World Wide Web Conferences Steering Committee, Republic and Canton of Geneva, CHE (2018). https://doi.org/10.1145/3178876.3186150
24. Liu, F., Chen, H., Cheng, Z., Liu, A., Nie, L., Kankanhalli, M.: Disentangled multimodal representation learning for recommendation. In: 2021 IEEE International Conference on Multimedia and Expo (ICME), Virtual, pp. 1–11. IEEE (2021). https://doi.org/10.1109/TMM.2022.3217449
25. Liu, F., Cheng, Z., Sun, C., Wang, Y., Nie, L., Kankanhalli, M.: User diverse preference modeling by multimodal attentive metric learning. In: Proceedings of the 27th ACM International Conference on Multimedia, MM 2019, pp. 1526–1534. Association for Computing Machinery, New York (2019). https://doi.org/10.1145/3343031.3350953
26. Liu, S., Chen, Z., Liu, H., Hu, X.: User-video co-attention network for personalized micro-video recommendation. In: The World Wide Web Conference, WWW '19, pp. 3020–3026. Association for Computing Machinery, New York (2019). https://doi.org/10.1145/3308558.3313513
27. Locatello, F., et al.: Challenging common assumptions in the unsupervised learning of disentangled representations. In: Chaudhuri, K., Salakhutdinov, R. (eds.) Proceedings of the 36th International Conference on Machine Learning, June 2019, vol. 97, pp. 4114–4124. Proceedings of Machine Learning Research, PMLR (20219). https://proceedings.mlr.press/v97/locatello19a.html
28. Luo, C., Zhan, J., Xue, X., Wang, L., Ren, R., Yang, Q.: Cosine normalization: using cosine similarity instead of dot product in neural networks. In: Kůrková, V., Manolopoulos, Y., Hammer, B., Iliadis, L., Maglogiannis, I. (eds.) ICANN 2018. LNCS, vol. 11139, pp. 382–391. Springer, Cham (2018). https://doi.org/10.1007/978-3-030-01418-6_38

29. Ma, J., Zhou, C., Cui, P., Yang, H., Zhu, W.: Learning disentangled representations for recommendation. In: Proceedings of the 33rd International Conference on Neural Information Processing Systems, Red Hook, NY, USA. Curran Associates Inc. (2019)
30. MacQueen, J.: Some methods for classification and analysis of multivariate observations. In: Proceedings of the Fifth Berkeley Symposium on Mathematical Statistics and Probability, Oakland, CA, USA, vol. 1, pp. 281–297 (1967)
31. Nauta, M., et al.: From anecdotal evidence to quantitative evaluation methods: a systematic review on evaluating explainable AI. ACM Comput. Surv. **55**(13s) (2023). https://doi.org/10.1145/3583558
32. Ni, J., Li, J., McAuley, J.: Justifying recommendations using distantly-labeled reviews and fine-grained aspects. In: Proceedings of the 2019 Conference on Empirical Methods in Natural Language Processing and the 9th International Joint Conference on Natural Language Processing (EMNLP-IJCNLP), Hong Kong, China, November 2019, pp. 188–197. Association for Computational Linguistics (2019). https://doi.org/10.18653/v1/D19-1018. https://aclanthology.org/D19-1018
33. Ontañón, S.: An overview of distance and similarity functions for structured data. Artif. Intel. Rev. **53**(7), 5309–5351 (2020). https://doi.org/10.1007/s10462-020-09821-w
34. Radford, A., et al.: Learning transferable visual models from natural language supervision. In: Meila, M., Zhang, T. (eds.) Proceedings of the 38th International Conference on Machine Learning, , Virtual, July 2021, vol. 139, pp. 8748–8763. Proceedings of Machine Learning Research, PMLR (2021). https://proceedings.mlr.press/v139/radford21a.html
35. Rendle, S., Freudenthaler, C., Gantner, Z., Schmidt-Thieme, L.: BPR: Bayesian personalized ranking from implicit feedback. In: Proceedings of the Twenty-Fifth Conference on Uncertainty in Artificial Intelligence, UAI 2009, Arlington, Virginia, USA, pp. 452–461. AUAI Press (2009)
36. Roelleke, T.: Foundations of IR models. In: Information Retrieval Models. Synthesis Lectures on Information Concepts, Retrieval, and Services, p. 76. Springer, Cham (2013). https://doi.org/10.1007/978-3-031-02328-6_2
37. Wang, Q., Wei, Y., Yin, J., Wu, J., Song, X., Nie, L.: DualGNN: dual graph neural network for multimedia recommendation. IEEE Trans. Multimedia **25**, 1074–1084 (2023). https://doi.org/10.1109/TMM.2021.3138298
38. Wang, X., Chen, H., Zhou, Y., Ma, J., Zhu, W.: Disentangled representation learning for recommendation. IEEE Trans. Pattern Anal. Mach. Intell. **45**(1), 408–424 (2023). https://doi.org/10.1109/TPAMI.2022.3153112
39. Wang, X., Chen, H., Zhu, W.: Multimodal disentangled representation for recommendation. In: 2021 IEEE International Conference on Multimedia and Expo (ICME), Virtual, pp. 1–6. IEEE (2021). https://doi.org/10.1109/ICME51207.2021.9428193
40. Wang, Y., Wang, L., Li, Y., He, D., Liu, T.Y.: A theoretical analysis of NDCG type ranking measures. In: Shalev-Shwartz, S., Steinwart, I. (eds.) Proceedings of the 26th Annual Conference on Learning Theory, Princeton, NJ, USA, 12–14 Jun 2013, vol. 30, pp. 25–54. Proceedings of Machine Learning Research. PMLR (2013). https://proceedings.mlr.press/v30/Wang13.html
41. Wei, Y., Wang, X., Nie, L., He, X., Chua, T.S.: Graph-refined convolutional network for multimedia recommendation with implicit feedback. In: Proceedings of the 28th ACM International Conference on Multimedia, MM 2020, pp. 3541–3549. Association for Computing Machinery, New York (2020). https://doi.org/10.1145/3394171.3413556

42. Wei, Y., Wang, X., Nie, L., He, X., Hong, R., Chua, T.S.: MMGCN: multi-modal graph convolution network for personalized recommendation of micro-video. In: Proceedings of the 27th ACM International Conference on Multimedia, MM 2019, pp. 1437–1445. Association for Computing Machinery, New York (2019). https://doi.org/10.1145/3343031.3351034
43. Wu, Z., Xiong, Y., Yu, S.X., Lin, D.: Unsupervised feature learning via non-parametric instance discrimination. In: 2018 IEEE/CVF Conference on Computer Vision and Pattern Recognition, Salt Lake City, UT, USA, pp. 3733–3742. IEEE Computer Society (2018). https://doi.org/10.1109/CVPR.2018.00393
44. Zhang, S.F., Zhai, J.H., Xie, B.J., Zhan, Y., Wang, X.: Multimodal representation learning: advances, trends and challenges. In: 2019 International Conference on Machine Learning and Cybernetics (ICMLC), pp. 1–6. Association for Computing Machinery, New York (2019). https://doi.org/10.1109/ICMLC48188.2019.8949228
45. Zheng, Y., Gao, C., Li, X., He, X., Li, Y., Jin, D.: Disentangling user interest and conformity for recommendation with causal embedding. In: Proceedings of the Web Conference 2021, WWW '21, pp. 2980–2991. Association for Computing Machinery, New York (2021). https://doi.org/10.1145/3442381.3449788
46. Zhou, H., Zhou, X., Zeng, Z., Zhang, L., Shen, Z.: A comprehensive survey on multimodal recommender systems: taxonomy, evaluation, and future directions (2023). https://doi.org/10.48550/ARXIV.2302.04473. https://arxiv.org/abs/2302.04473
47. Ziegler, C.N., McNee, S.M., Konstan, J.A., Lausen, G.: Improving recommendation lists through topic diversification. In: Proceedings of the 14th International Conference on World Wide Web, WWW 2005, pp. 22–32. Association for Computing Machinery, New York (2005). https://doi.org/10.1145/1060745.1060754

Eliminating Contextual Bias in Aspect-Based Sentiment Analysis

Ruize An[1], Chen Zhang[1], and Dawei Song[1,2](✉)

[1] Beijing Institute of Technology, Beijing, China
{rz.an,czhang,dwsong}@bit.edu.cn
[2] The Open University, Milton Keynes, UK

Abstract. Pretrained language models (LMs) have made remarkable achievements in aspect-based sentiment analysis (ABSA). However, it is discovered that these models may struggle in some particular cases (e.g., to detect sentiments expressed towards targeted aspects with only implicit or adversarial expressions). Since it is hard for models to align implicit or adversarial expressions with their corresponding aspects, the sentiments of the targeted aspects would largely be impacted by the expressions towards other aspects in the sentence. We name this phenomenon as contextual bias. To tackle the problem, we propose a flexible aspect-oriented debiasing method (ARDE) to eliminate the harmful contextual bias without the need of adjusting the underlying LMs. Intuitively, ARDE calibrates the prediction towards the targeted aspect by subtracting the bias towards the context. Favorably, ARDE can get theoretical support from counterfactual reasoning theory. Experiments are conducted on SemEval benchmark, and the results show that ARDE can empirically improve the accuracy on contextually biased aspect sentiments without degrading the accuracy on unbiased ones. Driven by recent success of large language models (LLMs, e.g., ChatGPT), we further uncover that even LLMs can fail to address certain contextual bias, which yet can be effectively tackled by ARDE.

Keywords: aspect-based sentiment analysis · counterfactual inference · implicit sentiment

1 Introduction

Aspect-based sentiment analysis (ABSA) aims to predict the sentiment expressed towards a particular aspect in a given sentence. Recent advances have preferably employed pretrained language models (LMs) and achieved remarkable gains. Built upon LMs, memory networks [2,33,36], convolutional networks [14,45], attention mechanisms [12,18], linguistic structures [4,34,46], and input transformations [20,25] have been introduced for aspect-oriented finetuning.

While these aspect-oriented finetuning approaches can largely lift the performance, they may struggle with the so-called contextual bias problem in some

N. Goharian et al. (Eds.): ECIR 2024, LNCS 14608, pp. 90–107, 2024.
https://doi.org/10.1007/978-3-031-56027-9_6

Table 1. The accuracy of an existing LM-based ABSA model, namely AspectMarker, on the SemEval Laptop dataset, showing how the model's performance drops in implicit and adversarial cases.

Method	Normal	Implicit	Adversarial
AspectMarker [20]	81.25	71.43	72.25
Δ	−0.00	−9.82	−9.00

particular cases. For example, in a review "*the food here is just great, and the waiter should be more friendly*", the sentiment towards the aspect *waiter* is negative indicated by the implicit expression "*should be more friendly*". However, it can be misjudged as positive due to the explicit context *just great.* As another example, in "*the food here is not bad, but the waiter is awful*", the sentiment towards the aspect *food* is positive but can be misjudged as negative due to the adversarial expressions used. Specifically, although *not just bad* suggests a positive sentiment regarding the *food*, the evident context *awful* in reference to the *waiter* might mislead the overall sentiment assessment towards the *food* as negative. In these cases, the sentiment judgements are largely impacted by the implicit and adversarial expressions towards contextual aspects. We term the afore-discussed phenomena as ***contextual bias***, which can cause remarkable performance degradation in sentiment analysis as demonstrated in Table 1.

Figure 1 provides a quantitative analysis on the impact of contextual bias. Given a review "*My friend had a burger and I had these not wonderful blueberry pancakes*", the sentiment of the aspect *blueberry pancakes* is negative. However, the predicted probability towards it locates more densely at a neutral sentiment. Indeed, the contextual probability for the aspect *blueberry pancakes* leads to a contextual bias that makes its predicted probability move towards a neutral sentiment instead of a negative one.

In this paper, we propose an aspect-oriented debiasing method (ARDE) that is aimed to eliminate the harmful contextual bias without any intrusive adjustments to the existing aspect-oriented LM finetuning approaches. Specifically, ARDE operates at the LM inference stage and contains three crucial steps. Firstly, for an already finetuned LM for ABSA, ARDE obtains the sentiment distribution towards an aspect through LM inference as normal. Then, it gets the sentiment distribution towards the context by LM inference with the aspect-oriented information eliminated. Finally, it induces the calibrated sentiment distribution by conditionally subtracting the bias from the original prediction. Besides, the training-agnostic property of ARDE enables universal pluggability to almost all ABSA models.

It is also important to stress that ARDE can be viewed as a counterfactual-related instantiation of causal inference [22]. In the language of counterfactuals, the identified contextual bias is a sort of confounding bias and can be counterfactually derived even though it has not ever been seen [27]. As discussed in Sect. 3.4, the proposed debiasing approach naturally corresponds to spurious

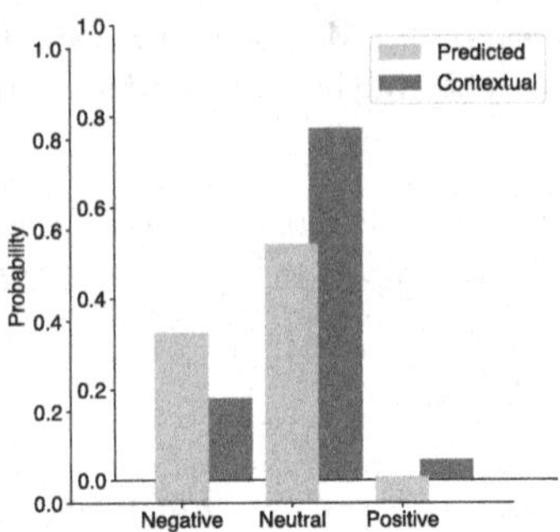

Fig. 1. The predicted probability versus the contextual probability of Aspect-Marker [20] on a case, where the predicted one is biased by the contextual one. The contextual probability is obtained by removing the marker around the target aspect at the input end.

correlation decoupling in the counterfactual theory [22], showing the theoretical soundness of ARDE.

We conduct experiments on the widely used SemEval benchmark [26]. Thanks to previous studies, we can easily distinguish the implicit aspect sentiments [15] and adversarial aspect sentiments [41] from the normal ones. The experimental results demonstrate that ARDE can improve the accuracy on contextually biased aspect sentiments with negligible affect on the unbiased ones.

As a further exploration, we investigate whether the more recently emerged large language models (LLMs, e.g., ChatGPT), which have led to performance breakthroughs in a diverse range of downstream tasks [28], also suffer from the contextual bias problem in ABSA. Our preliminary results indicate that even ChatGPT (at the time when the experiment was carried out) failed to handle the problem, which can be largely alleviated by incorporating ARDE.

Our main contributions can be summarized as follows:

- We discover that existing finetuned LMs can struggle with the contextual bias problem in particular cases, e.g., implicit and adversarial aspect sentiments.
- We design a flexible aspect-oriented debiasing method, ARDE, to eliminate the contextual bias, which is training-agnostic and pluggable to almost all ABSA models.
- We prove that ARDE is theoretically sound from a causal inference perspective and empirically effective on a commonly used benchmark.
- We uncover that LLMs can fail to circumvent contextual bias, but the problem can be effectively tackled by ARDE. To our best knowledge, this is also the very first trial of examining the ability of ChatGPT in the ABSA task.

2 Related Work

2.1 Aspect-Based Sentiment Analysis

In recent years, LMs such as BERT [7] and RoBERTa [17], have played a crucial role in NLP. Based on LMs, large performance improvements have been achieved in ABSA [20,29,30,42]. Taking the advantage of LMs, ABSA can be treated as a sequence pair classification task [9,31] or a reading comprehension task [3, 21]. Moreover, aspect-oriented dependency tree is used to further improve the performance of LMs [5,40,46,47]. Most recently, ABSA is resolved as part of a triplet extraction task to realize a more complete solution [23,39,48,49]. Our work generally falls in this line of approach but specifically concentrates on the contextual bias problem.

2.2 Implicit Aspect Sentiment Analysis

Supervised contrastive pretraining is used to align the representation of implicit sentiment expressions with the corresponding sentiments, and distinguish the implicit slices from the explicit ones in the SemEval benchmark [15]. Furthermore, a knowledge graph is produced to supplement the implicit sentiment expressions and a novel implicit sentiment model is proposed to combine the knowledge enhancements and context features [43]. In addition, structured generation is used for aspect sentiment quadruple extraction to detect implicit sentiment expressions more effectively with a newly-designed quadruple predictor and an encoder-decoder model [24]. Our work utilizes implicit aspect sentiments as one type of contextual bias, and they are also used as instances to measure the performance of our approach in addressing the contextual bias problem.

2.3 Adversarial Aspect Sentiment Analysis

It has been recognized that LMs suffer from significant performance drops on adversarial aspect sentiments [41], and various methods [6,11,19,44] are proposed to improve the robustness of ABSA models. Similarly, a dual-feature extraction module is used to extract aspect-related and aspect-unrelated features, while an aspect-feature distillation module is used to eliminate the interference of aspect-unrelated words [16]. Likewise, our work considers adversarial aspect sentiments as another type of contextual bias and leverages them as a major testbed.

2.4 Debiasing in NLP

Debiasing has been considered as important to improve model robustness in NLP, and a range of methods are proposed for debiasing [10,27,35]. Among them, the idea of counterfactual has inspired several debiasing studies [8,13,32,38]. As for ABSA, the bias often refers to the fact that some aspects are more associated with some sentiments rather than others [35]. Such aspect bias can be alleviated by a no-aspect template [1]. In the work [37] that is most related to ours, implicit aspect sentiments are thought to hurt the model robustness and should be intervened by a complex and training-specific instrumental variable model. Differently, our work goes beyond the implicit aspect sentiments, and presents a straightforward and training-agnostic method based on the counterfactual theory to eliminate the contextual bias.

3 Methodology

3.1 Problem Definition

When presented with a sentence $x = (x_1, \ldots, a, \ldots, x_n)$ (where n denotes the sentence length) and a specific aspect $a = (a_1, \ldots, a_m)$ (where m denotes the aspect's length) within the sentence, a fine-tuned Language Model, denoted as $(\mathcal{P}, \mathcal{M})$ for Aspect-Based Sentiment Analysis (ABSA), is essential. This model is tasked with providing a predictive distribution $\bar{y}$ encompassing sentiments (e.g., positive, negative, neutral). Here, $\mathcal{P}$ and $\mathcal{M}$ represent a three-way classifier and a pretrained backbone, respectively. The objective is to align the predicted distribution with the ground truth one-hot distribution y as closely as possible, irrespective of the presence or absence of contextual bias.

3.2 Aspect-Oriented Finetuning

Aspect-oriented finetuning of LMs for ABSA can be reduced to two major paradigms: adjusting 1) the input structure of LMs (i.e., $\mathcal{I}$) or 2) the output feature of LMs (i.e., $\mathcal{O}$) for aspect-oriented information. Typical input-based methods include the aspect paired and aspect marked structures [20,42], while typical output-based methods are based on the aspect averaged and aspect weighted features [5,47]. These methods are used to make LMs pay mpre attention to the corresponding expressions of the target aspect, and are used in our work for finetuning baselines. An overview of these methods are given in Fig. 2a.

For abstraction, aspect-oriented finetuning can be represented typically as minimizing the cross-entropy loss:

$$\mathcal{L} = -y \log \mathcal{P} \circ \mathcal{O} \circ \mathcal{M} \circ \mathcal{I}(x, a)$$

where $\mathcal{I}$ or $\mathcal{O}$ insert aspect-unaware or aspect-oriented transformations on either the input end or the output end. $\circ$ means sequential function composition.

Aspect Paired Input Structure [42]. This input structure is proved to make LMs concentrate more on the targeted aspect by appending the targeted aspect to the sentence:

$$\mathcal{I}(x, a) = [\mathsf{CLS}]\ x_1\ \cdots\ x_n\ [\mathsf{SEP}]\ a_1\ \cdots\ a_m\ [\mathsf{SEP}]$$

Aspect Marked Input Structure [20]. This input structure is realized by annotating the targeted aspect by placing two markers around the aspect:

$$\mathcal{I}(x, a) = [\mathsf{CLS}]\ x_1\ \cdots\ [\mathsf{M}]\ a_1\ \cdots\ a_m\ [\mathsf{M}] \cdots\ x_n\ [\mathsf{SEP}]$$

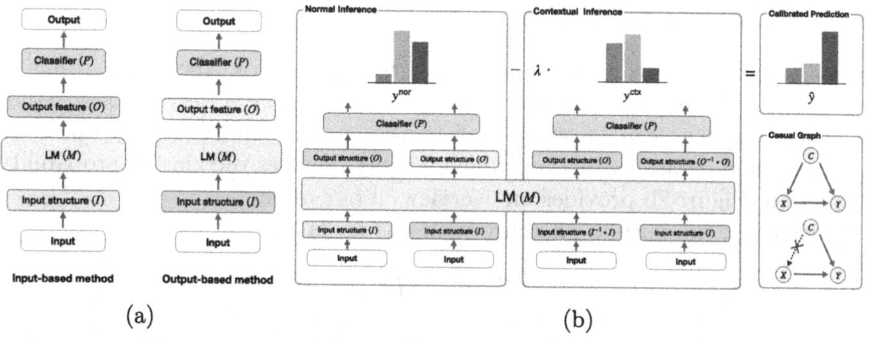

(a) (b)

Fig. 2. (a) An overview of aspect-oriented finetuning paradigms. The grey boxes stand for either aspect-unaware input structures or output features. (b) An overview of ARDE and its connections to counterfactual theory. (Color figure online)

For both input-based methods, the output feature is the last hidden state corresponding to [CLS]:

$$\mathcal{O}(\mathbf{h}, x, a) = \mathbf{h}_{[\mathsf{CLS}]}$$

where $\mathbf{h}$ generally indicates the last hidden states.

Aspect Averaged Output Feature [5]. This output feature means taking the average over the last hidden states of the targeted aspect:

$$\mathcal{O}(\mathbf{h}, x, a) = \sum_{k}^{m} 1/m \cdot \mathbf{h}_{a_k}$$

Aspect Weighted Output Feature [47]. This output features is derived from the hypothesis that the hidden states locating close to the targeted aspect are more important than the ones locating further away:

$$\mathcal{O}(\mathbf{h}, x, a) = \sum_{k}^{n} d(x_k, a_1, a_m) \cdot \mathbf{h}_{x_k}$$

where d is a function measuring the proximity from any token to the target aspect and it takes the form of:

$$d(x_k, a_1, a_m) = 1 - \min(|x_k, a_1|, |x_k, a_m|)/n$$

where, with abuse of notation, $|p, q|$ denotes the absolute position distance between two tokens.

For both output-based methods, the input structure is the naive sentence:

$$\mathcal{I}(x, a) = [\mathsf{CLS}]\ x_1\ \cdots\ x_n\ [\mathsf{SEP}]$$

3.3 Aspect-Oriented Debiasing

Normally the inference with the above finetuned LMs for ABSA will result in errors due to the existence of the contextual bias. As we observe in Fig. 1 that the contextual bias can be attributed to the context directly, we propose an aspect-oriented debiasing method which flexibly removes the bias by probability subtraction. Figure 2b provides an overview of our method.

Specifically, the normal inference with LM leads to:

$$y^{\mathsf{nor}} = \mathcal{P} \circ \mathcal{O} \circ \mathcal{M} \circ \mathcal{I}(x, a)$$

In contrast, the contextual inference, where the aspect-oriented information is eliminated and only the context is kept, results in:

$$y^{\mathsf{ctx}} = \mathcal{P} \circ \mathcal{O}^{-1} \circ \mathcal{O} \circ \mathcal{M} \circ \mathcal{I}^{-1} \circ \mathcal{I}(x, a)$$

where $\mathcal{I}^{-1}$ or $\mathcal{O}^{-1}$ is used as inverse function to respectively eliminate the effect of $\mathcal{I}$ or $\mathcal{O}$. Then we can reach a calibrated prediction without bias as:

$$\hat{y} = y^{\mathsf{nor}} - \lambda \cdot y^{\mathsf{ctx}}$$

It is noteworthy that even the subtraction can give negative probability, we can always use it since the LM inference is only concerned with relative magnitudes.

Aspect Paired Input Structure. For the Aspect Marked Input Structure, the inverse input structure function is achieved by removing the appended targeted aspect:

$$I^{-1} \circ I(x, a) = [\mathsf{CLS}]\ x_1\ \cdots\ x_n\ [\mathsf{SEP}]$$

Aspect Marked Input Structure. Similarly, the inverse input structure function for Aspect Marked Input Structure is implemented by getting rid of the markers around the targeted aspect, which results in an aspect-eliminated input.

Aspect Averaged Output Feature. The inverse output feature function for this feature expands the averaging range from the targeted aspect to the whole sentence:

$$\mathcal{O}^{-1} \circ \mathcal{O}(\mathbf{h}, x, a) = \sum_{k}^{n} 1/n \cdot \mathbf{h}_{x_k}$$

Here, the expansion safely erases the aspect-related information.

Aspect Weighted Output Feature. The inverse output feature function for this feature drops the proximity weights. Together with the original output feature function, it behaves exactly the same as the one positioned above.

3.4 Connections to Counterfactual Theory

On the theoretical side, we find that ARDE nicely aligns with the counterfactual theory, which essentially tells that the confounding bias can be counter factually conceptualized even though it has never be seen, thus the confounding bias can be described and potentially eliminated [22,27].

In case of ABSA, the contextual bias is exactly a kind of confounding bias which can be depicted with a causal graph as in Fig. 2b (right). In the causal graph, X represents the targeted aspect and its associated sentiment expression, Y represents the ground truth sentiment, and C represents other aspects and their associated sentiment expressions. Two linguistic facts in the causal graph are: 1) X and C are correlated since some aspects would be mentioned concurrently; and 2) C empirically has an impact on Y as sentiment expressions are interconnected and is illustrated in Fig. 1. The fork structure $X \leftarrow C \rightarrow Y$ admits C as a confounding bias, which leaves a spurious correlation between X and Y through C.

Fortunately, as the contextual bias can be derived with our tactics, the spurious correlation can be cut off with the subtraction at the probability level.

4 Experiments

4.1 Datasets and Metrics

Our experiments are conducted on the SemEval benchmark. There are generally two domains of data from the benchmark, i.e., one from laptop and the other from restaurant. Each domain consists of both training and test sets. The training part is used to finetune the backbone LMs, and the test part is used to evaluate the model performance. Previous studies have reannotated the test sets to provide implicit test data held out from the original test data [15] and additional adversarial test data [41]. The summary is described in Table 2. We adopt accuracy (Acc.) and F1 scores (F1) as standard measures for evaluation.

Table 2. The summary of data.

Dataset		Positive	Neutral	Negative	Total
Laptop	train	987	460	866	2313
	nor. test	341	169	128	638
	imp. test	36	111	28	175
	adv. test	883	407	587	1877
Restaurant	train	2164	633	805	3602
	nor. test	728	196	196	1120
	imp. test	76	137	54	267
	adv. test	1953	473	1104	3530

4.2 Baselines and Implementation

We compare ARDE to the aspect-oriented finetuning baselines, which utilize Bert and Roberta as the backbone LMs. These models are finetuned with learning rate 0.00005 and batch size 64, which are decided by grid searching. The training objective is to minimize the cross-entropy loss with L_2 regularization. We conduct a grid search for the parameter λ over a range from 0 to 1 with a step size of 0.05. This finetuning-searching-predicting process is iterated for five times, and the final results are obtained by averaging the outcomes. For each model, both the mean ($\triangle$) and standard deviation (σ) of the improvements in accuracy and F1 scores are included as part of the results.

Table 3. The results on normal and implicit aspect sentiments (Bert).

Method	Laptop				Restaurant			
	Nor. Acc.	Nor. F1	Imp. Acc.	Imp. F1	Nor. Acc.	Nor. F1	Imp. Acc.	Imp. F1
Aspect Paired Input Structure	78.50	74.04	64.91	58.41	84.32	77.30	64.72	62.81
w/ ARDE	78.75	74.44	65.71	58.95	84.37	77.37	64.79	62.87
$\triangle$	+0.25	+0.40	+0.80	+0.54	+0.05	+0.07	+0.07	+0.06
σ	0.29	0.50	1.06	0.69	0.07	0.09	0.15	0.12
Aspect Marked Input Structure	78.94	75.03	70.63	65.03	84.18	77.44	62.77	61.24
w/ ARDE	79.03	75.14	70.74	65.12	84.43	77.96	63.30	61.51
$\triangle$	+0.09	+0.11	+0.11	+0.09	+0.25	+0.52	+0.53	+0.27
σ	0.19	0.22	0.23	0.18	0.13	0.45	0.30	0.37
Aspect Averaged Output Feature	78.56	74.78	71.43	65.45	85.36	79.36	66.37	63.66
w/ ARDE	79.09	75.43	72.57	66.29	85.66	79.97	67.19	64.48
$\triangle$	+0.53	+0.65	+1.14	+0.84	+0.30	+0.61	+0.82	+0.82
σ	0.26	0.33	0.63	0.57	0.11	0.21	0.49	0.51
Aspect Weighted Output Feature	78.90	75.41	72.00	65.83	84.29	77.80	63.00	60.40
w/ ARDE	79.09	75.61	72.11	65.93	84.70	78.46	63.67	61.11
$\triangle$	+0.10	+0.20	+0.11	+0.10	+0.59	+0.66	+0.67	+0.71
σ	0.15	0.14	0.23	0.18	0.18	0.27	0.44	0.42

4.3 Main Results

Results on Normal Aspect Sentiments. As shown in Table 3 and Table 4, we can see that ARDE would not affect the performance on normal aspect sentiments and could even bring certain boost. For example, Aspect Weighted Output Feature w/ARDE achieves 0.59 and 0.66 performance gains in terms of Acc. and F1 compared with Aspect Weighted Output Feature on Restaurant in Table 3, and the respective standard deviations of improvement in Acc. and F1 are 0.18 and 0.27. Meanwhile, Aspect Marked Input Structure w/ARDE gives 0.19 and 0.21 absolute performance gains over Aspect Marked Input Structure in terms of Acc. and F1 on Laptop in Table 4, and the standard deviations of improvement are 0.18 and 0.24.

Results on Implicit Aspect Sentiments. We find again from Table 3 and Table 4 that ARDE can enhance the performance more significantly on implicit aspect sentiments to a large extent. As shown in Table 3, Aspect Averaged Output

Table 4. The results on normal and implicit aspect sentiments (Roberta).

Method	Laptop				Restaurant			
	Nor. Acc.	Nor. F1	Imp. Acc.	Imp. F1	Nor. Acc.	Nor. F1	Imp. Acc.	Imp. F1
Aspect Paired Input Structure	83.01	80.17	80.45	76.81	87.05	80.53	69.21	68.03
w/ ARDE	83.04	80.21	80.45	76.81	87.12	80.73	69.51	68.41
Δ	+0.03	+0.04	+0.00	+0.00	+0.07	+0.20	+0.30	+0.38
σ	0.06	0.08	0	0	0.07	0.19	0.28	0.34
Aspect Marked Input Structure	83.86	81.05	83.09	78.95	87.36	81.53	70.56	69.35
w/ ARDE	84.05	81.26	83.43	79.40	87.53	81.79	71.01	69.81
Δ	+0.19	+0.21	+0.34	+0.45	+0.18	+0.26	+0.45	+0.46
σ	0.18	0.24	0.28	0.37	0.27	0.38	0.60	0.61
Aspect Averaged Output Feature	83.57	80.71	83.43	79.73	86.80	80.55	70.00	69.29
w/ ARDE	83.86	81.13	84.00	80.37	87.02	80.93	71.39	70.15
Δ	+0.28	+0.41	+0.57	+0.64	+0.21	+0.38	+0.98	+0.86
σ	0.18	0.23	0.63	0.66	0.14	0.29	0.81	0.78
Aspect Weighted Output Feature	83.17	80.09	78.74	73.48	86.84	80.49	70.04	68.91
w/ ARDE	83.20	80.11	78.86	73.57	86.95	80.67	70.26	69.00
Δ	+0.03	+0.03	+0.11	+0.09	+0.11	+0.18	+0.22	+0.21
σ	0.06	0.05	0.23	0.19	0.07	0.12	0.18	0.17

Table 5. The results on adversarial aspect sentiments (Bert).

Method	Laptop		Restaurant	
	Adv. Acc.	Adv. F1	Adv. Acc.	Adv. F1
Aspect Paired Input Structure	66.02	63.22	77.64	71.62
w/ ARDE	66.59	63.93	77.96	71.98
Δ	+0.57	+0.71	+0.32	+0.36
σ	0.23	0.36	0.22	0.26
Aspect Marked Input Structure	72.25	69.33	78.47	72.34
w/ ARDE	73.08	70.30	78.82	72.83
Δ	+0.83	+0.97	+0.35	+0.49
σ	1.22	1.66	0.32	0.48
Aspect Averaged Output Feature	72.32	69.35	77.89	72.48
w/ ARDE	73.35	70.77	78.53	73.32
Δ	+1.03	+1.42	+0.64	+0.84
σ	1.13	1.53	0.21	0.22
Aspect Weighted Output Feature	70.69	68.03	77.71	72.04
w/ ARDE	71.52	68.90	78.56	73.17
Δ	+0.83	+0.87	+0.85	+1.13
σ	0.62	0.79	0.62	0.92

Feature w/ ARDE improves the Acc. and F1 on Laptop by 1.14 and 0.84 over Aspect Averaged Output Feature, and the standard deviations of improvement are 0.63 and 0.57. The improvement can also be seen on Laptop with a 0.57 and 0.64 boost respectively on Acc. and F1 in Table 4, while the standard deviations of improvement are 0.63 and 0.66. The improvements outweigh those on normal aspect sentiments. The results retrospectively demonstrate that the implicit expressions can lead to contextual bias, which can be alleviated by ARDE.

Results on Adversarial Aspect Sentiments. From Table 5 and Table 6, we unearth a similar performance trend on adversarial aspect sentiments. As shown in Table 5, Aspect Weighted Output Feature w/ARDE improves the Acc. and F1

Table 6. The results on adversarial aspect sentiments (Roberta).

Method	Laptop		Restaurant	
	Adv. Acc.	Adv. F1	Adv. Acc.	Adv. F1
Aspect Paired Input Structure	75.64	73.10	81.69	75.37
w/ ARDE	75.97	73.42	81.83	75.58
Δ	+0.33	+0.32	+0.14	+0.22
σ	0.28	0.35	0.10	0.19
Aspect Marked Input Structure	77.44	74.96	81.77	76.12
w/ ARDE	77.75	75.29	82.06	76.49
Δ	+0.31	+0.33	+0.29	+0.37
σ	0.16	0.16	0.23	0.33
Aspect Averaged Output Feature	76.51	74.39	81.13	75.01
w/ ARDE	77.11	74.65	81.59	75.71
Δ	+0.61	+0.25	+0.46	+0.70
σ	0.29	1.18	0.12	0.17
Aspect Weighted Output Feature	75.23	72.62	80.21	74.20
w/ ARDE	75.65	73.11	80.57	74.76
Δ	+0.43	+0.49	+0.36	+0.55
σ	0.25	0.31	0.14	0.25

on Laptop respectively by 0.83 and 0.87 over Aspect Weighted Output Feature. Additionally, the standard deviations of improvement are 0.62 and 0.92. As for Aspect Averaged Output Feature w/ARDE in Table 6, the improvement of the Acc. and F1 on Restaurant are 0.46 and 0.70 over Aspect Averaged Output Feature and the associated standard deviations of improvement are 0.12 and 0.17. The results indicate that Aspect Weighted Output Feature with ARDE exhibits a higher level of stability in handling adversarial aspect sentiments. The results validate that the output-based methods can handle contextual bias more effectively and more stably than the input-based methods.

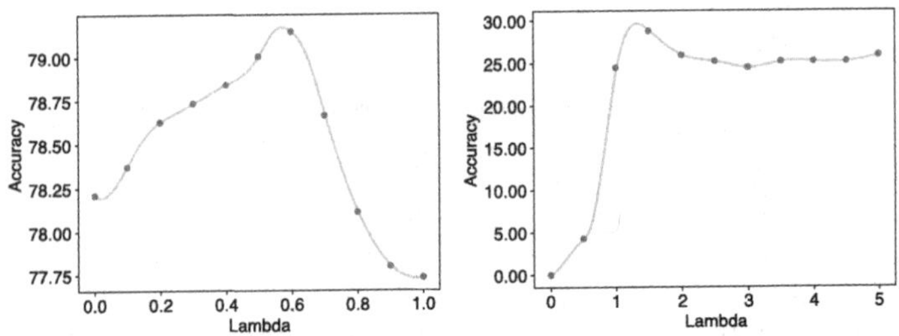

Fig. 3. The impact of λ.

Impact of λ. To investigate the impact of λ, we compare the results of Aspect Average Output Feature w/ARDE on Laptop normal aspect sentiments corresponding to different values of λ. We vary λ from 0 to 1 with a step of 0.05. From Fig. 3 (left), we can observe a first-increase-then-decrease phenomenon on Acc., suggesting there is an optimal λ value for ARDE.

Besides, it is not always sufficient for some particular cases to calibrate correctly (i.e., under-calibration) with λ valued from 0 to 1. Yet, λ valued larger than 1 could possibly lead to worse results due to over-calibration on adequately calibrated aspect sentiments. So, we further study the impact of λ ranging from 0 to 5 with a step of 0.5 on probably under-calibrated aspect sentiments, which are wrongly-predicted aspect sentiments (i.e., Acc. is 0 when λ is 0) from the concerned baselines. According to Fig. 3 (right), Aspect Average Output Feature w/ ARDE shows an Acc. peak of 37.5 at λ of 1.5, and it also performs well with λ valued from 1.5 to 5.0, maintaining an Acc. of at least 30.0. The results showcase that ARDE with larger λ can have a wider applicable range on more extreme cases, which can't be solved with λ valued from 0 to 1.

4.4 Case Study

Table 7. The attention study. The targeted aspects are underlined. The contextual aspects are underwaved. The groudtruth sentiments are [bracketed]. In these examples, the contextual aspects are paid with more attention, leading to contextual bias.

Example	Contextual Aspect	Targeted Aspect
[CLS] reasonably priced with very stale su ##shi . [SEP]	Pos.	Pos. [Neg.]
[CLS] coffee is a better deal than over ##pr ##ice ##d co ##si sandwiches . [SEP]	Neu.	Neu. [Pos.]
[CLS] go with the specials , and stay away from the salmon . [SEP]	Neg.	Neu. [Pos.]
[CLS] dessert ##s are almost credible : my personal favorite is their tar ##t of the day . [SEP]	Pos.	Pos. [Neg.]
[CLS] it is a not great size and amazing windows 8 included ! [SEP]	Pos.	Pos. [Neg.]
[CLS] :) great product , not great price , great delivery , and great service . [SEP]	Pos.	Pos. [Neg.]
[CLS] it is really easy to use but it is not quick to start up . [SEP]	Pos.	Pos. [Neg.]

Attention Study. We select some representative examples to show the attention pattern of contextual bias. Table 7 visualizes the attention of Aspect Average Output Feature without any aspect-oriented information given. We can see that contextual bias arises with varying attentions on different aspects. The unevenness of attention make the aspects that the model pays more attention to play a greater role in the contextual prediction. Therefore, when predicting the sentiments towards the aspects overlooked by the model, they are more likely to be influenced by the aspects of high presence, possibly leading predictions towards an opposite direction.

4.5 Results on ChatGPT

We also apply ARDE to a LLM, namely ChatGPT (v3.5-0315). Different from application of ARDE to aspect-oriented finetuning, we need to give ChatGPT a background about ABSA so that ChatGPT can output the predictive sentiment probability towards the targeted aspect. Then, to evaluate ARDE, we introduce ARDE to calibrate the sentiment probabilities produced from ChatGPT, and uncover the examples that simply can not be solved by ChatGPT but can be tackled by ARDE.

Table 8. ABSA Results with ChatGPT using sampled data on LapTop.

λ	0	0.1	0.2	0.3	0.4	0.5	0.6	0.7	0.8	0.9	1	1.1	1.2
Adv.Acc.	79.17	79.17	79.17	79.17	79.17	81.25	82.29	80.21	76.04	66.67	36.46	18.75	14.58

The evaluation in Table 8 on Laptop adversarial aspect sentiments shows that ARDE can effectively promote the performance on ChatGPT by adjusting the coefficient λ properly. That is, we find that ChatGPT tends to be confused by these particular cases. As shown in Table 9, we can use ARDE to remove the contextual bias, thus yielding correct answers.

Table 9. A Case Study on ChatGPT.

Example		Predicted Distribution		
Sentence	Aspect	Pos.	Neu.	Neg.
It has all the not expected features and more + plus a wide screen but more than roomy keyboard.	features	Normal		
		0.98	0.01	0.01
		Contextual		
		0.92	0.06	0.02
		Calibrated ($\lambda = 1.5$)		
		-0.4	-0.08	-0.02

5 Conclusions and Future Work

We have proposed ARDE, a training-agnostic and pluggable method to eliminate the contextual bias in aspect-based sentiment analysis. The approach has been shown theoretically sound from a causal inference perspective, and empirically effective. Another important discovery is that even LLMs can make mistakes when faced with contextual bias, and the problem can be largely alleviated by ARDE. Moving forward, we will broaden the application scope, e.g., to joint aspect extraction and sentiment analysis, and explore the inference costs. Currently, while using ChatGPT, we identified specific instances but were unable to obtain comprehensive results across the entire dataset. In the future, we aim to conduct further testing of ARDE within the state-of-the-art LLM framework using the complete dataset.

Acknowledgements. This work is funded in part by the Natural Science Foundation of China (grant no: 62376027) and Beijing Municipal Natural Science Foundation (grant no: 4222036 and IS23061).

References

1. Cao, J., Liu, R., Peng, H., Jiang, L., Bai, X.: Aspect is not you need: no-aspect differential sentiment framework for aspect-based sentiment analysis. In: Carpuat, M., de Marneffe, M., Ruíz, I.V.M. (eds.) Proceedings of the 2022 Conference of the North American Chapter of the Association for Computational Linguistics: Human Language Technologies, NAACL 2022, Seattle, WA, United States, 10–15 July 2022, pp. 1599–1609. Association for Computational Linguistics (2022). https://doi.org/10.18653/v1/2022.naacl-main.115
2. Chen, P., Sun, Z., Bing, L., Yang, W.: Recurrent attention network on memory for aspect sentiment analysis. In: Proceedings of the 2017 Conference on Empirical Methods in Natural Language Processing, pp. 452–461 (2017)
3. Chen, S., Wang, Y., Liu, J., Wang, Y.: Bidirectional machine reading comprehension for aspect sentiment triplet extraction. In: Thirty-Fifth AAAI Conference on Artificial Intelligence, AAAI 2021, Thirty-Third Conference on Innovative Applications of Artificial Intelligence, IAAI 2021, The Eleventh Symposium on Educational Advances in Artificial Intelligence, EAAI 2021, Virtual Event, 2–9 February 2021, pp. 12666–12674. AAAI Press (2021). https://ojs.aaai.org/index.php/AAAI/article/view/17500
4. Chen, X., Sun, C., Wang, J., Li, S., Si, L., Zhang, M., Zhou, G.: Aspect sentiment classification with document-level sentiment preference modeling. In: Proceedings of the 58th Annual Meeting of the Association for Computational Linguistics, pp. 3667–3677 (2020)
5. Dai, J., Yan, H., Sun, T., Liu, P., Qiu, X.: Does syntax matter? a strong baseline for aspect-based sentiment analysis with roberta. In: Toutanova, K., et al. (eds.) Proceedings of the 2021 Conference of the North American Chapter of the Association for Computational Linguistics: Human Language Technologies, NAACL-HLT 2021, Online, 6–11 June 2021, pp. 1816–1829. Association for Computational Linguistics (2021). https://doi.org/10.18653/v1/2021.naacl-main.146
6. Deng, P., Yuan, J., Zhao, Y., Qin, B.: Zero-shot aspect-level sentiment classification via explicit utilization of aspect-to-document sentiment composition. CoRR abs/2209.02276 (2022). https://doi.org/10.48550/arXiv.2209.02276
7. Devlin, J., Chang, M., Lee, K., Toutanova, K.: BERT: pre-training of deep bidirectional transformers for language understanding. In: Burstein, J., Doran, C., Solorio, T. (eds.) Proceedings of the 2019 Conference of the North American Chapter of the Association for Computational Linguistics: Human Language Technologies, NAACL-HLT 2019, Minneapolis, MN, USA, 2–7 June 2019, vol. 1 (Long and Short Papers), pp. 4171–4186. Association for Computational Linguistics (2019). https://doi.org/10.18653/v1/n19-1423
8. Feng, F., Zhang, J., He, X., Zhang, H., Chua, T.: Empowering language understanding with counterfactual reasoning. In: Zong, C., Xia, F., Li, W., Navigli, R. (eds.) Findings of the Association for Computational Linguistics: ACL/IJCNLP 2021, Online Event, 1–6 August 2021. Findings of ACL, vol. ACL/IJCNLP 2021, pp. 2226–2236. Association for Computational Linguistics (2021). https://doi.org/10.18653/v1/2021.findings-acl.196
9. Gao, L., Wang, Y., Liu, T., Wang, J., Zhang, L., Liao, J.: Question-driven span labeling model for aspect-opinion pair extraction. In: Thirty-Fifth AAAI Conference on Artificial Intelligence, AAAI 2021, Thirty-Third Conference on Innovative Applications of Artificial Intelligence, IAAI 2021, The Eleventh Symposium on Educational Advances in Artificial Intelligence, EAAI 2021, Virtual Event, 2–9 February 2021, pp. 12875–12883. AAAI Press (2021). https://ojs.aaai.org/index.php/AAAI/article/view/17523

10. He, M., Chen, X., Hu, X., Li, C.: Causal intervention for sentiment de-biasing in recommendation. In: Proceedings of the 31st ACM International Conference on Information & Knowledge Management, pp. 4014–4018 (2022)
11. Hou, X., et al.: Graph ensemble learning over multiple dependency trees for aspect-level sentiment classification. In: Toutanova, K., et al. (eds.) Proceedings of the 2021 Conference of the North American Chapter of the Association for Computational Linguistics: Human Language Technologies, NAACL-HLT 2021, Online, 6–11 June 2021, pp. 2884–2894. Association for Computational Linguistics (2021). https://doi.org/10.18653/v1/2021.naacl-main.229
12. Huang, B., Ou, Y., Carley, K.M.: Aspect level sentiment classification with attention-over-attention neural networks. In: Thomson, R., Dancy, C., Hyder, A., Bisgin, H. (eds.) SBP-BRiMS 2018. LNCS, vol. 10899, pp. 197–206. Springer, Cham (2018). https://doi.org/10.1007/978-3-319-93372-6_22
13. Kaushik, D., Hovy, E.H., Lipton, Z.C.: Learning the difference that makes a difference with counterfactually-augmented data. In: 8th International Conference on Learning Representations, ICLR 2020, Addis Ababa, Ethiopia, 26–30 April 2020. OpenReview.net (2020). https://openreview.net/forum?id=Sklgs0NFvr
14. Li, X., Bing, L., Lam, W., Shi, B.: Transformation networks for target-oriented sentiment classification. In: Gurevych, I., Miyao, Y. (eds.) Proceedings of the 56th Annual Meeting of the Association for Computational Linguistics, ACL 2018, Melbourne, Australia, 15–20 July 2018, vol. 1: Long Papers, pp. 946–956. Association for Computational Linguistics (2018). https://doi.org/10.18653/v1/P18-1087. https://aclanthology.org/P18-1087/
15. Li, Z., Zou, Y., Zhang, C., Zhang, Q., Wei, Z.: Learning implicit sentiment in aspect-based sentiment analysis with supervised contrastive pre-training. In: Moens, M., Huang, X., Specia, L., Yih, S.W. (eds.) Proceedings of the 2021 Conference on Empirical Methods in Natural Language Processing, EMNLP 2021, Virtual Event/Punta Cana, Dominican Republic, 7–11 November 2021, pp. 246–256. Association for Computational Linguistics (2021). https://doi.org/10.18653/v1/2021.emnlp-main.22
16. Liu, R., Cao, J., Sun, N., Jiang, L.: Aspect feature distillation and enhancement network for aspect-based sentiment analysis. In: Amigó, E., Castells, P., Gonzalo, J., Carterette, B., Culpepper, J.S., Kazai, G. (eds.) SIGIR 2022: The 45th International ACM SIGIR Conference on Research and Development in Information Retrieval, Madrid, Spain, 11–15 July 2022, pp. 1577–1587. ACM (2022). https://doi.org/10.1145/3477495.3531938
17. Liu, Y., et al.: Roberta: a robustly optimized BERT pretraining approach. CoRR abs/1907.11692 (2019). http://arxiv.org/abs/1907.11692
18. Ma, D., Li, S., Zhang, X., Wang, H.: Interactive attention networks for aspect-level sentiment classification. In: Sierra, C. (ed.) Proceedings of the Twenty-Sixth International Joint Conference on Artificial Intelligence, IJCAI 2017, Melbourne, Australia, 19–25 August 2017, pp. 4068–4074. ijcai.org (2017). https://doi.org/10.24963/ijcai.2017/568
19. Ma, F., Zhang, C., Song, D.: Exploiting position bias for robust aspect sentiment classification. In: Zong, C., Xia, F., Li, W., Navigli, R. (eds.) Findings of the Association for Computational Linguistics: ACL/IJCNLP 2021, Online Event, 1–6 August 2021. Findings of ACL, vol. ACL/IJCNLP 2021, pp. 1352–1358. Association for Computational Linguistics (2021). https://doi.org/10.18653/v1/2021.findings-acl.116

20. Ma, F., Zhang, C., Zhang, B., Song, D.: Aspect-specific context modeling for aspect-based sentiment analysis. In: Lu, W., Huang, S., Hong, Y., Zhou, X. (eds.) Natural Language Processing and Chinese Computing - 11th CCF International Conference, NLPCC 2022, Guilin, China, 24–25 September 2022, Proceedings, Part I. Lecture Notes in Computer Science, vol. 13551, pp. 513–526. Springer, Heidelberg (2022). https://doi.org/10.1007/978-3-031-17120-8_40
21. Mao, Y., Shen, Y., Yu, C., Cai, L.: A joint training dual-mrc framework for aspect based sentiment analysis. In: Thirty-Fifth AAAI Conference on Artificial Intelligence, AAAI 2021, Thirty-Third Conference on Innovative Applications of Artificial Intelligence, IAAI 2021, The Eleventh Symposium on Educational Advances in Artificial Intelligence, EAAI 2021, Virtual Event, 2–9 February 2021, pp. 13543–13551. AAAI Press (2021). https://ojs.aaai.org/index.php/AAAI/article/view/17597
22. Pearl, J.: Causal inference in statistics: an overview (2009)
23. Peng, H., Xu, L., Bing, L., Huang, F., Lu, W., Si, L.: Knowing what, how and why: a near complete solution for aspect-based sentiment analysis. In: The Thirty-Fourth AAAI Conference on Artificial Intelligence, AAAI 2020, The Thirty-Second Innovative Applications of Artificial Intelligence Conference, IAAI 2020, The Tenth AAAI Symposium on Educational Advances in Artificial Intelligence, EAAI 2020, New York, NY, USA, 7–12 February 2020, pp. 8600–8607. AAAI Press (2020). https://ojs.aaai.org/index.php/AAAI/article/view/6383
24. Peper, J., Wang, L.: Generative aspect-based sentiment analysis with contrastive learning and expressive structure. In: Goldberg, Y., Kozareva, Z., Zhang, Y. (eds.) Findings of the Association for Computational Linguistics: EMNLP 2022, Abu Dhabi, United Arab Emirates, 7–11 December 2022, pp. 6089–6095. Association for Computational Linguistics (2022). https://aclanthology.org/2022.findings-emnlp.451
25. Phan, M.H., Ogunbona, P.O.: Modelling context and syntactical features for aspect-based sentiment analysis. In: Proceedings of the 58th Annual Meeting of the Association for Computational Linguistics, pp. 3211–3220 (2020)
26. Pontiki, M., Galanis, D., Pavlopoulos, J., Papageorgiou, H., Androutsopoulos, I., Manandhar, S.: SemEval-2014 task 4: aspect based sentiment analysis. In: Proceedings of the 8th International Workshop on Semantic Evaluation (SemEval 2014), pp. 27–35. Association for Computational Linguistics, Dublin (2014). https://doi.org/10.3115/v1/S14-2004. https://aclanthology.org/S14-2004
27. Qian, C., Feng, F., Wen, L., Ma, C., Xie, P.: Counterfactual inference for text classification debiasing. In: Proceedings of the 59th Annual Meeting of the Association for Computational Linguistics and the 11th International Joint Conference on Natural Language Processing, vol. 1: Long Papers, pp. 5434–5445 (2021)
28. Qin, C., Zhang, A., Zhang, Z., Chen, J., Yasunaga, M., Yang, D.: Is chatgpt a general-purpose natural language processing task solver? CoRR abs/2302.06476 (2023). https://doi.org/10.48550/arXiv.2302.06476
29. Rietzler, A., Stabinger, S., Opitz, P., Engl, S.: Adapt or get left behind: Domain adaptation through BERT language model finetuning for aspect-target sentiment classification. In: Calzolari, N., et al. (eds.) Proceedings of The 12th Language Resources and Evaluation Conference, LREC 2020, Marseille, France, 11–16 May 2020, pp. 4933–4941. European Language Resources Association (2020). https://aclanthology.org/2020.lrec-1.607/

30. Song, Y., Wang, J., Jiang, T., Liu, Z., Rao, Y.: Attentional encoder network for targeted sentiment classification. CoRR abs/1902.09314 (2019). http://arxiv.org/abs/1902.09314
31. Sun, C., Huang, L., Qiu, X.: Utilizing BERT for aspect-based sentiment analysis via constructing auxiliary sentence. In: Burstein, J., Doran, C., Solorio, T. (eds.) Proceedings of the 2019 Conference of the North American Chapter of the Association for Computational Linguistics: Human Language Technologies, NAACL-HLT 2019, Minneapolis, MN, USA, 2–7 June 2019, vol. 1 (Long and Short Papers), pp. 380–385. Association for Computational Linguistics (2019). https://doi.org/10.18653/v1/n19-1035
32. Sun, T., Wang, W., Jing, L., Cui, Y., Song, X., Nie, L.: Counterfactual reasoning for out-of-distribution multimodal sentiment analysis. In: Proceedings of the 30th ACM International Conference on Multimedia, pp. 15–23 (2022)
33. Tang, D., Qin, B., Liu, T.: Aspect level sentiment classification with deep memory network. In: Su, J., Carreras, X., Duh, K. (eds.) Proceedings of the 2016 Conference on Empirical Methods in Natural Language Processing, EMNLP 2016, Austin, Texas, USA, 1–4 November 2016, pp. 214–224. The Association for Computational Linguistics (2016). https://doi.org/10.18653/v1/d16-1021
34. Tang, H., Ji, D., Li, C., Zhou, Q.: Dependency graph enhanced dual-transformer structure for aspect-based sentiment classification. In: Proceedings of the 58th Annual Meeting of the Association for Computational Linguistics, pp. 6578–6588 (2020)
35. Wang, B., Shen, T., Long, G., Zhou, T., Chang, Y.: Eliminating sentiment bias for aspect-level sentiment classification with unsupervised opinion extraction. In: Findings of the Association for Computational Linguistics: EMNLP 2021, pp. 3002–3012 (2021)
36. Wang, S., Mazumder, S., Liu, B., Zhou, M., Chang, Y.: Target-sensitive memory networks for aspect sentiment classification. In: Proceedings of the 56th Annual Meeting of the Association for Computational Linguistics, vol. 1: Long Papers (2018)
37. Wang, S., Zhou, J., Sun, C., Ye, J., Gui, T., Zhang, Q., Huang, X.: Causal intervention improves implicit sentiment analysis. In: Calzolari, N., et al. (eds.) Proceedings of the 29th International Conference on Computational Linguistics, COLING 2022, Gyeongju, Republic of Korea, 12–17 October 2022, pp. 6966–6977. International Committee on Computational Linguistics (2022). https://aclanthology.org/2022.coling-1.607
38. Wei, T., Feng, F., Chen, J., Wu, Z., Yi, J., He, X.: Model-agnostic counterfactual reasoning for eliminating popularity bias in recommender system. In: Proceedings of the 27th ACM SIGKDD Conference on Knowledge Discovery & Data Mining, pp. 1791–1800 (2021)
39. Wu, Z., Ying, C., Zhao, F., Fan, Z., Dai, X., Xia, R.: Grid tagging scheme for aspect-oriented fine-grained opinion extraction. CoRR abs/2010.04640 (2020). https://arxiv.org/abs/2010.04640
40. Wu, Z., Chen, Y., Kao, B., Liu, Q.: Perturbed masking: parameter-free probing for analyzing and interpreting BERT. In: Jurafsky, D., Chai, J., Schluter, N., Tetreault, J.R. (eds.) Proceedings of the 58th Annual Meeting of the Association for Computational Linguistics, ACL 2020, Online, 5–10 July 2020, pp. 4166–4176. Association for Computational Linguistics (2020). https://doi.org/10.18653/v1/2020.acl-main.383

41. Xing, X., Jin, Z., Jin, D., Wang, B., Zhang, Q., Huang, X.: Tasty burgers, soggy fries: probing aspect robustness in aspect-based sentiment analysis. In: Webber, B., Cohn, T., He, Y., Liu, Y. (eds.) Proceedings of the 2020 Conference on Empirical Methods in Natural Language Processing, EMNLP 2020, Online, 16–20 November 2020, pp. 3594–3605. Association for Computational Linguistics (2020). https://doi.org/10.18653/v1/2020.emnlp-main.292
42. Xu, H., Liu, B., Shu, L., Yu, P.S.: BERT post-training for review reading comprehension and aspect-based sentiment analysis. In: Burstein, J., Doran, C., Solorio, T. (eds.) Proceedings of the 2019 Conference of the North American Chapter of the Association for Computational Linguistics: Human Language Technologies, NAACL-HLT 2019, Minneapolis, MN, USA, 2–7 June 2019, vol. 1 (Long and Short Papers). pp, 2324–2335. Association for Computational Linguistics (2019). https://doi.org/10.18653/v1/n19-1242
43. Xu, M., Wang, D., Feng, S., Yang, Z., Zhang, Y.: KC-ISA: an implicit sentiment analysis model combining knowledge enhancement and context features. In: Calzolari, N., et al. (eds.) Proceedings of the 29th International Conference on Computational Linguistics, COLING 2022, Gyeongju, Republic of Korea, 12–17 October 2022, pp. 6906–6915. International Committee on Computational Linguistics (2022). https://aclanthology.org/2022.coling-1.601
44. Xu, W., Li, X., Deng, Y., Bing, L., Lam, W.: Peerda: data augmentation via modeling peer relation for span identification tasks. CoRR abs/2210.08855 (2022). https://doi.org/10.48550/arXiv.2210.08855
45. Xue, W., Li, T.: Aspect based sentiment analysis with gated convolutional networks. In: Gurevych, I., Miyao, Y. (eds.) Proceedings of the 56th Annual Meeting of the Association for Computational Linguistics, ACL 2018, Melbourne, Australia, 15–20 July 2018, vol. 1: Long Papers, pp. 2514–2523. Association for Computational Linguistics (2018). https://doi.org/10.18653/v1/P18-1234. https://aclanthology.org/P18-1234/
46. Zhang, C., Li, Q., Song, D.: Aspect-based sentiment classification with aspect-specific graph convolutional networks. In: Inui, K., Jiang, J., Ng, V., Wan, X. (eds.) Proceedings of the 2019 Conference on Empirical Methods in Natural Language Processing and the 9th International Joint Conference on Natural Language Processing, EMNLP-IJCNLP 2019, Hong Kong, China, 3–7 November 2019, pp. 4567–4577. Association for Computational Linguistics (2019). https://doi.org/10.18653/v1/D19-1464
47. Zhang, C., Li, Q., Song, D.: Syntax-aware aspect-level sentiment classification with proximity-weighted convolution network. In: Piwowarski, B., Chevalier, M., Gaussier, É., Maarek, Y., Nie, J., Scholer, F. (eds.) Proceedings of the 42nd International ACM SIGIR Conference on Research and Development in Information Retrieval, SIGIR 2019, Paris, France, 21–25 July 2019, pp. 1145–1148. ACM (2019). https://doi.org/10.1145/3331184.3331351
48. Zhang, C., Li, Q., Song, D., Wang, B.: A multi-task learning framework for opinion triplet extraction. In: Cohn, T., He, Y., Liu, Y. (eds.) Findings of the Association for Computational Linguistics: EMNLP 2020, Online Event, 16–20 November 2020. Findings of ACL, vol. EMNLP 2020, pp. 819–828. Association for Computational Linguistics (2020). https://doi.org/10.18653/v1/2020.findings-emnlp.72
49. Zhang, C., Ren, L., Ma, F., Wang, J., Wu, W., Song, D.: Structural bias for aspect sentiment triplet extraction. In: Calzolari, N., et al. (eds.) Proceedings of the 29th International Conference on Computational Linguistics, COLING 2022, Gyeongju, Republic of Korea, 12–17 October 2022, pp. 6736–6745. International Committee on Computational Linguistics (2022). https://aclanthology.org/2022.coling-1.585

Learning Action Embeddings for Off-Policy Evaluation

Matej Cief[1,2](✉), Jacek Golebiowski[3], Philipp Schmidt[3], Ziawasch Abedjan[4], and Artur Bekasov[5]

[1] Brno University of Technology, Brno, Czech Republic
[2] Kempelen Institute of Intelligent Technologies, Bratislava, Slovakia
matej.cief@kinit.sk
[3] Amazon, Berlin, Germany
[4] Leibniz University Hannover, Hanover, Germany
[5] Amazon, London, UK

Abstract. Off-policy evaluation (OPE) methods allow us to compute the expected reward of a policy by using the logged data collected by a different policy. However, when the number of actions is large, or certain actions are under-explored by the logging policy, existing estimators based on inverse-propensity scoring (IPS) can have a high or even infinite variance. Saito and Joachims [13] propose marginalized IPS (MIPS) that uses action *embeddings* instead, which reduces the variance of IPS in large action spaces. MIPS assumes that good action embeddings can be defined by the practitioner, which is difficult to do in many real-world applications. In this work, we explore *learning* action embeddings from logged data. In particular, we use intermediate outputs of a trained reward model to define action embeddings for MIPS. This approach extends MIPS to more applications, and in our experiments improves upon MIPS with pre-defined embeddings, as well as standard baselines, both on synthetic and real-world data. Our method does not make assumptions about the reward model class, and supports using additional action information to further improve the estimates. The proposed approach presents an appealing alternative to DR for combining the low variance of DM with the low bias of IPS.

Keywords: off-policy evaluation · multi-armed bandits · large action space · representation learning · recommender systems

1 Introduction

The *multi-armed bandit* (MAB) framework is commonly used to model the interaction between recommender systems or search engines with their users [21]. In MAB, an *agent* performs *actions* and receives *rewards*, where each reward is sampled from an unknown reward distribution conditioned on the action. The

M. Cief—Work done during an internship at Amazon.

N. Goharian et al. (Eds.): ECIR 2024, LNCS 14608, pp. 108–122, 2024.
https://doi.org/10.1007/978-3-031-56027-9_7

goal is to learn a *policy* for the agent that maximizes the cumulative reward over multiple iterations. In a *contextual* MAB, the reward distribution is also conditioned on a *context* that is observed by the agent, and is used as an additional input to the policy function.

Bandit policies are commonly learned *online*, where we run a policy in production and update it iteratively using the observed rewards [21]. However, deploying untested policies to production is risky. An alternative is to leverage the logged data from historical customer interactions to learn a policy *offline*. The fundamental problem of offline learning is *off-policy evaluation* (OPE), where we try to estimate the expected reward of a new policy without deploying it. The disadvantage of OPE is that the computed value is only an *estimate* of the true value of the policy, and the success of offline learning depends on how accurate this estimate is. When the policy used to gather logged data has a low probability of choosing a subset of actions, existing estimators based on *inverse propensity scoring* [IPS, 10] can be inaccurate [11]. This problem is especially potent when the number of actions is large, as is often the case in recommender systems [19].

Embedding contexts into a lower-dimensional space can improve the off-policy estimates when dealing with a large number of contexts [21]. Similarly, Saito and Joachims [13] propose to embed the actions, and to use the embeddings in *marginalized* IPS (MIPS). Saito and Joachims use additional action information to define the embeddings. For example, in fashion recommendation, where an action is a particular fashion item to recommend, they use the price, the brand and the category of an item as embedding dimensions. For many problems of interest defining action embeddings by hand will be difficult, and/or will require expert knowledge.

Motivated by the two considerations above, we propose methods for *learning* or *fine-tuning* the action embeddings, aiming to improve the performance of MIPS. Our key hypothesis is that an action embedding that is useful for reward prediction will also be useful for MIPS. In particular, we propose to use the intermediate outputs of a trained reward model as action embeddings for MIPS, instead of using them for direct reward prediction as in DM.

We study this approach on both synthetic and real-world datasets, demonstrating that it consistently outperforms MIPS with pre-defined action embeddings, as well as common baselines like standard IPS, *direct method* (DM), and *doubly robust* estimation [DR, 3]. We demonstrate that the method is not sensitive to reward model misspecification: a linear regression reward model produces useful action embeddings even when the true reward function itself is non-linear. We propose methods for utilizing additional action information when it is available, and show that learning/fine-tuning action embeddings is especially useful when the additional action information is high-dimensional, or when the information affecting the reward is incomplete. Finally, we provide a theoretical analysis of our method, showing that it can be interpreted as kernelized DM.

2 Background and Related Work

The interaction of a recommender system with the user starts with the system receiving the context information (e.g. user's purchase history) $x \in \mathcal{X} \subseteq \mathbb{R}^{d_{\mathcal{X}}}$ drawn i.i.d. from an unknown distribution $p(x)$. The system then chooses an action (e.g. a ranked set of products) $a \sim \pi(a \mid x)$ from the set of available actions $\mathcal{A}$ according to the system's policy π. Finally, the user interacts with the displayed products via clicks or purchases, and the system receives a reward $r \in [r_{min}, r_{max}]$ sampled from a reward distribution $p(r \mid x, a)$. This process repeats for n rounds, where in round t the system observes a context x_t, draws an action a_t according to $\pi(a \mid x_t)$, and observes a reward $r_t \sim p(r \mid x_t, a_t)$. All interactions are recorded in a *logged dataset* $\mathcal{D} = \{(x_t, a_t, r_t)\}_{t=1}^n$. The policy π_0 that collects $\mathcal{D}$ is called a *logging policy.* The goal of *off-policy evaluation* [OPE, 3,8,13,15,16,20] is to develop $\hat{V}$, an estimator of the policy *value* $V(\pi) = \mathbb{E}_{p(x)\pi(a|x)p(r|x,a)}[r]$ such that $\hat{V}(\pi) \approx V(\pi)$.

Inverse propensity scoring [IPS, 10] is a commonly used estimator that works by re-weighting the rewards in $\mathcal{D}$ based on how likely the corresponding actions are to be selected under the target policy:

$$\hat{V}_{\text{IPS}}(\pi) = \frac{1}{n} \sum_{t=1}^{n} \frac{\pi(a_t \mid x_t)}{\pi_0(a_t \mid x_t)} r_t. \tag{1}$$

Many state-of-the-art OPE methods are based on IPS [3,8,15,16,20]. A key assumption of IPS, however, is that the logging policy π_0 and the target policy π have *common support*, i.e. $\pi(a \mid x) > 0 \implies \pi_0(a \mid x) > 0$ [13]. When dealing with large action spaces, it can be difficult to select a logging policy that puts non-zero probability on *all* actions, yet does not degrade the customer experience. Even if the two policies have common support, a low probability of selecting an action under the logging policy results in a large weight $\pi(a_t|x_t)/\pi_0(a_t|x_t)$, which increases the variance of the IPS estimate. To reduce the variance of IPS, we can either clip the importance weights according to a tunable clipping parameter [17] or self-normalize them [SNIPS, 18], but the former biases the estimator and the latter can still result in a relatively high variance. To deal with continuous space, Kallus and Zhou [7] use a kernel function to calculate propensities for IPS. We also use a similar technique to estimate the propensities of learned embeddings, and we show that the propensity estimation used in MIPS can be interpreted as kernel regression in Sect. 3.3.

In their work Saito and Joachims [13] instead assume that useful *action embeddings* are available to the estimator, and that we can pool information across similar actions to improve the estimates. Suppose we have an action embedding $e \in \mathcal{E} \subseteq \mathbb{R}^{d_{\mathcal{E}}}$ for each action a. Saito and Joachims show that if the logging policy π_0 has common *embedding support*, i.e. $p(e \mid \pi, x) > 0 \implies p(e \mid \pi_0, x) > 0$, and the action a has no *direct effect* on the reward r, i.e. $a \perp r \mid x, e$, then the proposed *marginalized* IPS (MIPS) estimator

$$\hat{V}_{\text{MIPS}}(\pi) = \frac{1}{n}\sum_{t=1}^{n} \frac{p(e_t \mid \pi, x_t)}{p(e_t \mid \pi_0, x_t)} r_t \tag{2}$$

is unbiased. Experiments by Saito and Joachims demonstrate that in practice MIPS has a lower variance than IPS, and produces estimates with a lower MSE. Our goal in this work is to develop a method for *learning* the action embeddings from logged data to further improve MIPS, and to make it applicable to problems where pre-defined embeddings are not available.

The *direct method* (DM) learns the model of the expected reward for the context-action pair, and uses it to estimate the expected reward of a policy. DM has a low variance, but suffers from a high bias when the model is misspecified. *Doubly robust* estimator [DR, 3,5,20] combines DM and IPS: it uses the learned model of the reward as a control variate. In other words, it computes a non-parametric estimation on the reward model residuals. If the expected reward is correctly specified, DR can achieve a lower variance than IPS, and a lower bias than DM. Several extensions of DR have been proposed [15,20], but, to the best of our knowledge, none of them aim to improve the performance of DR in large action spaces.

Two methods for off-policy evaluation with large action spaces have been developed in parallel with our work [9,14]. Peng et al. [9] propose to cluster similar actions, and replace individual action propensity scores with those of action clusters. Saito et al. [14] train a two-step regression model, hypothesizing that similar actions share a cluster effect on the reward, but also have their own residual effect.

3 Methods

In this work we propose to use the reward signal to learn the embeddings of actions for the MIPS estimator, which would estimate the propensities of these embeddings, and use them instead of the original action propensities to re-weight the observed rewards.

3.1 Motivation

We first formalize the main assumption that motivates the proposed method.

Assumption 1 (Action Embedding Space). *For every set of actions $\mathcal{A}$, there exists a lower-dimensional embedding space $\mathcal{E} \subseteq \mathbb{R}^{d_\mathcal{E}}$ so that every action $a \in \mathcal{A}$ can be mapped to an embedding $e \in \mathcal{E}$ while $p(r \mid a, x) \approx p(r \mid e, x)$ holds for every context x. We denote $d_\mathcal{E} < |\mathcal{A}|$ to be the number of dimensions of embedding space $\mathcal{E}$.*

Table 1. Mean squared error of the estimators on a synthetic experiment while varying the number of actions. Reporting mean and standard error averaged over 50 runs.

$\lvert\mathcal{A}\rvert$	IPS	DM	Learned MIPS
50	0.63 ± 0.05	0.64 ± 0.05	$\mathbf{0.56 \pm 0.05}$
100	0.68 ± 0.04	2.13 ± 0.11	$\mathbf{0.58 \pm 0.05}$
200	0.61 ± 0.04	7.16 ± 0.19	$\mathbf{0.55 \pm 0.04}$
500	0.71 ± 0.05	27.9 ± 0.32	$\mathbf{0.56 \pm 0.05}$
1000	0.70 ± 0.04	62.3 ± 0.31	$\mathbf{0.55 \pm 0.03}$

This assumption holds when $d_{\mathcal{E}} = |\mathcal{A}|$ as the actions can be mapped as one-hot encoded representations. But we hypothesize that in practice some actions are "similar", and that the action space can be represented more efficiently. For example, if certain types of customers mostly clicked on shoes in the logged data, we expect to observe a high reward for a shoe product, even if we have never recommended this particular product for this type of customer before.

Based on Assumption 1, the environment can be represented as a graphical model $a \to e \to r \leftarrow x$. This can be learned the same way as learning a reward model for DM, but instead of using the model for reward prediction, we use the intermediate model outputs as action embeddings. As model misspecification is one of the biggest issues with DM [5], it is often difficult to predict the reward end-to-end. But even when the true reward function is complex, and the model class is not rich enough to recover it, we hypothesize that the model will learn a useful $a \to e$ mapping that can be used in MIPS as defined in Eq. (2).

To test our hypothesis, we design the following toy example[1]. Let $\mathcal{D} = \{(x_t, a_t, r_t)\}_{t=1}^n$ for $n = 1000$ be the logged dataset, where a context $x \in [-1,1]^{d_\mathcal{X}}$, $d_\mathcal{X} = 5$, is drawn from the standard normal distribution, action $a \sim \pi_0$ is chosen from set $\mathcal{A}$ by the unconditional logging policy $\pi_0(a) = \nu_a / \sum_{a' \in \mathcal{A}} \nu_{a'}$, where $\nu_a \sim \text{Exp}(1)$ is drawn from an exponential distribution, and the reward is generated by $r = f_a(x) + \eta$, where $f_a(x) = 1/(1+e^{-(x^\top \theta_a + \mu_a)})$ is a logistic model for action a with randomly initialized parameters $\theta_a \sim \mathcal{N}(0,1)^{d_\mathcal{X}}$, $\mu_a \sim \mathcal{N}(0,1)$, and $\eta \sim \mathcal{N}(0, 0.1)$ is random noise. We use the uniform random target policy. For DM, we fit a linear regression $\hat{f}_a(x)$ to estimate r. Of course, linear regression is not expressive enough to fit the non-linear f_a, but it learns the correct *order* of the contexts, meaning $\forall x_1, x_2 \in \mathcal{X}, \hat{f}_a(x_1) < \hat{f}_a(x_2) \Leftrightarrow f_a(x_1) < f_a(x_2)$. We take these learned embeddings and use them in the MIPS estimator [12] and refer to this method as *Learned MIPS*. We vary the number of actions $|\mathcal{A}| \in \{50, 100, 200, 500, 1000\}$, while keeping the size of the logged data fixed. For each $|\mathcal{A}|$ we synthesize 50 reward functions, and for each reward function we generate 15 logged datasets to evaluate the methods. The results are summarized in Table 1.

[1] Code to reproduce this and further experiments is available at https://github.com/amazon-science/ope-learn-action-embeddings.

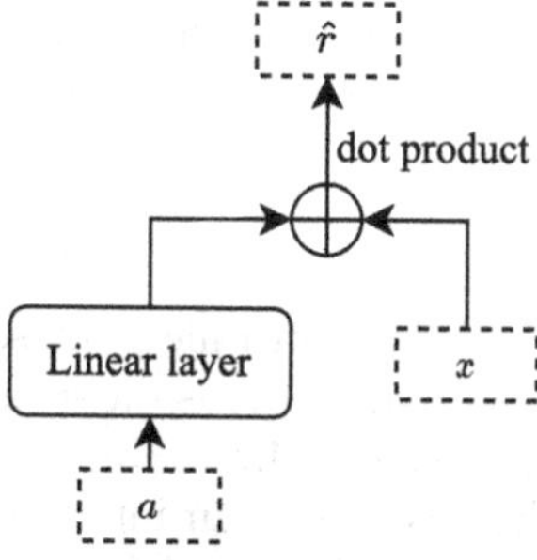

Fig. 1. The model we use to learn the action embeddings. We use a simple linear model in this work, but any model class can be used. Input a can be a one-hot encoded representation of the action identity, a pre-defined action embedding, or a concatenated vector of both.

As the number of samples per action decreases, the learned embeddings become increasingly inaccurate for *direct* reward prediction (see the increasing error of DM in Table 1), but they still reflect the structure of the problem, and we can build an accurate model-free estimator on top of them (Learned MIPS outperforms both DM and IPS). This simple example demonstrates the potential of the proposed method as a novel way to combine the strengths of model-based and model-free off-policy estimation methods.

3.2 Algorithm

As shown in the previous section, we can learn action embeddings by fitting a separate linear regression model $\hat{f}_a(x)$ for each action a. In general, as in DM, we can use any model class. For example, we can fit a deep neural network, and use the intermediate outputs as action embeddings. In this work, we use a linear model visualized in Fig. 1. In our preliminary experiments, more complex neural network models demonstrated comparable or worse performance. We hypothesize that this is due to the high expressivity (and hence high variance) of neural networks, which is counter to our goal of reducing estimator variance. We leave further empirical study and theoretical analysis of this phenomenon for future work.

The model takes the action identity a, and computes a reward estimate $\hat{r}$ as a dot product between the action embedding and the context vector x. We train the model to minimize the MSE between $\hat{r}$ and the reward r observed in the logged data. The learned embedding for action a is the output of the linear layer. As we show in Sect. 4, we can also include additional action information (e.g. content-based features of the corresponding product), which can further improve the performance of the estimator.

Having learned the action embeddings, we follow MIPS [13] and fit a logistic regression model to estimate $p(a \mid e)$. This estimate is used to compute the embedding propensities ($p(e \mid \pi, x)$ and $p(e \mid \pi_0, x)$) in Eq. (2), noting that

$p(e \mid \pi, x) = p(a \mid e)\pi(a \mid x)$. We discuss this step in more detail in the next section, where we show a connection between our method and kernelized DM.

3.3 Connection to DM

Learned MIPS resembles DM in that both methods solve the OPE task by first training a reward predictor. The two methods differ, however, in how they apply said predictor. In this section, we show a direct connection between learned MIPS and DM. In particular, we show that we can interpret the proposed method as DM that uses *kernel regression* with a learned feature embedding as the reward predictor.

To simplify the notation, we consider a non-contextual estimation task: given logged data $\mathcal{D} = \{a_t, r_t\}_{t=1}^n$ gathered with a policy $\pi_0(a)$, we want to estimate the value of a policy $\pi(a)$. In this setting, the DM estimate is simply

$$\hat{V}_{\mathrm{DM}}(\pi) = \sum_{a \in \mathcal{A}} \pi(a)\hat{r}(a), \tag{3}$$

where $\hat{r}(a)$ is the model of expected reward for action a that we learn using $\mathcal{D}$ as the training data. We now define $\tilde{\mathcal{D}} = \{a_t, e_t, r_t\}_{t=1}^n$, where $e_t \sim p(e \mid a_t)$ is the corresponding action embedding. The MIPS estimate is then

$$\hat{V}_{\mathrm{MIPS}}(\pi) = \frac{1}{n}\sum_{t=1}^{n} \mathbb{E}_{p(a|e_t)}\left[\frac{\pi(a)}{\pi_0(a)}\right] r_t = \frac{1}{n}\sum_{t=1}^{n}\sum_{a \in \mathcal{A}} \frac{\pi(a)}{\pi_0(a)} p(a \mid e_t) r_t \tag{4}$$

$$= \sum_{a \in \mathcal{A}} \pi(a) \frac{1}{n}\sum_{t=1}^{n} \frac{p(a \mid e_t)}{\pi_0(a)} r_t = \sum_{a \in \mathcal{A}} \pi(a)\tilde{r}(a), \tag{5}$$

where $p(a \mid e_t)$ is estimated as in Eq. (7). In other words, the MIPS estimate matches the DM estimate with a different expected reward model:

$$\tilde{r}(a) = \frac{1}{n}\sum_{t=1}^{n} \frac{p(a \mid e_t)}{\pi_0(a)} r_t. \tag{6}$$

We now try to understand the nature of $\tilde{r}$. Intuitively, $p(a \mid e)$ will be higher when the embedding of a is "closer" to e according to some distance measure. In particular, let $e \sim \mathcal{N}(f(a), \sigma^2 I)$ be the embedding distribution, where $f(a)$ is the learned deterministic embedding. As in MIPS, we estimate $p(a \mid e)$ by fitting a probabilistic classifier to the dataset $\{a_t, e_t\}_{t=1}^n$. Saito and Joachims use logistic regression, but here we consider *linear discriminant analysis* [6, LDA,Section 4.3], a comparable model. LDA fits a normal distribution to each class, hence we expect it to recover $\mathcal{N}(f(a), \sigma^2 I)$. We also expect it to recover $\pi_0(a)$ as the class prior, because we sampled $a_t \sim \pi_0(a)$ to produce the training data $\{a_t, e_t\}_{t=1}^n$. The predictive probability of LDA is then

$$p(a \mid e) \propto \mathcal{N}(e; f(a), \sigma^2 I)\pi_0(a). \tag{7}$$

Combining Eqs. (6) and (7) we get

$$\tilde{r}(a) \propto \frac{1}{n}\sum_{t=1}^{n} \mathcal{N}(e_t; f(a), \sigma^2 I) r_t, \tag{8}$$

where

$$\mathcal{N}(e_t; f(a), \sigma^2 I) \propto \exp\left(-\frac{1}{2\sigma^2}(e_t - f(a))^\top (e_t - f(a))\right). \tag{9}$$

This confirms our intuition: if we use LDA as the classifier, the expected reward $\tilde{r}(a)$ in Eq. (5) is a weighted combination of all rewards in the logged data, where samples with e_t "closer" to the action embedding $f(a)$ have a higher weight. For the normal embedding distribution, the weight is determined by a squared Euclidean distance between vectors e_t and $f(a)$, but other distributions will induce different distance functions. In fact, Eq. (8) is equivalent to *kernel regression* [2, Chapter 10] in a learned embedding space determined by f, with a kernel determined by the embedding distribution.

While the above holds when using LDA as the classifier, in our experiments we follow Saito and Joachims and use the logistic regression classifier. Both of these methods fit linear decision boundaries and give similar results in practice ([4]; [6, Chapter 4]), which means that our interpretation likely holds for logistic regression-based MIPS as well. As DM, our method does not provide theoretical guarantees on the bias/variance for a general reward model class, but, as we show in the next section, demonstrates good performance in practice.

4 Empirical Evaluation

In this section we extend our proof-of-concept experiments in Sect. 3 to a more complex synthetic dataset, and conduct experiments on a real-world fashion recommendation dataset. In the synthetic experiments, we evaluate two classes of methods based on whether the method uses additional action information or not. We find that our methods outperform all the standard baselines and, in some cases, improve upon the true embeddings used to generate the reward.

4.1 Synthetic Experiments

For our empirical evaluation, we follow the setup by Saito and Joachims [13]. We generate synthetic data by first sampling $d_{\mathcal{X}}$-dimensional context vectors x from the standard normal distribution. We also sample $d_{\mathcal{E}}$-dimensional categorical action embeddings $e \in \mathcal{E}$ from the following distribution

$$p(e \mid a) = \prod_{k=1}^{d_{\mathcal{E}}} \frac{\exp \alpha_{a,e_k}}{\sum_{e' \in \mathcal{E}_k} \exp \alpha_{a,e'}}, \tag{10}$$

where $\{\alpha_{a,e_k}\}$ is a set of parameters sampled independently from the standard normal distribution. Each dimension of $\mathcal{E}$ has a cardinality of 10. We then synthesize the expected reward as

$$q(x, e) = \sum_{k=1}^{d_\mathcal{E}} \eta \cdot (x^\top M x_{e_k} + \theta_x^\top x + \theta_e^\top x_{e_k}), \tag{11}$$

where M is the parameter matrix and θ_x, θ_e are parameter vectors sampled uniformly in $[-1, 1]$. x_{e_k} is a parameter vector corresponding to the k-th dimension of the action embedding and is sampled from the standard normal distribution. Parameter η_k specifies the importance of the k-th dimension and is drawn from a Dirichlet distribution $\eta \sim \text{Dir}(\alpha)$ with its parameters $\alpha = (\alpha_1, \dots, \alpha_{d_\mathcal{E}})$ sampled from a uniform distribution $[0, 1)$. With this setup, the action embeddings are fully informative, and the action identity has no *direct effect* on the reward. On the other hand, if action embeddings have some missing dimensions, they do not fully capture the action's effect on the reward. In this case, we hypothesize that we can learn better embeddings, and we test this in one of the experiments.

We also follow Saito and Joachims [13] in how we synthesize the logging policy π_0. We apply the softmax function to $q(x, a) = \mathbb{E}_{p(e|a)}[q(x, e)]$ as

$$\pi_0(a \mid x) = \frac{\exp \beta \cdot q(x, a)}{\sum_{a' \in \mathcal{A}} \exp \beta \cdot q(x, a')}, \tag{12}$$

where β is the parameter that controls the entropy and the optimality of the logging policy. A large positive value leads to an optimal, near-deterministic logging policy, while 0 leads to a uniform-random policy. We define the target policy π as an ε-greedy policy

$$\pi(a \mid x) = (1 - \varepsilon) \cdot \mathbb{1}\left\{a = \underset{a' \in \mathcal{A}}{\arg\max}\ a(x, a')\right\} + \frac{\varepsilon}{|\mathcal{A}|}, \tag{13}$$

where $\varepsilon \in [0, 1]$ controls the exploration rate of the target policy. Similar to Saito and Joachims [13], we set $\beta = -1$ and $\varepsilon = 0.05$ in our experiments.

To generate a batch of logged data, we follow an iterative process to generate n samples $\mathcal{D} = \{x_t, a_t, e_t, r_t\}_{t=1}^n$. In each iteration, we sample a context x from the standard normal distribution ($d_\mathcal{X} = 10$), and a discrete action a from π_0 according to Eq. (12). Given an action a, we then sample a categorical action embedding e using Eq. (10). Finally, we sample the reward from a normal distribution with mean $q(x, e)$ defined in Eq. (11) and standard deviation $\sigma = 2.5$ [13].

We assume three variants of our method, based on which action representation is passed to the model:

- *Learned MIPS OneHot* only sees one-hot encoded action identities.
- *Learned MIPS FineTune* only sees pre-defined action embeddings.
- *Learned MIPS Combined* sees both of the above concatenated into a single vector.

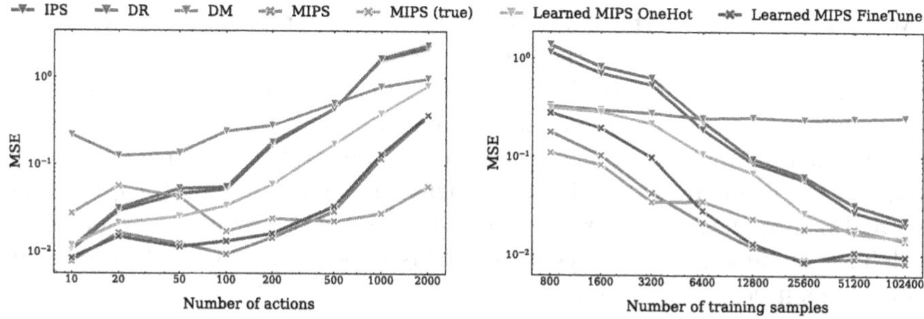

Fig. 2. Synthetic experiments varying the number of actions and training samples. To better distinguish comparable methods, those marked with ▽ only use action identities, and those marked with × also use pre-defined action embeddings. *Learned MIPS OneHot* outperforms all standard baselines. When we have enough data for every action, it performs just as well as IPS. As the variance grows with fewer samples per action, its error approaches the one of DM using the same model. The pre-defined embeddings ($d_{\mathcal{E}} = 3$) have low bias and variance; hence our methods can not improve upon them. Shaded areas around the lines are standard errors (almost invisible).

We compare our methods to all common estimators, including DM, IPS, and DR. To gauge the quality of the learned embeddings, we specifically compare to MIPS with pre-defined embeddings, and *MIPS (true)*, which uses the true propensity scores of the pre-defined embeddings, which are not available in practice. For all experiments we report the MSE and standard errors averaged over 100 different subsampled datasets.

Can We Improve Upon Standard Baselines Without Using Pre-defined Embeddings? We evaluate our performance against all standard estimators in two sets of experiments with data generated according to the procedure described in Sect. 4.1. All pre-defined embedding dimensions are preserved, hence these embeddings are *fully-informative*, i.e. $q(x, e) = q(x, e, a)$. We fix the number of pre-defined embedding dimensions to $d_{\mathcal{E}} = 3$. In the first experiment, we vary the number of actions from 10 to 2,000 with a fixed sample size of $n = 20{,}000$. In the second experiment, we vary the number of samples in the training data $n = [800, 1600, \ldots, 102400]$ and leave the number of actions fixed at $|\mathcal{A}| = 100$. The results of these two experiments are presented in Fig. 2.

Learned MIPS OneHot outperforms all other methods using action identities. We can observe the difference between the two methods that use the same model, *Learned MIPS OneHot* and *DM*. Using learned embedding propensities in MIPS results in a smaller bias compared to using them in direct reward prediction. The error of *Learned MIPS* keeps shrinking with more data. While IPS and DR do not pool information across actions, MIPS is using the embedding propensities influenced by every action, resulting in a better bias-variance trade-off. When the number of actions is small enough, and we have enough data, IPS estimates already have a low variance, and we do not observe any significant improvements.

We also report the performance of the methods that use pre-defined embeddings. Perhaps unsurprisingly, the learned embeddings perform worse than pre-defined embeddings, and even *Learned MIPS FineTune* does not improve over them. This is because the pre-defined embeddings compress the information about the action well. In further experiments, we study more realistic benchmarks, where embeddings are higher-dimensional or do not fully capture the reward. Both of these cases are common in practice, for example, when using pre-defined embeddings extracted from LLMs or when extending the setup to lists of items in learning to rank.

Can We Improve Upon the Pre-defined Embeddings? In the previous experiment, fine-tuned embeddings did not improve upon the pre-defined embeddings. The reason is that the pre-defined embeddings have low bias and low variance. This is rarely the case in practice. In this experiment, we show that we can learn better embeddings when the pre-defined embeddings have a high number of embedding dimensions or when they do not capture the full effect on the reward. We evaluate how the bias of the pre-defined embeddings influences our methods and whether it is better to learn the embeddings from action identities or fine-tune them from pre-defined embeddings in this setting. We progressively add more bias similarly to Saito and Joachims [13]. We fix the number of dimensions in the action embedding to $d_{\mathcal{E}} = 20$, and after the reward is generated, we hide a certain number of dimensions so the estimators do not have access to fully-informative embeddings. The results are shown in Fig. 3. As the embeddings worsen, our methods that learn or fine-tune them outperform MIPS. We can see the bias-variance trade-off between *Learned MIPS OneHot* and *Learned MIPS FineTune*. When the number of unobserved dimensions is small, the variance reduction gained from pre-defined embeddings is greater than induced bias. As we increase the number of unobserved dimensions, the bias of pre-defined embeddings can get even higher than the variance of IPS. In practice, we often do not know how biased the pre-defined embeddings are. Therefore, we introduce *Learned MIPS Combined* and observe performance improvements when using this method for the majority of the values of the unobserved dimensions. When the information in the embedding is complete (we do not hide any dimensions), providing additional dimensions as an action identity only makes the problem more difficult (*Learned MIPS FineTune* outperforms *Learned MIPS Combined* with more embedding dimensions). But as soon as some of the information in the embedding is omitted and the common *embedding support* assumption [13] is violated, the action identity information is very useful, and *Learned MIPS Combined* uses it to model the direct effect of the action missing from the pre-defined embedding. As we show in the next section, it may be difficult to gauge in practice how much information is missing in the real-world embedding and which method would yield the best performance. We leave the optimal method selection for future work.

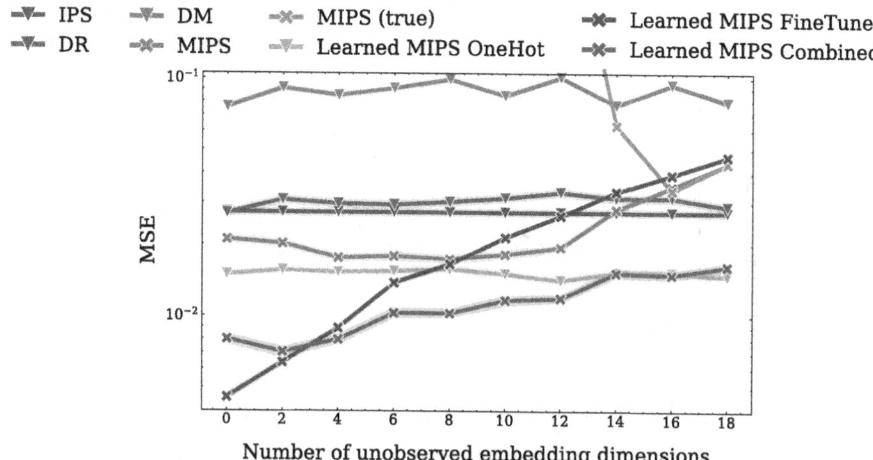

Fig. 3. Synthetic experiments varying the number of unobserved dimensions. As we progressively hide more dimensions, pre-defined embeddings get more biased and methods using them get less accurate. Combining *Learned MIPS OneHot* and *FineTune* yields the most robust results when the bias of pre-defined embeddings is unknown. Shaded areas around the lines are standard errors (almost invisible).

4.2 Real-World Data

We follow the experimental setup of Saito and Joachims [13] to evaluate how the estimators perform on a real-world bandit dataset. In particular, we use a fashion recommendation dataset that was collected on a large-scale fashion e-commerce platform, and comes in the form of $\mathcal{D} = \{(x_t, a_t, e_t, r_t)\}_{t=1}^n$ tuples. The dataset was collected during an A/B test under two policies: a uniform policy π_0 and a Thompson sampling policy π. The size of the action space is $|\mathcal{A}| = 240$, and every action (a fashion item) comes with a pre-defined 4-dimensional action embedding e, which includes the item's price, brand, and hierarchical category information. We use $n = 10000$ sub-sampled observations from the "ALL" campaign in the dataset to be directly comparable to the prior work [9,13].

In line with Saito and Joachims [13], we repeat the experiment over 150 bootstrap samples, normalize the estimators' MSE relative to the MSE of IPS (i.e. divide the method's MSEs for a given data sample by the MSE of IPS for the same sample), and compute the cumulative distribution function (CDF) of these MSEs. We report the results in Fig. 4. We compare against another competitive baseline *MIPS(w/ SLOPE)* that greedily improves the performance by dropping embedding dimensions [13]. *Learned MIPS* methods outperform IPS in about 75% of runs, more often than any other baseline. The next best-performing method is MIPS (w/ SLOPE), which outperforms IPS in about 65% of runs. For DM, we use the same model as in *Learned MIPS OneHot*. These results support our hypothesis that even though DM fails to model the reward, the embeddings it learns in the process can still be useful for model-free estimation.

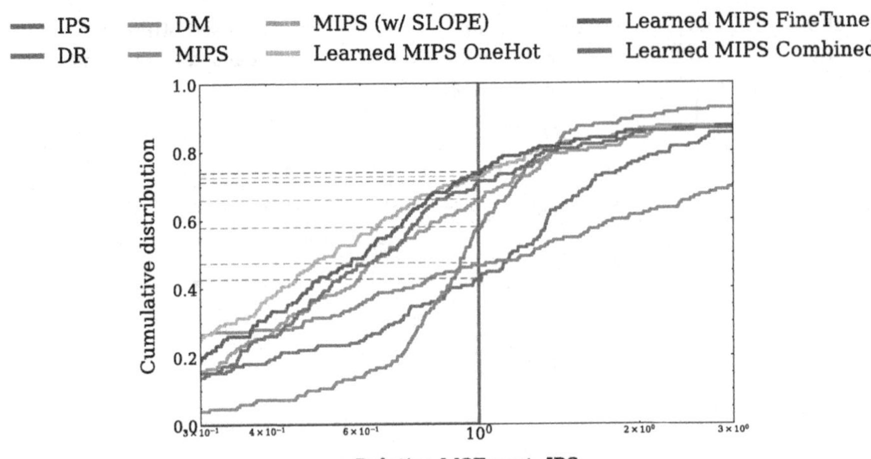

Fig. 4. CDF of relative MSEs w.r.t IPS on the real-world dataset. The intersection of a method's curve with the IPS curve tells us the proportion of experiments in which the method performs better than IPS.

5 Conclusion

In this work, we propose methods that assume a structured action space in the contextual bandit setting, and reduce the variance of model-free off-policy estimators by *learning* action embeddings from the reward signal. We extend MIPS to settings where pre-defined embeddings are not available, or are difficult to define. At the same time, we show that the method can improve upon the pre-defined embeddings even when they are available. The proposed method outperforms all standard baselines on synthetic and real-world fashion recommendation data, even if the reward model itself is inaccurate. An interesting future direction is to study the embeddings learned by more complex classes of reward models, such as neural networks. In the current experiments, we estimate the action distribution over the embeddings $p(a \mid e)$ using a discriminative model as proposed by Saito and Joachims [13]. To enable explicit bias-variance control in future work, it would be interesting to experiment with estimating these weights with a generative classifier and a *prescribed* form of $p(a \mid e)$.

We would also like to extend our methods to the learning-to-rank setting, as list-level embeddings can dramatically reduce the combinatorial complexity of the problem. This can be done by concatenating the item-level embeddings of a single list into a longer embedding and applying our method to it. We assume that in the same way our model exploits the similarity relationships between individual items, it would also learn to embed the browsing customer behavior impacting the final reward (e.g., the items at the end of the list are observed less frequently). This would eliminate the need for explicitly modeling these behaviors by click models [1]. Another way to use our method in LTR, as it works with any model class, is to use some of the well-known list-wise LTR methods and use its intermediate output as the learned action embedding.

Acknowledgement. We want to thank Mohamed Sadek for his contributions to the codebase. The research conducted by Matej Cief (also with slovak.AI) was partially supported by TAILOR, a project funded by EU Horizon 2020 under GA No. 952215, https://doi.org/10.3030/952215.

References

1. Chuklin, A.: Click Models for Web Search, vol. 7, no. 3, pp. 1–115 (2015)
2. Dhrymes, P.J.: Topics in Advanced Econometrics: Probability Foundations, vol. 1. Springer, Heidelberg (1989). https://doi.org/10.1007/978-1-4612-4548-3
3. Dudík, M., Erhan, D., Langford, J., Li, L.: Doubly robust policy evaluation and optimization. Stat. Sci. **29**(4), 485–511 (2014). ISSN 0883-4237, 2168-8745. https://doi.org/10.1214/14-sts500. https://projecteuclid.org/journals/statistical-science/volume-29/issue-4/Doubly-Robust-Policy-Evaluation-and-Optimization/10.1214/14-STS500.full
4. Efron, B.: The efficiency of logistic regression compared to normal discriminant analysis. J. Am. Stat. Assoc. **70**(352), 892–898 (1975)
5. Farajtabar, M., Chow, Y., Ghavamzadeh, M.: More robust doubly robust off-policy evaluation. In: Proceedings of the 35th International Conference on Machine Learning, pp. 1447–1456. PMLR (2018). https://proceedings.mlr.press/v80/farajtabar18a.html. iSSN: 2640-3498
6. Hastie, T., Tibshirani, R., Friedman, J.: The Elements of Statistical Learning. Springer Series in Statistics. Springer, New York (2009). https://doi.org/10.1007/978-0-387-84858-7
7. Kallus, N., Zhou, A.: Policy evaluation and optimization with continuous treatments. In: Proceedings of the Twenty-First International Conference on Artificial Intelligence and Statistics, pp. 1243–1251. PMLR (2018). https://proceedings.mlr.press/v84/kallus18a.html. iSSN: 2640-3498
8. Metelli, A.M., Russo, A., Restelli, M.: Subgaussian and differentiable importance sampling for off-policy evaluation and learning. In: Advances in Neural Information Processing Systems, vol. 34, pp. 8119–8132. Curran Associates, Inc. (2021). https://proceedings.neurips.cc/paper/2021/hash/4476b929e30dd0c4e8bdbcc82c6ba23a-Abstract.html
9. Peng, J., et al.: Offline policy evaluation in large action spaces via outcome-oriented action grouping. In: Proceedings of the ACM Web Conference 2023, WWW 2023, pp. 1220–1230. Association for Computing Machinery, New York (2023). ISBN 978-1-4503-9416-1. https://doi.org/10.1145/3543507.3583448. https://dl.acm.org/doi/10.1145/3543507.3583448
10. Robins, J.M., Rotnitzky, A., Zhao, L.P.: Estimation of regression coefficients when some regressors are not always observed. J. Am. Stat. Assoc. **89**(427), 846–866 (1994). ISSN 0162-1459. https://doi.org/10.1080/01621459.1994.10476818
11. Sachdeva, N., Su, Y., Joachims, T.: Off-policy bandits with deficient support. In: Proceedings of the 26th ACM SIGKDD International Conference on Knowledge Discovery & Data Mining, KDD 2020, pp. 965–975. Association for Computing Machinery, New York (2020). ISBN 978-1-4503-7998-4. https://doi.org/10.1145/3394486.3403139. https://dl.acm.org/doi/10.1145/3394486.3403139

12. Saito, Y., Aihara, S., Matsutani, M., Narita, Y.: Open bandit dataset and pipeline: towards realistic and reproducible off-policy evaluation (2021). https://doi.org/10.48550/arXiv.2008.07146. arXiv:2008.07146 [cs, stat]
13. Saito, Y., Joachims, T.: Off-policy evaluation for large action spaces via embeddings. In: Proceedings of the 39th International Conference on Machine Learning, pp. 19089–19122. PMLR (2022). https://proceedings.mlr.press/v162/saito22a.html. iSSN: 2640–3498
14. Saito, Y., Ren, Q., Joachims, T.: Off-policy evaluation for large action spaces via conjunct effect modeling. In: Proceedings of the 40th International Conference on Machine Learning, pp. 29734–29759. PMLR (2023). https://proceedings.mlr.press/v202/saito23b.html. iSSN: 2640–3498
15. Su, Y., Dimakopoulou, M., Krishnamurthy, A., Dudik, M.: Doubly robust off-policy evaluation with shrinkage. In: Proceedings of the 37th International Conference on Machine Learning, pp. 9167–9176. PMLR (2020). https://proceedings.mlr.press/v119/su20a.html. iSSN: 2640–3498
16. Su, Y., Wang, L., Santacatterina, M., Joachims, T.: CAB: continuous adaptive blending for policy evaluation and learning. In: Proceedings of the 36th International Conference on Machine Learning, pp. 6005–6014. PMLR (2019). https://proceedings.mlr.press/v97/su19a.html. iSSN: 2640–3498
17. Swaminathan, A.: Counterfactual Evaluation and Learning From Logged User Feedback. Ph.D. thesis, Cornell University, Ithaca, NY, United States (2017). https://ecommons.cornell.edu/handle/1813/51557
18. Swaminathan, A., Joachims, T.: The self-normalized estimator for counterfactual learning. In: Advances in Neural Information Processing Systems, vol. 28. Curran Associates, Inc. (2015). https://proceedings.neurips.cc/paper/2015/hash/39027dfad5138c9ca0c474d71db915c3-Abstract.html
19. Swaminathan, A., et al.: Off-policy evaluation for slate recommendation. In: Advances in Neural Information Processing Systems, vol. 30. Curran Associates, Inc. (2017). https://proceedings.neurips.cc/paper_files/paper/2017/hash/5352696a9ca3397beb79f116f3a33991-Abstract.html
20. Wang, Y.X., Agarwal, A., Dudik, M.: Optimal and adaptive off-policy evaluation in contextual bandits. In: Proceedings of the 34th International Conference on Machine Learning, pp. 3589–3597. PMLR (2017). https://proceedings.mlr.press/v70/wang17a.html. iSSN: 2640–3498
21. Zhou, L.: A Survey on Contextual Multi-armed Bandits (2016). https://doi.org/10.48550/arXiv.1508.03326. arXiv:1508.03326 [cs]

Lightweight Modality Adaptation to Sequential Recommendation via Correlation Supervision

Hengchang Hu[1(✉)], Qijiong Liu[2], Chuang Li[1], and Min-Yen Kan[1]

[1] National University of Singapore, Singapore, Singapore
{huhengc,lichuang,kanmy}@comp.nus.edu.sg
[2] The Hong Kong Polytechnic University, Hong Kong, China
jyonn.liu@connect.polyu.hk

Abstract. In Sequential Recommenders (SR), *encoding* and *utilizing* modalities in an end-to-end manner is costly in terms of modality encoder sizes. Two-stage approaches can mitigate such concerns, but they suffer from poor performance due to modality forgetting, where the sequential objective overshadows modality representation. We propose a lightweight knowledge distillation solution that preserves both merits: retaining modality information and maintaining high efficiency. Specifically, we introduce a novel method that enhances the learning of embeddings in SR through the supervision of modality correlations. The supervision signals are distilled from the original modality representations, including both (1) holistic correlations, which quantify their overall associations, and (2) dissected correlation types, which refine their relationship facets (honing in on specific aspects like color or shape consistency).To further address the issue of modality forgetting, we propose an asynchronous learning step, allowing the original information to be retained longer for training the representation learning module. Our approach is compatible with various backbone architectures and outperforms the top baselines by 6.8% on average. We empirically demonstrate that preserving original feature associations from modality encoders significantly boosts task-specific recommendation adaptation. Additionally, we find that larger modality encoders (e.g., Large Language Models) contain richer feature sets which necessitate more fine-grained modeling to reach their full performance potential.

Keywords: Recommender System · Multimodal Recommendation · Knowledge Distillation

1 Introduction

Recommender systems reduce information overload by helping users select their preferred next items. The Sequential Recommender (SR) paradigm specifically focuses on learning sequential relationships among the user's historical items, where the models range from GRU [13], Transformer [21] to Graph Neural Network [49] based methods. The paradigm of using unique identities (IDs for short)

N. Goharian et al. (Eds.): ECIR 2024, LNCS 14608, pp. 123–139, 2024.
https://doi.org/10.1007/978-3-031-56027-9_8

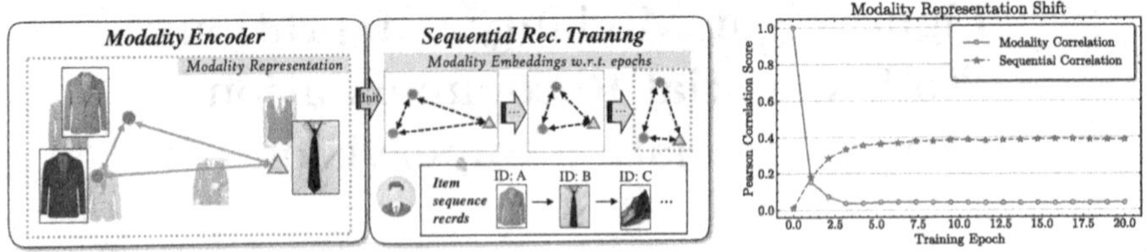

Fig. 1. (left) Diagram of adaptation from original modality representation to modality embedding over training epochs. (right) Experiment on the Beauty dataset with GRU-based models, highlighting the issue of modality forgetting.

to represent distinct items has been well-studied [23,40]. Until recently, there has been a shift towards using modality to represent items [52], with the emergence of sophisticated modality encoders such as Large Language Models (LLM) [46,53] and large-scale image encoder models [33]. These models represent modality features with vectors, discerning inherent modality correlations among items.

Nonetheless, directly utilizing encoded modality features to represent items presents a gap. SRs are specifically designed to optimize item representations for *sequential correlations* (i.e., selection order) modeling, making them ill-suited for leveraging the original modal representations (containing intrinsic *modality correlations*) created by modality encoders. Let us consider sequence selection in the fashion domain (Fig. 1): people usually pick a `suit` followed by a `tie`, rather than choosing two visually similar `suits`. Modality encoders can highlight their modality correlations (such as similarity), but they are easily overshadowed in SR training, as SR is tuned to capture item sequential relations.

It is thus essential to tune and align modality knowledge to the SR systems. While the end-to-end training of modality encoders and SR is a common practice [6], as it maintains the original features of items, it would be less feasible as the emergence of large-scale modality encoders such as LLMs. Two-stage training [5,35] was proposed as a more efficient method that decouples the pretraining phase of the modality encoders and the downstream recommendation phase, which only takes cached modality representations as embedding input.

However, the embedding suffers from *modality forgetting* in the recommendation phase. This issue is demonstrated clearly in Fig. 1 (right). By training a modality-enriched SR model, we compare the epoch-updated modality embeddings (E) with two other standard representation spaces: the original modality representations (M, including only modality correlations), and the item ID embeddings (V, that contains only sequential correlations) in a pure ID-based model.0 Assessing pairwise similarities in these spaces by Pearson score [4], we find that as training progresses, the item correlations within E align more with V but drift away from M (the modality correlation rapidly deteriorates to nearly zero). As modality embeddings readily over-adapt to sequential correlations, it reduces their distinct advantage over plain IDs [52].

Given the concerns above, we introduce a Knowledge Distillation [14] framework for modality-enriched SR (KDSR) that embraces both merits: it retains the modality original features from end-to-end training, while maintaining the effectiveness of two-stage training. Specifically, it contains two components: a teacher model distills correlations from modality encoders, and a student model training embeddings supervised by these correlations. We further propose designs to solve a few inherent challenges: Since a unified correlation can only represent an overall relationship between two items, we design a differentiated holistic and dissected correlation operation to offer the modalities' interrelations capturing diverse perspectives, such as shape and color. Furthermore, to allow for some discrepancies from noise, we employ a soft-matching mechanism to align teacher's and student's holistic correlations, and a vector quantization mechanism to convert dissected correlations into discrete codebook representations. In this context, dissected correlations are transformed into a set of distinct *codes*, simplifying complex relationships while retaining granular insights.

We conducted experiments across five distinct scenarios with image and text modalities. Our proposed KDSR effectively addresses the dilution of modal information, outperforming the top baselines by 6.8%, on average. Additionally, we introduce an asynchronous training step strategy that prevents the early loss of original modal details during training, further offering an enhanced adaptation. Our approach is compatible with different encoders and SR backbones. Notably, our method integrates well with all large-scale modality encoders, like Swin [33] and LLaMA [46]. As these encoders expand in size, our quantization technique is able to yield deeper insights, with an adequate embedding dimension required to capture these more fine-grained correlations.

Our contributions are summarized as: (i) We are the first to spotlight rapid modal information dilution challenges, and propose a knowledge distillation framework to preserve the original modal information. (ii) We offer a lightweight solution that combines coarse- and fine-grained correlation supervision in embedding learning. (iii) We conduct comprehensive experiments, which show our solution can keep pace with the evolving modality encoders, while being practical in real multi-modal use.

2 Related Work

Modality-enriched recommender systems typically utilize a modality encoder to represent item content knowledge, atop which a recommender model is applied for user modeling [29] or collaborative filtering [47]. One line of research is the joint training of the modality encoder and recommender [20,34,36,48]. Another line takes a two-stage training paradigm [17], which firstly pretrains the modality encoder, and then makes extracted item representations detached from the modality encoder for downstream recommendation training. Such a two-stage design allows for more efficient use of resources, as large-scale modality encoders can be employed independently. Works in different domains (such as PREC [30] in news recommendation, UVCAN [31] in micro-video recommendation, and CLEP [37] in music recommendation) have demonstrated the effectiveness of the application of extracted modality knowledge without updating the modality encoders. Our proposed KDSR method takes advantage of the efficient training of the two-stage design and the inherent modality correlation from the end-to-end paradigm.

Knowledge distillation [14] is introduced to transfer valuable information from a large and complex model (i.e., "teacher model") to a smaller and more efficient one ("student"). Knowledge distillation techniques are categorized into supervised [28,57], semi-supervised [1,44,58] and self-supervised ones [24,39,50]. It has been widely used in various domains, including computer vision [2,7], natural language processing [19,41], and speech [25,32]. In the recommendation domain, Tan *et al.* [45] first introduced knowledge distillation for SR. A portion of the previous work is aimed at improving efficiency [3,26], while the majority is focused on enhancing inference quality [50,56]. However, an inadequately trained teacher can misguide, causing slow convergence leading to entrapment in local optima. In this paper, we design a self-supervised pipeline by introducing pre-calculated modality correlations.

3 Methodology

Problem Formulation. Conventional sequential recommendation aims to predict the next item v (from an overall set of N items) that a user u is likely to consume, based on their historical interactions $\{v_1, \ldots, v_m\}$. Extending this, Modality-enriched Sequential Recommendation (MSR) considers item modalities to refine the interrelationships among items within the sequence. Specifically, the objective is to acquire a function $f : \{(v_1, a_1, b_1), \ldots, (v_m, a_m, b_m)\} \rightarrow v$, where a_i/b_i indicates the modality features (image/text) of item v_i respectively.

In this section, we present a novel method for optimizing modality usage in MSR via correlation supervision. We first outline the framework and then give the details of how we implement the correlation supervision in two phases: correlation distillation and approximation.

3.1 Overall Framework of Knowledge Distillation for MSR

Most existing MSR methods adhere to an Encoding–Utilization paradigm, as shown in Fig. 2. For simplification and clarity, we use the symbol a in place of either a or b when referring to a specific modality type in the following text.

Feature Encoding. Item ID features v are initially represented as integer values and can be converted into low-dimensional, dense real-value vectors $\mathbf{e}_v$ using a randomly initialized embedding table. Modality feature encoding comprises two key modules: (1) the feature representation module, which represents the original features with vectors $a \rightarrow \mathbf{m}_a$ via modality encoders [9,38] pretrained on extensive datasets; and (2) the embedding initialization module, which adapts these pretrained representations $\mathbf{m}_a \rightarrow \mathbf{e}_a$ for recommendation tasks.

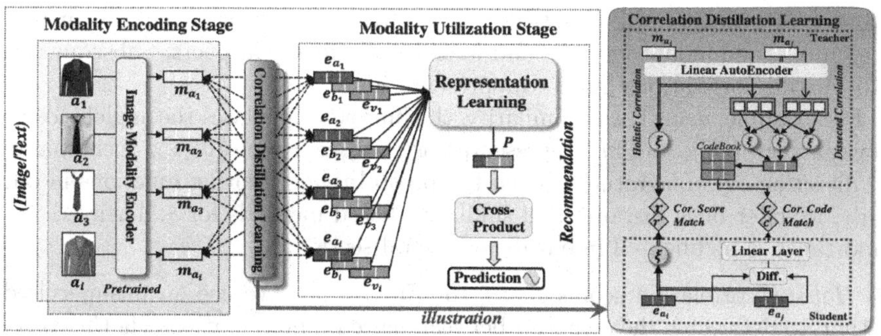

Fig. 2. (left) Our KDSR framework. Green, orange, and blue denote image modality, text modality, and ID features, respectively. (right) Detail of Correlation Distillation Learning on the image (green) modality. (Color figure online)

Feature Utilization. Several prior studies have focused on designing networks to model modality feature interactions and generate the overall sequence representation $\mathbf{P}$. Using the *representation learning* function g, this is given by:

$$\mathbf{P} = g((\mathbf{e_{v_1}}, \mathbf{e_{a_1}}, \mathbf{e_{b_1}}), (\mathbf{e_{v_2}}, \mathbf{e_{a_2}}, \mathbf{e_{b_2}}), ..., (\mathbf{e_{v_m}}, \mathbf{e_{a_m}}, \mathbf{e_{b_m}})) \quad (1)$$

Prediction. To match the learned sequence representation $\mathbf{P}$ with candidate items $\mathbf{e}_v$, we adopt the wide-used dot product $\hat{y} = \mathbf{P} \cdot \mathbf{e}_v^\top$. During training, the model measures the differences between the ground-truth y and the predicted score $\hat{y}$ using the cross-entropy loss [21], denoted as $\mathcal{L}^{RS}(y, \hat{y})$.

Knowledge Distillation. As in the introduction, during training, $\mathbf{e}_a$ tends to deviate from $\mathbf{m}_a$, risking the loss of knowledge from the original feature a. To solve it, we employ a teacher–student paradigm for Correlation Distillation Learning (CDL), as illustrated in the central portion of Fig. 2 (l). The teacher model, during the feature encoding phase, distills correlations among item modalities and supervises the student model in the subsequent feature utilization stage. Formally, it predicts item pair correlations ($T(v_i, v_j) = r_{ij}$) relative to their modality representations, while the student model forecasts $S(v_i, v_j) = \hat{r}_{ij}$ using fine-tuned modality embeddings. CDL's targets to minimize the disparity between r_{ij} and $\hat{r}_{ij}$, which we detail next.

3.2 Bridging the Gap: Correlation Distillation Learning

As shown in Fig. 2 (right), we employ both coarse- and fine-grained supervision signals. The teacher distills both signals, and the student approximates them.

Modality Correlation Distillation. The original m_i derived from the modality encoder is often high-dimensional and laden with redundant data. For discerning refined and distinctive features from their vector representations, we employ a linear autoencoder $\tilde{\mathbf{m}}_i = AE(\mathbf{m}_{a_i})$ to convert $\mathbf{m}$ into a condensed $\tilde{\mathbf{m}}$ first to eliminate redundant features and focus on distinctive aspects. Subsequently, CDL distills correlations into two representations.

For example, using cosine similarity, the function calculates the angle's cosine between vectors m_i and m_j, represented as $\xi(m_i, m_j) = \cos(\theta_{m_i,m_j})$, emphasizing direction over magnitude. Given the pivotal role of scoring functions in our study, we offer both distance and similarity-based methods for ξ. Their critical importance is highlighted in our experimental section.

(1) Holistic correlation scores $T_s : \xi(\tilde{\mathbf{m}}_i, \tilde{\mathbf{m}}_j) \rightarrow r_{ij}$. We use an unsupervised method to determine the disparity between paired representations via the correlation scoring function ξ, yielding a continuous value r. For instance, with cosine similarity, $\xi(\mathbf{m}_i, \mathbf{m}_j) = \cos(\theta_{m_i,m_j})$ computes the cosine of the angle between vectors $\mathbf{m}_i$ and $\mathbf{m}_j$, prioritizing direction over magnitude. Given the pivotal role of the scoring function in our study, our experimental results showcase the significance of both distance and similarity-based approaches to ξ.

(2) Dissected correlation codes $T_c : \pi(\tilde{\mathbf{m}}_i, \tilde{\mathbf{m}}_j) \rightarrow c_{ij}$. Considering just vector-wise operations ξ may be overly simplistic; we distill fine-grained discrete correlation signals c through a more detailed process π. Given that modality vectors capture diverse features like color and shape, an in-depth approach is also vital to model their intrinsic relationships. For example, the pairs (`black-suit`, `black-tie`) and (`black-suit`, `yellow-suit`) might both exhibit high correlation, but their attributes differ. Therefore, pinpointing these unique correlation patterns into distinct codes is essential for such generalization. This process involves two main steps: correlation vectorization and vector quantization.

In the first step, rather than representing the correlation with a single scalar value r_{ij}, we vectorize their correlations in a vector $\mathbb{r}_{ij}$. Drawing inspiration from Product Quantization [18], we treat each sub-vector from $\tilde{\mathbf{m}}$ as an individual, meaningful segment (where $\tilde{\mathbf{m}}$ from the autoencoder offers distinct, significant features). Specifically, we split $\tilde{\mathbf{m}}_i$ into D sub-vectors, represented as $[\tilde{\mathbf{m}}_{i,1}; ...; \tilde{\mathbf{m}}_{i,D}]$. Then we use the scoring function ξ to measure the correlations for each sub-vector in $\tilde{\mathbf{m}}_i$ against its counterpart in $\tilde{\mathbf{m}}_j$. These scores are then concatenated to form correlation vectors as $\mathbb{r}_{ij} = [\xi(\tilde{\mathbf{m}}_{i,1}, \tilde{\mathbf{m}}_{j,1}); ...; \xi(\tilde{\mathbf{m}}_{i,D}, \tilde{\mathbf{m}}_{j,D})]$. As such, it provides a more nuanced set of relationships between pairs of sub-vectors within the vectors.

In the second step, to condense the information and encapsulate the correlation pattern, we use Vector Quantization (VQ) [15] to convert the correlation vectors $\mathbb{r}_{ij}$ into correlation codes c_{ij}. This approach retains signal while accommodating for minor errors in the imperfect alignment between the encoder's modality spaces and the SR embedding space. We start by generating a codebook composed of x vectors, where x is a predetermined number of codes. Each *codeword* in this codebook denotes a distinct correlation type. They are initialized randomly, and then optimized with training data to better match the vector distribution of $\mathbb{r}$. Each vector is then mapped to its nearest codeword, based on Euclidean distance (a common practice for quantized code learning [15]). This process transforms $\mathbb{r} \in \mathbb{R}^{N \times N \times D}$ to $c \in \mathbb{R}^{N \times N \times 1}$, where each entry signifies an index value between 1 and x, representing the closest codeword. While there are other quantization methods which we tried (see Sect. 4.2), this method validated well empirically.

With the aforementioned steps, similar correlation patterns are mapped to the same codeword. This process also distinguishes subtle patterns, such as *"similar color, different shape"* versus *"similar shape, different color"*, through distinct codewords, highlighting more nuanced relationships.

Distilled Correlation Approximation. To ensure correlation consistency between embeddings E in the modality utilization stage and modality representation M in the modality encoding stage, the student network needs to mirror the outcomes of the more complex teacher model. Specifically, we assess item correlation through holistic proximity and directional relations, which directly pair with holistic correlation scores T_s and dissected correlation codes T_c.

(1) *Holistic* correlation learning. In the student model, the correlation between items v_i and v_j is derived from their fine-tuned embeddings $\mathbf{e}_{a_i}$ and $\mathbf{e}_{a_j}$:

$$\hat{r}_{ij} = \xi\left(g_\phi(\mathbf{e}_{a_i}), g_\phi(\mathbf{e}_{a_j})\right) \tag{2}$$

where g_ϕ is a transformation linear layer. In recommendation settings, item similarity is not as crucial as in image classification tasks. Thus, we adjust the sensitivity of modality supervision by tweaking the temperature parameter τ in a soft match equation, where the loss function is defined as:

$$\mathcal{L}_S^{KD} = \frac{1}{N^2} \sum_{i=1,j=1}^{N,N} ||\sigma\left(r_{ij}/\tau\right), \sigma\left(\hat{r}_{ij}/\tau\right)||_{l2} \tag{3}$$

Here r and r' are predictions from the teacher and student models, respectively, and $l2$ is the squared error loss. Raising τ smooths the probability distribution, reducing disparities. Conversely, decreasing the temperature sharpens the distribution, amplifying the distinctions between probabilities.

(2) *Dissected* correlation learning. In the teacher model, VQ is employed to differentiate correlation patterns. However, it also incurs increased complexity. Alternatively, we use a simpler, computationally-efficient direct subtraction mechanism to simulate inter-modal correlation patterns in student models; i.e., terming different correlation facets as directional relations in embedding spaces. Specifically, for any item pair, the student model first determines their directional relationship via a subtraction vector, subsequently predicting a probability distribution p across x classes. This prediction logit p aligns with the correlation code probabilities supplied by the teacher model. Mathematically,

$$\tilde{e}_{ij} = |\mathbf{e}_{a_i} - \mathbf{e}_{a_j}|, \quad p = W_{out}(ReLU(W_c(\tilde{e}_{ij}))) \tag{4}$$

Here, $W_c \in \mathcal{R}^{d \times d}$ and $W_{out} \in \mathcal{R}^{d \times x}$ are two linear transformation weight matrices, and $ReLU$ is the activation function. The teacher's output c_{ij} is expressed as a binary distribution, where the index of the predicted correlation code corresponds to 1; otherwise, 0. The loss function for learning can then be formulated as the cross-entropy loss between the student's and teacher's output distributions:

$$\mathcal{L}_C^{KD} = -\frac{1}{N^2} \sum_{i=1,j=1}^{N,N} \sum_{t=1}^{x} 1(c_{ij} = t) \log (p_{ij,t}) \tag{5}$$

3.3 Prediction and Training Strategy

As depicted in Fig. 2, sequences of image, text, and ID embeddings—denoted as $\{\mathbf{e}_{a_i}\}$, $\{\mathbf{e}_{b_i}\}$, and $\{\mathbf{e}_{v_i}\}$—are channeled into representation learning modules. For instance, we utilize MMSR [16] as our representation learning module, leveraging a heterogeneous GNN for information exchange. Initializing $\mathbf{e}_a$, $\mathbf{e}_b$, and $\mathbf{e}_v$ as distinct node types and using sequential transitions as edges, we obtain a sequence of hidden states $[\mathbf{h}_1, ..., \mathbf{h}_m]$ after l layers of graph propagation.

The final sequence representation $\mathbf{P}$ integrates both long- and short-term interests. For the long-term ones, the mean is calculated as $h^l = avg(\mathbf{h}_i)_{i=1}^m$. For the short-term interests, we concentrate on the sequence's last q items in the sequence, applying attention mechanisms. The final $\mathbf{P}$ is calculated by:

$$\alpha_i = W_2(ReLU(W_1 \mathbf{h}_i)),\ h^s = \sum_{i=m-q+1}^{m} (\alpha_i \mathbf{h}_i),\ \text{and} \quad \mathbf{P} = W_3[h^s; h^l], \tag{6}$$

with transformation matrices $W_1 \in \mathbb{R}^{d \times d}$ and $W_2 \in \mathbb{R}^{d \times 1}$, the attention weights, $\alpha \in \mathbb{R}^{q \times 1}$, emphasize recent items. Combining short-term interest h^s with long-term h^l ones yields the final sequence representation $\mathbf{P}$ and the matrix $W_3 \in \mathbb{R}^{2d \times d}$ projects this combined representation.

Finally, to supervise the embedding learning using correlation, we combine the supervision loss with prediction loss, resulting in the training objective:

$$\mathcal{L} = \mathcal{L}^{RS}(y, \hat{y}) + \lambda_1 \mathcal{L}_S^{KD} + \lambda_2 \mathcal{L}_C^{KD} \tag{7}$$

Asynchronous Training. We also explore adaptation operations that prioritize preserving the original domain's information while training new domain parameters. Specifically, we use an asynchronous update training strategy [55] for modality embedding parameters and other representation learning parameters. The learning rates for embedding parameters begin at 0.1ϵ and linearly increase to ϵ over the first η epochs, with ϵ being the learning rate for other parameters and η is the pre-set parameter for stopping the asynchronous training. The effectiveness of this training strategy is validated in our ablation study, detailed in Sect. 4.2.

3.4 Complexity Analysis

Compared to standard two-stage methods, our approach comes with the additional cost of CDL. Using efficient scoring functions like cosine similarity (costing $O(d)$ with d-dimensional vectors for each item pair), CDL adds a cost of $O(L^2 d)$

per sequence during training, where L denotes the average sequence length. Compared to backbones such as SASRec, which uses transformer cores and requires $O(L^2dH)$ (with H being the number of the attention head) for representation learning, CDL adds an additional computational increase of $1/(1+H)$, largely negligible compared to end-to-end training costs. In end-to-end training, the main cost comes from tuning the modality encoder, which has a very large parameter size (seen in Sect. 4.3), compared to the parameter size for training SR. During the inference phase, the student model operates independently, so its computational complexity is **the same** as standard two-stage models.

4 Experiment

Datasets. Consistent with prior MSR work [11,54], we used the Amazon review [10] and MovieLens [8] datasets for our experiments, both of which provide item descriptions and images. For Amazon data, to ensure our approach's versatility in various scenarios, we selected four broad per-category datasets, namely Beauty, Clothing, Sport, and Toys. In these datasets, each review rating (of products or movies) signifies a positive user–item interaction, as commonly accepted in prior studies [11,12,54]. To ensure a fair comparison with existing methods, we implemented core-5 filtering, a method that iteratively filters the dataset until each user and item has at least five interaction records (Table 1).

Table 1. Dataset statistics after preprocessing.

	Beauty	Clothing	Sports	Toys	ML-1M
User #	22,363	39,387	35,598	19,412	6,040
Item #	12,101	23,033	18,357	11,924	3,416
Inter. #	198,502	278,677	296,337	167,597	999,611
Avg. Len. #	8.88	7.12	8.46	8.79	165.50
Sparsity	99.93%	99.97%	99.95%	99.93%	95.16%

Baseline Models. We select three baseline categories for comparison. **(A)** ID-based SR models: GRU4Rec [13], using Gated Recurrent Units; SASRec [21], employing self-attention; and SR-GNN [49], a graph-based model considering user-item and item-item relations. **(B)** MSR models with frozen features: MMSRec [43], which uses features as supervised signals, and modified SASRec models (SASRM) integrating static modality and ID features for representation learning. **(C)** MSR models with fine-tuned features: NOVA [27] and DIF-SR [51], applying non-invasive fusion methods, and Trans2D [42] and MMSR [16], which employ holistic fusion to merge items' modality and ID features.

Evaluation Protocol. Following conventional SR evaluations, we split each user's sequence: the first 80% for training and the last 20% for testing, where each user has at least one test data point. Performance is gauged using two common ranking metrics: hit ratio (HR, or H@k) and mean reciprocal rank (MRR, or M@k), with higher values denoting superior efficacy in top k predictions.

Parameter Settings. For a fair comparison, all models use a 128-dimensional embedding and 512 batch size. Learning rates spanned logarithmically from 10^{-1} to 10^{-4}, and L_2 regularization ranged from $\{0, 10^{-5}\}$. Dropout ratios were set from 0 to 0.9. The Adam optimizer [22] was used. Each test was averaged over five runs for consistency. τ is set to 1.5. We set the optimized quantization split $D = 8$ and code number $x = 100$. We adjusted the (λ_1,λ_2) settings to (1, 0.5) for the Amazon dataset and (0.5, 0.1) for the ML-1M dataset.

4.1 Performance Comparison

As seen in Table 2, even when leveraging the best baseline as the representation learning module, our KDSR consistently surpasses top benchmarks.

Table 2. Overall performance (%). Bold denotes peak average; underline highlights best baselines. † marks KDSR's statistical significance (p-value < 0.05) against the best baseline. KDSR builds on the best per-dataset baseline as the backbone, marked as $\star$ (MMSR) or $\diamond$ (SASRM).

	Metric	GRU4Rec	SASRec	SR-GNN	MMSRec	SASRM	NOVA	DIF-SR	Trans2D	MMSR	KDSR
Beauty	H@5	5.6420	6.1900	4.1483	5.2851	6.3078	4.2219	6.5789	6.0191	7.1563	**7.4156**†*
	M@5	3.1110	3.2165	2.2123	3.0391	3.5983	2.1785	4.0735	3.4387	4.4429	**4.6355**†*
	H@20	12.7217	14.0681	10.2351	11.9568	13.7529	10.7978	14.0137	13.2214	14.1470	**14.675**†*
	M@20	3.7714	3.9668	2.7911	3.6539	4.2998	2.8160	4.7983	3.9460	5.0433	**5.2444**†*
Cloth.	H@5	1.3340	1.5885	0.8547	1.3995	1.9660	1.2937	1.5524	1.3929	1.8684	**2.4368**†*
	M@5	0.6765	0.7820	0.4555	0.7841	0.9992	0.6503	0.7961	0.6682	1.1365	**1.1647**†*
	H@20	3.8111	3.9574	2.7528	3.8329	4.8163	3.4866	4.0571	4.0683	4.4136	**4.9499**†*
	M@20	0.9418	1.0339	0.6251	1.0384	1.2683	0.8783	1.0530	1.0391	1.3344	**1.3851**†*
Sport	H@5	2.4388	2.9549	2.0742	2.8996	3.0306	2.1539	2.5145	2.7168	3.2657	**3.5008**†*
	M@5	1.2696	1.5858	1.0790	1.5391	1.6625	1.1271	1.3469	1.4235	1.9846	**1.9934***
	H@20	6.6430	7.2208	5.4376	6.8258	7.3683	5.8062	7.0774	6.9453	7.7466	**7.8126**†*
	M@20	1.6947	2.0357	1.4349	1.8650	2.0682	1.5648	1.9214	1.7058	2.2826	**2.3066***
Toys	H@5	3.8663	5.0902	2.7329	4.1151	4.6208	3.7899	5.2363	4.1908	6.1159	**6.6027**†*
	M@5	2.0022	2.7536	1.4878	2.5188	2.7823	1.9641	3.1944	2.2370	3.8987	**4.2785**†*
	H@20	10.0727	11.8668	6.7452	9.9061	9.5824	9.0609	12.0284	10.5082	12.1192	**12.3361**†*
	M@20	2.7267	3.4228	1.8655	3.0021	3.2572	2.4502	3.8777	2.9298	4.3551	**4.7925**†*
ML-1M	H@5	12.0854	10.6164	10.0049	14.0942	15.4602	13.8782	14.2002	13.2905	12.5701	**15.5644**$^{\dagger\diamond}$
	M@5	6.3590	5.1311	5.1460	7.6364	8.8079	7.5799	7.4881	7.1383	6.6671	**8.8109**$^{\dagger\diamond}$
	H@20	29.3807	29.1878	25.8203	31.2147	32.9212	31.2121	33.1632	30.844	30.0923	**33.1966**$^{\dagger\diamond}$
	M@20	7.9991	6.8462	6.6340	9.4173	10.4825	9.2362	9.3023	8.3520	8.3287	**10.4995**$^{\dagger\diamond}$

Among ID-based models, SASRec stands out, emphasizing the effectiveness of attention mechanisms. Among MSR models incorporating modality features, the fine-tuned modality approach (MMSR) works better on datasets with shorter sequences (like Beauty, Sport, and Toy). However, for longer sequences as in the ML-1M dataset, the frozen modality method (SASRM) prevails. This underscores that the fine-tuned approach can capture the inherent modal information in shorter sequences, but maintaining this information becomes problematic in longer sequences, rendering the frozen feature approach more favorable. Meanwhile, in fusion-based models, holistic methods [16] like MMSR or Trans2D outperform early/late fusion approaches [16] like NOVE and DIF-SR on shorter Amazon datasets, with MMSR being particularly effective. However, for longer sequences, the advantage shifts away from these holistic methods. This indicates that fusing modality information into items *during* sequential modeling benefits shorter sequences, but it can complicate and diminish results for longer ones.

For our KDSR, there was a notable increase in HR (average of ↑10.2%), compared to a modest rise in MRR (average of ↑3.4%). This was particularly evident in the Sport and ML-1M datasets. The difference indicates that while harnessing original modality correlation can boost recommendation overall precision (i.e., HR), its influence on ranking improvement (i.e., MRR) might be limited.

Compatibility Study. Observing that optimal representation varied across learning backbones for tasks with long and short sequences, we further explored CDL's compatibility with different backbones, as shown in Fig. 3.

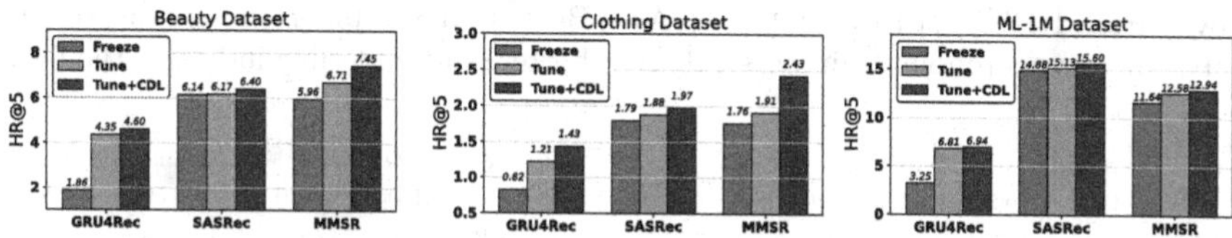

Fig. 3. Compatibility study with different representation learning backbones.

Across the three *representation learning* backbones, a clear trend emerges: *frozen* embeddings are less effective than *tuning*, and our CDL enhances embedding tuning adaptability. For GRU4Rec, frozen embeddings yield notably lower scores, given the recurrent models' demand for more adaptive embedding inputs. In contrast, SASRec, which concentrates on using attention scores to signify pairwise correlations among items, exhibits the least fluctuation in training strategies among the three backbones, as even original representations yield meaningful attention calculation. MMSR shines on the Clothing dataset with the shortest average sequence, but falters with longer sequences from ML-1M. In ML-1M, although all backbones improve slightly with CDL, the gains are modest. This could be because supervising correlations in lengthy sequences may apply overly tight constraints, considering the exponential growth of item correlation pairs with sequence length, which in turn can impede effective embedding learning in recommendation systems.

4.2 Modality Correlation Study

Ablation Within the CDL Module. To investigate the efficacy of components in the student model, we conducted an ablation study, as illustrated in Table 3.

The soft match in KDSR, when compared to a hard match (w/o τ), shows improved performance. This improvement is more evident in H@20 but less in H@5, especially in the Beauty dataset where the soft match actually decreases performance. This suggests that a hard match might compromise recommendation diversity, increasing overall hit rates but negatively affecting the top recommendations. Using single correlation supervision signals, the holistic correlation score signal (w/o $\mathcal{L}_C^{KD}$) outperforms the dissected correlation code (w/o $\mathcal{L}_S^{KD}$). Yet, combining dissected and holistic signals further enhances performance compared to using holistic ones solely, particularly in H@20. However, there is a slight drop in H@5 for the Sports dataset. This implies that holistic signals mainly boost recommendation accuracy, while dissected signals add to diversity, as more fine-grained guidance. As diversity increases, performance under broader top rankings will increase more. Last but not least, omitting asynchronous training (AT) or the dual supervision signals mildly impacts performance. We believe they have interdependent roles in preserving the original modality information.

Table 3. Ablation Study with modules removed from KDSR, in three datasets. AT denotes the asynchronous training strategy.

Model	Beauty		Clothing		Sport	
	H@5	H@20	H@5	H@20	H@5	H@20
KDSR	7.416	14.675	2.437	4.950	3.501	7.813
w/o τ	7.460	14.340	2.369	4.861	3.466	7.422
w/o $\mathcal{L}_C^{KD}$	7.407	14.292	2.367	4.860	3.470	7.641
w/o $\mathcal{L}_S^{KD}$	7.395	14.215	2.291	4.853	3.422	7.715
w/o $\mathcal{L}_{C\&S}^{KD}$	7.045	13.951	2.269	4.716	3.391	7.623
w/o AT	7.394	14.493	2.388	4.852	3.418	7.709

Table 4. Quantization Functions over Beauty datasets. *Italics* indicate the better, and <u>underline</u> indicates the best.

ξ	VQ		k-Means	
	H@5	H@20	H@5	H@20
Euc	7.121	*14.657*	*7.201*	14.631
Man	*7.174*	*14.569*	7.165	14.537
Dot	7.117	*14.519*	*7.118*	14.507
Cos	<u>*7.245*</u>	<u>*14.752*</u>	7.224	14.743

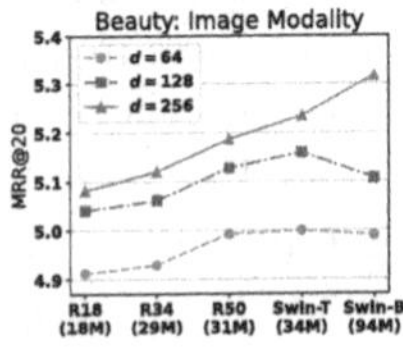

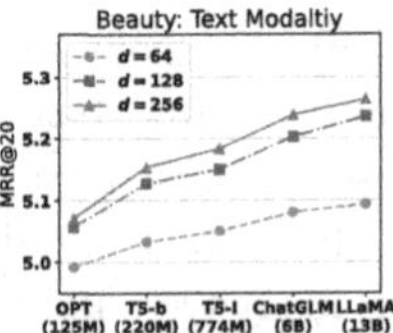

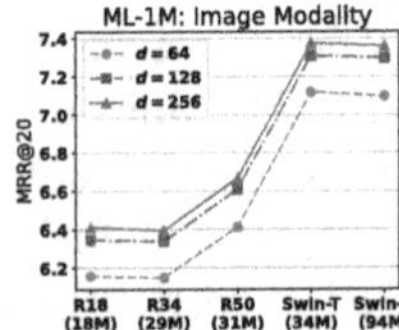

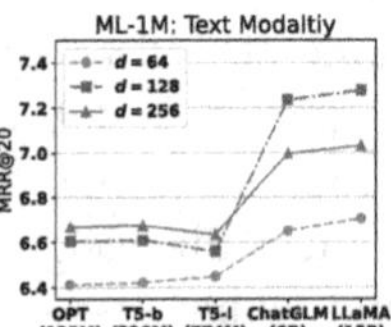

Fig. 4. Evaluation on two datasets equipped with MMSR backbone and different modality encoders (image and text) with various parameter sizes. R* indicates the ResNet [9] with different sizes. Swin-T/B are the transformer-based models [33] in two sizes. T5-b/l indicates the base and large version of T5 [38]. ChatGLM [53] and LLaMA-13B [46] are two recently introduced LLMs.

Correlation Distillation Study. Capturing the correlation supervision signal is pivotal in our method, and we examine this through two lenses: quantization methods (Vector Quantization and k-Means) and scoring functions (Euclidean/ Manhattan distance, and Dot/Cosine similarity).

In evaluations using H@5 and 20 (Table 4), VQ and k-Means have distinct performances. While both are unsupervised quantization approaches, VQ notably surpasses k-Means in noise resilience. We attribute this to VQ's adaptable codebook training from scratch, contrasting with k-Means' reliance on random initial centroid placements. The inherent refined VQ representations thrive in a wider range of contexts, evidenced by VQ's leading H@20 scores. Conversely, in scenarios like H@5 where precision is crucial, k-Means may occasionally (*Euc* and *Dot*) yield more precise outcomes. Meanwhile, cosine similarity emerges as the most compatible scoring function for both VQ and k-Means. These results may arise from cosine similarity's resilience with sparse data, as distance-based metrics can be swayed by noise. Furthermore, the distinction between Cosine and Dot arises from the dot product's sensitivity to vector magnitudes, making it prone to distortions from noise and outliers. Conversely, cosine similarity is normalized, which presents more consistent and robust results.

4.3 Influence of Modality Encoder Sophistication

As modality encoders become more sophisticated in terms of parameters, they become better equipped to capture nuanced modality features. To assess our method's proficiency in grasping correlations amid these refined features, we executed controlled experiments (in Fig. 4), tweaking the encoder for each modality.

Overall, for both long (ML-1M) and short (Beauty) sequence tasks, modality encoders with richer parameters extract more information, generating more accurate dissected correlations resulting in improved embedding learning. If a modality's embedding dimension (d) is too small, it might struggle to both represent the original modality and support sequential modeling. For instance, in the Beauty dataset discussing image encoders, when the embedding dimension $d = 64$ or 128, overall performance declines slightly as the encoder parameter size increases. But this does not happen when $d = 256$. The optimal modality embedding size varies across scenarios. In ML-1M, for text representations extracted by LLM (i.e., ChatGLM or LLaMA), the embedding dimension $d = 256$ performs worse than $d = 128$. This is likely because longer sequences will cause an exponential increase in item relation pairs. This will lead to more numerous correlation supervisions which may result in the embedding overfitting to the CDL task, leading to suboptimal outcomes.

5 Conclusion

As encoded modality features remain underleveraged in two-stage SRs, and current modality encoders grow in complexity, we introduce a lightweight adaptation strategy. Concerning modality forgetting during SR training, we leverage self-supervised knowledge distillation. It first comprehensively distills both fine- and coarse-grained correlation signals from encoded modality representations, and then enhances embedding learning through these signal supervision. Experiments underscore that while holistic correlations augment accuracy, dissected correlations enhance diversity. Together they ensure a balanced recommendation.

In this study, we present an efficient approach to distill knowledge from the output of the modality encoder's last layer, leaving room for integrating deeper encoder layers in future research. On the technical front, our current distilled knowledge is in the format of triplets, i.e., (v_i, v_j, r_{ij}). Further adopting structured distillation formats, like graphs, may also be a compelling next step. Additionally, while we emphasize intra-modality supervision, we believe it is promising to study joint supervision across varied modality types, enhancing multi-modal SR tasks.

References

1. Cao, Y., et al.: Semi-supervised knowledge distillation for tiny defect detection. In: 2022 IEEE 25th International Conference on Computer Supported Cooperative Work in Design (CSCWD), pp. 1010–1015. IEEE (2022)
2. Chen, X., Cao, Q., Zhong, Y., Zhang, J., Gao, S., Tao, D.: Dearkd: data-efficient early knowledge distillation for vision transformers. In: Proceedings of the IEEE/CVF Conference on Computer Vision and Pattern Recognition, pp. 12052–12062 (2022)
3. Chen, X., Zhang, Y., Xu, H., Qin, Z., Zha, H.: Adversarial distillation for efficient recommendation with external knowledge. ACM Trans. Inf. Syst. (TOIS) **37**(1), 1–28 (2018)
4. Cohen, I., et al.: Pearson correlation coefficient. Noise Reduct. Speech Process. 1–4 (2009)
5. Covington, P., Adams, J., Sargin, E.: Deep neural networks for youtube recommendations. In: Proceedings of the 10th ACM Conference on Recommender Systems, pp. 191–198 (2016)
6. Elsayed, S., Brinkmeyer, L., Schmidt-Thieme, L.: End-to-end image-based fashion recommendation. In: Corona Pampin, H.J., Shirvany, R. (eds.) RECSYS 2022, vol. 981, pp. 109–119. Springer, Heidelberg (2022). https://doi.org/10.1007/978-3-031-22192-7_7
7. Gao, Q., Zhao, Y., Li, G., Tong, T.: Image super-resolution using knowledge distillation. In: Jawahar, C.V., Li, H., Mori, G., Schindler, K. (eds.) ACCV 2018. LNCS, vol. 11362, pp. 527–541. Springer, Cham (2019). https://doi.org/10.1007/978-3-030-20890-5_34
8. Harper, F.M., Konstan, J.A.: The movielens datasets: history and context. ACM Trans. Interact. Intell. Syst. (TIIS) **5**(4), 1–19 (2015)
9. He, K., Zhang, X., Ren, S., Sun, J.: Deep residual learning for image recognition. In: Proceedings of the IEEE Conference on Computer Vision and Pattern Recognition, pp. 770–778 (2016)
10. He, R., McAuley, J.: Ups and downs: modeling the visual evolution of fashion trends with one-class collaborative filtering. In: Proceedings of the 25th International Conference on World Wide Web, pp. 507–517 (2016)
11. He, R., McAuley, J.: VBPR: visual bayesian personalized ranking from implicit feedback. In: Proceedings of the AAAI Conference on Artificial Intelligence, vol. 30 (2016)

12. He, X., Deng, K., Wang, X., Li, Y., Zhang, Y., Wang, M.: LightGCN: simplifying and powering graph convolution network for recommendation. In: Proceedings of the 43rd International ACM SIGIR Conference on Research and Development in Information Retrieval, pp. 639–648 (2020)
13. Hidasi, B., Karatzoglou, A., Baltrunas, L., Tikk, D.: Session-based recommendations with recurrent neural networks (2016)
14. Hinton, G., Vinyals, O., Dean, J.: Distilling the knowledge in a neural network. arXiv preprint arXiv:1503.02531 (2015)
15. Hou, Y., He, Z., McAuley, J., Zhao, W.X.: Learning vector-quantized item representation for transferable sequential recommenders. In: Proceedings of the ACM Web Conference 2023, pp. 1162–1171 (2023)
16. Hu, H., Guo, W., Liu, Y., Kan, M.Y.: Adaptive multi-modalities fusion in sequential recommendation systems. In: Proceedings of the 32nd ACM International Conference on Information & Knowledge Management (2023)
17. Hu, H., Pan, L., Ran, Y., Kan, M.Y.: Modeling and leveraging prerequisite context in recommendation. In: Proceedings of the 16th ACM Conference on Recommender Systems, Context-Aware Recommender System Workshop (2022)
18. Jegou, H., Douze, M., Schmid, C.: Product quantization for nearest neighbor search. IEEE Trans. Pattern Anal. Mach. Intell. **33**(1), 117–128 (2010)
19. Jiao, X., et al.: Tinybert: distilling bert for natural language understanding. arXiv preprint arXiv:1909.10351 (2020)
20. Kang, W.C., Fang, C., Wang, Z., McAuley, J.: Visually-aware fashion recommendation and design with generative image models. In: Proceedings of the 2017 IEEE International Conference on Data Mining (ICDM), pp. 207–216. IEEE (2017)
21. Kang, W.C., McAuley, J.: Self-attentive sequential recommendation. In: Proceedings of the 2018 International Conference on Data Mining (ICDM), pp. 197–206. IEEE (2018)
22. Kingma, D.P., Ba, J.: Adam: a method for stochastic optimization (2015)
23. Koren, Y., Bell, R., Volinsky, C.: Matrix factorization techniques for recommender systems. Computer **42**(8), 30–37 (2009)
24. Lee, S.H., Kim, D.H., Song, B.C.: Self-supervised knowledge distillation using singular value decomposition. In: Proceedings of the European Conference on Computer Vision (ECCV), pp. 335–350 (2018)
25. Lee, Y., Jang, K., Goo, J., Jung, Y., Kim, H.: Fithubert: going thinner and deeper for knowledge distillation of speech self-supervised learning (2022)
26. Lian, D., Wang, H., Liu, Z., Lian, J., Chen, E., Xie, X.: Lightrec: a memory and search-efficient recommender system. In: Proceedings of The Web Conference 2020, pp. 695–705 (2020)
27. Liu, C., Li, X., Cai, G., Dong, Z., Zhu, H., Shang, L.: Noninvasive self-attention for side information fusion in sequential recommendation. In: Proceedings of the AAAI Conference on Artificial Intelligence, vol. 35, pp. 4249–4256 (2021)
28. Liu, D., Cheng, P., Dong, Z., He, X., Pan, W., Ming, Z.: A general knowledge distillation framework for counterfactual recommendation via uniform data. In: Proceedings of the 43rd International ACM SIGIR Conference on Research and Development in Information Retrieval, pp. 831–840 (2020)
29. Liu, F., Cheng, Z., Sun, C., Wang, Y., Nie, L., Kankanhalli, M.: User diverse preference modeling by multimodal attentive metric learning. In: Proceedings of the 27th ACM International Conference on Multimedia, pp. 1526–1534 (2019)
30. Liu, Q., Zhu, J., Dai, Q., Wu, X.: Boosting deep CTR prediction with a plug-and-play pre-trainer for news recommendation. In: Proceedings of the 29th International Conference on Computational Linguistics, pp. 2823–2833 (2022)

31. Liu, S., Chen, Z., Liu, H., Hu, X.: User-video co-attention network for personalized micro-video recommendation. In: The World Wide Web Conference, pp. 3020–3026 (2019)
32. Liu, Y., et al.: End-to-end speech translation with knowledge distillation (2019)
33. Liu, Z., et al.: Swin transformer: hierarchical vision transformer using shifted windows. In: Proceedings of the IEEE/CVF International Conference on Computer Vision, pp. 10012–10022 (2021)
34. Liu, Z., Ma, Y., Schubert, M., Ouyang, Y., Xiong, Z.: Multi-modal contrastive pre-training for recommendation. In: Proceedings of the 2022 International Conference on Multimedia Retrieval, pp. 99–108 (2022)
35. McAuley, J., Targett, C., Shi, Q., Van Den Hengel, A.: Image-based recommendations on styles and substitutes. In: Proceedings of the 38th International ACM SIGIR Conference on Research and Development in Information Retrieval, pp. 43–52 (2015)
36. Oramas, S., Nieto, O., Sordo, M., Serra, X.: A deep multimodal approach for cold-start music recommendation. In: Proceedings of the 2nd Workshop on Deep Learning for Recommender Systems, pp. 32–37 (2017)
37. Park, M., Lee, K.: Exploiting negative preference in content-based music recommendation with contrastive learning. In: Proceedings of the 16th ACM Conference on Recommender Systems, pp. 229–236 (2022)
38. Raffel, C.: Exploring the limits of transfer learning with a unified text-to-text transformer. J. Mach. Learn. Res. **21**(1), 5485–5551 (2020)
39. Rajasegaran, J., Khan, S., Hayat, M., Khan, F.S., Shah, M.: Self-supervised knowledge distillation for few-shot learning (2021)
40. Rendle, S., Freudenthaler, C., Gantner, Z., Schmidt-Thieme, L.: BPR: bayesian personalized ranking from implicit feedback. arXiv preprint arXiv:1205.2618 (2012)
41. Sanh, V., Debut, L., Chaumond, J., Wolf, T.: Distilbert, a distilled version of bert: smaller, faster, cheaper and lighter (2019)
42. Singer, U., et al.: Sequential modeling with multiple attributes for watchlist recommendation in e-commerce. In: Proceedings of the Fifteenth ACM International Conference on Web Search and Data Mining, pp. 937–946 (2022)
43. Song, K., Sun, Q., Xu, C., Zheng, K., Yang, Y.: Self-supervised multi-modal sequential recommendation. arXiv preprint arXiv:2304.13277 (2023)
44. Su, M., Gu, G., Ren, X., Fu, H., Zhao, Y.: Semi-supervised knowledge distillation for cross-modal hashing. IEEE Trans. Multimedia **25**, 28–35 (2021)
45. Tan, Y.K., Xu, X., Liu, Y.: Improved recurrent neural networks for session-based recommendations. In: Proceedings of the 1st Workshop on Deep Learning for Recommender Systems, pp. 17–22 (2016)
46. Touvron, H., et al.: Llama: open and efficient foundation language models. arXiv preprint arXiv:2302.13971 (2023)
47. Wei, Y., Wang, X., Nie, L., He, X., Hong, R., Chua, T.S.: Mmgcn: multi-modal graph convolution network for personalized recommendation of micro-video. In: Proceedings of the 27th ACM International Conference on Multimedia, pp. 1437–1445 (2019)
48. Wu, C., Wu, F., Qi, T., Huang, Y.: Empowering news recommendation with pre-trained language models. In: Proceedings of the 44th International ACM SIGIR Conference on Research and Development in Information Retrieval, pp. 1652–1656 (2021)
49. Wu, S., Tang, Y., Zhu, Y., Wang, L., Xie, X., Tan, T.: Session-based recommendation with graph neural networks. In: Proceedings of the AAAI Conference on Artificial Intelligence, vol. 33, pp. 346–353 (2019)

50. Xia, X., Yin, H., Yu, J., Wang, Q., Xu, G., Nguyen, Q.V.H.: On-device next-item recommendation with self-supervised knowledge distillation. In: Proceedings of the 45th International ACM SIGIR Conference on Research and Development in Information Retrieval, pp. 546–555 (2022)
51. Xie, Y., Zhou, P., Kim, S.: Decoupled side information fusion for sequential recommendation. In: Proceedings of the 45th International ACM SIGIR Conference on Research and Development in Information Retrieval, pp. 1611–1621 (2022)
52. Yuan, Z., et al.: Where to go next for recommender systems? id-vs. modality-based recommender models revisited. arXiv preprint arXiv:2303.13835 (2023)
53. Zeng, A., et al.: GLM-130b: an open bilingual pre-trained model. In: Proceedings of the Eleventh International Conference on Learning Representations (ICLR) (2023)
54. Zhang, J., Zhu, Y., Liu, Q., Wu, S., Wang, S., Wang, L.: Mining latent structures for multimedia recommendation. In: Proceedings of the 29th ACM International Conference on Multimedia, pp. 3872–3880 (2021)
55. Zhang, S., Choromanska, A.E., LeCun, Y.: Deep learning with elastic averaging SGD. Adv. Neural. Inf. Process. Syst. **28**, 685–693 (2015)
56. Zhang, Y., Xu, X., Zhou, H., Zhang, Y.: Distilling structured knowledge into embeddings for explainable and accurate recommendation. In: Proceedings of the 13th International Conference on Web Search and Data Mining, pp. 735–743 (2020)
57. Zhao, B., Cui, Q., Song, R., Qiu, Y., Liang, J.: Decoupled knowledge distillation. In: Proceedings of the IEEE/CVF Conference on Computer Vision and Pattern Recognition, pp. 11953–11962 (2022)
58. Zhou, Y., Chen, H., Lin, H., Heng, P.-A.: Deep semi-supervised knowledge distillation for overlapping cervical cell instance segmentation. In: Martel, A.L., et al. (eds.) MICCAI 2020. LNCS, vol. 12261, pp. 521–531. Springer, Cham (2020). https://doi.org/10.1007/978-3-030-59710-8_51

Revealing the Hidden Impact of Top-N Metrics on Optimization in Recommender Systems

Lukas Wegmeth(✉), Tobias Vente, and Lennart Purucker

Intelligent Systems Group, University of Siegen, Siegen, Germany
{lukas.wegmeth,tobias.vente}@uni-siegen.de, purucker@cs.uni-freiburg.de

Abstract. The hyperparameters of recommender systems for top-n predictions are typically optimized to enhance the predictive performance of algorithms. Thereby, the optimization algorithm, e.g., grid search or random search, searches for the best hyperparameter configuration according to an optimization-target metric, like *nDCG* or *Precision*. In contrast, the optimized algorithm, e.g., *Alternating Least Squares Matrix Factorization* or *Bayesian Personalized Ranking*, internally optimizes a different loss function during training, like *squared error* or *cross-entropy*. To tackle this discrepancy, recent work focused on generating loss functions better suited for recommender systems. Yet, when evaluating an algorithm using a top-n metric during optimization, another discrepancy between the optimization-target metric and the training loss has so far been ignored. During optimization, the top-n items are selected for computing a top-n metric; ignoring that the top-n items are selected from the recommendations of a model trained with an entirely different loss function. Item recommendations suitable for optimization-target metrics could be outside the top-n recommended items; hiddenly impacting the optimization performance. Therefore, we were motivated to analyze whether the top-n items are optimal for optimization-target top-n metrics. In pursuit of an answer, we exhaustively evaluate the predictive performance of 250 *selection strategies* besides selecting the top-n. We extensively evaluate each *selection strategy* over twelve implicit feedback and eight explicit feedback data sets with eleven recommender systems algorithms. Our results show that there exist *selection strategies* other than top-n that increase predictive performance for various algorithms and recommendation domains. However, the performance of the top $\sim 43\%$ of *selection strategies* is not significantly different. We discuss the impact of our findings on optimization and re-ranking in recommender systems and feasible solutions. The implementation of our study is publicly available.

Keywords: recommender systems · re-ranking · optimization · autorecsys · hyperparameter · top-n · evaluation

1 Introduction

Top-n recommendations, i.e., recommending ranked item lists, are probably the most common task for recommender systems nowadays. To tackle this task,

N. Goharian et al. (Eds.): ECIR 2024, LNCS 14608, pp. 140–156, 2024.
https://doi.org/10.1007/978-3-031-56027-9_9

recommender systems developers often apply machine learning algorithms, e.g., nearest neighbor or matrix factorization approaches [37]. The models produced by such algorithms are then used to predict personalized ranked lists. These recommendations are then commonly evaluated in terms of predictive performance with optimization-target metrics like the *nDCG* or *Precision* [19,20,53,54]. This performance is influenced by the data set, the algorithm, and its hyperparameters. Hyperparameter optimization techniques like grid search, random search, or Bayesian optimization are commonly applied to improve recommendation performance by determining the best hyperparameter values for an algorithm [3,12,21,30,39,42,49,55,56].

No matter which optimization technique is applied, it is vital to correctly approximate the predictive accuracy of a set of hyperparameters, as this influences future decisions during the optimization or when deploying the model in production. However, the predictive performance according to optimization-target metrics is not necessarily optimized during the training of recommender systems. The research community knows this discrepancy between training loss metrics, e.g., *squared error*, and optimization-target metrics, like *nDCG*. Their problem is finding a training loss metric that accurately represents the optimization-target metric such that the trained weights are optimized correctly. To this end, recent work focused on empirically proven loss functions or learning better loss functions [25,38,48].

In theory, the aforementioned discrepancy also exists in evaluation metrics, e.g., *nDCG@10*. By definition, *nDCG@10* is the result of calculating the *nDCG* of the top 10 predicted items, ignoring all items ranked below the top 10. Metrics like *nDCG@10* strongly differ from training loss functions that evaluate model performance without the concept of a ranking or relevance threshold. Consequently, we hypothesize that, due to the aforementioned discrepancy between the metrics, evaluating only the top-n selection is insufficient to optimize for the highest predictive accuracy of the recommender system. In other words, evaluating only the top-n predicted items but ignoring all others might not always yield the highest possible predictive accuracy of a trained model.

Therefore, we aim to determine whether there are cases where selecting items other than the top-n results in higher predictive accuracy. If there were such cases, it would imply that there is a hidden impact of top-n metrics on optimization performance due to the assumption that evaluating the top-n recommended items results in the highest predictive accuracy w.r.t. the evaluated hyperparameters. The aforementioned implication motivated us to conduct the exploratory study that we present in this paper. In detail, our study aims to answer the following questions:

RQ1 Does the selection of items other than the top-n during the evaluation of recommender systems yield improved predictive accuracy for specific algorithms, domains, or data sets?

RQ2 If there are cases where selecting items other than the top-n improves predictive accuracy, is there a *significant* impact of top-n metrics on optimization?

To answer the research questions, we conduct a study of the performance of nine recommendation algorithms on twelve implicit feedback and eight explicit feedback data sets. **Our contribution** is the first large-scale study and discussion of the, so far, hidden impact of top-n metrics on optimization in recommender systems. Our results prove that there are cases where selecting items other than the top-n yields increased performance, answering **RQ1**. However, we indicate that the impact is likely insignificant, answering **RQ2**. Therefore, we reveal the hidden impact of top-n metrics on optimization in recommender systems and provide evidence that researchers do not have to worry about it being a confounding factor in the evaluation of recommender systems using traditional collaborative filtering algorithms. Our long-term goal is to clear doubts about hidden problems in evaluating recommender systems and to raise awareness in the community.

Our implementation is publicly available on our GitHub repository[1] and contains documentation for the reproducibility of our experiments.

2 Related Work

To the best of our knowledge, there exists no analysis of the quality of the top-n selection of ranked lists. However, our work is related to and motivated by recent work on hyperparameter optimization, training loss metrics, and re-ranking.

The optimization of hyperparameters requires some score that approximates the highest predictive accuracy of a model. If the score fails to do that, the optimization strategy runs into the risk of either optimizing for undesired criteria, e.g. in Bayesian optimization, or the results can not be interpreted correctly, e.g. in grid search and random search. Recommender systems research that reports evaluation scores usually obtains these scores on optimized hyperparameters [9,44,50]. Additionally, there are efforts to transfer automated machine learning techniques to recommender systems [2,8,23,27,45,57], and one of the core problems of automated machine learning is automated hyperparameter optimization [13].

Related to optimization, the discrepancy between training loss and optimization-target metrics has been explored by the recommender systems community [6,47]. The problem can be tackled either by engineering loss metrics that fit well to the desired accuracy metric [38,48] or by learning the loss metric automatically [25]. In contrast, we thoroughly analyze the validity of selecting only the top-n elements for evaluation.

To achieve secondary optimization goals that are different from increasing predictive accuracy, e.g., removing popularity bias, re-ranking techniques are commonly applied [1,15,22,28,34]. Re-ranking techniques assume that the underlying ranked list is predicted by an implicit feedback recommender system optimized for predictive accuracy. However, our analysis focuses on re-ranking predicted items during the optimization of a model to better approximate the

[1] https://code.isg.beel.org/scoring-optimizer.

predictive accuracy of a set of hyperparameters. Our approach is more similar to works that analyze strategies that randomly sample recommendations from the top predicted items [24], or define a relevance cutoff for recommended items [5].

The Probability Ranking Principle (PRP) [40] assumes that an optimal recommender algorithm ranks items in order of probability of relevance to the user [33]. In contrast, our work investigates whether a global ranking order of items is optimal on average for all users. Hence, if the best possible global ranking order of an algorithm's predictions contains exactly the top-n predicted items, then the PRP holds on average for all users. However, our work aims to research the impact of top-n metrics on optimization.

3 Method

At the core of this paper is the evaluation of different *selection strategies*. A recommender system predicts a ranked list of items P for each user. We define a *selection strategy* as an approach to select any n items from P, with $n \leq |P|$, for the evaluation of the recommender system's accuracy with a threshold-based metric like *nDCG@n*. Hence, the number of possible *selection strategies* is $\binom{|P|}{n}$. In the traditional evaluation of recommender systems, e.g. with *nDCG@10* (i.e., $n = 10$), the commonly used *selection strategy* is choosing the top 10 items from P. We call this *selection strategy* the *top-n selection strategy.* To illustrate an alternative to the *top-n selection strategy*, assume we randomly select 10 items from P and use this *non-top-n selection strategy* to compute the *nDCG@10.* Such an approach was, for example, proposed in research on the effectiveness of randomly sampling ranked items [24].

Our work focuses on finding a *non-top-n selection strategy* with a higher *nDCG@n* than the *top-n selection strategy.* Additionally, we define the subset K, with $K \subseteq P$, as the subset that contains the top-$|K|$ elements of P, which we need later.

The experiments presented in this paper were executed in a highly parallel manner on a cluster where each node has 256 GB RAM and a total of 64 cores from two AMD EPYC 7452 CPUs.

3.1 Selection Strategy Design Decisions

For our exploratory study, we set $|K| = 10$ and $n = 5$ to investigate the resulting 252 distinct *selection strategies.* In this paragraph, we provide our motivation for this choice. For context, the *top-n selection strategy* in this setting is choosing the top 5 predicted items from K. To answer **RQ1**, technically, we only need to show that there exists at least one *non-top-n selection strategy* with higher performance than the *top-n selection strategy* for any combination of $|K|$ and n. However, to sufficiently answer **RQ2**, we want to acquire more exhaustive results than for, e.g. $|K| = 2$ and $n = 1$, or $|K| = 11$ and $n = 10$. Further, evaluating the lowest-ranked items does not provide additional information, e.g., when $|K|$ is close to or equal to $|P|$. Moreover, if *selection strategies* that choose the lowest-ranked items perform well, the evaluated algorithm likely fails to learn correctly from the data. Additionally, the number of possible *selection strategies* increases exponentially with $|K|$ if it significantly differs to n. In the complete study, including data preprocessing, fitting, and predicting, the parameters $|K|$ and n have the most considerable influence on computational resource requirements for our exploratory study. With $|K| = 10$ and $n = 5$, analyzing one data set requires about one CPU year on average with our evaluation setup. Furthermore, $n = 5$ is commonly chosen for recommender systems evaluations. Considering all the above points, we selected $|K| = 10$ and $n = 5$ for this study.

3.2 Data Sets and Algorithms

We analyzed twelve implicit and eight explicit feedback data sets from six different domains with nine recommendation algorithms plus two baselines from two recommender systems libraries. Four of the recommendation domains, *shopping*, *music*, *movies*, and *articles*, are represented by three or more data sets each. The remaining data sets are from the domains *social* and *locations*. The implicit feedback data sets are: *Adressa One Week* [17], *Citeulike-a* [46], *Cosmetics-Shop*[2], *Globo* [31,41], *Gowalla* [10], *Hetrec-Lastfm* [7], *Nowplaying-rs* [36], *Retailrocket*[3], *Sketchfab*[4], *Spotify-Playlists* [35], *Yelp*[5], and *Yoochoose*[6]. The explicit feedback data sets are: *Amazon CDs&Vinyl* [32], *Amazon Musical Instruments* [32], *Amazon Video Games* [32], *CiaoDVD*[7], *Jester3* [16], *MovieLens-1M* [18], *MovieLens-100k* [18], and *MovieTweetings* [11]. For these explicit feedback data sets, we treat a rating that is $> 60\%$ of the maximum rating as an interaction according to standard practice [4,26,29]. Furthermore, we prune all data sets such that all users and items have at least five interactions, commonly called five-core pruning [43,51,52]. We do this to reduce the impact of cold start cases since the used algorithms can not predict cold start scenarios. Table 1 contains the data set statistics for the preprocessed data sets. We used the libraries *Implicit* [14] and *LensKit* [12] for their implementation of the recommender algorithms. We used all algorithms from these libraries that natively support implicit feedback for a total of nine algorithms plus two baselines. The algorithms from *Implicit* are *Alternating Least Squares*, *Logistic Matrix Factorization*, *Bayesian Personalized Ranking*, and *Item-Item Nearest Neighbors* with distance metrics *Cosine Similarity*, *TF-IDF*, and *BM25*. The algorithms from *LensKit* are *Implicit Matrix Factorization*, *User-User Nearest Neighbors*, *Item-Item Nearest Neighbors*, *Most Popular*, and *Random*.

[2] https://rees46.com/.
[3] https://www.kaggle.com/datasets/retailrocket/ecommerce-dataset.
[4] https://github.com/EthanRosenthal/rec-a-sketch.
[5] https://www.yelp.com/dataset.
[6] https://www.kaggle.com/datasets/chadgostopp/recsys-challenge-2015.
[7] https://guoguibing.github.io/librec/datasets.html.

3.3 Experimental Pipeline

We perform five-fold cross-validation by randomly splitting the interactions of each user into three separate sets with fixed sizes: training (60%), validation (20%), and test (20%). All recommenders were optimized on the validation set with random search for two hours. The range of hyperparameter values in the configuration space is set to be around the default values given by the library. Finally, we exhaustively evaluate all 252 *selection strategies* for both the validation and test data. We focus on and present the results from evaluating *selection strategies* on the test data. However, we also obtain the results from evaluating *selection strategies* on the validation data to analyze the generalization capabilities of *selection strategies*.

Table 1. Data set statistics after five-core pruning. Split between the implicit (first part) and explicit (second part) feedback data sets.

Name	#Interactions	#Users	#Items	Avg.#Int./User	Avg.#Int./Item	Sparsity	Domain
Adressa One Week	2,020,328	146,635	2,441	13.78	827.66	99.44%	Articles
Citeulike-a	200,180	5,536	15,429	36.16	12.97	99.77%	Articles
Cosmetics-Shop	4,949,482	230,248	44,009	21.5	112.47	99.95%	Shopping
Globo	2,482,163	157,926	11,832	15.72	209.78	99.87%	Articles
Gowalla	2,018,421	64,115	164,532	31.48	12.27	99.98%	Locations
Hetrec-Lastfm	71,355	1,859	2,823	38.38	25.28	98.64%	Music
Nowplaying-rs	2,447,318	64,392	95,277	38.01	25.69	99.96%	Music
Retailrocket	240,938	22,178	17,803	10.86	13.53	99.94%	Shopping
Sketchfab	547,477	25,655	15,274	21.34	35.84	99.86%	Social
Spotify-Playlists	8,718,742	15,146	337,256	575.65	25.85	99.83%	Music
Yelp	3,999,684	268,658	109,340	14.89	36.58	99.99%	Locations
Yoochoose	10,195,058	1,283,296	27,995	7.94	364.17	99.97%	Shopping
Amazon CDs&Vinyl	1,241,336	98,228	66,979	12.64	18.53	99.98%	Shopping
Amazon Musical Instruments	176,631	21,420	8,642	8.25	20.44	99.9%	Shopping
Amazon Video Games	362,996	40,793	14,497	8.9	25.04	99.94%	Shopping
CiaoDVD	23,467	1,582	1,788	14.83	13.12	99.17%	Movies
Jester3	777,118	34,884	140	22.28	5,550.84	84.09%	Social
MovieLens-1M	835,789	6,038	3,307	138.42	252.73	95.81%	Movies
MovieLens-100k	81,697	943	1,203	86.64	67.91	92.8%	Movies
MovieTweetings	563,309	20,643	8,810	27.29	63.94	99.69%	Movies

4 Results

Our results are presented in the following order. We first show **aggregated** evaluation results of all twelve implicit and eight explicit data sets with all eleven algorithms. Next, we present a closer look using a **domain-specific** subset of the results. Zooming further into this subset, we provide a more detailed, **data set-specific**, view of the behavior of different *selection strategies*. Moreover, we analyze the **generalization capabilities** of *selection strategies*. Finally, we perform **statistical tests** on the significance of *selection strategies* to understand the impact of potential solutions. All results are split between implicit and explicit feedback data sets to compare them.

Though we aggregated results as much as possible, we can only show exemplary results for some parts of the analysis in this paper due to space constraints. The results we can not show in this paper do not provide any additional insight and do not contradict the presented results. For example, all data presented in this paper feature results exclusively on the *nDCG* metric. However, we also performed the same evaluation procedure on the *Precision* metric and found no difference compared to the results of the *nDGC* metric. This applies to algorithms, domains, and data sets as well. To confirm, the interested reader may refer to the complete set of results of the whole evaluation procedure, which are stored in our public repository. Any additional material is therefore made available only for completeness and reproducibility purposes.

Aggregated Results. Figure 1 provides an aggregated view of our results on all twelve implicit and eight explicit data sets split between implicit and explicit feedback with all eleven algorithms. We observe that there are some data sets and algorithms for which the best *nDCG* performance is not achieved with the *top-n selection strategy*. However, the distribution of the data points also shows that the *top-n selection strategy* is *the best on average*. Furthermore, with the baseline plot, we observe that the random and popularity baselines behave as expected. When recommending random elements, there is a high probability that the *top-n selection strategy* is sub-optimal due to the amount of other possible *selection strategies*. Recommending items based on popularity using the *top-n selection strategy* is also sub-optimal in most cases, simply confirming that the most popular recommendations are not automatically the best.

In all cases where a *non-top-n selection strategy* performs better than the *top-n selection strategy*, we could search for the best *non-top-n selection strategy* to increase performance. The tested recommendation algorithms appear to be stable in terms of relative performance since the difference in performance in both directions is marginal at less than 0.4% for implicit feedback and less than 1.5% for explicit feedback, indicating that finding the best *selection strategy* only has a marginal performance impact.

Domain-Specific Results. To further analyze the observations, we focus on the *articles* and *movies* domains, which contain only implicit and explicit data sets, respectively, in Fig. 2. Notably, there are cases with different data sets in these domains where the best *non-top-n selection strategy* is better than the *top-n selection strategy*. Therefore, the aggregated and domain-specific plots indicate that the relative performance of the best *selection strategies* may be specific to an algorithm or data set rather than a recommendation domain. Additionally, the average difference in the relative performance of data sets changes for different recommender system algorithms.

Data Set-Specific Results. To better understand the scope of the previous observations, we have to analyze not only the best *non-top-n selection strategy* but all of them. We, therefore, take an exemplary look at the exhaustive evaluation of *selection strategies* for the *Adressa One Week* data set from the *articles* domain on the *Alternating Least Squares* algorithm for implicit feedback. Fur-

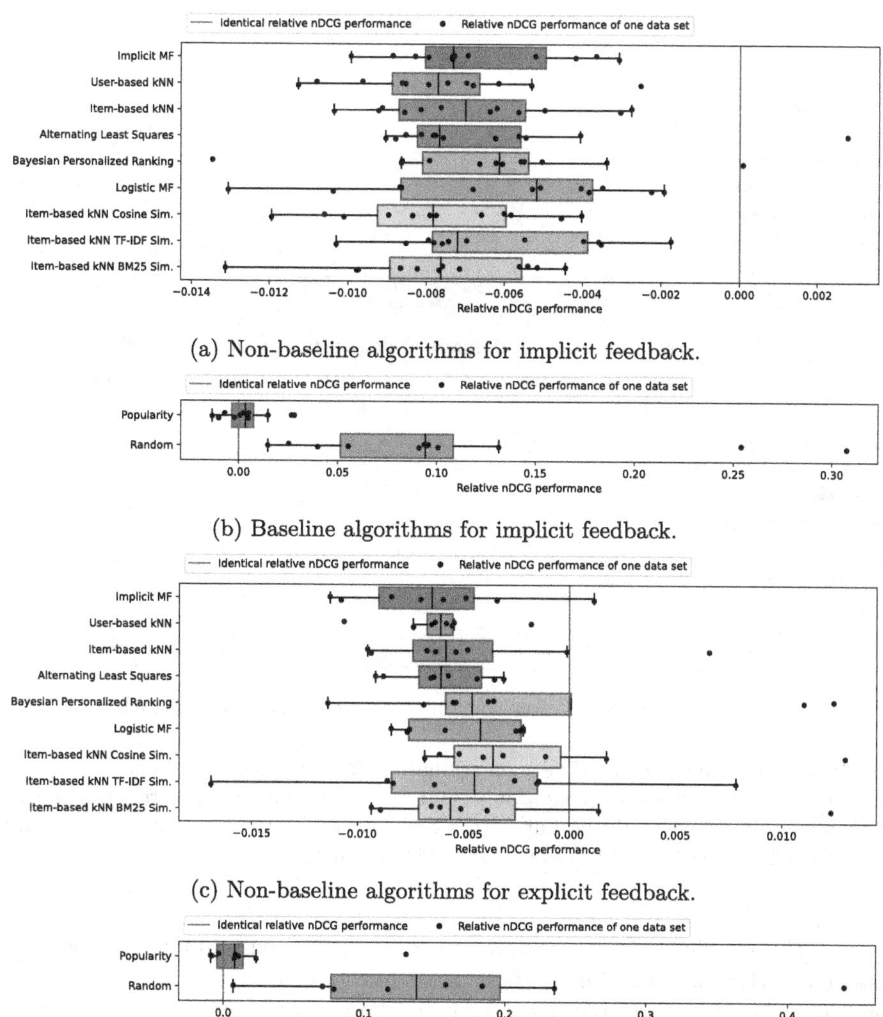

(a) Non-baseline algorithms for implicit feedback.

(b) Baseline algorithms for implicit feedback.

(c) Non-baseline algorithms for explicit feedback.

(d) Baseline algorithms for explicit feedback.

Fig. 1. The relative *nDCG* performance of the best *non-top-n selection strategy* versus the *top-n selection strategy* evaluated on the test set. A point to the right of the red line indicates that the best *non-top-n selection strategy* is better than the *top-n selection strategy*.

thermore, we do the same for the *MovieLens-100k* data set from the *movies* domain on the *Item-Item Nearest Neighbors* algorithm for explicit feedback.

Figure 3 visualizes the results of the exhaustive evaluation through box plots. If the algorithm ranks the items correctly, we expect a consistent movement toward a lower median for lower-ranked elements, but there is no such trend.

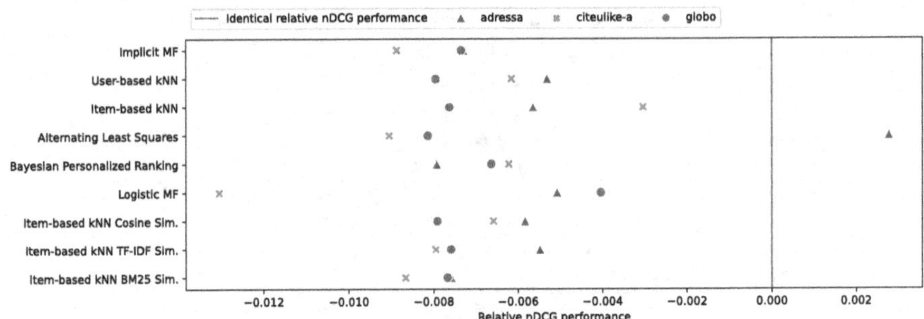

(a) *Articles* recommendation domain (3 data sets, implicit feedback).

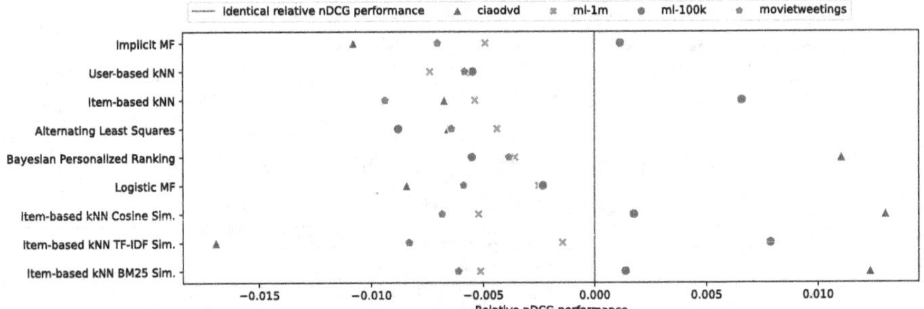

(b) *Movies* recommendation domain (4 data sets, explicit feedback).

Fig. 2. This Figure is a focused view on evaluating data sets that represent the *articles* (2a) and *movies* (2b) recommendation domains. The shown data points are a subset of Figs. 1a and 1c, respectively, but with a focus on the data sets. The data sets are indicated with markers and colors to distinguish them better.

In fact, the interquartile range shows that the model only appears to be consistently performing well for the first predicted item. However, there are items for which the plot shows increased average performance compared to a higher-ranked element. Additionally, we already know from Figs. 2a and 2b, specifically for these examples, that there is at least one *non-top-n selection strategy* that improves over the *top-n selection strategy*. Figure 3 reveals the elements chosen in these strategies, indicated by the points to the right of the red vertical line, e.g., the line indicating the performance of the *top-n selection strategy*.

Generalization Capabilities. Since we can show that there are *selection strategies* that are better than the *top-n selection strategy*, we would need a way to find them reliably. All of the results so far are evaluated on the test set, e.g., the data split that we did not know during training. As a result, if we want to find the best *selection strategy* in a real-world scenario, we would need to search for the best *selection strategy* on the validation set and hope that it generalizes to the test set. To that end, we performed the same exhaustive evaluation on the validation set to observe the generalization capability of the

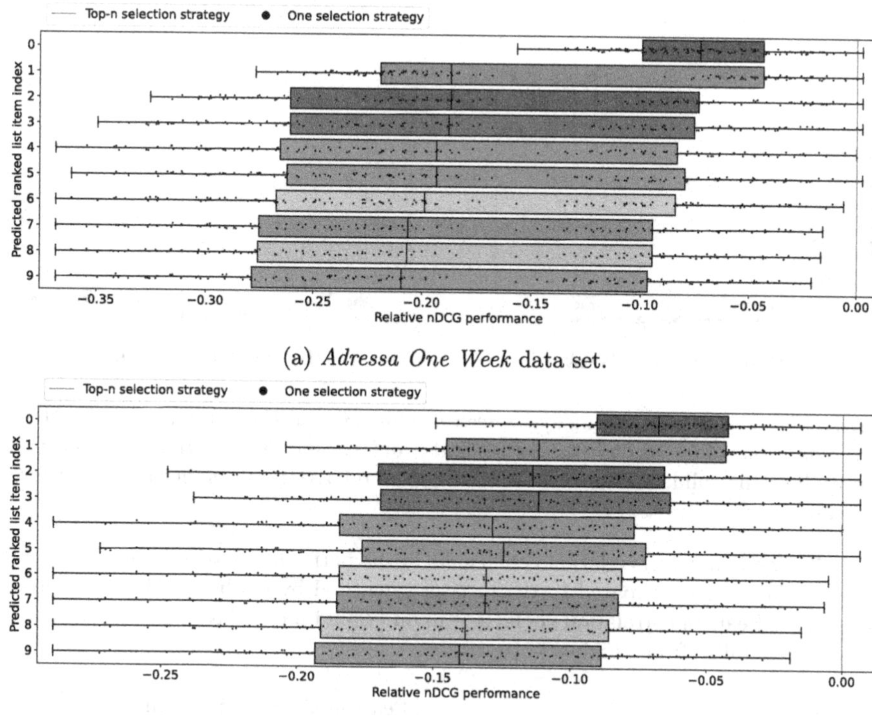

(a) *Adressa One Week* data set.

(b) *MovieLens-100k* data set.

Fig. 3. The exhaustive evaluation of 252 *selection strategies* with the *nDCG* metric on the data set *Adressa One Week* with the *Logistic Matrix Factorization* algorithm (3a) and the data set *MovieLens-100k* with the *Alternating Least Squares* algorithm (3b). Each *selection strategy* picks a different combination of 5 items from the top 10 predicted items. The performance of a *selection strategy* is averaged over all users in the test data set. The y-axis refers to the index of the items in the predicted ranked list, e.g., element 0 is the highest-ranked and the most relevant item. A black dot refers to a *selection strategy* that contains said item, e.g., all *selection strategies* represented by black dots in a row contain the item according to the index stated on the y-axis.

selection strategies. Continuing with the previously chosen example data set and recommendation algorithm, we show the generalization capability in Fig. 4. With the generalization plot, we already see a trend regarding the generalization capability of *selection strategies*. To measure the generalization capability directly, we additionally calculated the Pearson correlation coefficient of the performance of *selection strategies* over the validation and the test. A higher Pearson correlation coefficient indicates higher generalization capability. The Pearson correlation coefficient averaged over all data sets per algorithm is shown in Table 2.

Statistical Significance. Finally, we tested for the statistical difference in the performance of *selection strategies* over all data sets but per algorithm. We applied the Friedman test to confirm that different *selection strategies* result

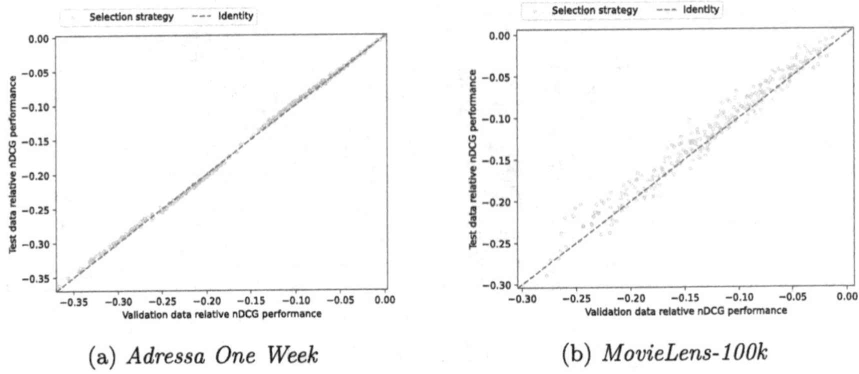

(a) *Adressa One Week* (b) *MovieLens-100k*

Fig. 4. The relative performance of *selection strategies* on the validation and test set for the data sets *Adressa One Week* (4a) and *MovieLens-100k* (4b). If a point is on the identity line, the validation and test set have equal relative performance and, therefore, generalize.

Table 2. The Pearson correlation coefficients of the performance of *selection strategies* of the validation set compared to the test set separated by implicit and explicit feedback sets. A Pearson correlation coefficient of one would mean that *selection strategies* perfectly generalize from validation to test.

Algorithm	Pearson Corr. Coeff. Implicit	Pearson Corr. Coeff. Explicit
Implicit MF	0.998	0.988
User-based kNN	0.999	0.990
Item-based kNN	0.998	0.994
Alternating Least Squares	0.999	0.986
Bayesian Personalized Ranking	0.995	0.966
Logistic MF	0.995	0.983
Item-based kNN Cosine Sim	0.999	0.990
Item-based kNN TF-IDF Sim	0.998	0.992
Item-based kNN BM25 Sim	0.999	0.992
Random	0.004	−0.064
Popularity	0.994	0.995

in statistically significant performance differences ($p < 0.05$). Then, we applied the Nemenyi post hoc test to obtain the critical difference. The result shows that for all non-baseline algorithms, the best $\sim 43\%$ *selection strategies* are not significantly different in performance. We note, however, that the theory behind the Nemenyi test may not hold up due to the high number of compared methods. Nevertheless, we believe it to be a worthwhile indicator to answer our research questions and beyond.

5 Discussion

First, we consider our →research questions. To answer **RQ1**, we evaluated all possible *selection strategies* that choose 5 out of the top 10 predicted items. We accept **RQ1** since we found that $\sim 0.01\%$ of *non-top-n selection strategies* result in higher performance than the *top-n selection strategy.* However, we can not identify specific criteria that lead to this effect, e.g. it occurred in different types of algorithms and data sets from different domains. To answer **RQ2**, we additionally tested the statistical significance of *selection strategies* based on their performance. The tests indicate that most *selection strategies* are not significantly different. Moreover, the maximum performance gain is marginal at less than 0.4% for implicit feedback data sets and less than 1.5% for explicit feedback data sets, as shown in the aggregated results (Fig. 1).

Searching for the Best Selection Strategy. The average Pearson correlation coefficient of > 0.99 for implicit feedback data sets and > 0.96 for explicit feedback sets for all non-baseline algorithms shows that *selection strategies* generalize their performance from validation to test. Therefore, it is possible to reliably search and find the best *selection strategy* on the validation set. However, the search strategy we used for the evaluation, e.g., exhaustively evaluating all possible options, is not cost-efficient. We have also shown that the *top-n selection strategy* is *the best on average.* Therefore, searching for the best *selection strategy* in practice is only feasible with a highly efficient search strategy or when the potential for a minor improvement outweighs the cost of the search. In future work, finding the best *selection strategy* could be accomplished with an efficient search algorithm, e.g., a greedy search. Alternatively, our results indicate that we do not have to observe items too far away from the top-n, e.g., we searched *selection strategies* with 5 items out of a sample range of 10 items, but 8 items may have already been enough to find the optimum. As a result, an exhaustive search on fewer combinations may be feasible.

The Impact on Hyperparameter Optimization. Hyperparameter optimization techniques rely on the developer to correctly approximate the predictive accuracy of a model to find the best hyperparameters. If the *top-n selection strategy* does not achieve that, due to being sub-optimal in terms of performance, we should avoid using it. However, our statistical tests indicate that optimizing the *selection strategy* likely has no practical impact since the evaluated performance with the majority of *selection strategies* is not significantly different. In turn, this means we may use most *selection strategies* that choose 5 out of 10 items in practice.

The Impact on Re-ranking. Re-ranking algorithms change which items are recommended, e.g. they apply a different *selection strategy.* Often, this means that predictive accuracy performance, e.g. *nDCG* performance, is sacrificed to gain increased performance in secondary metrics, e.g. *diversity* or *fairness.* Our results show that, when the re-ranking algorithm is confined to elements close to the top-n, the re-ranked predictions may actually not be significantly different in terms of predictive accuracy.

Conclusion. We reveal the hidden impact of top-n metrics on optimization in recommender systems and show that it is insignificant. As a result, there is no practical benefit in optimizing *selection strategies*. Conclusively, our exploratory study cleared any doubts on this confounding factor for the evaluation and reproducibility of traditional collaborative filtering algorithms.

Acknowledgement. The OMNI cluster of the University of Siegen was used to compute the results presented in this paper.

References

1. Abdollahpouri, H., Burke, R., Mobasher, B.: Managing popularity bias in recommender systems with personalized re-ranking. arXiv preprint arXiv:1901.07555 (2019)
2. Anand, R., Beel, J.: Auto-surprise: an automated recommender-system (autorecsys) library with tree of parzens estimator (TPE) optimization. In: Proceedings of the 14th ACM Conference on Recommender Systems (RecSys 2020), pp. 585–587. Association for Computing Machinery, New York (2020). https://doi.org/10.1145/3383313.3411467
3. Anelli, V.W., et al.: Elliot: a comprehensive and rigorous framework for reproducible recommender systems evaluation. In: Diaz, F., Shah, C., Suel, T., Castells, P., Jones, R., Sakai, T. (eds.) The 44th International ACM SIGIR Conference on Research and Development in Information Retrieval, Virtual Event (SIGIR 2021), Canada, 11–15 July 2021, pp. 2405–2414. ACM (2021). https://doi.org/10.1145/3404835.3463245
4. Barkan, O., Hirsch, R., Katz, O., Caciularu, A., Koenigstein, N.: Anchor-based collaborative filtering. In: Proceedings of the 30th ACM International Conference on Information & Knowledge Management (CIKM 2021), pp. 2877–2881. Association for Computing Machinery, New York (2021). https://doi.org/10.1145/3459637.3482056
5. Beel, J., Dinesh, S.: Real-world recommender systems for academia: the pain and gain in building, operating, and researching them. In: Mayr, P., Frommholz, I., Cabanac, G. (eds.) Proceedings of the Fifth Workshop on Bibliometric-enhanced Information Retrieval (BIR) Co-located with the 39th European Conference on Information Retrieval (ECIR 2017), Aberdeen, UK, 9th April 2017. CEUR Workshop Proceedings, vol. 1823, pp. 6–17. CEUR-WS.org (2017). https://ceur-ws.org/Vol-1823/paper1.pdf
6. Bruch, S., Wang, X., Bendersky, M., Najork, M.: An analysis of the softmax cross entropy loss for learning-to-rank with binary relevance. In: Proceedings of the 2019 ACM SIGIR International Conference on Theory of Information Retrieval (ICTIR 2019), pp. 75–78. Association for Computing Machinery, New York (2019). https://doi.org/10.1145/3341981.3344221
7. Cantador, I., Brusilovsky, P., Kuflik, T.: Second workshop on information heterogeneity and fusion in recommender systems (hetrec2011). In: Proceedings of the Fifth ACM Conference on Recommender Systems (RecSys 2011), pp. 387–388. Association for Computing Machinery, New York (2011). https://doi.org/10.1145/2043932.2044016

8. Chen, B., Zhao, X., Wang, Y., Fan, W., Guo, H., Tang, R.: A comprehensive survey on automated machine learning for recommendations. ACM Trans. Recomm. Syst. (2023). https://doi.org/10.1145/3630104
9. Chen, H., et al.: Denoising self-attentive sequential recommendation. In: Proceedings of the 16th ACM Conference on Recommender Systems (RecSys 2022). pp. 92–101. Association for Computing Machinery, New York (2022). https://doi.org/10.1145/3523227.3546788
10. Cho, E., Myers, S.A., Leskovec, J.: Friendship and mobility: user movement in location-based social networks. In: Proceedings of the 17th ACM SIGKDD International Conference on Knowledge Discovery and Data Mining (KDD 2011), pp. 1082–1090. Association for Computing Machinery, New York (2011). https://doi.org/10.1145/2020408.2020579
11. Dooms, S., De Pessemier, T., Martens, L.: MovieTweetings: a movie rating dataset collected from twitter. In: Workshop on Crowdsourcing and Human Computation for Recommender Systems, Held in Conjunction with the 7th ACM Conference on Recommender Systems, p. 2 (2013)
12. Ekstrand, M.D.: Lenskit for python: next-generation software for recommender systems experiments. In: Proceedings of the 29th ACM International Conference on Information and Knowledge Management (CIKM 2020), pp. 2999–3006. Association for Computing Machinery, New York (2020). https://doi.org/10.1145/3340531.3412778
13. Feurer, M., Klein, A., Eggensperger, K., Springenberg, J., Blum, M., Hutter, F.: Efficient and robust automated machine learning. In: Cortes, C., Lawrence, N., Lee, D., Sugiyama, M., Garnett, R. (eds.) Advances in Neural Information Processing Systems, vol. 28. Curran Associates, Inc. (2015)
14. Frederickson, B.: Fast python collaborative filtering for implicit datasets (2018). https://githubcom/benfred/implicit
15. Ge, Y., et al.: Understanding echo chambers in e-commerce recommender systems. In: Proceedings of the 43rd International ACM SIGIR Conference on Research and Development in Information Retrieval (SIGIR 2020), pp. 2261–2270. Association for Computing Machinery, New York (2020). https://doi.org/10.1145/3397271.3401431
16. Goldberg, K., Roeder, T., Gupta, D., Perkins, C.: Eigentaste: a constant time collaborative filtering algorithm. Inf. Retr. **4**(2), 133–151 (2001). https://doi.org/10.1023/A:1011419012209
17. Gulla, J.A., Zhang, L., Liu, P., Özgöbek, O., Su, X.: The adressa dataset for news recommendation. In: Proceedings of the International Conference on Web Intelligence (WI 2017), pp. 1042–1048. Association for Computing Machinery, New York (2017). https://doi.org/10.1145/3106426.3109436
18. Harper, F.M., Konstan, J.A.: The movielens datasets: History and context. ACM Trans. Interact. Intell. Syst. **5**(4), 1–19 (2015). https://doi.org/10.1145/2827872
19. Herlocker, J.L., Konstan, J.A., Terveen, L.G., Riedl, J.T.: Evaluating collaborative filtering recommender systems. ACM Trans. Inf. Syst. **22**(1), 5–53 (2004). https://doi.org/10.1145/963770.963772
20. Hernández del Olmo, F., Gaudioso, E.: Evaluation of recommender systems: a new approach. Exp. Syst. Appl. **35**(3), 790–804 (2008). https://doi.org/10.1016/j.eswa.2007.07.047

21. Jankiewicz, P., Kyrashchuk, L., Sienkowski, P., Wójcik, M.: Boosting algorithms for a session-based, context-aware recommender system in an online travel domain. In: Proceedings of the Workshop on ACM Recommender Systems Challenge (RecSys Challenge 2019). Association for Computing Machinery, New York (2019). https://doi.org/10.1145/3359555.3359557
22. Jannach, D., Adomavicius, G.: Price and profit awareness in recommender systems. arXiv preprint arXiv:1707.08029 (2017)
23. Joglekar, M.R., et al.: Neural input search for large scale recommendation models. In: Proceedings of the 26th ACM SIGKDD International Conference on Knowledge Discovery and Data Mining (KDD 2020), pp. 2387–2397. Association for Computing Machinery, New York (2020). https://doi.org/10.1145/3394486.3403288
24. Langer, S., Beel, J.: Apache lucene as content-based-filtering recommender system: 3 lessons learned. In: Mayr, P., Frommholz, I., Cabanac, G. (eds.) Proceedings of the Fifth Workshop on Bibliometric-enhanced Information Retrieval (BIR) co-located with the 39th European Conference on Information Retrieval (ECIR 2017), Aberdeen, 9 April 2017. CEUR Workshop Proceedings, vol. 1823, pp. 85–92. CEUR-WS.org (2017). https://ceur-ws.org/Vol-1823/paper8.pdf
25. Li, Z., Ji, J., Ge, Y., Zhang, Y.: Autolossgen: automatic loss function generation for recommender systems. In: Proceedings of the 45th International ACM SIGIR Conference on Research and Development in Information Retrieval (SIGIR 2022), pp. 1304–1315. Association for Computing Machinery, New York (2022). https://doi.org/10.1145/3477495.3531941
26. Liang, D., Krishnan, R.G., Hoffman, M.D., Jebara, T.: Variational autoencoders for collaborative filtering. In: Proceedings of the 2018 World Wide Web Conference (WWW 2018), pp. 689–698. International World Wide Web Conferences Steering Committee, Republic and Canton of Geneva, CHE (2018). https://doi.org/10.1145/3178876.3186150
27. Liu, H., Zhao, X., Wang, C., Liu, X., Tang, J.: Automated embedding size search in deep recommender systems. In: Proceedings of the 43rd International ACM SIGIR Conference on Research and Development in Information Retrieval (SIGIR 2020), pp. 2307–2316. Association for Computing Machinery, New York (2020). https://doi.org/10.1145/3397271.3401436
28. Liu, W., et al.: Neural re-ranking in multi-stage recommender systems: a review. In: Raedt, L.D. (ed.) Proceedings of the Thirty-First International Joint Conference on Artificial Intelligence (IJCAI-22), pp. 5512–5520. International Joint Conferences on Artificial Intelligence Organization (2022). https://doi.org/10.24963/ijcai.2022/771
29. Melchiorre, A.B., Rekabsaz, N., Ganhör, C., Schedl, M.: Protomf: prototype-based matrix factorization for effective and explainable recommendations. In: Proceedings of the 16th ACM Conference on Recommender Systems (RecSys 2022), pp. 246–256. Association for Computing Machinery, New York (2022). https://doi.org/10.1145/3523227.3546756
30. Michiels, L., Verachtert, R., Goethals, B.: Recpack: an(other) experimentation toolkit for top-n recommendation using implicit feedback data. In: Proceedings of the 16th ACM Conference on Recommender Systems (RecSys 2022). pp. 648–651. Association for Computing Machinery, New York (2022). https://doi.org/10.1145/3523227.3551472
31. Moreira, G.D.S.P., Jannach, D., Cunha, A.M.D.: Contextual hybrid session-based news recommendation with recurrent neural networks. IEEE Access **7**, 169185–169203 (2019)

32. Ni, J., Li, J., McAuley, J.: Justifying recommendations using distantly-labeled reviews and fine-grained aspects. In: Proceedings of the 2019 Conference on Empirical Methods in Natural Language Processing and the 9th International Joint Conference on Natural Language Processing (EMNLP-IJCNLP), pp. 188–197. Association for Computational Linguistics, Hong Kong (2019). https://doi.org/10.18653/v1/D19-1018
33. Pang, L., Ai, Q., Xu, J.: Beyond probability ranking principle: modeling the dependencies among documents. In: Proceedings of the 14th ACM International Conference on Web Search and Data Mining (WSDM 2021), pp. 1137–1140. Association for Computing Machinery, New York (2021). https://doi.org/10.1145/3437963.3441662
34. Pei, C., et al.: Personalized re-ranking for recommendation. In: Proceedings of the 13th ACM Conference on Recommender Systems (RecSys 2019), pp. 3–11. Association for Computing Machinery, New York (2019). https://doi.org/10.1145/3298689.3347000
35. Pichl, M., Zangerle, E., Specht, G.: Towards a context-aware music recommendation approach: what is hidden in the playlist name? In: 2015 IEEE International Conference on Data Mining Workshop (ICDMW), pp. 1360–1365 (2015). https://doi.org/10.1109/ICDMW.2015.145
36. Poddar, A., Zangerle, E., Yang, Y.H.: #nowplaying-rs: a new benchmark dataset for building context-aware music recommender systems. In: Proceedings of the 15th Sound and Music Computing Conference. Limassol, Cyprus (2018). https://mac.citi.sinica.edu.tw/~yang/pub/poddar18smc.pdf, code at https://github.com/asmitapoddar/nowplaying-RS-Music-Reco-FM
37. Portugal, I., Alencar, P., Cowan, D.: The use of machine learning algorithms in recommender systems: a systematic review. Expert Syst. Appl. **97**, 205–227 (2018)
38. Rendle, S., Freudenthaler, C., Gantner, Z., Schmidt-Thieme, L.: BPR: Bayesian personalized ranking from implicit feedback. In: Proceedings of the Twenty-Fifth Conference on Uncertainty in Artificial Intelligence (UAI 2009), pp. 452–461. AUAI Press, Arlington (2009)
39. Rendle, S., Krichene, W., Zhang, L., Koren, Y.: Revisiting the performance of ials on item recommendation benchmarks. In: Proceedings of the 16th ACM Conference on Recommender Systems (RecSys 2022), pp. 427–435. Association for Computing Machinery, New York (2022). https://doi.org/10.1145/3523227.3548486
40. Robertson, S.: The probability ranking principle in IR. J. Document. **33**(4), 294–304 (1977). https://doi.org/10.1108/eb026647
41. de Souza Pereira Moreira, G., Ferreira, F., da Cunha, A.M.: News session-based recommendations using deep neural networks. In: Proceedings of the 3rd Workshop on Deep Learning for Recommender Systems. ACM (2018). https://doi.org/10.1145/3270323.3270328
42. Sun, B., Wu, D., Shang, M., He, Y.: Toward auto-learning hyperparameters for deep learning-based recommender systems. In: Bhattacharya, A., et al. (eds.) Database Systems for Advanced Applications, pp. 323–331. Springer, Cham (2022). https://doi.org/10.1007/978-3-031-00126-0_25
43. Sun, F., et al.: Bert4rec: sequential recommendation with bidirectional encoder representations from transformer. In: Proceedings of the 28th ACM International Conference on Information and Knowledge Management (CIKM 2019), pp. 1441–1450. Association for Computing Machinery, New York (2019). https://doi.org/10.1145/3357384.3357895

44. Tang, H., Liu, J., Zhao, M., Gong, X.: Progressive layered extraction (PLE): a novel multi-task learning (MTL) model for personalized recommendations. In: Proceedings of the 14th ACM Conference on Recommender Systems (RecSys 2020), pp. 269–278. Association for Computing Machinery, New York (2020). https://doi.org/10.1145/3383313.3412236
45. Vente, T., Ekstrand, M., Beel, J.: Introducing lenskit-auto, an experimental automated recommender system (autorecsys) toolkit. In: Proceedings of the 17th ACM Conference on Recommender Systems (RecSys 2023), pp. 1212–1216. Association for Computing Machinery, New York (2023). https://doi.org/10.1145/3604915.3610656
46. Wang, H., Chen, B., Li, W.J.: Collaborative topic regression with social regularization for tag recommendation. In: Proceedings of the Twenty-Third International Joint Conference on Artificial Intelligence (IJCAI 2013), pp. 2719–2725. AAAI Press (2013)
47. Wang, X., Li, C., Golbandi, N., Bendersky, M., Najork, M.: The lambdaloss framework for ranking metric optimization. In: Proceedings of the 27th ACM International Conference on Information and Knowledge Management (CIKM 2018), pp. 1313–1322. Association for Computing Machinery, New York (2018)
48. Weston, J., Yee, H., Weiss, R.J.: Learning to rank recommendations with the k-order statistic loss. In: Proceedings of the 7th ACM Conference on Recommender Systems (RecSys 2013), pp. 245–248. Association for Computing Machinery, New York (2013). https://doi.org/10.1145/2507157.2507210
49. Xu, L., et al.: Recent advances in RecBole: extensions with more practical considerations. arXiv preprints arXiv:2211.15148 (2022)
50. Yang, C., Hou, Y., Song, Y., Zhang, T., Wen, J.R., Zhao, W.X.: Modeling two-way selection preference for person-job fit. In: Proceedings of the 16th ACM Conference on Recommender Systems (RecSys 2022), pp. 102–112. Association for Computing Machinery, New York (2022). https://doi.org/10.1145/3523227.3546752
51. Yue, Z., He, Z., Zeng, H., McAuley, J.: black-box attacks on sequential recommenders via data-free model extraction. In: Proceedings of the 15th ACM Conference on Recommender Systems (RecSys 2021), pp. 44–54. Association for Computing Machinery, New York (2021). https://doi.org/10.1145/3460231.3474275
52. Yue, Z., Zeng, H., Kou, Z., Shang, L., Wang, D.: Defending substitution-based profile pollution attacks on sequential recommenders. In: Proceedings of the 16th ACM Conference on Recommender Systems (RecSys 2022), pp. 59–70. Association for Computing Machinery, New York (2022). https://doi.org/10.1145/3523227.3546770
53. Zangerle, E., Bauer, C.: Evaluating recommender systems: survey and framework. ACM Comput. Surv. **55**(8) (2022). https://doi.org/10.1145/3556536
54. Zhang, R., Bao, H., Sun, H., Wang, Y., Liu, X.: Recommender systems based on ranking performance optimization. Front. Comp. Sci. **10**(2), 270–280 (2015)
55. Zhao, W.X., et al.: Recbole 2.0: towards a more up-to-date recommendation library. In: Proceedings of the 31st ACM International Conference on Information and Knowledge Management, pp. 4722–4726 (2022)
56. Zhao, W.X., et al.: Recbole: towards a unified, comprehensive and efficient framework for recommendation algorithms. In: CIKM, pp. 4653–4664. ACM (2021)
57. Zheng, R., Qu, L., Cui, B., Shi, Y., Yin, H.: Automl for deep recommender systems: a survey. ACM Trans. Inf. Syst. **41**(4) (2023). https://doi.org/10.1145/3579355

Simulated Task Oriented Dialogues for Developing Versatile Conversational Agents

Xi Wang[1(✉)], Procheta Sen[2], Ruizhe Li[3], and Emine Yilmaz[1]

[1] University College London, London, UK
{xi-wang,emine.yilmaz}@ucl.ac.uk
[2] University of Liverpool, Liverpool, UK
procheta.sen@liverpool.ac.uk
[3] University of Aberdeen, Aberdeen, UK
ruizhe.li@abdn.ac.uk

Abstract. Task-Oriented Dialogue (TOD) Systems are increasingly important for managing a variety of daily tasks, yet often underperform in unfamiliar scenarios due to limitations in existing training datasets. This study addresses the challenge of generating robust and versatile TOD systems by transforming instructional task descriptions into natural user-system dialogues to serve as enhanced pre-training data. We explore three strategies for synthetic dialogue generation: crowdsourcing, encoder-decoder models, and in-context learning with large language models. The evaluation of these approaches, based on a comprehensive user study employing 10 different metrics, reveals the top quality of the dialogues generated by learning an encoder-decoder model as per human evaluation. Notably, employing this synthetic dialogue further improves the performance of advanced TOD models, especially in unfamiliar domains, with improvements spanning 5.5% to as much as 20.9% in combined evaluation scores. Our findings advocate for the use of specialised, task-oriented knowledge bases and step-wise dialogue generation techniques to advance the capabilities and generalizability of TOD systems.

1 Introduction

Task-Oriented Dialogue (TOD) Systems have recently proven valuable in helping users with various daily tasks like restaurant bookings [10,16]. For these systems to be effective, they must understand tasks and offer relevant suggestions. A key challenge is arming TOD models with extensive knowledge for effective responses across multiple tasks. Researchers tackle this by fine-tuning models with rich, large-scale datasets [36]. However, a notable gap exists in available datasets that cover a broad range of tasks and details. Existing datasets, such as MultiWoZ [40], Frames [11], and SGD [30], mainly focus on common scenarios like travel. While SGD covers 16 domains, it lacks comprehensive instructional content crucial for task-specific adaptation. One solution is creating enriched, larger-scale datasets, but this faces two main hurdles. First, the scarcity of clean, structured, domain-specific knowledge. A notable and recent advancement in

N. Goharian et al. (Eds.): ECIR 2024, LNCS 14608, pp. 157–172, 2024.
https://doi.org/10.1007/978-3-031-56027-9_10

this regard is Task2KB [25], which compiles task instructions and a wealth of associated information from WikiHow[1] – an online platform offering detailed guides for diverse tasks. However, a second challenge remains: the significant resource investment required to develop high-quality, large-scale TOD datasets. Traditional human labelling methods are both time-consuming and often result in noisy or unreliable data [6]. Therefore, it is essential to leverage existing publicly available task-oriented knowledge bases, like Task2KB, to develop rich, reliable and diverse task-oriented dialogue datasets, so as to further benefit the advancement of TOD models.

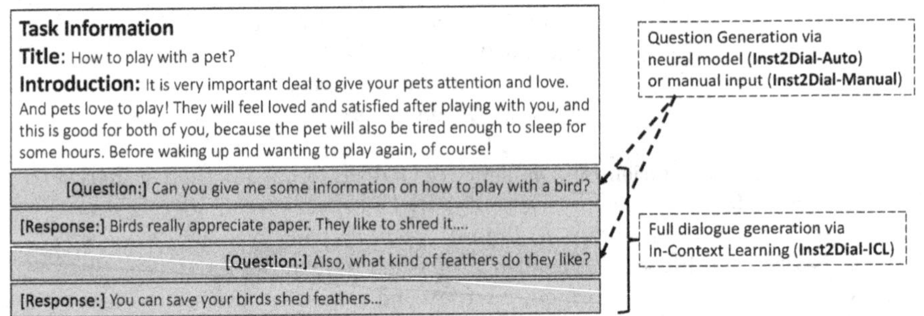

Fig. 1. An illustrative example of synthetic dialogues that can be generated using different methodologies.

Concurrent with recent advancements in generative models, such as GPT-3.5/4 and Flan-T5, research has demonstrated the feasibility of generating high-quality dialogue data using descriptive text as input, such as Wikipedia passages [9]. In light of these findings, our study aims to transform instructional task descriptions into natural user-system dialogues. These dialogues intend to serve as training data to enhance both the quality of responses generated and the generalizability of state-of-the-art TOD models. Specifically, we systematically explored three strategies for generating synthetic data using task-related instructional information. These strategies encompass a) *step-wise dialogue generation employing encoder-decoder models*, b) *crowdsourcing* and c) *in-context learning* with large language models. In Fig. 1, we exemplify the use of the three strategies with the corresponding resulting datasets, Inst2Dial-Auto/Manual/ICL, which generate questions and use instructions as responses (Inst2Dial-Auto&Manual) or generate full dialogue at once (Inst2Dial-ICL).

To provide a comprehensive evaluation of these methodologies, we conducted both offline assessments and user studies. The evaluation results, based on a user study employing 10 different metrics-tailored to whether additional task information was provided-demonstrate that step-wise dialogue generation with encoder-decoder models consistently outperforms the other two strategies. Specifically, synthetic dialogues produced through such a strategy have been shown to enhance the performance of state-of-the-art models, especially in domains with

[1] https://wikihow.com.

limited knowledge. Improvements ranged from a minimum of 5.5% to as much as 20.9% in the combined evaluation score.

This study presents three pivotal contributions to the field of Task-Oriented Dialogue (TOD) systems. **Firstly**, we introduce novel approaches for simulating task-oriented dialogues by leveraging instructional documents from WikiHow, thereby creating rich training datasets. **Secondly**, we implement rigorous quality control mechanisms to ensure the generated dialogues are both contextually relevant and of high quality. **Thirdly**, we demonstrate the practical application of large-scale simulated dialogues (full Inst2Dial-Auto) by utilizing them as pre-training data, which results in significant performance improvements in state-of-the-art TOD models.

2 Related Work

Task-Oriented Dialogue (TOD) systems, a subdomain of conversational systems, act in the role of task completion assistant with a requirement of comprehensive task knowledge [19]. Traditional techniques have aimed to improve various components, such as dialogue state tracking [39], action prediction [3] and response generation [7,13]. Recent end-to-end approaches started leveraging advanced language models as backbones for enhanced performance in natural language understanding and generation [5,14,38]. In line with the advancements in conversational systems, such as UBAR [38] and JSA-TOD [5], many conversational datasets have emerged. These include the Schema-Guided Dialogue (SGD) [30], MultiWoZ [40] and RiSAWOZ [27] datasets. However, these datasets are primarily limited by restricted domain coverage (as exemplified by MultiWoZ's coverage of only eight domains and SGD's extension to a current maximum of 16) and insufficient instructional information, impairing the effectiveness of TOD models trained on them.

To address these limitations, various research efforts have either augmented existing datasets or employed human engagement to create more inclusive dialogue datasets. However, these approaches are often contained by their considerable financial implications and intricate design requirements [9]. Meanwhile, there exists a thread of work leveraging simulation techniques for dialogue generation [9,20,34]. Classical approaches predominantly utilise rule-based methods [23,34]. A contemporary instance is [9], which employed a BERT-like architecture to generate dialogues based on given texts, culminating in the release of the WikiDialog corpus. Each dialogue in this corpus is synthesised from a corresponding Wikipedia passage. For task-oriented dialogues, Mohapatra et al., [20] fine-tuned a GPT-2 model [28] and applied it to a certain dialogue context to generate simulated dialogues, aiming to improve the performance of TOD models, particularly in a low-resource setting. While their work aligns closely with our own contributions, it fully relies on existing task completion scenarios within the dataset, thereby neglecting the rich instructional data contained in external knowledge bases. This oversight results in the persistence of limited domain coverage, an issue we previously identified. In summary, extant research has not leveraged step-wise instructional content to develop synthetic dialogues, a gap that our work aims to fill, especially considering recent advancements in generative models.

3 INST2DIAL Synthetic Dialogue Development

In this section, we discuss three strategies for generating synthetic dialogues, leveraging rich task-specific instructional content from an external knowledge base, Task2KB [25].

Problem Description. Formally, each task t is characterised by its title τ^t, introduction i^t and a series of k instructional steps $S^t = \{s_1^t, s_2^t, ..., s_k^t\}$. The process of generating synthetic dialogues is modelled as $\hat{d}^t = f(\tau^t, i^t, S^t)$. For readability, subsequent descriptions will omit the task index t. We conceptualise $f(\cdot)$ as either a question generator paired with selected instructions as responses or as a full-dialogue generator using the instructions as input. Next, we proceed to detail the three strategies we propose to implement $f(\tau^t, i^t, S^t)$.

3.1 INST2DIAL-Auto

In this approach, we employ an advanced encoder-decoder model to generate a full conversation. This model uses instructional steps for task completion as responses while automatically generating pertinent questions. Unlike in open-domain scenarios, the question generator is specifically designed to ask task-relevant questions that help progress task completion. The generation process is divided into three stages: (1) neural question generator learning, (2) formulating the input for dialogue generation, and (3) the actual dialogue generation.

Neural Question Generator Learning. The aim of this module is to fine-tune an encoder-decoder model, such as Flan-T5 [8], to produce high-quality questions for synthetic task-oriented dialogues. Inspired by the methodology in [9], we employ sequential masking on questions within existing dialogues to create input-output pairs (i.e., dialogue inpainting). In this format, the input contains masked text designated to be filled with a generated question. For a dialogue d consisting of two sequential Question-Answer (QA) pairs, $d = \{q_0, a_0, q_1, a_1\}$, we derive two input-output pairs as follows:

$$\textbf{input}: [MASK][SEP]a_0 \rightarrow \textbf{output}: \hat{q}_0$$
$$\textbf{input}: q_0[SEP]a_0[SEP][MASK][SEP]a_1 \rightarrow \textbf{output}: \hat{q}_1$$

Here, $[MASK]$ marks where the generated question should be inserted, and $[SEP]$ separates questions and answers. The model is then trained to predict suitable questions for such structured inputs. To tailor the question generator for task completion, we propose training on specialised TOD datasets rather than commonly-used open-domain datasets [2,9].

Input Formation. In the next stage, we focus on input formation using our trained task-oriented question generator. The goal is to capture the logical progression of a task in dialogue form. To this end, we use step-wise task instructions

S as answers to prospective user questions, forming a chain of linked responses with missing questions for a task t. We also introduce strategies to balance dialogue length and information content. Specifically, we investigate two methods: (1) employing topic sentences from each step description, and (2) choosing the most specific sentence as determined by a text specificity predictor, speciteller [18]. After a thorough evaluation, we find that using topic sentences leads to higher-quality synthetic dialogues – they are both fluent and informative with less noise. We publicly make available both implementations in our GitHub repository for the details.

Dialogue Generation. Finally, we turn to generating the missing questions in our prepared dialogues from the previous stage, represented as $d = \{\square, a_0, \square, a_1, ..., \square, a_{|d|}\}$. We introduce three strategies, Single-QA, Last-QA and Full-QA, that take different input contexts into account. All strategies initiate the dialogue using the task's title τ and introduction i from the Task2KB knowledge base [33] as the opening QA pair to set the context. Subsequent dialogue is generated incrementally, with each new input influenced by prior ones. The input formatting is as follows:

$$\textbf{input 1}: q^{\tau}[SEP]a^{i}[SEP][MASK][SEP]a_0$$
$$\textbf{input 2}: q^{\tau}[SEP]a^{i}[SEP](\hat{q}_0[SEP]a_0)[SEP][MASK][SEP]a_1$$
$$\textbf{input n}: ...$$

The second input incorporates the question generated from the first, serving as an extended context. We differentiate between the three strategies by varying the scope of the input: Single-QA includes only the immediate preceding answer, Last-QA adds the last QA pair, and Full-QA incorporates all previous QA pairs. Upon optimising these generation methods, we produce a comprehensive set of automated task-oriented dialogues (INST2DIAL-Auto) covering a broad array of tasks within a large-scale knowledge base, Task2KB.

3.2 INST2DIAL-Manual

Next, we introduce our second approach for generating synthetic task-oriented dialogues, Crowdsourced Question Generation (i.e., INST2DIAL-Manual). Unlike the first strategy, which relies on a learned generative model, this method leverages human efforts for dialogue generation, particularly focusing on generating high-quality questions that integrate seamlessly into full dialogues with step-by-step instructional responses.

To achieve this, we designed a user study, creating a custom interface that enables crowd workers to generate the INST2DIAL-Manual dataset. In Fig. 2, we present an example task, "Install a Rear View Camera", to demonstrate the interface used by the crowd workers. The interface is split into two sections. The left side provides detailed guidelines for the task, while the right side is designed for worker input. As can be seen in Fig. 2, crowd workers are instructed to perform two actions for each step of a task: (1) formulate a relevant question concerning the task, and (2) select an appropriate response from the instructional material. Due to space limitations, Fig. 2 only provides the first step (in

Ask a series of questions and **Select** the corresponding answers.
Instructions:
This user study is to give a sequence of question-answering pairs that related to a given task. Such a task in this study can be something like "How to roast a chicken?". For each question-answering pair, it is related to one of the methods or steps while addressing a given task. A full list of such methods or steps will also be presented. We also show which method/step you are in while addressing the question-answering pair.
For each method/step, you need to ask a question and select the part of the offered text that can answer the corresponding question. Therefore, each question-answering task can be split into two steps: **(1) Ask an appropriate question; (2) select the matched answer from a list of sentences.** Eventually, we would like to see the questions and answers can be joined together, which results in a complete conversational dialogue (i.e., questions are sequentially related in completing the task.).
1. Ask an appropriate question while addressing a task:
For the first stage, you will be settled into a context that you are addressing a particular task, like "exploring how to roast a chicken". A task overview will also be offered to describe the corresponding task. What you need to do is asking a **question** that *can be answered by a part of an also offered task instruction (i.e. **answer**).*
The question needs to be:
1. A complete and grammar correct question.
2. A question that can be directly answered by part of the task instruction (next to the question text box).
3. Including more details of the task, such as "what do I need to do after seasoning the chicken?" for the "roast chicken" task.
4. Relevant, and no rewards will be awarded if the question is irrelevant to the given task.
5. Sequential-related questions. We prefer questions that are sequentially related to the previously asked questions. Such as "How long does it take for roasting the chicken after I have finished the seasoning step?" for the task "How to roast a chicken".
2. Select answer from the instruction:
Next, you need to select which part of the offered text (next to the question text box) is the corresponding answer by selecting sentences.
1. The selected sentences should be directly relevant to the asked question or partly answer the question.
2. Again, no rewards will be awarded if the selected answer is not relevant to the question.
Task
Install a Rear View Camera
Task Overview
A rear view camera, also known as a backup camera, lets you see what's Behind your vehicle without having to look backwards. Though the device comes standard with many new car models, you can add a rear-view camera to your vehicle if it didn't come with one.
Method / Steps
Purchase the Necessary Equipment
You are at this method/step.
Installing the camara cables
Putting In Your Monitor
Mounting the Camera
Question
Input a question can be answered by current method/step description.
Check
Answer List
Title: Purchase the necessary equipment.
☐ Buy a mountable backup camera for your specific device.
☐ For safety, make sure you purchase a device specifically designed to be rear-view camera.
☐ Purchasing one made for your specific vehicle will make it easier to install than a standard aftermarket camera.

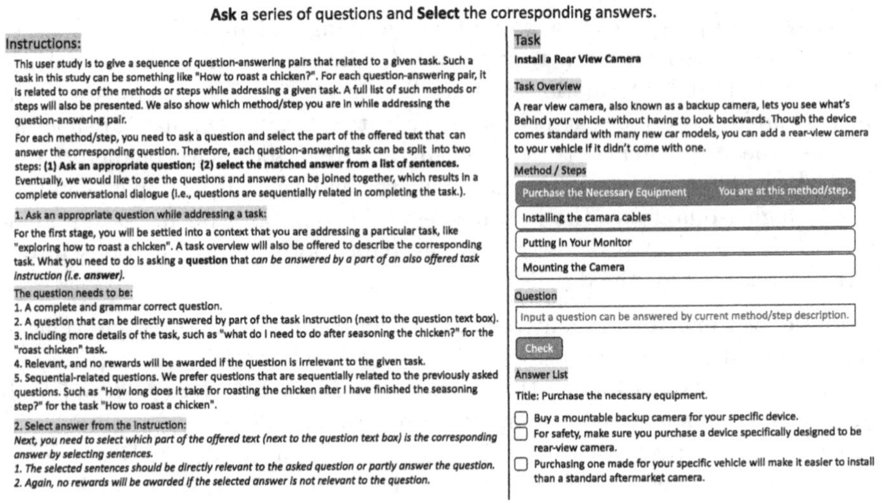

Fig. 2. The interface for crowd workers. A complete instruction is in the left, which stands next to an example interface that allows workers to write a question for the first step of an example task (i.e., install a rear-view camera) and select the corresponding sentence-wise answer.

the bottom right) of the full instructions available to workers. Post data collection, we implement quality control measures, involving manual labelling by domain experts. They assess each dialogue on three criteria: the relevance and meaningfulness of the questions, the alignment of the questions with the context, and the compatibility between questions and answers. Dialogues that fail any of these tests are excluded to ensure a high-quality dataset.

3.3 INST2DIAL-ICL

In addition to the previous methods, we also explore the potential of advanced language models, like GPT [21], for generating synthetic dialogues through in-context learning [37]. This approach has demonstrated strong performance in various generative tasks. We craft a specialised prompt that incorporates task-related information, including the task title, introduction and step-by-step instructions. The model employed for this study is GPT-3.5, and the corresponding prompt is as follows:

> You help in generating a mix-initiative synthetic multi-turn task-oriented dialogue, each utterance starts with [user] or [system], while [user] initiates the conversations, and they take turns (with a similar number of turns to the number of steps) in giving utterances, by leveraging the task instructions as input but without explicitly mentioning the steps. Example input: [task title: □, instruction: □], Output: {[user] □, [system] □, ..., [system] □}.

□ denotes placeholder information, omitted here due to space constraints. Using Task2KB, this approach allows us to automatically produce complete dialogues for each task without the need for heavy model training or human intervention.

To summarise, this section outlines three strategies employed in this study to produce synthetic task-oriented dialogues. These strategies aim to enhance TOD models and ultimately improve user experience through more accurate and contextually relevant system responses. Details for the implementation of each method will be covered in the subsequent section.

4 Experimental Setup

In this section, we outline the experimental framework designed to implement and assess the three strategies for generating synthetic datasets and their effectiveness in enhancing task-oriented dialogue systems. We commence with an investigation into the optimization of the INST2DIAL-Auto strategy, guided by the following research question:

RQ1: Which strategy learns the best question generator (INST2DIAL-Auto) for the generation of task-oriented dialogues?

In line with the methodology articulated in [9], our first step involves conducting a user study to evaluate the quality of the synthetic dialogues we generate, especially given the absence of ground-truth benchmarks. Subsequently, we use these high-quality synthetic dialogues as model pre-training data, thereby boosting more robust model performance. With this approach, we intend to address the subsequent research questions:

RQ2: Which among the three strategies for synthetic dialogue generation yields the highest quality of dialogues as per human evaluation?

RQ3: Does the top-performing strategy with the resulting synthetic dialogues also contribute to improving state-of-the-art TOD models?

To answer these questions, we have crafted a comprehensive set of experiments for each of the three strategies with code and resources publicly[2]:

Table 1. Statistic of datasets for learning question generators.

Type	Dataset	Train	Valid	Test
# Dialogues	QReCC	11,020	1,409	1,409
	ORConvQA	8,766	980	1,542
	MultiWoZ	8,437	1,000	1,000
# QA pairs	QReCC	52,481	6,816	6,817
	ORConvQA	22,760	2,450	4,029
	MultiWoZ	16,352	2,085	2,133

Table 2. The Perplexity score of running various language models on multiple conversational datasets.

Models	ORConvQA	QReCC	MultiWoZ
T5-base	4.6423	3.7980	3.8165
Flan-T5-base	**4.5446**	**3.7214**	3.6611
Flan-T5-large	5.3126	3.8204	**3.6058**

[2] https://github.com/wangxieric/task2kb-resource.

INST2DIAL-Auto: At first, we explore the performance differences between open-domain and task-oriented dialogues for fine-tuning Language Models (LMs) as question generators. We use T5-base, Flan-T5-base, and Flan-T5-large models, fine-tuning them on two open-domain datasets – ORConvQA [26] and QReCC [1] – as well as one task-oriented dataset, MultiWoZ. Table 1 provides a summary of these datasets, highlighting similarities in dialogue count but variations in conversation length (measured by the number of QA turns). QReCC encompasses longer conversations compared to the other two datasets, which have similar turns of QAs. Then, we fine-tune these models as question generators using a default learning rate of 1e-4, the widely-used Adam optimizer [17] and cross-entropy loss. Our evaluation proceeds in several steps: We first assess the overall performance of each fine-tuned model across the three datasets. We then evaluate the effectiveness of these models specifically for question generation in task-oriented dialogues, aiming to highlight any performance disparities. Finally, we examine the impact of context input types-Single-QA, Last-QA, and Full-QA-each of which incorporates historical utterances differently. This aspect of the evaluation is particularly focused on the MultiWoZ dataset and the inputs for final generation (TOC-Auto) to explore the influence of varying input lengths.

INST2DIAl-Manual: To collect INST2DIAL-Manual, we conduct a user study using the Amazon Mechanical Turk platform[3]. To ensure a high quality of written English, we restricted participation to individuals residing in English-speaking countries, as identified by the relevant Wikipedia page[4]. Further refining our participant tool, we required each worker to have successfully completed at least 2,000 previous tasks (known as HITs) with an approval rate exceeding 95%. To guide question construction, we implemented specific guidelines into the "Check" button functionality. These guidelines mandate each question should begin with one of the updated 5Ws ('how', 'what', 'when', 'where', or 'why'), contain at least five words, and conclude with a question mark. After crafting a question, workers are obliged to select the corresponding answer from the task's step description before proceeding to the next step or submitting their final responses. To minimize the risk of low-quality or copied inputs, we deactivated the copy-paste feature within the text box designated for question formulation and collected a minimum of two dialogues for each task. For their contributions, workers were compensated at a rate of US$0.10 per completed dialogue.

INST2DIAL-ICL: For the generation of the INST2DIAL-ICL dataset, we followed the prompt configuration detailed in Sect. 3.3. We utilized the pretrained GPT-3.5 model, specifically the gpt-3.5-turbo version, in conjunction with OpenAI's Chat Completion API. We adhered to the API's default settings for the dialogue generation process.

To address RQ2, we evaluate the three types of synthetic dialogues generated from a shared random sample of 100 tasks in 19 categories from Task2KB. Conducted on Amazon Mechanical Turk, our user study employs 10 criteria

[3] https://www.mturk.com/.

[4] https://en.wikipedia.org/wiki/English-speaking_world.

Interestingness

The overall engagement level of the dialogue.

Inquisitiveness

Did the user actively seek relevant information or did they appear uninterested in the details?

Task Completeness

Was the task they set out to accomplish fully completed by the end of the conversation?

Task Relatedness

The relevancy of the conversation with respect to the task

Question-Answer Relatedness

Assess how aptly the system's responses match the user's questions.

Informativeness

The depth, richness, and usefulness of the information the system provides in its responses.

Fluency

Did both the user and the system communicate in a coherent and smooth manner?

Humanness (User)

Did the user exhibit qualities typical of human interaction Or did they seem more like an automated entity?

Mechanicalness (System)

Did the user exhibit qualities typical of human interaction Or did they seem more like an automated entity?

***Misinformation** (If task information given)

Misinformation refers to the accuracy of the information provided by the system.

Fig. 3. Dialogue Evaluation Aspects

based on a recent dialogue quality framework [32]. These 10 evaluative aspects are detailed in Fig. 3. Participants evaluate dialogues in two scenarios: with and without task instructions. The 'misinformation' metric is applied only when instructions are available. Each dialogue undergoes six independent evaluations-three per scenario-and workers receive US$0.10 per evaluation.

Table 3. Generated questions linked to an example response selected from a task step description. Key information are highlighted in bold. Sentences that are not proper questions are marked with a [×] symbol.

Task Title		How to learn music theory online?
LLM	Dataset	Generated Dialogue Examples
Flan-T5-Base	**QReCC**	You can **learn music theory online** for **free**? [×]
	ORConvQA	How to learn **music theory online**?
	MultiWoZ	Can you recommend a good place to find a **music theory lesson**?
Flan-T5-Large	**QReCC**	if you're looking for **a low cost way to learn music theory**? [×]
	ORConvQA	You can find a tutor or teacher to teach you? [×]
	MultiWoZ	What is **the best way** to find a **free course**?
Example Response		*Online learning* is **a great way** to find a **lesson** taught by a professional **without having to pay the cost**."

5 Results

In this section, we present the experimental results and conduct a thorough analysis that answers the first two research questions.

Optimise Question Generator for INST2DIAL-Auto (RQ1): To optimise the configuration for our learned question generator, we initially evaluate various encoder-decoder models across three datasets: ORConvQA, QReCC and MultiWoZ. The results of this evaluation can be found in Table 2. Notably, the

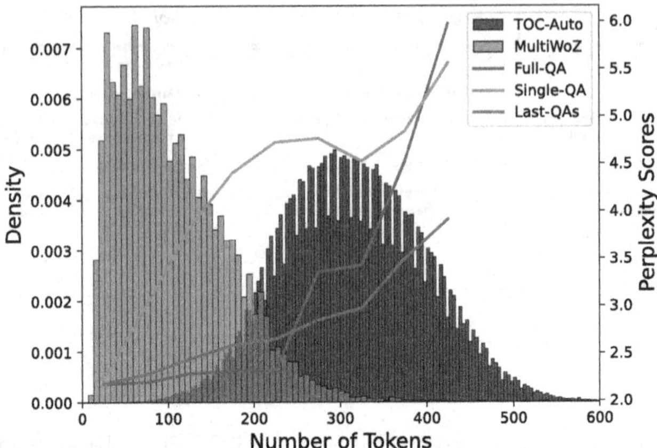

Fig. 4. Evaluation of single-QA, Pair-QAs and Full-QAs on MultiWoZ with distinct input lengths.

Flan-T5 model [8] outperforms T5 [29] with lower perplexity scores, underscoring the effectiveness of language models fine-tuned on instructional resources or chain-of-thought documents. To elucidate the disparities among questions generated by different language models, we provide an illustrative example, using a response as input and comparing various generated questions. These are detailed in Table 3. In our observations, models trained on open-domain dialogues struggle to generate contextually accurate and proper questions. In addition, the Flan-T5-Large model, which scored best in our initial evaluation, after trained on MultiWoZ, adeptly captures critical response elements like 'a great way', 'lesson' and 'without having to pay the cost', resulting in highly relevant questions. Therefore, we employ the Flan-T5-Large model, fine-tuned on the MultiWoZ dataset, for our subsequent dialogue generation experiments. We also assess model performance under varying conditions by examining average perplexity scores as the length and complexity of conversational history change. Figure 4 visualizes this, juxtaposing three question-generating setups: Single-QA, Last-QA, and Full-QA. Our findings indicate that all three approaches experience a performance decline in generating questions for longer conversations. Specifically, Single-QA consistently lags behind, primarily due to its disregard for conversational history. Full-QA excels in shorter dialogues, but its performance decreases as the input length increases. Last-QA proves to be the most resilient, maintaining stable performance irrespective of input length. To sum up, in response to Research Question 1 (RQ1), the Last-QA approach, when used in conjunction with the Flan-T5-Large model fine-tuned on the MultiWoZ dataset, achieves an effective balance between historical context retention and adaptability to longer dialogues.

Quality of Synthetic Dialogues as per Human Evaluation (RQ2): To evaluate the quality of dialogues generated through three distinct strategies –

Inst2Dial-Auto/Manual/ICL – we conducted a comprehensive user study, collecting feedback on 100 sampled dialogues from each group. Specifically, each dialogue was evaluated by at least 3 crowd workers as per 10 evaluation criteria (see Fig. 3). These criteria were applied both with and without the provision of task instructions. Figure 5 displays the average scores for these metrics, which range from 1 to 4, for each dataset. Upon examining the evaluations under both scenarios, with and without task instructions, it became evident that participants gave more varied scores when additional contextual information was available. Across all three strategies, the generated dialogues were found to be interesting, fluent, and task-relevant. However, when metrics such as task completeness, question-answer relevance, and informativeness were considered, Inst2Dial-Auto consistently outperformed the other two datasets. This was particularly true for the informativeness metric when complete information was provided for comparison. Interestingly, Inst2Dial-ICL scored the lowest in terms of accurate information dissemination, as reflected by its misinformation scores – a finding that aligns with the observed hallucination issue of LLMs [35]. Overall, Inst2Dial-Auto emerged as the most satisfactory dataset under both evaluation conditions. Therefore, in response to RQ2, Inst2Dial-Auto yields the most satisfactory dialogues according to user feedback, compared to the other two synthetic dialogue types.

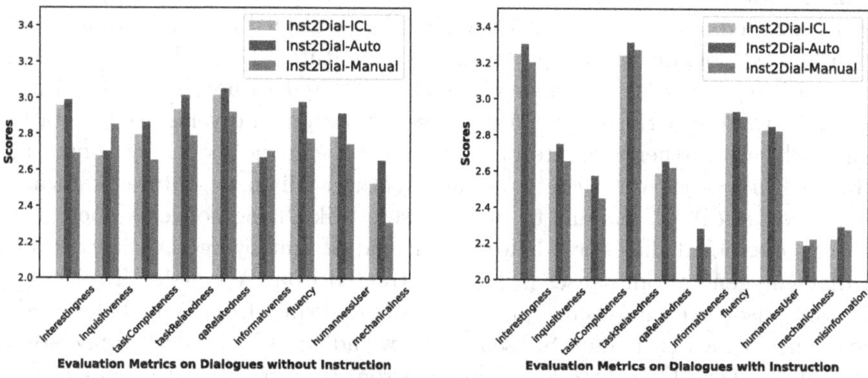

Fig. 5. Dialogue Quality Comparison on Essential Aspects as per User Feedback.

6 Application

Based on the evaluation results from the previous section, the INST2DIAL-Auto dataset – generated using a fine-tuned encoder-decoder model – emerged as the best-performing strategy. This leads us to RQ3, which investigates the impact of using INST2DIAL-Auto on SOTA TOD models. While incorporating synthetic data into the training set could improve model performance, it also risks complicating dialogue structuring and demands additional engineering efforts. To mitigate these challenges, we opt to fine-tune a pre-existing language model,

DistilGPT2 [31], on our synthetic dataset to encode task-specific knowledge. We then integrate this fine-tuned model into two recent advanced neural TOD models to assess the impact on performance. Additionally, we introduce a knowledge-augmented loss function specifically designed for fine-tuning Language Models (LMs) to effectively generate task-oriented responses on the full INST2DIAL-Auto dataset with all tasks, 251,433 in total, from Task2KB. The loss function for response generation ($\mathcal{L}_{RG}$) is defined as $\mathcal{L}_{RG} = -\sum_{t=1}^{T} \log p(r_t|q_t, r < t, q^{title}, a^i)$, where q_t denotes the question asked during the t-th conversational turn and $r < t$ represents the modelling of prior conversations. We employ the Adam optimiser [17] with a learning rate of $2e^{-}5$. For empirical validation, the fine-tuned LMs serve as substitute backbones for two state-of-the-art TOD models: **(1) UBAR** [38], an advanced model that extends SimpleTOD [14] and Soloist [24], giving it complete access to full dialogues, beliefs, database states and system acts for each conversational turn. **(2) JSA-TOD** [5], which employs a recent joint stochastic approximation algorithm [22] for semi-supervised learning, concentrating on leveraging both labelled and unlabelled dialogue data.

We assess the performance of the trained DistilGPT2 across multiple training settings: full supervision, few-shot learning and limited domain knowledge setting. For full supervision, we adhere to the publicly accessible implementations of baseline models, using fully labelled training data. In the few-shot learning context, we train UBAR and JSA-TOD models on randomly sample 10% of the training data. Notably, for JSA-TOD, we use a semi-supervised setup, comprising 3% labelled and 7% unlabeled dialogue data sampled from the training set. Upon limited domain knowledge setup, we substitute 10% of the training data with domain-specific dialogues, such as those related to hotels or restaurants, and then evaluate performance on a test set containing a diverse array of tasks. Conversely, we further evaluate our fine-tuned backbone model by comparing it to an LM trained on a contemporary synthetic dialogue dataset known as WikiDialog (wdl) [9]. Similar to our dataset, WikiDialog converts documents into conversational dialogues; however, instead of employing task instructions for dialogue generation, it uses passages extracted from Wikipedia. To ensure a fair comparison, we sample an equal number of dialogues from WikiDialog as the size of ours, which is smaller. Subsequently, we adhere to an identical procedure in training a DistilGPT2 model for comparison. In particular, we evaluate various UBAR and JSA-TOD implementations on the MultiWoZ datasets, version 2.0 [4] and 2.1 [12], respectively, as they reported in their paper. For a comprehensive evaluation, we rely on key metrics associated with the MultiWoZ dataset: informativeness, success rate, BLEU scored and combined performance metrics.

In Table 4, we present the experimental results. First, under a mixed-domain setup, we note tangible enhancements in both UBAR and JSA-TOD models when employing DistilGPT2 trained on INST2DIAL-Auto in a few-shot learning context. These gains do diminish when juxtaposed with the full supervision baseline, a phenomenon attributable to the well-understood issue of catastrophic forgetting [15]. Nevertheless, the use of INST2DIAL-Auto continues to offer observable benefits. Subsequently, we scrutinize the performance of these models in sce-

Table 4. The experimental results of UBAR and JSA-TOD on MultiWoZ. 're', 'wdl' and 'our' refer to the use of initial DistilGPT2, the one trained on Wikidialog and our synthetic data, respectively. The improvement ratio is compared to the reproduced model on combined scores.

UBAR					
Setups	**Inform**	**Success**	**BLEU**	**Combined**	**Impr. %**
Few Shot Setting with Mixture Domains					
Few Shot + re	50.55	37.94	10.88	55.12	-
Few Shot + wdl	53.95	41.94	11.49	59.44	7.8%
Few Shot + our	**54.85**	**42.34**	**11.59**	**60.19**	**9.1%**
Full Supv. + re	87.69	75.88	**14.87**	96.65	-
Full Supv. + wdl	89.99	78.58	14.83	99.11	2.5%
Full Supv. + our	**90.19**	**79.08**	14.83	**99.46**	**2.9%**
Domain Generalisability					
Few Shot (hotel) + re	41.74	25.33	10.81	44.34	-
Few Shot (hotel) + wdl	45.75	29.43	11.78	49.36	11.3%
Few Shot (hotel) + our	**48.25**	**32.13**	**11.86**	**52.05**	**17.4%**
Few Shot (train) + re	51.65	32.23	10.79	52.73	-
Few Shot (train) + wdl	48.15	28.63	10.99	49.38	-6.3%
Few Shot (train) + our	**56.45**	**37.73**	**11.56**	**58.65**	**11.2%**
Few Shot (attraction) + re	47.45	31.53	11.13	50.62	-
Few Shot (attraction) + wdl	51.65	32.93	9.92	52.21	3.1%
Few Shot (attraction) + re	**55.16**	**39.04**	**11.52**	**58.62**	**15.8%**
Few Shot (restaurant) + re	47.75	29.13	10.69	49.13	-
Few Shot (restaurant) + wdl	51.75	34.53	12.17	55.31	12.6%
Few Shot (restaurant) + our	**54.95**	**37.74**	**12.33**	**58.68**	**19.4%**
Few Shot (taxi) + re	37.14	18.62	8.58	36.46	-
Few Shot (taxi) + wdl	45.65	24.12	8.98	43.87	20.3%
Few Shot (taxi) + our	**45.75**	**24.22**	**9.10**	**44.09**	**20.9%**

JSA-TOD					
Setups	**Inform**	**Success**	**BLEU**	**Combined**	**Impr.%**
Few Shot Setting with Mixture Domains					
Few Shot + re	53.10	36.30	**14.86**	59.56	
Few Shot + wdl	55.80	37.40	13.38	59.98	0.7%
Few Shot + our	**57.80**	**40.40**	12.96	**62.06**	**4.2%**
Full Supv. + re	86.30	75.90	19.08	100.18	
Full Supv. + wdl	83.40	73.40	18.65	97.05	-3.1%
Full Supv. + our	**87.51**	**76.20**	**19.87**	**101.73**	**1.5%**
Domain Generalisability					
Few Shot (hotel) + re	28.90	11.40	11.09	31.24	
Few Shot (hotel) + wdl	**29.50**	13.40	11.12	32.57	4.2%
Few Shot (hotel) + our	28.10	**15.60**	**12.06**	**33.91**	**8.5%**
Few Shot (train) + re	28.40	11.80	11.00	31.10	
Few Shot (train) + wdl	27.40	10.80	12.61	31.71	2.0%
Few Shot (hotel) + our	**28.80**	**14.60**	**12.64**	**34.34**	**10.4%**
Few Shot (attraction) + re	27.60	**13.90**	8.17	28.92	
Few Shot (attraction) + wdl	25.90	11.30	11.00	29.60	2.4%
Few Shot (attraction) + our	**28.50**	12.30	**10.11**	**30.51**	**5.5%**
Few Shot (restaurant) + re	28.90	17.20	12.65	35.70	
Few Shot (restaurant) + wkl	27.40	14.80	**13.25**	34.35	-3.8%
Few Shot (restaurant) + our	**33.60**	**24.00**	13.02	**41.82**	**17.1%**
Few Shot (taxi) + re	23.90	9.80	**8.35**	25.20	
Few Shot (taxi) + wkl	26.10	9.50	7.96	25.76	2.2%
Few Shot (taxi) + our	**27.30**	**11.80**	8.26	**27.81**	**10.4%**

narios with restricted domain knowledge. Here, we identify marked and consistent performance uplifts when integrating our INST2DIAL-Auto dataset, as substantiated by the data in Table 4. Thus, we affirm that employing INST2DIAL-Auto can augment the capabilities of pre-trained language models when deployed in task-oriented dialogue systems. This elevation in performance is particularly pronounced in both few-shot learning and out-of-domain application scenarios.

Next, we assess the utility of encoding task-specific instructions as opposed to utilizing open-domain resources like Wikipedia. Firstly, under both few-shot and full supervision paradigms, a DistilGPT2 model trained on WikiDialog fails to consistently outperform the baseline, particularly for the JSA-TOD model. In fact, it results in a 3.1% decrease in the combined score under full supervision. Secondly, when juxtaposing DistilGPT2 models fine-tuned on both WikiDialog and INST2DIAL-Auto to test domain generalizability, INST2DIAL-Auto emerges as demonstrably more effective, driving both models to achieve superior results. To address **RQ3**, our conclusions indicate that the INST2DIAL-Auto dataset is highly effective in enhancing the ability of task-oriented dialogue models to accurately interpret and respond to task-specific information across diverse categories. This results in noticeably improved conversational responses.

7 Conclusions

In this study, we explored three innovative strategies for generating synthetic dialogues with the aim of enhancing Task-Oriented Dialogue (TOD) models. The first approach leverages a sophisticated neural question generator within an optimized pipeline to produce the Inst2Dial-Auto dataset. For the other two datasets, Inst2Dial-Manual/ICL, we deployed a carefully designed user study and in-context learning prompts, respectively. Our empirical evaluation, rooted in human evaluation metrics, revealed that dialogues produced via a finely-tuned question generator-Inst2Dial-Auto-consistently yielded the highest quality. When applied to state-of-the-art TOD models, this dataset contributed to substantial improvements, most notably in scenarios with limited domain knowledge, registering a minimum uplift of 5.5% in combined evaluation scores.

Acknowledgement. This research is supported by the Alan Turing Institute under the EPSRC grant [EP/N510129/1] and the EPSRC Fellowship titled "Task Based Information Retrieval" [EP/P024289/1].

References

1. Anantha, R., Vakulenko, S., Tu, Z., Longpre, S., Pulman, S., Chappidi, S.: Open-domain question answering goes conversational via question rewriting. In: Proceedings of NAACL (2021)
2. Bao, J., et al.: A synthetic data generation framework for grounded dialogues. In: Proceedings of ACL (2023)
3. Boyer, K., Ha, E.Y., Phillips, R., Wallis, M., Vouk, M., Lester, J.: Dialogue act modeling in a complex task-oriented domain. In: Proceedings of SIGDIAL (2010)
4. Budzianowski, P., et al.: MultiWOZ - a large-scale multi-domain Wizard-of-Oz dataset for task-oriented dialogue modelling. In: Proceedings of EMNLP (2018)
5. Cai, Y., Liu, H., Ou, Z., Huang, Y., Feng, J.: Advancing semi-supervised task oriented dialog systems by JSA learning of discrete latent variable models. In: Proceedings of SIGDIAL (2022)
6. Chen, D., Yu, Z.: Sources of noise in dialogue and how to deal with them. In: Proceedings of SIGDIAL (2023)
7. Chen, X., Xu, J., Xu, B.: A working memory model for task-oriented dialog response generation. In: Proceedings of ACL (2019)
8. Chung, H.W., et al.: Scaling instruction-finetuned language models. arXiv preprint arXiv:2210.11416 (2022)
9. Dai, Z., et al.: Dialog inpainting: turning documents into dialogs. In: Proceedings of ICML (2022)
10. De Cicco, R., Silva, S.C.L.d.C.e., Alparone, F.R.: "It's on its way": chatbots applied for online food delivery services, social or task-oriented interaction style? J. Foodserv. Bus. Res. **24**(2), 140–164 (2021)
11. El Asri, L., et al.: Frames: a corpus for adding memory to goal-oriented dialogue systems. In: Proceedings of SIGdial (2017)
12. Eric, M., et al.: Multiwoz 2.1: a consolidated multi-domain dialogue dataset with state corrections and state tracking baselines. In: Proceedings of LREC (2020)
13. Hosseini-Asl, E., McCann, B., Wu, C.S., Yavuz, S., Socher, R.: A simple language model for task-oriented dialogue. In: Proceedings of NeurIPS (2020)
14. Hosseini-Asl, E., McCann, B., Wu, C., Yavuz, S., Socher, R.: A simple language model for task-oriented dialogue. In: Proceedings of NeurIPS (2020)
15. Hu, W., et al.: Overcoming catastrophic forgetting for continual learning via model adaptation. In: Proceeding of ICLR (2019)

16. Jin, D., Kim, S., Hakkani-Tur, D.: Can i be of further assistance? using unstructured knowledge access to improve task-oriented conversational modeling. In: Proceedings of DialDoc (2021)
17. Kingma, D.P., Ba, J.L.: Adam: a method for stochastic optimization. In: Proceedings of ICLR (2015)
18. Li, J.J., Nenkova, A.: Fast and accurate prediction of sentence specificity. In: Proceedings of AAAI (2015)
19. Madotto, A., Wu, C.S., Fung, P.: Mem2seq: effectively incorporating knowledge bases into end-to-end task-oriented dialog systems. In: Proceedings of ACL (2018)
20. Mohapatra, B., Pandey, G., Contractor, D., Joshi, S.: Simulated chats for building dialog systems: learning to generate conversations from instructions. In: Proceedings of EMNLP (2021)
21. OpenAI: GPT-4 technical report. arXiv preprint arXiv:2303.08774 (2023)
22. Ou, Z., Song, Y.: Joint stochastic approximation and its application to learning discrete latent variable models. In: Proceedings of UAI (2020)
23. Papangelis, A., Wang, Y.C., Molino, P., Tur, G.: Collaborative multi-agent dialogue model training via reinforcement learning. In: Proceedings of SIGDIAL (2019)
24. Peng, B., Li, C., Li, J., Shayandeh, S., Liden, L., Gao, J.: SOLOIST: building task bots at scale with transfer learning and machine teaching. Trans. Assoc. Comput. Linguist. **9**, 824–907 (2021)
25. Procheta, S., Xi, W., Ruiqing, X., Emine, Y.: Task2kb: a public task-oriented knowledge base. In: Proceedings of AAAI (2023)
26. Qu, C., Yang, L., Chen, C., Qiu, M., Croft, W.B., Iyyer, M.: Open-retrieval conversational question answering. In: Proceedings of SIGIR (2020)
27. Quan, J., Zhang, S., Cao, Q., Li, Z., Xiong, D.: Risawoz: a large-scale multi-domain wizard-of-oz dataset with rich semantic annotations for task-oriented dialogue modeling. In: Proceedings of EMNLP (2020)
28. Radford, A., et al.: Language models are unsupervised multitask learners. OpenAI blog **1**(8), 9 (2019)
29. Raffel, C., et al.: Exploring the limits of transfer learning with a unified text-to-text transformer. J. Mach. Learn. Res. **21**(1), 5485–5551 (2020)
30. Rastogi, A., Zang, X., Sunkara, S., Gupta, R., Khaitan, P.: Towards scalable multi-domain conversational agents: the schema-guided dialogue dataset. In: Proceedings of AAAI (2020)
31. Sanh, V., Debut, L., Chaumond, J., Wolf, T.: Distilbert, a distilled version of BERT: smaller, faster, cheaper and lighter. arXiv preprint arXiv:1910.01108 (2019)
32. See, A., Roller, S., Kiela, D., Weston, J.: What makes a good conversation? how controllable attributes affect human judgments. In: Proceedings of NAACL-HLT (2019)
33. Sen, P., Wang, X., Xu, R., Yilmaz, E.: Task2kb: a public task-oriented knowledge base. In: Proceedings of AAAI (2023)
34. Shah, P., et al.: Building a conversational agent overnight with dialogue self-play. arXiv preprint arXiv:1801.04871 (2018)
35. Shuster, K., Poff, S., Chen, M., Kiela, D., Weston, J.: Retrieval augmentation reduces hallucination in conversation. In: Proceedings of EMNLP (Findings) (2021)
36. Srivastava, M., Lu, Y., Peschon, R., Li, C.: Pretrain-finetune based training of task-oriented dialogue systems in a real-world setting. In: Proceedings of NAACL (2021)
37. Wei, J., et al.: Chain-of-thought prompting elicits reasoning in large language models. In: Proceedings of NeurIPS (2022)

38. Yang, Y., Li, Y., Quan, X.: Ubar: towards fully end-to-end task-oriented dialog system with gpt-2. In: Proceedings of AAAI (2021)
39. Ye, F., Wang, X., Huang, J., Li, S., Stern, S., Yilmaz, E.: Metaassist: robust dialogue state tracking with meta learning. In: Proceedings of EMNLP (2022)
40. Zang, X., Rastogi, A., Sunkara, S., Gupta, R., Zhang, J., Chen, J.: Multiwoz 2.2 : a dialogue dataset with additional annotation corrections and state tracking baselines. arXiv preprint arXiv:2007.12720 (2020)

Investigating the Effects of Sparse Attention on Cross-Encoders

Ferdinand Schlatt(✉), Maik Fröbe, and Matthias Hagen

Friedrich-Schiller-Universität Jena, Jena, Germany
ferdinand.schlatt@uni-jena.de

Abstract. Cross-encoders are effective passage and document re-rankers but less efficient than other neural or classic retrieval models. A few previous studies have applied windowed self-attention to make cross-encoders more efficient. However, these studies did not investigate the potential and limits of different attention patterns or window sizes. We close this gap and systematically analyze how token interactions can be reduced without harming the re-ranking effectiveness. Experimenting with asymmetric attention and different window sizes, we find that the query tokens do not need to attend to the passage or document tokens for effective re-ranking and that very small window sizes suffice. In our experiments, even windows of 4 tokens still yield effectiveness on par with previous cross-encoders while reducing the memory requirements by at least 22%/59% and being 1%/43% faster at inference time for passages/documents. Our code is publicly available (https://github.com/webis-de/ECIR-24).

Keywords: Cross-encoder · Re-ranking · Windowed attention · Cross-attention

1 Introduction

Pre-trained transformer-based language models (PLMs) are important components of modern retrieval and re-ranking pipelines as they help to mitigate the vocabulary mismatch problem of lexical systems [61,84]. Especially cross-encoders are effective [52,55,57,81] but less efficient than bi-encoders or other classic machine learning-based approaches with respect to inference run time, memory footprint, and energy consumption [68]. The run time issue is particularly problematic for practical applications as searchers often expect results after a few hundred milliseconds [2]. To increase the efficiency but maintain the effectiveness of cross-encoders, previous studies have, for instance, investigated reducing the number of token interactions by applying sparse attention patterns [44,70].

Sparse attention PLMs restrict the attention of most tokens to local windows, thereby reducing token interactions and improving efficiency [72]. Which tokens have local attention is a task-specific decision. For instance, cross-encoders using

N. Goharian et al. (Eds.): ECIR 2024, LNCS 14608, pp. 173–190, 2024.
https://doi.org/10.1007/978-3-031-56027-9_11

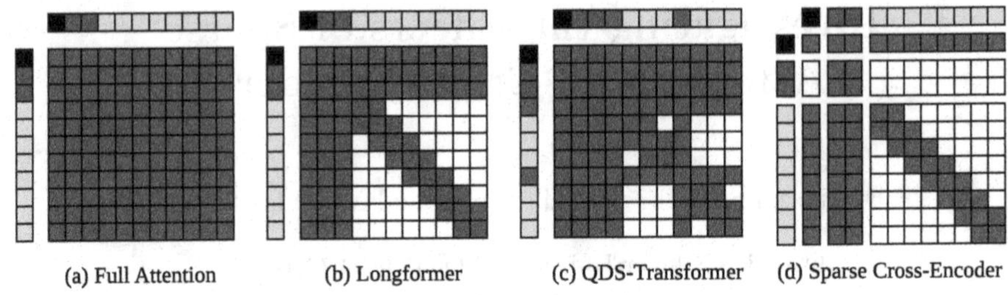

Fig. 1. Previous cross-encoder attention patterns (a, b, and c) and our newly proposed sparse cross-encoder (d). The marginal boxes denote input tokens (black: [CLS], blue: query, yellow: passage/document, red: start of sentence). The inner green boxes indicate token attention. Our new pattern considers the sub-sequences separately (indicated by the added spacing) and is asymmetric. (Color figure online)

the Longformer model [3] apply normal global attention to query tokens but local attention to document tokens. The underlying idea is that a document token does not require the context of the entire document to determine whether it is relevant to a query—within a document, most token interactions are unnecessary, and a smaller context window suffices (cf. Fig. 1(b) for a visualization).

Previously, sparse attention has been applied to cross-encoders to be able to re-rank long documents without cropping or splitting them [44,70]. However, the impact of sparsity on effectiveness has not been investigated in detail. To close this gap, we explore the limits of sparse cross-encoders and try to clarify which token interactions are (un)necessary. As mechanisms to reduce token interactions, we investigate local attention window sizes and disabling attention between sub-sequences (e.g., between the query and the passage or document). Our analyses are based on the following two assumptions.

(1) Cross-encoders create contextualized embeddings that encode query and passage or document semantics. We hypothesize that the contextualized embeddings of passage or document tokens do not actually need to encode fine-grained semantics. Rather, an overall "gist" in the form of small local context windows is sufficient to estimate relevance well.

(2) Cross-encoders allow queries and documents to exchange information via symmetric attention. We hypothesize that full symmetry is unnecessary as we view the query–document relevance relationship as asymmetric: for ranking, it should suffice to determine whether a result is relevant to a query, not vice versa. To further reduce token interactions, we propose a novel configurable asymmetric cross-attention pattern with varying amounts of interaction between [CLS], query, and passage or document tokens (cf. Fig. 1(d)).

In experiments on re-ranking tasks from the TREC Deep Learning tracks [26–29] and the TIREx benchmark [36], our new model's effectiveness is consistently on par with previous (sparse) cross-encoder models, even with a local window size of only four tokens and asymmetric attention. Further efficiency analyses for sequences with 174/4096 tokens show that our sparsification techniques reduce the memory requirements by at least 22%/59% and yield inference times that are at least 1%/43% faster.

2 Related Work

One strategy for using PLMs in ranking is to separate the encoding of queries and documents (or passages) [45,46,61,67]. Such bi-encoder models are efficient retrievers as the document encodings can be indexed, so only the query needs to be encoded at retrieval time. However, the good efficiency of bi-encoders comes at reduced effectiveness compared to cross-encoders [40], especially in out-of-domain scenarios [66]. Consequently, bi-encoders are often used for first-stage retrieval, while the more effective but less efficient cross-encoders serve as second-stage re-rankers [52,55,57,81]. We focus on improving the efficiency of cross-encoders while maintaining their effectiveness.

One strategy to make cross-encoders more efficient is to reduce the model size via knowledge distillation [41,42,50]. During knowledge distillation, a smaller and more efficient model aims to mimic a larger "teacher" model. The distilled models can often match the effectiveness of the teacher model at only a fraction of the computational costs [79], indicating that PLMs can be "overparameterized" for re-ranking [40]. We follow a similar idea and try to reduce token interactions in cross-encoders using sparse attention techniques to substantially lower computational costs at comparable effectiveness.

Previously, sparse attention PLMs aimed to increase the processable input length [72]. Instead of full attention across the entire input, a sparse PLM restricts attention to neighboring tokens. For instance, the Sparse Transformer [12] uses a block-sparse kernel, splitting the input into small blocks where tokens only attend to tokens in their block. Additional strided attention allows for attention between blocks. The Longformer [3], BigBird [83], and ETC [1] use a different approach. Tokens can attend to a fixed window of neighboring tokens, with additional global attention tokens that can attend to the entire sequence.

For efficient windowed self-attention, the Longformer, BigBird, and ETC use block-sparse matrices. However, block-sparse techniques often make concessions on the flexibility of window sizes or incur additional overhead that does not affect the resulting predictions. We compare several previously proposed windowed self-attention implementations and find them inefficient in terms of time or space compared to the reduction in operations. Therefore, we implement a custom CUDA kernel and compare it to other implementations (cf. Sect. 4.2).

Sparse attention PLMs have also been applied to document re-ranking. For example, the Longformer without global attention was used as a cross-encoder to re-rank long documents [70]. However, effectiveness was not convincing as later-appearing document tokens were unable to attend to query tokens. The QDS-Transformer [44] fixed this problem by correctly applying global attention to query tokens and achieved better retrieval effectiveness than previous cross-encoder strategies that split documents [30,82]. While the QDS-Transformer was evaluated with different window sizes, the effectiveness results were inconclusive.

A model fine-tuned on a window size of 64 tokens was tested with smaller (down to 16 tokens) and larger window sizes (up to 512)—always yielding worse effectiveness. We hypothesize that models specifically fine-tuned for smaller window sizes will be as effective as models fine-tuned for larger window sizes.

Besides analyzing models fine-tuned for different window sizes, we hypothesize that token interactions between some input sub-sequences are unnecessary. For example, bi-encoder models show that independent contextualization of query and document tokens can be effective [45,46,61,67]. However, the symmetric attention mechanisms of previous sparse PLM architectures do not accommodate asymmetric attention patterns. We develop a new cross-encoder variant that combines windowed self-attention from sparse PLMs with asymmetric cross-attention. Cross-attention allows a sequence to attend to an arbitrary other sequence and is commonly used in transformer architectures for machine translation [73], computer vision [35,59,71], and in multi-modal settings [43].

3 Sparse Asymmetric Attention Using Cross-Encoders

We propose a novel sparse asymmetric attention pattern for re-ranking documents (and passages) with cross-encoders. Besides combining existing windowed self-attention and cross-attention ideas, our pattern also flexibly allows for asymmetric query–document interactions (e.g., allowing a document to attend to the query but not vice versa). To this end, we partition the input sequence into the [CLS] token, query tokens, and document tokens, with customizable attention between these groups and local attention windows around document tokens.

Figure 1 depicts our and previous cross-encoder attention patterns. In full attention, each token can attend to every other token. Instead, Longformer-based cross-encoders apply windowed self-attention to document tokens to which the QDS-Transformer adds global attention tokens per sentence. Our pattern is similar to the Longformer but deactivates attention from query tokens to [CLS] and document tokens. But, [CLS] and document tokens still have access to query tokens. Our hypothesis is that cross-encoders do not need symmetric query–document attention for re-ranking as a one-sided relationship suffices.

3.1 Preliminaries

A cross-encoder predicts a relevance score for a query–document pair (q, d) as follows. Given q and d as token sequences $q_1 \dots q_m$ and $d_1 \dots d_n$ and adopting BERT-style encoding [33], a concatenation of a special classification token [CLS], then q, then a separator token [SEP], then d, and then another [SEP] token is passed through a transformer encoder. The output of the last transformer layer is an $s \times h$ matrix O, with $s = m + n + 3$ being the total sequence length and h being the embedding dimensionality. Each column of O is an embedding vector of a token from the input sequence. There is one such O-matrix per transformer layer, but only the [CLS] token embedding of the last transformer layer (first column of that layer's O-matrix) is used as input to a final linear transformation that then predicts the relevance score of d for q.

The transformer encoder internally uses a dot-product attention mechanism [73]. For a single transformer layer, three separate linear transformations map the embedding matrix O' of the previous layer to three vector-lookup matrices Q, K, and V. An $s \times s$ attention probability matrix that contains the probabilities of a token attending to another is obtained by softmaxing the $\sqrt{h}$-normalized product QK^T. The attention probabilities are then used as weights for the vector-lookup matrix V to obtain the layer's output embedding matrix:

$$O = \text{Attention}(Q, K, V) = \text{softmax}\left(\frac{QK^T}{\sqrt{h}}\right) V .$$

3.2 Windowed Self-attention

Windowed self-attention was proposed for more efficient sparse PLM architectures [1,3,83]. The idea is that a token does not attend to the entire input sequence of length s, but only to a local window of w tokens, with $w \ll s$ (e.g., $w = 4$ means that a token attends to $2 \cdot 4 + 1$ tokens: to the 4 tokens before itself, to itself, and to the 4 tokens after itself). For a window size w, windowed self-attention changes the dot-products of the transformer attention mechanism to windowed variants $\boxdot_w$ and $\odot_w$:

$$O = \text{Attention}_w(Q, K, V) = \text{softmax}\left(\frac{Q \boxdot_w K^T}{\sqrt{h}}\right) \odot_w V \text{, with}$$

$$Q_{(s\times h)} \boxdot_w K^T_{(h\times s)} \rightarrow A_{(s\times 2w+1)}, \qquad \text{where } a_{i,j} = \sum_{l=1}^{h} q_{i,l} \cdot k_{l,i+j-w-1} \,,$$

$$P_{(s\times 2w+1)} \odot_w V_{(s\times h)} \rightarrow O_{(s\times h)}, \qquad \text{where } o_{i,l} = \sum_{j=1}^{2w+1} p_{i,j} \cdot v_{i+j-w-1,l} \,.$$

Thus, $\boxdot_w$ outputs a band matrix subset of the standard matrix–matrix multiplication, stored in a space-efficient form (non-band entries omitted), and $\odot_w$ multiplies a space-efficiently stored band matrix and a standard matrix. To ensure correctness, we zero-pad windows exceeding the sequence bounds: when either $i + j - w \leq 0$ or $i + j - w > s$, we set $k_{l,i+j-w-1} = v_{i+j-w-1,l} = 0$.

For a visual impression of windowed attention, consider the lower right document-to-document attention matrix in Fig. 1(d). Only the diagonal band is computed: in Fig. 1(d) for $w = 1$.

Compared to full self-attention, in theory, windowed self-attention reduces the space complexity from $\mathcal{O}(s^2)$ to $\mathcal{O}(s \cdot (2w + 1))$ and the (naïve) computational complexity from $\mathcal{O}(s^2 \cdot h)$ to $\mathcal{O}(s \cdot (2w + 1) \cdot h)$. However, fully achieving these improvements is difficult in practice. Previous windowed self-attention implementations avoided writing hardware-specific kernels and made concessions regarding flexibility, time efficiency, or space efficiency [3,83]. Therefore, we implement our own CUDA kernel for windowed self-attention; Sect. 4.2 compares our implementation's efficiency to previous implementations.

3.3 Cross-Attention

Cross-attention is a type of attention where a token sequence does not attend to itself, as in self-attention, but to a different sequence. We use cross-attention to configure attention between different token types. Instead of representing a cross-encoder's input as a single sequence, we split it into three disjoint subsequences: the [CLS] token, the query tokens, and the document tokens (Fig. 1(d) visually represents this for our proposed pattern by splitting the marginal vectors; the [SEP] tokens are part of "their" respective subsequence). Each subsequence can then have its own individual attention function $\text{Attention}(Q, K, V)$.

We split the vector-lookup matrices column-wise into [CLS], query, and document token-specific submatrices Q_c, Q_q, Q_d, etc. These matrices are pre-computed and shared between the different attention functions for efficiency. Restricting attention between subsequences then means to call the attention function for a subsequence's Q-matrix and the respective K- and V-matrices of the attended-to subsequences. For example, to let a document attend to itself and the query, the function call is $\text{Attention}(Q_d, [K_q, K_d], [V_q, V_d])$, where $[\cdot, \cdot]$ denotes matrix concatenation by columns (i.e., $[M, M']$ yields a matrix whose "left" columns come from M and the "right" columns from M').

3.4 Locally Windowed Cross-Attention

The above-described cross-attention mechanism using concatenation is not directly applicable in our case, as we want to apply windowed self-attention to document tokens and asymmetric attention to query tokens. Instead of concatenating the matrices K and V, our mechanism uses tuples $\mathcal{K}$ and $\mathcal{V}$ of matrices and a tuple $\mathcal{W}$ of window sizes to assign different attention window sizes $w \in \mathcal{W}$ to each attended-to subsequence. As a result, we can combine windowed self-attention with asymmetric attention based on token types. Formally, given j-tuples $\mathcal{K}$ and $\mathcal{V}$ of matrices K_i and V_i and a j-tuple $\mathcal{W}$ of window sizes w_i, our generalized windowed cross-attention mechanism works as follows:

$$\text{Attention}_{\mathcal{W}}(Q, \mathcal{K}, \mathcal{V}) = \sum_{i=1}^{j} P_i \odot_{w_i} V_i \,, \text{ where}$$

$$[P_1, \ldots, P_j] = \text{softmax}\left(\frac{[A_1, \ldots, A_j]}{\sqrt{h}}\right) \quad \text{and} \quad A_i = Q \boxdot_{w_i} K_i \,.$$

Our proposed attention pattern (visualization in Fig. 1(d)) is then formally defined as follows. The [CLS] token has full attention over all subsequences (Eq. 1; for notation convenience, we use $w = \infty$ to refer to full self-attention), the query tokens can only attend to query tokens (Eq. 2), and the document tokens can attend to all subsequences but use windowed self-attention on their own subsequence (Eq. 3):

$$O_c = \text{Attention}_{(\infty,\infty,\infty)}(Q_c, (K_c, K_q, K_d), (V_c, V_q, V_d)), \tag{1}$$

$$O_q = \text{Attention}_{(\infty)}(Q_q, (K_q), (V_q)), \tag{2}$$

$$O_d = \text{Attention}_{(\infty,\infty,w)}(Q_d, (K_c, K_q, K_d), (V_c, V_q, V_d)). \tag{3}$$

3.5 Experimental Setup

We fine-tune various models using the Longformer and our proposed attention pattern with window sizes $w \in \{\infty, 64, 16, 4, 1, 0\}$ (∞: full self-attention). We start from an already fine-tuned and distilled cross-encoder model[1] which also serves as our baseline [62]. We additionally fine-tune a QDS-Transformer model with its default $w = 64$ window for comparison. All models are fine-tuned for 100,000 steps with 1,000 linear warm-up steps and a batch size of 32 (16 document pairs) with margin MSE loss using MS MARCO-based knowledge distillation triples [40]. For documents, we extend the models fine-tuned on passages using positional interpolation [10] and further fine-tune them on document pairs from MS MARCO Document [54] for 20,000 steps using RankNet loss [8]. Negative documents are sampled from the top 200 documents retrieved by BM25 [65]. We use a learning rate of $7 \cdot 10^{-6}$, an AdamW optimizer [51], and a weight decay of 0.01. We truncate passages and documents to a maximum sequence length of 512 and 4096 tokens, respectively. All models were implemented in PyTorch [58] and PyTorchLightning [34] and fine-tuned on a single NVIDIA A100 40GB GPU.

We evaluate the models on the TREC 2019–2022 Deep Learning (DL) passage and document retrieval tasks [26–29] and the TIREx benchmark [36]. For each TREC DL task, we re-rank the top 100 passages/documents retrieved by BM25 using `pyserini` [49]. For TIREx, we use the official first-stage retrieval files retrieved by BM25 and ChatNoir [4,60] and also re-rank the top 100 documents. We measure nDCG@10 and access all corpora and tasks via `ir_datasets` [53], using the default text field for passages and documents.

To evaluate time and space efficiency, we generate random data with a query length of 10 tokens and passage/document lengths from 54 to 4086 tokens. For the QDS-Transformer, we set global sentence attention at every 30th token, corresponding to the average sentence length in MS MARCO documents. We use the largest possible batch size per model, but up to a maximum of 100.

4 Empirical Evaluation

We compare our sparse cross-encoder's re-ranking effectiveness and efficiency to full attention and previous sparse cross-encoder implementations. We also examine the impact of different small window sizes and of our attention deactivation pattern—analyses that provide further insights into how cross-encoders work.

[1] https://huggingface.co/cross-encoder/ms-marco-MiniLM-L-6-v2.

4.1 Effectiveness Results

In-domain Effectiveness Table 1 reports the nDCG@10 of various models with different attention patterns and window sizes on the TREC Deep Learning passage and document re-ranking tasks. We group Full Attention and Longformer models into the same category because they have the same pattern but different window sizes in our framework. We fine-tune separate models for passage and document re-ranking (cf. Sect. 3.5) except models with $w = \infty$. Their lack of efficiency prevents training on long sequences, and we only fine-tune them on passages but include their MaxP scores [30] for documents (in gray).

Since we hypothesize that sparse attention will not substantially affect the re-ranking effectiveness, we test for significant equivalence instead of differences. Therefore, we cannot use the typical t-test, but instead use a paired TOST procedure (two one-sided t-tests [69]; $p < 0.003$, multiple test correction [47]) to determine if the difference between two models is within ± 0.02. We deem ± 0.02 a reasonable threshold for equivalence since it is approximately the difference between the top two models in the different TREC Deep Learning tasks.

Table 1. Re-ranking effectiveness as nDCG@10 on TREC Deep Learning [26–29]. The highest score per task is given in bold. Scores obtained using a MaxP strategy are grayed out. † denotes significant equivalence within ± 0.02 (paired TOST [69], $p < 0.003$), compared to Full Attention $w = \infty$ for passage tasks and Longformer $w = 64$ for document tasks.

Task		Full Att./Longformer						Sparse Cross-Encoder						QDS
$w =$		∞	64	16	4	1	0	∞	64	16	4	1	0	64
Passage	2019	.724	.719†	.725†	.719	.714	.694	.722	.717	.724	**.728**	.715	.696	.720†
	2020	.674	.681†	.680	**.684**	.676	.632	.666	.672	.661	.665	.649	.605	.682
	2021	**.656**	.653	.650	.645	.629	.602	**.656**	.650	.639	.647	.625	.593	**.656**†
	2022	**.496**	.494†	.487	.486	.481	.441	.490	.492†	.479	.484	.471	.427	.495†
	Avg	.619	.619†	.616†	.615†	.607	.572	.615†	.615†	.607	.612†	.596	.560	**.620**†
Document	2019	.658	.683	.678	.667	.689	.663	.638	.672	.685	.669	.692	.646	**.697**
	2020	0 .622	.640	.639	**.661**	.655	.644	.636	.638	.650	.642	.657	.638	.639
	2021	.678	.671	.681	**.683**	**.683**	.629	.677	.681	.681	.670	.679	.644	.676
	2022	.424	.425	.431	.425	.409	.389	.421	**.446**	.443	.417	.424	.405	.428
	Avg	.575	.582	.586†	.587	.584†	.556	.573	.590	**.594**	.577	.589	.561	.587†

We consider two different reference models for the passage and document re-ranking tasks. The Full Attention cross-encoder has complete information access in the passage re-ranking setting and serves as the reference model for the passage tasks. Since the models without windowed attention ($w = \infty$) only process a limited number of tokens in the document re-ranking setting, we use the standard Longformer ($w = 64$) as the reference model for document tasks.

We first examine the effectiveness of the QDS-Transformer. In contrast to the original work [44], it does not improve re-ranking effectiveness despite having more token interactions. The reference models are statistically equivalent to the QDS-Transformer within ±0.02 for both passage and document re-ranking.

Next, we examine the effect of independent query contextualization on effectiveness. We compare the reference models for the passage and document tasks with our sparse cross-encoder with window size $w = \infty$ and $w = 64$, respectively. This comparison is a type of ablation test, as the two models being compared have identical configurations, except that our sparse cross-encoders independently contextualize the query. Our model is statistically equivalent within ±0.02 to the reference model on average across all passage tasks. The same does not hold for the document task, but our sparse cross-encoder is slightly more effective, achieving an 0.008 higher nDCG@10. We conclude that independent query contextualization only marginally affects re-ranking effectiveness.

Finally, we examine the effect of decreasing window size on effectiveness. For the Longformer, the window sizes 64, 16, and 4 are all significantly equivalent within ±0.02 on average across passage tasks. Even reducing the window size to just a single token, meaning a passage token can only attend to its immediate left and right neighboring tokens, reduces effectiveness by only 0.012 but is no longer statistically equivalent. The results are similar for the document tasks, with the window sizes 16 and 1 being statistically equivalent to the reference model. Furthermore, deactivating attention for passage or document tokens to its own subsequence ($w = 0$) does not yield statistically equivalent effectiveness but only drops effectiveness by 0.047 and 0.026 nDCG@10, respectively.

The effect of decreasing window sizes is similar for our sparse cross-encoder model. Window sizes 64 and 4 are statistically equivalent for passage tasks. Window sizes 16 and 1 are not statistically equivalent but only drop effectiveness by 0.012 and 0.023 nDCG@10, respectively. On the document tasks, our sparse cross-encoder models with smaller window sizes are never statistically equivalent within ±0.02 compared to the reference Longformer with window size 64. However, window sizes 64, 16, and 1 slightly improve over the reference model, and window size 4 is slightly less effective.

In summary, both independent query contextualization and windowed self-attention do not substantially affect re-ranking effectiveness, confirming our initial assumptions. That is, symmetric modeling of the query–passage relationship is unnecessary, and very small window sizes suffice to determine a passage's or document's relevance to the query. Interestingly, we find a window size of 0 to still feature competitive effectiveness. In this case, a document (or passage) token cannot attend to other tokens from its sub-sequence, making it similar to a lexical or bag-of-words model. We leave a more in-depth investigation into the implications of these results for future work.

Out-of-Domain Effectiveness. Table 2 reports nDCG@10 on all out-of-domain tasks from the TIREx [36] benchmark of our sparse cross-encoder compared to two other cross-encoders of various sizes: monoT5 [56] and monoBERT [55]. Our model uses a window size of 4 tokens, and all models use a maximum sequence length of 512 token except for our sparse cross-encoder trained on documents, which has access to a maximum of 4096 tokens. Overall, the out-of-domain

Table 2. Re-ranking effectiveness as nDCG@10 on TIREx [36]. The average document length per corpus and first-stage (FS) effectiveness are listed for context. We report micro-averaged scores across all queries from a corpus and macro-average these in the "Average" row. The highest score per corpus is given in bold. Our sparse cross-encoder models use a window size of 4.

Corpus	Doc. Len.	FS	monoT5			monoBERT		Sparse CE	
			Base	Large	3b	Base	Large	512	4096
Antique [37]	49.9	.510	.505	.527	.537	.507	.484	**.540**	.174
Args.me [5,6]	435.5	**.405**	.305	.338	.392	.314	.371	.313	.180
CW09 [14–17]	1132.6	.178	.186	.182	.201	.192	.134	.198	**.212**
CW12 [5,6,21,22]	5641.7	**.364**	.260	.266	.279	.263	.251	.312	.338
CORD-19 [78]	3647.7	.586	.688	.636	.603	**.690**	.625	.673	.642
Cranfield [19,20]	234.8	.008	.006	.007	.007	.006	.006	**.009**	.003
Disks4+5 [74–77]	749.3	.429	.516	.548	**.555**	.514	.494	.487	.293
GOV [23–25]	2700.5	.266	.320	.327	**.351**	.318	.292	.316	.292
GOV2 [9,13,18]	2410.3	.467	.486	.513	**.514**	.489	.474	.503	.460
MED. [38,39,63,64]	309.1	**.366**	.264	.318	.350	.267	.298	.237	.180
NFCorpus [7]	364.6	.268	.295	.296	**.308**	.295	.288	.284	.151
Vaswani	51.3	.447	.306	.414	.458	.321	**.476**	.436	.163
WaPo	713.0	.364	.451	**.492**	.476	.449	.438	.434	.296
Average	–	.358	.353	.374	**.387**	.356	.356	.365	.260

re-ranking effectiveness of all cross-encoders is lower than in-domain. Only the monoT5 large and 3b variants and our sparse cross-encoder trained on passages can improve the ranking of the first-stage retrieval on average across all corpora.

Our model trained solely on passages (512-token sequence length) features competitive effectiveness despite having substantially fewer parameters. On average over all corpora, it slightly outperforms both the base and large monoBERT variants and the base variant of monoT5. We emphasize that our model only has about 24 million parameters, making it around four times smaller than the base variant of monoBERT, nine times smaller than the base variant of monoT5, and fourteen times smaller than monoBERT-large. It is slightly less effective than the monoT5 large variant, and the largest model, monoT5 3b, is the most effective.

In contrast, our model trained on documents is substantially less effective in out-of-domain retrieval. Across all corpora, it is 0.105 nDCG@10 less effective than our model trained on passages. However, it can take advantage of its longer context length on the corpora containing long documents. For example, on the ClueWeb corpora it is the most effective of all cross-encoder models and features competitive effectiveness on CORD-19, GOV, and GOV2.

4.2 Efficiency Results

Finally, we study how the various attention patterns affect efficiency. We first compare our custom windowed matrix multiplication kernel with previous implementations. We then compare the efficiency of our proposed cross-encoder model to the reference cross-encoder, Longformer, and QDS-Transformer.

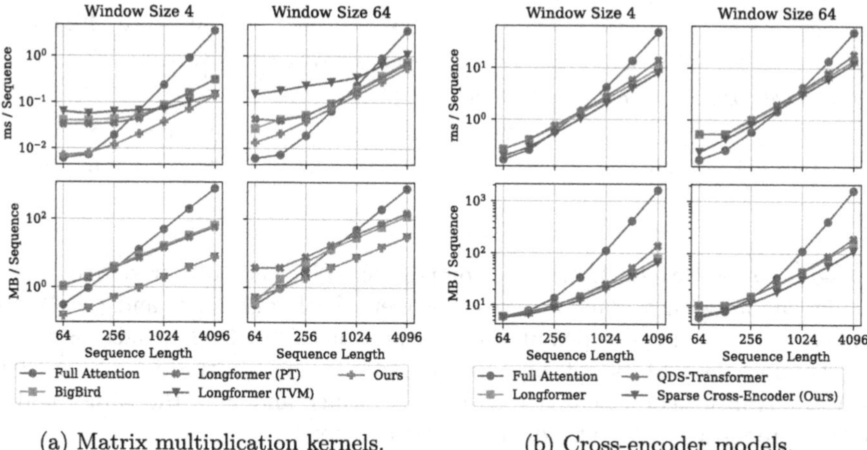

(a) Matrix multiplication kernels. (b) Cross-encoder models.

Fig. 2. Comparison of windowed matrix multiplication kernels (a) and sparse cross-encoder models (b) in terms of efficiency. Time (ms/Document) and space (GB) efficiency are reported for window sizes $w \in \{4, 64\}$. All plots use a logarithmic scale with base 2 on the x-axis and base 10 on the y-axis.

Windowed Matrix Multiplication. Figure 2a compares our windowed matrix multiplication kernel with PyTorch's built-in matrix multiplication kernel (Full Attention), Longformer's TVM-based [11] implementation, and the two PyTorch-based block-sparse implementations from BigBird and Longformer. All implementations are intended as drop-in replacements for matrix multiplication but use different ways to interface with CUDA and add varying levels of overhead.

We observe similar behavior as Beltagy et al. [3] for a window size of 64 tokens. Both block-sparse matrix implementations are time-efficient but sacrifice space efficiency. The opposite is true for the TVM-based kernel. For window size 4, the TVM implementation fairs better, achieving similar time efficiency as the PyTorch-based kernels for short sequence lengths and outperforming them at longer sequences. At the same time, it upholds its space efficiency. However, previous kernels are vastly slower compared to full matrix multiplication for shorter sequences. Our custom kernel achieves optimal space efficiency and is faster than all previous windowed matrix multiplication kernels for both window sizes. Compared to full matrix multiplication, our kernel is only slower for the edge case when the window size exceeds the sequence length.

Table 3. Time and space efficiency of cross-encoder models, including our sparse cross-encoder without our kernel and independent query contextualization. Relative differences to baseline models (underlined) are given in parentheses.

Unit	Full Att.	Longf.	QDS.	Sp. CE	Sp. CE	~~Kernel~~	~~Query~~
$w =$	∞	64	64	64	4	4	4
Query length 10, Passage length 164							
ms	368	980 (+166%)	995 (+170%)	527 (+43%)	364 (−1%)	404 (+10%)	403 (+10%)
MB	9	15 (+67%)	15 (+67%)	9 (+0%)	7 (−22%)	8 (−11%)	7 (−22%)
Query length 10, Document length 4086							
ms	49 (+250%)	14	18 (+29%)	12 (−14%)	8 (−43%)	9 (−36%)	8 (−43%)
MB	1608 (+905%)	160	192 (+20%)	111 (−31%)	66 (−59%)	84 (−48%)	66 (−59%)

Cross-Encoders Models. Lastly, we compare the efficiency of our sparse cross-encoder model with a standard cross-encoder, the Longformer, and the QDS-Transformer. We use the default implementations from Huggingface [80] but omit BigBird, as it does not support task-specific global attention. Note that the QDS-Transformer is based on the Longformer and uses the same model architecture with a different global attention pattern.

Figure 2b gives a visual overview of the efficiency of the models for windows sizes 4 and 64. Table 3 reports the time and memory used per sequence for passages of length 164 and documents of length 4086. These lengths correspond to the average of the longest passage/document per top-100 ranking of a TREC Deep Learning query, i.e., the setup simulates re-ranking a batch of 100 sequences. For ablation analyses, Table 3 additionally reports the efficiency of our model without our custom kernel and independent query contextualization.

Figure 2b shows our model outperforms the other two sparse cross-encoders for time and space efficiency for both window sizes. The QDS-Transformer is the least efficient due to its additional global sentence attention. The Longformer lies between the QDS-Transformer and our sparse cross-encoder. The efficiency improvements can be attributed to two sources. The first is our improved windowed matrix multiplication kernel, and the second is our cross-attention mechanism. The Listformer uses a similar mechanism but extracts the matrices required for global attention in each transformer layer. Our sparse cross-encoder model splits the sub-sequences once and reuses the extracted matrices for all layers, avoiding repeating the expensive extraction and splitting step.

Table 3 underlines the efficiency improvements of our model. With a 64 token window size, our sparse cross-encoder is almost twice as fast and uses 40% less memory than the Longformer on passages. On documents, the difference is less pronounced but still substantial. Our model is 14% faster and uses 31% less memory. However, our sparse cross-encoder achieves the largest efficiency improvements when reducing the window size. Compared to the Longformer with a 64-token window size, our sparse cross-encoder with a 4-token window size is 63% faster and uses 53% less memory on passages. On documents, our model is 43% faster and uses 59% less memory. Despite the different window sizes, we deem this a fair comparison because the Longformer was previously not successfully used for re-ranking with smaller window sizes. It acts as the previous sparse cross-encoder efficiency standard.

Ablation tests show that our custom kernel and independent query contextualization both contribute to our model's improved efficiency. Using a Pytorch-based block-sparse windowed matrix multiplication kernel, our model is less time and space-efficient and loses between 9% and 11% percent of its time and space-efficiency improvements. Independent query contextualization only has a marginal effect on space efficiency and a noticeable effect on time efficiency only for passages. The query is generally not long enough compared to passages or documents to substantially impact efficiency in practice.

Comparing our sparse cross-encoder to the standard cross-encoder reveals that there is still room for improvement. Time and space efficiency on documents is orders of magnitude better, and our model uses 22% less memory on passages. But, regarding inference time, our model is on par with the standard cross-encoder for passages. The root cause is that the cross-attention incurs additional overhead. Each sub-sequence uses its own attention function. Multiple smaller attention functions are executed sequentially, while full attention uses a single large attention function for the entire sequence. Recent work on fused-attention kernels [31,32,48] has shown that moving the entire attention function to the GPU substantially improves efficiency. At the time of writing, fused-attention kernels do not support asymmetric attention patterns. We leave investigating their applicability to our model to future work.

5 Conclusion

We have investigated the impact of sparse attention on the re-ranking effectiveness of cross-encoders by combining windowed self-attention and token-specific cross-attention to analyze (1) decreasing context sizes for document tokens and (2) deactivating attention from the query to the [CLS] and document tokens.

In passage and document re-ranking experiments, we find a window size down to four tokens to be as effective as larger window sizes or full attention (significantly equivalent effectiveness within ± 0.02 nDCG@10 compared to previous cross-encoders), and we find that deactivating attention from the query to the [CLS] and document tokens does not impact effectiveness. At the same time, combining the sparsification techniques substantially improves efficiency of passage and document re-ranking. For these efficiency improvements, our custom CUDA kernel and asymmetric cross-attention play substantial roles but the largest gains are achieved using small window sizes.

Sparse attention thus is a viable option for decreasing computational effort without substantially affecting effectiveness. To further increase efficiency, integrating asymmetric cross-attention and windowed self-attention into newly developed fused attention kernels [31,32,48] seems to be a promising direction. The flexibility and efficiency of our custom attention pattern also allow for further research into the direction of listwise re-ranking by passing multiple documents to the cross-encoder at once.

Acknowledgments. This work has received funding from the European Union's Horizon Europe research and innovation programme under grant agreement No 101070014 (OpenWebSearch.EU, https://doi.org/10.3030/101070014).

References

1. Ainslie, J., et al.: ETC: encoding long and structured inputs in transformers. In: Proceedings of EMNLP 2020, pp. 268–284 (2020). https://doi.org/10.18653/v1/2020.emnlp-main.19
2. Arapakis, I., Bai, X., Cambazoglu, B.B.: Impact of response latency on user behavior in web search. In: Proceedings of SIGIR 2014, pp. 103–112 (2014). https://doi.org/10.1145/2600428.2609627
3. Beltagy, I., Peters, M.E., Cohan, A.: Longformer: The Long-Document Transformer. arXiv (2020). https://doi.org/10.48550/arXiv.2004.05150
4. Bevendorff, J., Stein, B., Hagen, M., Potthast, M.: Elastic ChatNoir: search engine for the ClueWeb and the common crawl. In: Proceedings of ECIR 2018, pp. 820–824 (2018). https://doi.org/10.1007/978-3-319-76941-7_83
5. Bondarenko, A., et al.: Overview of Touché 2022: argument retrieval. In: Proceedings of CLEF 2022, pp. 311–336 (2022). https://doi.org/10.1007/978-3-031-13643-6_21
6. Bondarenko, A., et al.: Overview of Touché 2021: argument retrieval. In: Proceedings of CLEF 2021, pp. 450–467 (2021). https://doi.org/10.1007/978-3-030-85251-1_28
7. Boteva, V., Ghalandari, D.G., Sokolov, A., Riezler, S.: A full-text learning to rank dataset for medical information retrieval. In: Proceedings of ECIR 2016, pp. 716–722 (2016)
8. Burges, C.J.C.: From RankNet to LambdaRank to LambdaMART: An Overview. Technical report, Microsoft Research (2010)
9. Büttcher, S., Clarke, C.L.A., Soboroff, I.: The TREC 2006 terabyte track. In: Proceedings of TREC 2006, NIST Special Publication, vol. 500-272 (2006)
10. Chen, S., Wong, S., Chen, L., Tian, Y.: Extending Context Window of Large Language Models via Positional Interpolation. arXiv (2023). https://doi.org/10.48550/arXiv.2306.15595
11. Chen, T., et al.: TVM: an automated end-to-end optimizing compiler for deep learning. In: Proceedings of OSDI 2018, pp. 579–594 (2018)
12. Child, R., Gray, S., Radford, A., Sutskever, I.: Generating Long Sequences with Sparse Transformers. arXiv (2019). https://doi.org/10.48550/arXiv.1904.10509
13. Clarke, C.L.A., Craswell, N., Soboroff, I.: Overview of the TREC 2004 terabyte track. In: Proceedings of TREC 2004, NIST Special Publication, vol. 500-261 (2004)
14. Clarke, C.L.A., Craswell, N., Soboroff, I.: Overview of the TREC 2009 web track. In: Voorhees, E.M., Buckland, L.P. (eds.) Proceedings of TREC 2009, NIST Special Publication, vol. 500-278 (2009)
15. Clarke, C.L.A., Craswell, N., Soboroff, I., Cormack, G.V.: Overview of the TREC 2010 web track. In: Voorhees, E.M., Buckland, L.P. (eds.) Proceedings of TREC 2010, NIST Special Publication, vol. 500-294 (2010)

16. Clarke, C.L.A., Craswell, N., Soboroff, I., Voorhees, E.M.: Overview of the TREC 2011 web track. In: Voorhees, E.M., Buckland, L.P. (eds.) Proceedings of TREC 2011, NIST Special Publication, vol. 500-296 (2011)
17. Clarke, C.L.A., Craswell, N., Voorhees, E.M.: Overview of the TREC 2012 web track. In: Voorhees, E.M., Buckland, L.P. (eds.) Proceedings of TREC 2012, NIST Special Publication, vol. 500-298 (2012)
18. Clarke, C.L.A., Scholer, F., Soboroff, I.: The TREC 2005 terabyte track. In: Proceedings of TREC 2005, NIST Special Publication, vol. 500-266 (2005)
19. Cleverdon, C.: The cranfield tests on index language devices. In: ASLIB Proceedings, pp. 173–192, MCB UP Ltd. (Reprinted in Readings in Information Retrieval, Karen Sparck-Jones and Peter Willett, editors, Morgan Kaufmann, 1997) (1967)
20. Cleverdon, C.W.: The significance of the cranfield tests on index languages. In: Bookstein, A., Chiaramella, Y., Salton, G., Raghavan, V.V. (eds.) Proceedings of the 14th Annual International ACM SIGIR Conference on Research and Development in Information Retrieval, Chicago, Illinois, USA, 13–16 October 1991 (Special Issue of the SIGIR Forum), pp. 3–12 (1991)
21. Collins-Thompson, K., Bennett, P.N., Diaz, F., Clarke, C., Voorhees, E.M.: TREC 2013 web track overview. In: Proceedings of TREC 2013, NIST Special Publication, vol. 500-302 (2013)
22. Collins-Thompson, K., Macdonald, C., Bennett, P.N., Diaz, F., Voorhees, E.M.: TREC 2014 web track overview. In: Proceedings of TREC 2014, NIST Special Publication, vol. 500-308 (2014)
23. Craswell, N., Hawking, D.: Overview of the TREC 2002 web track. In: Proceedings of TREC 2002, NIST Special Publication, vol. 500-251 (2002)
24. Craswell, N., Hawking, D.: Overview of the TREC 2004 web track. In: Proceedings of TREC 2004, NIST Special Publication, vol. 500-261 (2004)
25. Craswell, N., Hawking, D., Wilkinson, R., Wu, M.: Overview of the TREC 2003 web track. In: Proceedings of TREC 2003, NIST Special Publication, vol. 500-255, pp. 78–92 (2003)
26. Craswell, N., Mitra, B., Yilmaz, E., Campos, D.: Overview of the TREC 2020 deep learning track. In: Proceedings TREC 2020, NIST Special Publication, vol. 1266 (2020)
27. Craswell, N., Mitra, B., Yilmaz, E., Campos, D., Lin, J.: Overview of the TREC 2021 deep learning track. In: Proceedings TREC 2021, NIST Special Publication, vol. 500-335 (2021)
28. Craswell, N., Mitra, B., Yilmaz, E., Campos, D., Voorhees, E.M.: Overview of the TREC 2019 deep learning track. In: Proceedings TREC 2019, NIST Special Publication, vol. 500-331 (2019)
29. Craswell, N., Mitra, B., Yilmaz, E., Campos, D., Voorhess, J.L.E.M., Soboroff, I.: Overview of the TREC 2021 deep learning track. In: Proceedings TREC 2021, NIST Special Publication, vol. 500-338 (2022)
30. Dai, Z., Callan, J.: Deeper text understanding for IR with contextual neural language modeling. In: Proceedings of SIGIR 2019, pp. 985–988 (2019). https://doi.org/10.1145/3331184.3331303
31. Dao, T.: FlashAttention-2: Faster Attention with Better Parallelism and Work Partitioning. arXiv (2023). https://doi.org/10.48550/arXiv.2307.08691
32. Dao, T., Fu, D.Y., Ermon, S., Rudra, A., Ré, C.: FlashAttention: fast and memory-efficient exact attention with IO-awareness. In: Proceedings of NeurIPS 2022, pp. 16344–16359 (2022)

33. Devlin, J., Chang, M.W., Lee, K., Toutanova, K.: BERT: pre-training of deep bidirectional transformers for language understanding. In: Proceedings of NAACL-HLT 2019, pp. 4171–4186 (2019). https://doi.org/10.18653/v1/N19-1423
34. Falcon, W.: The PyTorch Lightning team: PyTorch Lightning (2023). https://doi.org/10.5281/zenodo.7859091
35. Feng, C., Wang, X., Zhang, Y., Zhao, C., Song, M.: CASwin transformer: a hierarchical cross attention transformer for depth completion. In: Proceedings of ITSC 2022, pp. 2836–2841 (2022). https://doi.org/10.1109/ITSC55140.2022.9922273
36. Fröbe, M., et al.: The information retrieval experiment platform. In: Proceedings of SIGIR 2023, pp. 2826–2836 (2023). https://doi.org/10.1145/3539618.3591888
37. Hashemi, H., Aliannejadi, M., Zamani, H., Croft, W.B.: ANTIQUE: a non-factoid question answering benchmark. In: Proceedings of ECIR 2020, pp. 166–173 (2020)
38. Hersh, W.R., Bhupatiraju, R.T., Ross, L., Cohen, A.M., Kraemer, D., Johnson, P.: TREC 2004 genomics track overview. In: Proceedings of TREC 2004, NIST Special Publication, vol. 500-261 (2004)
39. Hersh, W.R., Cohen, A.M., Yang, J., Bhupatiraju, R.T., Roberts, P.M., Hearst, M.A.: TREC 2005 genomics track overview. In: Proceedings of TREC 2005, NIST Special Publication, vol. 500-266 (2005)
40. Hofstätter, S., Althammer, S., Schröder, M., Sertkan, M., Hanbury, A.: Improving Efficient Neural Ranking Models with Cross-Architecture Knowledge Distillation. arXiv (2021). https://doi.org/10.48550/arXiv.2010.02666
41. Hofstätter, S., Lin, S.C., Yang, J.H., Lin, J., Hanbury, A.: Efficiently teaching an effective dense retriever with balanced topic aware sampling. In: Proceedings of SIGIR 2021, pp. 113–122 (2021). https://doi.org/10.1145/3404835.3462891
42. Hofstätter, S., Zlabinger, M., Hanbury, A.: Interpretable & time-budget-constrained contextualization for re-ranking. In: Proceedings of ECAI 2020, pp. 513–520 (2020). https://doi.org/10.3233/FAIA200133
43. Ilinykh, N., Dobnik, S.: Attention as grounding: exploring textual and cross-modal attention on entities and relations in language-and-vision transformer. In: Findings of ACL 2022, pp. 4062–4073 (2022). https://doi.org/10.18653/v1/2022.findings-acl.320
44. Jiang, J.Y., Xiong, C., Lee, C.J., Wang, W.: Long document ranking with query-directed sparse transformer. In: Findings of EMNLP 2020, pp. 4594–4605 (2020). https://doi.org/10.18653/v1/2020.findings-emnlp.412
45. Karpukhin, V., et al.: Dense passage retrieval for open-domain question answering. In: Proceedings of EMNLP 2020, pp. 6769–6781 (2020). https://doi.org/10.18653/v1/2020.emnlp-main.550
46. Khattab, O., Zaharia, M.: ColBERT: efficient and effective passage search via contextualized late interaction over BERT. In: Proceedings of SIGIR 2020, pp. 39–48 (2020). https://doi.org/10.1145/3397271.3401075
47. Lauzon, C., Caffo, B.: Easy multiplicity control in equivalence testing using two one-sided tests. Am. Stat. **63**, 147–154 (2009). ISSN 0003-1305. https://doi.org/10.1198/tast.2009.0029
48. Lefaudeux, B., et al.: xFormers: A modular and hackable Transformer modelling library (2022). https://github.com/facebookresearch/xformers
49. Lin, J., Ma, X., Lin, S.C., Yang, J.H., Pradeep, R., Nogueira, R.: Pyserini: an easy-to-use python toolkit to support replicable IR research with sparse and dense representations. In: Proceedings of SIGIR 2021, pp. 2356–2362 (2021). https://doi.org/10.1145/3404835.3463238

50. Lin, S.C., Yang, J.H., Lin, J.: In-batch negatives for knowledge distillation with tightly-coupled teachers for dense retrieval. In: Proceedings of RepL4NLP 2021, pp. 163–173 (2021). https://doi.org/10.18653/v1/2021.repl4nlp-1.17
51. Loshchilov, I., Hutter, F.: Decoupled weight decay regularization. In: Proceedings of ICLR 2019 (2019)
52. MacAvaney, S., Nardini, F.M., Perego, R., Tonellotto, N., Goharian, N., Frieder, O.: Efficient document re-ranking for transformers by precomputing term representations. In: Proceedings of SIGIR 2020, pp. 49–58 (2020). https://doi.org/10.1145/3397271.3401093
53. MacAvaney, S., Yates, A., Feldman, S., Downey, D., Cohan, A., Goharian, N.: Simplified data wrangling with ir_datasets. In: Proceedings of SIGIR 2021, pp. 2429–2436 (2021). https://doi.org/10.1145/3404835.3463254
54. Nguyen, T., Rosenberg, M., Song, X., Gao, J., Tiwary, S., Majumder, R., Deng, L.: MS MARCO: a human generated MAchine reading COmprehension dataset. In: Proceedings of COCO@NeurIPS 2016 (2016)
55. Nogueira, R., Cho, K.: Passage Re-ranking with BERT. arXiv (2020). https://doi.org/10.48550/arXiv.1901.04085
56. Nogueira, R., Jiang, Z., Pradeep, R., Lin, J.: Document ranking with a pretrained sequence-to-sequence model. In: Findings of EMNLP 2020, pp. 708–718 (2020). https://doi.org/10.18653/v1/2020.findings-emnlp.63
57. Nogueira, R., Yang, W., Cho, K., Lin, J.: Multi-Stage Document Ranking with BERT. arXiv (2019). https://doi.org/10.48550/arXiv.1910.14424
58. Paszke, A., et al.: PyTorch: an imperative style, high-performance deep learning library. In: Proceedings of NeurIPS 2019, vol. 32 (2019)
59. Petit, O., Thome, N., Rambour, C., Themyr, L., Collins, T., Soler, L.: U-Net transformer: self and cross attention for medical image segmentation. In: Proceedings of MLMI@MICCAI 2021, pp. 267–276 (2021). https://doi.org/10.1007/978-3-030-87589-3_28
60. Potthast, M., et al.: ChatNoir: a search engine for the ClueWeb09 corpus. In: Proceedings of SIGIR 2012, p. 1004 (2012). https://doi.org/10.1145/2348283.2348429
61. Qu, Y., et al.: RocketQA: an optimized training approach to dense passage retrieval for open-domain question answering. In: Proceedings of NAACL-HLT 2021, pp. 5835–5847 (2021). https://doi.org/10.18653/v1/2021.naacl-main.466
62. Reimers, N., Gurevych, I.: Sentence-BERT: sentence embeddings using siamese BERT-networks. In: Proceedings of EMNLP-IJCNLP 2019, pp. 3980–3990 (2019). https://doi.org/10.18653/v1/D19-1410
63. Roberts, K., Demner-Fushman, D., Voorhees, E.M., Hersh, W.R., Bedrick, S., Lazar, A.J.: Overview of the TREC 2018 precision medicine track. In: Proceedings of TREC 2018, NIST Special Publication, vol. 500-331 (2018)
64. Roberts, K., et al.: Overview of the TREC 2017 precision medicine track. In: Proceedings of TREC 2017, NIST Special Publication, vol. 500-324 (2017)
65. Robertson, S.E., Walker, S., Jones, S., Hancock-Beaulieu, M., Gatford, M.: Okapi at TREC-3. In: Proceedings of TREC 1994, vol. 500-225, pp. 109–126 (1994)
66. Rosa, G., et al.: In Defense of Cross-Encoders for Zero-Shot Retrieval. arXiv (2022). https://doi.org/10.48550/arXiv.2212.06121
67. Santhanam, K., Khattab, O., Saad-Falcon, J., Potts, C., Zaharia, M.: ColBERTv2: effective and efficient retrieval via lightweight late interaction. In: Proceedings of NAACL-HLT 2022, pp. 3715–3734 (2022). https://doi.org/10.18653/v1/2022.naacl-main.272

68. Scells, H., Zhuang, S., Zuccon, G.: Reduce, reuse, recycle: green information retrieval research. In: Proceedings of SIGIR 2022, pp. 2825–2837 (2022). https://doi.org/10.1145/3477495.3531766
69. Schuirmann, D.J.: A comparison of the two one-sided tests procedure and the power approach for assessing the equivalence of average bioavailability. J. Pharmacokinet. Biopharm. **15**(6), 657–680 (1987). https://doi.org/10.1007/BF01068419
70. Sekulić, I., Soleimani, A., Aliannejadi, M., Crestani, F.: Longformer for MS MARCO document re-ranking task. In: Proceedings of TREC 2020, NIST Special Publication, vol. 1266 (2020)
71. Sui, X., et al.: CRAFT: cross-attentional flow transformer for robust optical flow. In: Proceedings of CVPR 2022, pp. 17581–17590 (2022). https://doi.org/10.1109/CVPR52688.2022.01708
72. Tay, Y., Dehghani, M., Bahri, D., Metzler, D.: Efficient transformers: a survey. ACM Comput. Surv. **55**, 109:1–109:28 (2023). https://doi.org/10.1145/3530811
73. Vaswani, A., et al.: Attention is all you need. In: Proceedings of NeurIPS 2017, pp. 5998–6008 (2017)
74. Voorhees, E.M.: NIST TREC Disks 4 and 5: Retrieval Test Collections Document Set (1996)
75. Voorhees, E.M.: Overview of the TREC 2004 robust track. In: Proceedings of TREC 2004, NIST Special Publication (2004)
76. Voorhees, E.M., Harman, D.: Overview of the seventh text retrieval conference (TREC-7). In: Proceedings of TREC 1998, NIST Special Publication (1998)
77. Voorhees, E.M., Harman, D.: Overview of the eight text retrieval conference (TREC-8). In: Proceedings of TREC 1999, NIST Special Publication (1999)
78. Wang, L.L., et al.: CORD-19: The Covid-19 Open Research Dataset. arXiv (2020). https://doi.org/10.48550/arXiv.2004.10706
79. Wang, W., Bao, H., Huang, S., Dong, L., Wei, F.: MiniLMv2: multi-head self-attention relation distillation for compressing pretrained transformers. In: Findings of ACL-IJCNLP 2021, pp. 2140–2151 (2021). https://doi.org/10.18653/v1/2021.findings-acl.188
80. Wolf, T., et al.: HuggingFace's Transformers: State-of-the-art Natural Language Processing. arXiv (2020). https://doi.org/10.48550/arXiv.1910.03771
81. Xiong, L., et al.: Approximate nearest neighbor negative contrastive learning for dense text retrieval. In: Proceedings of ICLR 2021 (2021)
82. Yan, M., et al.: IDST at TREC 2019 deep learning track: deep cascade ranking with generation-based document expansion and pre-trained language modeling. In: Proceedings of TREC 2019, NIST Special Publication, vol. 1250 (2019)
83. Zaheer, M., et al.: Big bird: transformers for longer sequences. In: Proceedings of NeurIPS 2020, pp. 17283–17297 (2020)
84. Zhang, Y., Long, D., Xu, G., Xie, P.: HLATR: Enhance Multi-stage Text Retrieval with Hybrid List Aware Transformer Reranking. arXiv (2022). https://doi.org/10.48550/arXiv.2205.10569

Reading Between the Frames: Multi-modal Depression Detection in Videos from Non-verbal Cues

David Gimeno-Gómez[1(✉)], Ana-Maria Bucur[1,2], Adrian Cosma[3], Carlos-David Martínez-Hinarejos[1], and Paolo Rosso[1,4]

[1] PRHLT Research Center, Universitat Politècnica de València, Valencia, Spain
{dagigo1,cmartine,prosso}@dsic.upv.es

[2] Interdisciplinary School of Doctoral Studies, University of Bucharest, Bucharest, Romania
ana-maria.bucur@drd.unibuc.ro

[3] Politehnica University of Bucharest, Bucharest, Romania
ioan_adrian.cosma@upb.ro

[4] ValgrAI Valencian Graduate School and Research Network of Artificial Intelligence, Valencia, Spain

Abstract. Depression, a prominent contributor to global disability, affects a substantial portion of the population. Efforts to detect depression from social media texts have been prevalent, yet only a few works explored depression detection from user-generated video content. In this work, we address this research gap by proposing a simple and flexible multi-modal temporal model capable of discerning non-verbal depression cues from diverse modalities in noisy, real-world videos. We show that, for in-the-wild videos, using additional high-level non-verbal cues is crucial to achieving good performance, and we extracted and processed audio speech embeddings, face emotion embeddings, face, body and hand landmarks, and gaze and blinking information. Through extensive experiments, we show that our model achieves state-of-the-art results on three key benchmark datasets for depression detection from video by a substantial margin. Our code is publicly available on GitHub (https://github.com/cosmaadrian/multimodal-depression-from-video).

Keywords: Affective Computing · Depression Detection · Multi-Modal

"There is only continual motion. If I rest, if I think inward, I go mad."

The Unabridged Journals of Sylvia Plath

1 Introduction

Depression is a leading cause of disability, with 5% of the adults worldwide suffering from it[1]. According to the World Health Organization (WHO), the prevalence of

[1] https://www.who.int/health-topics/depression. Accessed August 28th, 2023.

D. Gimeno-Gómez, A.-M. Bucur and A. Cosma—Equal contribution.
C.-D. Martínez-Hinarejos and P. Rosso—Equal supervision.

N. Goharian et al. (Eds.): ECIR 2024, LNCS 14608, pp. 191–209, 2024.
https://doi.org/10.1007/978-3-031-56027-9_12

depression was up to 25% during the first year of the COVID-19 pandemic[2]. Many efforts have been directed at detecting depression from text on social media sites such as Reddit [51,58,70] and Twitter [9,39,62]. This can be attributed to the relative ease of annotation (self-mentions [85,87], certain subreddits participation [31,52]) and abundance of available textual data that can be retrieved in a relevant context (i.e., as a thread). Although research on multi-modal depression detection from social media data has been performed using textual, visual, and online behavioral data, visual data is mostly limited to images [10,13,23,29,55,82]. However, many social media sites (YouTube, TikTok, Instagram) focus on user-generated video content and, currently, depression detection from in-the-wild video remains largely unexplored. The video modality poses an additional set of challenges for depression detection as it is significantly noisier, video length is highly variable (from shorts of a few seconds to half-hour vlogs in which the person talks directly to the camera), the face and body are not always visible, context rapidly changes, and audio quality is low, especially in amateur videos. Uploaded videos are mostly out of context and stand-alone as opposed to interactive discussions in social media threads. In this context, the task is to process subtle depression cues from the subject's behavior present in the video.

According to the DSM-5 criteria [3], major depressive disorder (depression) is manifested through symptoms such as depressed mood, sleep disturbance, appetite changes, loss of interest, etc. In addition, psychomotor changes (agitation or retardation) are a central feature of depression, intertwined with other symptoms, such as loss of energy, fatigue, and lack of concentration [8]. Some of the psychomotor particularities associated with depression are: reduced facial expressiveness [56], slower speech rate, longer pause time [83], and downward head tilt [24]. In addition, studying psychomotor changes in depression is essential because they can predict clinical response to medication in depression [68,83]. However, these cues have mainly been studied in laboratory settings [28,57] and significantly less so in in-the-wild videos. To the best of our knowledge, the only available benchmark dataset for depression detection using data from online platforms (i.e., YouTube) is the D-Vlog dataset [86].

In this work, we propose a multi-modal temporal model that processes multiple non-verbal cues across time to estimate depression from both noisy, in-the-wild, videos [86] and from controlled, laboratory recordings [28,57]. Our method is simple and flexible: by using appropriate high-level modality extractors, positional embeddings and modality conditioning vectors, this approach can be easily scaled to an arbitrary number of modalities. We surpass state of the art on three datasets: two controlled datasets (DAIC-WOZ [28] and E-DAIC [57]) and an in-the-wild dataset (D-Vlog [86]). Through extensive experiments, we show that using the appropriate data modalities and semantic embeddings is crucial in processing non-verbal cues from a relatively small amount of videos. This work makes the following contributions:

1. We propose a simple and flexible multi-modal architecture that can process non-verbal depression cues from an arbitrary number of modalities across time. Our model achieves state-of-the-art results on three key depression detection datasets, obtaining 0.78 F_1 on D-Vlog [86], 0.67 F_1 on DAIC-WOZ [28], and 0.56 F_1 on E-DAIC [57], surpassing the previous state of the art [42,44,64,67,76,77,86,91–93].

[2] https://www.who.int/news/item/02-03-2022-covid-19-pandemic-triggers-25-increase-in-prevalence-of-anxiety-and-depression-worldwide. Accessed August 28th, 2023.

2. We show that, for in-the-wild scenarios (e.g., for D-Vlog [86]), using appropriate high-level semantic embeddings is crucial, and we explore additional non-verbal cues informed by studies in psychology [8,24,56,83]: emotion-informed face embeddings, task-agnostic audio embeddings, body and hand landmarks, and eye movements (blinking and gaze). Using the additional modalities, our model exhibits a markedly better performance in in-the-wild depression detection.
3. We show that our model is interpretable by using Integrated Gradients [66], estimating the relevance of each modality across time for a particular subject. In this way, our method is a potentially valuable tool in preventive screening for psychologists.

2 Related Work

Depression Detection from Video. Depression detection from video content [34] follows the same high-level pipeline from general multi-modal video classification [47], which has been fueled by advances in multi-modal classification in images [1,36,63]. Similar to general multi-modal video classification, depression detection from video presumes the (automatic) extraction of temporal and spatial features and processing them with a classifier [47]. Some low-level features, extracted by using classical image and audio processing algorithms, are: loudness, spectral flux, phoneme duration, pitch slopes for audio and facial landmarks, facial action units [25], and histogram of oriented gradients features for video content [46,78]. High-level semantic features include facial expressions, smile intensity and duration, head movement, etc. [47]. Most current methods proposed for depression detection [67,91–93] use facial landmarks and low-level acoustic descriptors [86], disregarding the temporal information [44,86,93]. More recently, multi-modal large language models (LLMs) offer a promising paradigm for video classification, with approaches such as MiniGPT-4 [94] and PandaGPT [65]. In particular, some models are oriented towards mental health prediction from online text [81] with approaches such as MentalLLaMa [84]. However, in this work, we focus on processing a plethora of non-verbal cues, disregarding raw pixel data and transcripts, for which current LLMs are not suitable.

Benchmarking Datasets. Benchmark datasets for studying the non-verbal behavior of subjects with depression include mostly data collected in laboratory settings, e.g., the DAIC-WOZ [28], E-DAIC [57], Depression Severity Interviews Database [19], Audio-Visual Depressive Language Corpus [71], and the BlackDog Database [2]. These datasets consist of clinical interviews or footage with subjects performing different tasks (reading certain texts, counting, etc.). However, research in depression detection from video in general, in-the-wild, scenarios is limited. To the best of our knowledge, D-Vlog [86] is the only dataset containing in-the-wild user-generated videos.

Present Work. Inspired by psychological research [24,41,43,56,59], we employ additional features related to motor manifestations of depression, such as face and audio embeddings, hand and body landmarks, blinking, and gaze patterns. Our relatively simple architecture is able to process multiple modalities across different frame rates and processes only relevant parts of the video to perform classification. In contrast to previous approaches that mainly use global information [44,86,93], we design our architecture to handle both local, frame-level information, as well as temporal video dynamics.

3 Method

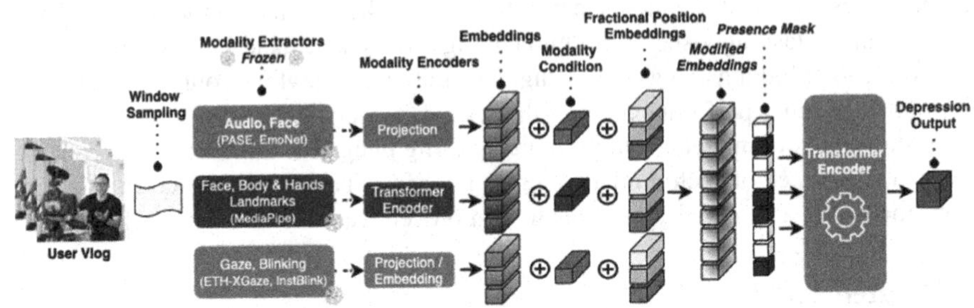

Fig. 1. The overall architecture of our proposed method. We extract high-level non-verbal cues using pretrained models, process them using a modality-specific encoder, condition the resulting embeddings with positional and modality embeddings, and process the sequence with a transformer encoder to perform the final classification.

3.1 Model Architecture

Overview. Given a dataset of vlogs with people directly talking to the camera, our aim is to train a model to estimate whether the subject in a video has depression or not. In this formulation, depression detection amounts to video-level classification. However, performing the classification of a high-level psychological affliction directly from pixels is infeasible without impractical amounts of labeled data. Consequently, in our work, we considered only high-level non-verbal cues, and discarded any textual information to remove any conversational topic bias [79]. The way most depression datasets are collected is based on direct mention of diagnosis [85,87] and by using only non-verbal cues, the model's performance is a more accurate depiction of its performance in realistic settings. Non-verbal cues are extracted using state-of-the-art pretrained models [14,53,69,80,88,89] and subsequently processed to perform classification. Figure 1 provides a high-level overview of our pipeline for depression classification from video. We further present a way to properly handle videos of different framerates and video lengths, and process an arbitrary amount of modalities.

Formally, given a dataset of depression labeled videos $\mathscr{D} = \{(V_1, \hat{y}_1), \dots (V_n, \hat{y}_n)\}$, where V_i is the video and $\hat{y}_i$ its corresponding label, the goal is to find the optimal parameters θ of a model f_θ that minimize the average cross entropy loss across the dataset: $\hat{\theta} = \arg\min_\theta \mathbb{E}_{1 \leq k \leq n}[\mathscr{L}(f_\theta(V_k), \hat{y}_k)]$.

Window Sampling. Since videos and vlogs in the wild have vastly different durations and framerates (see Fig. 3), with some videos exceeding 30 min, it is computationally unfeasible to process them directly in their entirety. Consequently, we operate on randomly sampled temporal windows $W_l^k \sim V_k$ from each video, with W_l^k of fixed time in seconds, and l a random temporal index in the video. Temporal windows can cover a different number of total frames, corresponding to each video's framerate. The audio is sampled at 100 frames per second [26], but video framerates from D-Vlog [86] vary between 6 and 30 frames per second. Through the window sampling approach, we assume that the mental health information in the video is time invariant.

Modality Extraction and Encoding. For each video, we considered multiple semantic modalities $\mathcal{M} = \{m_1, m_2 \dots\}$, that are extracted using a frozen pretrained model from both audio and visual information of the video at each frame. Consequently, the total number of frames for a video across all temporally concatenated modalities is the sum $T^k = T_1^k + \cdots + T_{|\mathcal{M}|}^k$. Since each modality has a different dimensionality d_{m_j}, we uniformize each modality output with an associated learnable modality encoder $E_{m_j} : \mathbb{R}^{d_{m_j}} \rightarrow \mathbb{R}^d$, operating at the frame-level, such that each modality has the same dimensionality d. Thus, the feature vector for a window W_l^k and a modality m_j is $x_{m_j} = E_{m_j}(W_l^k)$, with $x_{m_j} \in \mathbb{R}^{T_j \times d}$. Details about each modality encoder E_{m_j} are showcased in Sect. 3.2. Furthermore, each modality extractor can signal the presence of a modality in a frame, which allows us to construct a binary presence mask $M_j^k \in \mathbb{R}^{T_j^k}$. The presence mask is later used in attention masking for the final transformer encoder [72]. In practice, for in-the-wild videos, we first extracted and cached all modalities for each video, and sampled windows such that the subject is present in at least 50% of the frames in a window. This is because some videos were amateur-made and extremely noisy, and the subject's body, hands and face were not always visible; for instance, some people talk while driving their car and the steering wheel obscures the camera, or the subject walks out of the scene.

Modality Condition. We distinguish between the embeddings of different modalities by using an additional learned condition embedding $e_{m_j} \in \mathbb{R}^{|\mathcal{M}| \times d}$ that is added to each vector of its corresponding modality. This is similar to how BERT-type models [18] differentiate between sentences in tasks such as natural language inference.

Fractional Position Embeddings. Due to the mismatch between sampling rates, we temporally align the frames of acoustic and video-based modalities by using fractional positional embeddings [32] of the form $p_{m_j} \in \mathbb{R}^{\max_i(T_i) \times d}$. For a video, we construct a matrix of positional embeddings of size $\max_i(T_i)$, corresponding to the modality with the highest sample rate, usually audio. Afterwards, positions for a modality m_j are then uniformly indexed according to the ratio $r = \lfloor \max_i(T_i)/T_{m_j} \rfloor$. Fractional positional embeddings are based on sinusoidal positional embeddings [72] and are not learned during training. Frames across modalities have the same positional embedding at the same corresponding positions in time, irrespective of the sampling rate (see Fig. 2). Readers are referred to Harzig et al. [32] for more details.

An individual modified feature vector becomes $\bar{x}_{m_j} = x_{m_j} + e_{m_j} + p_{m_j}$. Finally, we process the concatenated modified feature vectors $X_l^k = \bar{x}_{m_1} \| \bar{x}_{m_2} \| \dots \| \bar{x}_{|\mathcal{M}|}$ alongside each corresponding presence mask $M_l^k = M_1 \| M_2 \| \dots \| M_{|\mathcal{M}|}$ from each modality of a sampled temporal window with a transformer encoder network [72] to obtain a prediction $y = f_\theta(X_l^k, M_l^k)$, with $X_l^k \in \mathbb{R}^{T_l^k \times |\mathcal{M}| \times d}$ and $M_k \in \mathbb{R}^{T_l^k \times |\mathcal{M}|}$. The final optimization formulation can be described as $\hat{\theta} = \arg\min_\theta \mathbb{E}_{(1 \leq k \leq n, 1 \leq l \leq |V_k|)}[\mathcal{L}(f_\theta(X_l^k, M_l^k), \hat{y}_k)]$.

Discussion. In our framework, modality fusion is done immediately after encoding each modality with the learnable encoders E_{m_j}, and the predominant part of the computation is performed with the main transformer encoder. This type of fusion can be considered

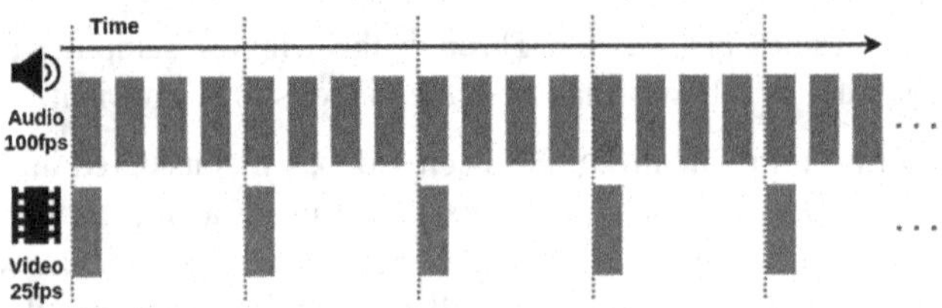

Fig. 2. Illustrative example of fractional positional embedding for temporally aligning video and audio sampling rates, similar to the work of Harzig et al. [32].

"early" fusion [63]. Furthermore, as defined by Vaswani et al. [72], the scaled dot-product attention is permutation invariant across the sequence without the addition of positional embeddings [75]. However, we employ two types of positional embeddings: learned modality conditions and fractional positional embeddings. In a single timestep t, the modality vectors $x^t_{m_j}$ are permutation invariant, as they share the same positional embedding. The order of the modalities does not affect the final computation, as the modalities are considered an unordered set, corresponding to the SetTransformer [38] formulation. Other approaches [44,86,93] mainly perform a global computation across a coarse temporal sampling of modalities, which is prone to lose the information of the finer non-verbal cues in manifestations of depression. However, through our window sampling, modality conditions and positional embeddings, the transformer attention operates both locally, inter-modality (at the same timestep), and globally, intra-modality (across multiple timesteps), to perform a successful depression classification. In this setting, the model performs a soft cross-attention between modalities. Furthermore, compared to other approaches in this area [91–93], which propose overly-complicated methods with modest results, our approach is simple, flexible, easy to reproduce, and it surpasses other methods on the three main datasets we evaluated.

3.2 Extracting and Encoding Non-verbal Cues

Psychological Motivation. Psychomotor manifestations are one of the main features of depression [8] found in the DSM-V criteria and self-assessment tools, such as Patient Health Questionnaire-9 (PHQ-9) or Beck Depression Inventory-Second Edition (BDI-II). The non-verbal facial cues associated with depression are reduced expressiveness [56], and fewer smiles [27]. In terms of body language, a downward head tilt [24], longer gait cycles, and a slower gait cadence [90] are common indicators. Regarding hand movements, individuals with depression may use fewer gestures [59] and exhibit impaired gesture performance [50]. A slower speech rate characterizes the voice, with longer pauses between words [83], and a decrease in loudness [74]. The eyes undergo subtle shifts with an increased blinking rate [43], reduced eye contact, and increased gaze angle [41]. Given these non-verbal changes in depressed individuals, we investigated how they interact and complement each other in depression detection. This work refers to major depressive disorder (MDD), not other related mood disorders, such as bipolar disorder, peripartum disorder, etc.

Audio-Visual Embeddings (AV). We extract semantic embeddings for the audio channels and the face gestures using pretrained models. For **audio embeddings**, we used the Pyannote Audio toolkit[3] [11, 12] to detect voice activities and extract 256-dimensional voice embeddings using the PASE+ model [49,53] in the time slots in which the person is speaking. The PASE+ model provides general embeddings [22,49] as it was trained in a task-agnostic manner in noisy and reverberant environments. For extracting emotion-informed **face embeddings**, we used EmoNet[4] [69], a model capable of estimating both discrete emotion classes and their corresponding valence and arousal [30,60] measures. We obtain a 256-dimensional embedding vector by taking the embedding from the last layer before the classification of EmoNet. For both types of embeddings, in our model we process the embeddings with a 1D batch normalization layer followed by a linear layer to project the embeddings into a 256-dimensional space.

Landmarks (LM). We extract relevant facial, body and hands keypoints using pre-trained networks. For **face landmarks**, we used the Face Alignment Network model [14] to extract 68 2-dimensional facial landmarks alongside their corresponding confidence score. For **body landmarks**, we used the MediaPipe[5] [6,80] toolkit to extract 33 3-dimensional landmarks. For **hand landmarks**, we used the hand landmark detector from the MediaPipe [80] toolkit to extract 21 3-dimensional landmarks for each hand. In our model, we process the landmarks using a 1D batch-normalization layer followed by a 2-layer transformer encoder that operates on the spatial dimension, obtaining the final embedding through average spatial pooling. Additionally, for encoding hand landmarks we distinguish between right and left hands by adding an embedding vector, similar to the modality condition described above.

Eye Information (EYES). We extract gaze and blinking information as additional modalities. For **gaze tracking** we used the official implementation[6] of the ETH-XGaze gaze estimator [89] to extract the gaze direction represented by 3 angle coordinates. For **blinking patterns**, we used the official implementation of the InstBlink model[7] [88], and labeled the video frames where the person is blinking (0 for not blinking and 1 for blinking). To encode gaze, we process the angles using a 1D batch normalization followed by a linear layer to project the angle vector into a 256-dimensional embedding. To encode blinking, at each frame we used one of two different learnable 256-dimensional embeddings, corresponding to blinking or not blinking states.

4 Experiments

4.1 Datasets

D-Vlog [86] is an in-the-wild dataset for multi-modal depression collected from user-generated YouTube vlogs. It comprises 961 video samples (159.2 h) with an average

[3] https://github.com/pyannote/pyannote-audio. Accessed August 28th, 2023.
[4] https://github.com/face-analysis/emonet. Accessed August 28th, 2023.
[5] https://github.com/google/mediapipe. Accessed August 28th, 2023.
[6] https://github.com/hysts/pytorch_mpiigaze_demo. Accessed August 28th, 2023.
[7] https://github.com/wenzhengzeng/MPEblink. Accessed August 28th, 2023.

duration of 9.9 min, containing 406 control and 555 depressed subjects. It is split into a training set with 647 samples, a validation set with 102 samples, and a test set with 212 samples. D-Vlog was manually annotated following overt mentions of depression symptoms, such as mentions of suicidal thoughts, mentions of depression medication, etc. In our work, in addition to studying the original audio-visual features provided by the authors, we explored the multiple non-verbal cues described in Subsect. 3.2. When extracting additional features, some of the videos were unavailable (deleted or made private by the owner), resulting in 861 videos. Specifically, 70, 8, and 23 videos were missing for training, validation, and test, respectively.

DAIC-WOZ [28] is a dataset composed of clinical interviews recorded in controlled settings to support the diagnosis of psychological distress conditions such as anxiety, depression, and post-traumatic stress disorder. It comprises 189 sessions with an average duration of 15.9 min, and it is highly imbalanced, containing 132 control and 57 depressed subjects. It is split into a training set with 107 samples, a validation set with 35 samples, and a test set with 47 samples. In addition to the transcript of the entire interview, COVAREP audio- [17] and OpenFace video-based [5] features are provided. Original recordings are not available for DAIC-WOZ for privacy reasons, and we used the original features provided by the authors. We used a total of 7 modalities: the acoustic COVAREP features, the 5 vocal tract resonance frequencies, the 68 3-dimensional facial landmarks, the facial action units [25], the gaze tracking, and the head pose landmarks. We did not use the text transcripts in our model. In our DAIC-WOZ experiment, we consider only the PHQ-8 binary labels. These binary labels are derived from the PHQ-8 score from each subject, with those who scored 10 or higher being labeled as depressed and the rest belonging to the control group.

E-DAIC [57] is the extended version of DAIC-WOZ [28]. It comprises a total of 275 sessions with an average duration of 16.2 min, and it is severely imbalanced, containing 209 control and 66 depressed subjects. It is split into a training set with 163 samples, a validation set with 56 samples, and a test set with 56 samples. In addition to the interview transcript, multiple audio-visual features are provided based on both expert knowledge-based methods [16,21] and deep learning representations [33]. Similar to DAIC-WOZ, we only used the features provided by the authors. Concretely, we used a total of 6 modalities: the 13 mel frequency cepstral coefficients [16] alongside their first and second derivatives, the 88 eGeMAPS measures [21] (covering information related to spectral, prosodic, and voice quality patterns), face embeddings extracted by a pre-trained ResNet-50 [33], and the OpenFace features [5], including gaze tracking, head pose landmarks and facial action units. We did not use the text transcripts in our model. Similar to DAIC-WOZ, we use the PHQ-8 binary labels as ground truth.

Discussion. We showcase the video durations for the datasets in Fig. 3. DAIC-WOZ and E-DAIC videos have a longer overall duration (15.9 and 16.2 min, respectively) than D-Vlog and the video lengths vary considerably less. Notably, in D-Vlog, the depression group's videos are longer than those of the control group, with averages of 10.5 and 8.9 min, respectively. In Fig. 4, we show the percentage of presence for each modality of interest in the D-Vlog dataset. Hands are the modality that is the least present in the vlogs. Even if the face and body are present in most of the videos, there are still many outliers in which the modalities are present less than 80% of the time. As expected, the voice is also commonly present in the videos, but less in the depression group. The fact that in-the-wild vlogs may contain different transitions or sequences without the subject's presence in the video motivated us to incorporate the presence mask in our architecture.

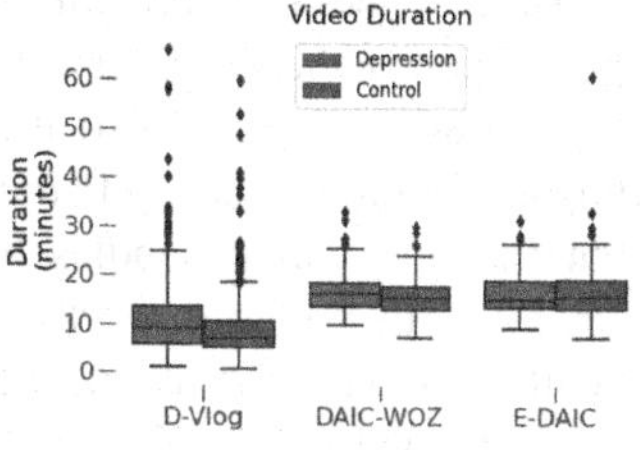

Fig. 3. Distributions of video durations for each of our benchmarking datasets.

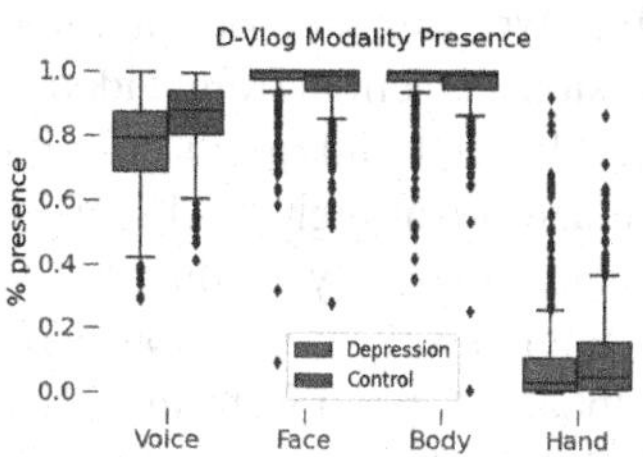

Fig. 4. Presence distributions for each of our considered modalities in D-Vlog.

4.2 Evaluation

Unlike other approaches, which either did not specify a formal evaluation protocol [91] or processed the features truncated to the average length of videos [86,92], we perform sequential evaluation across all temporal windows. For each non-overlapping window in a video $\{W_1, W_2, \ldots W_n\} \in V_k$ taken in order (containing at least 50% of modalities for D-Vlog), we perform depression classification and obtain a sequence of predictions: $\{y_1, y_2, \ldots y_n\}$, with $y_l = f_{\hat{\theta}}(W_l)$. The final decision is performed by voting on the decisions across windows: $y = \text{argmax}_{y_i} P(X = y_i)$. This evaluation protocol can be used for an early warning detection system [15] for suicide prevention by taking the decision at an earlier window $n' < n$. For instance, this procedure can be used in live, continuous video streams on platforms such as Facebook or Twitch, in which cases of live suicide have been unfortunately registered in the past [37].

4.3 Implementation Details

Experiments were conducted on 2 GeForce RTX 3060 GPUs with 12GB memory. In all our experiments, we used the AdamW optimizer [35] with a learning rate of 0.001 decaying through a cosine scheduler across 200 epochs, with a batch size of 8. On average, a training run took around 2 h. In initial hyper-parameter search experiments, we found a suitable configuration for each model. Depending on the number of modalities used, our model's parameters range between 8.4M to 15M. For D-Vlog, we used a context window size of 9 s and discarded window samples in which less than 50% of the frames did not contain the subject's face. For DAIC-WOZ and E-DAIC, we did not discard any window, and used a context window of 6 s. For D-Vlog and E-DAIC, the transformer encoder was composed of 8 layers with 8 32-dimensional attention heads, while for DAIC-WOZ we used 8 layers with 4 64-dimensional attention heads. Similar to D-Vlog (see Subsect. 3.2), for encoding the modalities in DAIC-WOZ and E-DAIC we defined a 2-layer transformer encoder when dealing with landmarks, and a linear projection when processing the other non-verbal features. For evaluation, we report mean and standard deviation across 5 training runs.

4.4 Comparison Methods

D-Vlog. We compared our proposed method to several state-of-the-art approaches for depression detection from video. In contrast with previous works, which use only low-level facial and audio descriptors [44,67,86,91–93], our model can also incorporate additional high-level facial and audio embeddings, gaze and blinking features, and hands and body landmarks. We also compare our method, which handles both local frame-level information and temporal video dynamics, to previous works that use global information from videos [44,86,93], and with previous time-aware methods [67,91,92]. We included two strong state-of-the-art methods [67,91]. Zheng et al. [91] proposed a temporal convolutional transformer that uses specialized medical and depression knowledge graphs. Tao et al. [67] proposed a spatio-temporal transformer architecture inspired by [20,40].

DAIC-WOZ. We compared our method to other works that approached the task from a non-verbal perspective, without considering the text transcripts. To show the capability of our proposed multi-modal architecture, we not only compared our method to those works that used hand-crafted features as our case [64,76,77], but also to those that explored deep-learning embedded representations [42]. Furthermore, although Wei et al. [76] introduced attention-based modules, most of these works were mainly based on CNNs combined with LSTMs [42,64,76].

E-DAIC. Although prior works [54,61,73,91] have been developed for this dataset, to the best of our knowledge, none of them explored the binary depression detection task from a non-verbal perspective, as all methods used the text transcript in their models. Consequently, we report the first results in this setting for E-DAIC.

Unlike our method, most previous works on D-Vlog only considered two modalities. Consequently, we introduce an additional strong baseline for general multi-modal classification for all three datasets. We used the Perceiver [36] based on an open-source implementation[8], as a general multi-modal encoder. However, different from our approach, the Perceiver is based on iterative attention between modalities on a lower-dimensional hidden latent sequence, which can suffer from gradient explosion [7,48] and loss of information. We constructed the Perceiver to be similar to our model in terms of the number of parameters and hidden dimensions.

5 Results

D-Vlog. In Table 1, we present the results of our method, compared to previous state-of-the-art approaches. As presented in Subsect. 4.1, since the release of the D-Vlog dataset, some YouTube videos became unavailable, leading to a reduction of the test

[8] https://github.com/lucidrains/perceiver-pytorch, Accessed August 28th, 2023.

Table 1. Results on the D-Vlog test set compared to existing approaches. Best results are highlighted in **bold** and second best results are underlined.

Method	Original Split	Additional Modalities	F_1	Precision	Recall
Yoon et al. [86]	✓	✗	0.64	0.65	0.66
Nguyen et al. [44]	✓	✗	0.64	0.66	0.64
Zheng et al. [91]	✓	✗	0.65	0.65	0.65
Zhou et al. [92]	✓	✗	0.66	0.66	0.67
Zhou et al. [93]	✓	✗	0.67	0.67	0.67
Tao et al. [67]	✓	✗	0.75	**0.73**	0.78
Perceiver [36]	✓	✗	0.74±0.01	0.67±0.03	0.84±0.07
Ours	✓	✗	**0.76±0.01**	0.67±0.02	**0.87±0.02**
Perceiver [36]	✗	✗	0.72±0.01	0.66±0.03	0.80±0.03
Perceiver [36]	✗	✓	0.73±0.01	0.68±0.08	0.81±0.15
Ours	✗	✗	0.73±0.01	0.62±0.01	**0.89±0.05**
Ours	✗	✓	**0.78±0.01**‡	**0.74±0.05**	0.84±0.06

Table 2. Performance of our method on D-Vlog using different modality combinations, compared to the Perceiver. Best results are highlighted in **bold** and second best results are underlined.

Method	Modality	F_1	Precision	Recall
Perceiver [36]	AV	0.69±0.06	0.72±0.07	0.69±0.13
	AV+EYES	0.73±0.01	0.68±0.08	0.81±0.15
	AV+LM	0.67±0.13	0.70±0.14	0.74±0.28
	AV+LM+EYES	0.64±0.17	**0.79± 0.07**	0.58±0.22
Ours	AV	0.72±0.05	0.73±0.03	0.72±0.11
	AV+EYES	0.75±0.04	0.72±0.03	0.79±0.12
	AV+LM	0.75±0.03	0.73±0.05	0.77± 0.06
	AV+LM+EYES	**0.78± 0.01**	0.74±0.05	**0.84± 0.06**

Table 3. Results on the DAIC-WOZ test set. Best results are highlighted in **bold** and second best results are underlined.

Method	Modality	F_1	Precision	Recall
S. Song et al. [64]	A	0.46	0.32	0.86
Ma et al. [42]	A	0.52	0.35	**1.00**
J. R. Williamson [77]	A	0.57	-	-
P-C. Wei et al. [76]	A	**0.61**	0.56	0.66
Perceiver [36]	A	0.33±0.26	0.30±0.28	0.45±0.21
Ours	A	0.49±0.05	**0.58±0.05**	0.47±0.05
S. Song et al. [64]	V	0.50	0.60	0.43
J. R. Williamson [77]	V	0.53	-	-
P-C. Wei et al. [76]	V	0.61	0.64	0.58
Perceiver [36]	V	**0.62± 0.05**	**0.67± 0.08**	**0.62± 0.05**
Ours	V	0.62±0.06	0.66±0.04	0.61±0.06
S. Song et al. [64]	AV	0.50	0.60	0.43
P-C. Wei et al. [76]	AV	0.61	**0.78**	0.50
Perceiver [36]	AV	0.58±0.13	0.75±0.05	0.59±0.11
Ours	AV	**0.67± 0.05**	0.68±0.04	**0.66± 0.06**

Table 4. Results on the E-DAIC test set. Best results are highlighted in **bold** and second best results are underlined.

Method	Modality	F_1	Precision	Recall
Perceiver [36]	A	0.36±0.17	0.66±0.12	0.41±0.09
	V	0.39±0.22	0.50±0.26	0.48±0.18
	AV	0.53±0.09	0.54±0.05	0.54±0.11
Ours	A	0.51±0.11	**0.71±0.05**	0.52±0.08
	V	0.50±0.08	0.58±0.05	0.49±0.07
	AV	**0.56±0.12**	0.59±0.07	**0.58±0.13**

split. To provide a fair comparison, we evaluate our model on both the original test split and the test split of currently available videos from which we extracted additional modalities. Our model achieves a 0.76 F_1 using the original low-level facial and audio features provided by D-Vlog authors from the original test split. However, when we evaluate our model using the original features on the reduced test split, the performance of the model is lower, with only a 0.73 F_1, pointing to the fact that the samples from the reduced test set are more challenging to classify than the original test split. Using the additional modalities (audio-visual, landmarks and eyes) inspired by psychological research, our model obtains the best performance of 0.78 F_1, surpassing the previous state-of-the-art method proposed by Tao et al. [67]. Moreover, compared to the Perceiver, a state-of-the-art method for multi-modal classification, our model achieves a considerable improvement of 5% F_1.

We provide the results of our method using different modalities in Table 2. Even if the Perceiver model is a strong baseline for multi-modal information processing, our proposed method surpasses it in all settings. Adding **LM** (face, body and hands landmarks) and **EYES** features (gaze and blinking) separately offers a slight improvement over the **AV** setting using only face and audio embeddings. Our model obtains the best performance with all modalities **AV+LM+EYES**, achieving an F_1 of 0.78. Results show that different non-verbal manifestations of depression are needed for accurate detection from video data. Our best-performing model uses emotional-informed face embeddings from EmoNet, audio embeddings, spatial information from the face, body and hand landmarks, and gaze and blinking patterns.

DAIC-WOZ. Even though the DAIC-WOZ dataset was extensively used for depression detection, most previous works incorporate text-based features extracted from the video transcripts, with state-of-the-art approaches achieving performances of around 0.96 F_1 [91]. However, this high performance is not due to a deep understanding of depression, but is rather due to spurious correlations between the text transcript and the depression label, as the dataset contains recordings of clinical interviews with predefined questions from PHQ-8. For this reason, we are using only non-verbal cues, and discard the text transcript, to provide a more realistic performance measurement for depression detection. The methods we considered in Table 3 do not incorporate textual features for depression modeling. Our model is competitive with other methods in video-only settings, but we achieve the best performance of 0.67 F_1 using both modalities. Compared to the Perceiver baseline, our method performs better in audio and audio-visual settings.

E-DAIC. Similar to DAIC-WOZ, the current state of the art for binary classification explored text-based features extracted from the transcripts, a considerably easier formulation than only using non-verbal cues, achieving high performance [91]. In Table 4, we report the first results addressing the task from a non-verbal perspective on binary classification for E-DAIC. Our model obtains a substantial improvement when combining both audio and visual modalities, and outperforms the Perceiver in all scenarios, showing that our architecture is highly efficient at integrating multiple modalities.

Explainability. Any model developed to work with data from humans, especially in clinical scenarios, needs to be interpretable and explainable, to increase trust and transparency, ethical and legal compliance, and to allow researchers and engineers to debug and perform error analysis. Consequently, we present a potential way of explaining multi-modal depression detection models from video using Integrated Gradients [66]. In Fig. 5, we show attribution scores for 6 modalities in a selected window from a video from D-Vlog, with higher values corresponding to a strong attribution towards a positive prediction. We show the attribution values across time, for the video frames (bottom) and the audio frames (top), sampled at a higher frequency. In this particular context window, movement through the body and hand landmarks and gaze are more indicative of the mental state of the subject, compared to audio or facial expressions. From around frame 150, the person is not visible in the video (while audio is still present), and consequently, the attribution scores for each missing modality are 0. This shows that our model can properly handle missing modalities. Our analysis provides the first steps in automatic interpretation and understanding of how different non-verbal cues contribute to the manifestations of depression.

Limitations and Ethical Considerations. Our approach has shown promising results in detecting depression through non-verbal behavior, but it is important to note that it

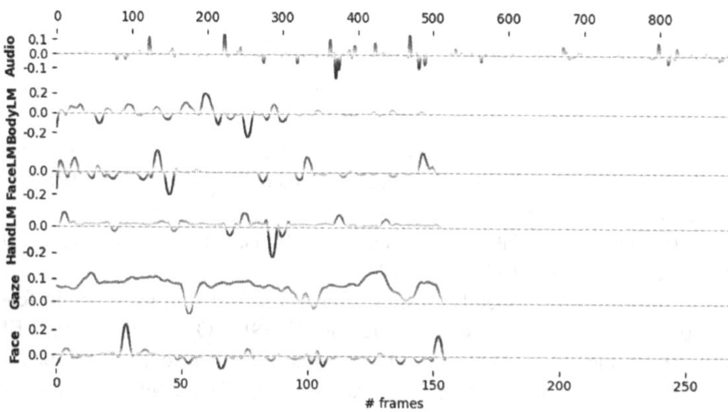

Fig. 5. Attribution scores per each modality across frames obtained with Integrated Gradients [66] on a selected window from a subject suffering from depression from D-Vlog. Higher values correspond to a strong attribution towards a positive prediction.

should not be used as a means of clinical diagnosis, which should only be done by a mental health professional. However, our method can be useful for initial depression screening by identifying psychomotor manifestations. In our experiments, we did not include any demographic information about the subjects; the results of our approach may differ when applied to other demographic groups [4,45]. Lastly, the data we used in our experiments were anonymized, and we made no attempt to contact the subjects.

6 Conclusion

In this work, we presented a simple and flexible multi-modal transformer architecture capable of detecting depression from multiple non-verbal cues from video. Following psychological studies [8,24,56,83], we explored additional high-level modalities and showed a markedly improvement in depression detection from in-the-wild videos. We obtained state-of-the-art results in three key benchmarking datasets for depression detection, both in-the-wild vlogs [86] and in recorded clinical interviews [28,57], surpassing previous works [42,44,64,67,76,77,86,91–93] by a considerable margin. Finally, we showed that our model is interpretable, and it can provide importance scores of each modality of a particular subject across time, making our method a viable solution for depression estimation, potentially in early-warning detection and suicide prevention in platforms with continuous video streams.

Acknowledgements. The work of David Gimeno-Gómez and Carlos-D. Martínez-Hinarejos was partially supported by Grant CIACIF/2021/295 funded by Generalitat Valenciana and by Grant PID2021-124719OB-I00 under project LLEER funded by MCIN/AEI/10.13039/501100011033/ and by ERDF, EU A way of making Europe. The work of Paolo Rosso was in the framework of the PID2021-124361OB-C31 research project funded by MCIN/AEI/10.13039/501100011033 and by ERDF, EU A way of making Europe.

References

1. Alayrac, J.B., et al.: Self-supervised multimodal versatile networks. Adv. Neural. Inf. Process. Syst. **33**, 25–37 (2020)
2. Alghowinem, S., Goecke, R., Wagner, M., Epps, J., Breakspear, M., Parker, G.: From joyous to clinically depressed: mood detection using spontaneous speech. In: FLAIRS Conference, vol. 19, pp. 141–146 (2012)
3. American Psychiatric Association: Diagnostic and statistical manual of mental disorders: DSM-5. Autor, Washington, DC, 5th edn (2013)
4. Bailey, A., Plumbley, M.D.: Gender bias in depression detection using audio features. In: 2021 29th European Signal Processing Conference (EUSIPCO), pp. 596–600. IEEE (2021)
5. Baltrušaitis, T., Robinson, P., Morency, L.P.: Openface: an open source facial behavior analysis toolkit. In: IEEE Winter Conference on Applications of Computer Vision (WACV), pp. 1–10 (2016)
6. Bazarevsky, V., Grishchenko, I., Raveendran, K., Zhu, T., Zhang, F., Grundmann, M.: Blazepose: on-device real-time body pose tracking. arXiv preprint arXiv:2006.10204 (2020)
7. Bengio, Y., Simard, P., Frasconi, P.: Learning long-term dependencies with gradient descent is difficult. IEEE Trans. Neural Networks **5**(2), 157–166 (1994)
8. Bennabi, D., Vandel, P., Papaxanthis, C., Pozzo, T., Haffen, E.: Psychomotor retardation in depression: a systematic review of diagnostic, pathophysiologic, and therapeutic implications. BioMed Res. Int. **2013** (2013)
9. Benton, A., Mitchell, M., Hovy, D.: Multitask learning for mental health conditions with limited social media data. In: Proceedings of the 15th Conference of the European Chapter of the Association for Computational Linguistics: Volume 1, Long Papers, Valencia, Spain, pp. 152–162. Association for Computational Linguistics (2017)
10. Birnbaum, M.L., et al.: Identifying signals associated with psychiatric illness utilizing language and images posted to Facebook. NPJ Schizophrenia **6**(1), 1–10 (2020)
11. Bredin, H., Laurent, A.: End-to-end speaker segmentation for overlap-aware resegmentation. In: Proceedings of Interspeech, pp. 3111–3115 (2021)
12. Bredin, H., et al.: Pyannote. Audio: neural building blocks for speaker diarization. In: IEEE International Conference on Acoustics, Speech and Signal Processing (ICASSP), pp. 7124–7128 (2020)
13. Bucur, A.M., Cosma, A., Rosso, P., Dinu, L.P.: It's just a matter of time: detecting depression with time-enriched multimodal transformers. In: Kamps, J., et al. (eds.) ECIR 2023. LNCS, vol. 13980, pp. 200–215. Springer, Cham (2023). https://doi.org/10.1007/978-3-031-28244-7_13
14. Bulat, A., Tzimiropoulos, G.: How far are we from solving the 2D & 3D face alignment problem? (and a dataset of 230,000 3D facial landmarks). In: Proceedings of the IEEE International Conference on Computer Vision, pp. 1021–1030 (2017)
15. Coppersmith, G., Leary, R., Crutchley, P., Fine, A.: Natural language processing of social media as screening for suicide risk. Biomed. Inform. Insights **10**, 1178222618792860 (2018)
16. Davis, S., Mermelstein, P.: Comparison of parametric representations for monosyllabic word recognition in continuously spoken sentences. IEEE Trans. Acoust. Speech Signal Process. **28**(4), 357–366 (1980)
17. Degottex, G., Kane, J., Drugman, T., Raitio, T., Scherer, S.: Covarep - a collaborative voice analysis repository for speech technologies. In: IEEE International Conference on Acoustics, Speech and Signal Processing (ICASSP), pp. 960–964 (2014)

18. Devlin, J., Chang, M.W., Lee, K., Toutanova, K.: BERT: pre-training of deep bidirectional transformers for language understanding. In: Proceedings of the 2019 Conference of the North American Chapter of the Association for Computational Linguistics: Human Language Technologies, Volume 1 (Long and Short Papers), Minneapolis, Minnesota, pp. 4171–4186. Association for Computational Linguistics (2019)
19. Dibeklioğlu, H., Hammal, Z., Cohn, J.F.: Dynamic multimodal measurement of depression severity using deep autoencoding. IEEE J. Biomed. Health Inform. **22**(2), 525–536 (2017)
20. Doersch, C., Gupta, A., Zisserman, A.: Crosstransformers: spatially-aware few-shot transfer. Adv. Neural. Inf. Process. Syst. **33**, 21981–21993 (2020)
21. Eyben, F., et al.: The Geneva minimalistic acoustic parameter set (GeMAPS) for voice research and affective computing. IEEE Trans. Affect. Comput. **7**(2), 190–202 (2016)
22. Fang, Z., Liu, Z., Hung, C.C., Sekhavat, Y.A., Liu, T., Wang, X.: Learning coordinated emotion representation between voice and face. Appl. Intell. **53**(11), 14470–14492 (2023)
23. Fernández-Barrera, I., Bravo-Bustos, S., Vidal, M.: Evaluating the social media users' mental health status during covid-19 pandemic using deep learning. In: International Conference on Biomedical and Health Informatics, vol. 14 (2022)
24. Fiquer, J.T., Boggio, P.S., Gorenstein, C.: Talking bodies: nonverbal behavior in the assessment of depression severity. J. Affect. Disord. **150**(3), 1114–1119 (2013)
25. Friesen, E., Ekman, P.: Facial action coding system: a technique for the measurement of facial movement. Palo Alto University, California, vol. 3, no. 2, p. 5 (1978)
26. Gales, M., Young, S.: The Application of Hidden Markov Models in Speech Recognition. Now Publishers Inc. (2008)
27. Girard, J.M., Cohn, J.F., Mahoor, M.H., Mavadati, S., Rosenwald, D.P.: Social risk and depression: evidence from manual and automatic facial expression analysis. In: 2013 10th IEEE International Conference and Workshops on Automatic Face and Gesture Recognition (FG), pp. 1–8. IEEE (2013)
28. Gratch, J., et al.: The distress analysis interview corpus of human and computer interviews. In: LREC, pp. 3123–3128 (2014)
29. Gui, T., et al.: Cooperative multimodal approach to depression detection in twitter. In: Proceedings of the AAAI Conference on Artificial Intelligence, vol. 33, pp. 110–117 (2019)
30. Gunes, H., Schuller, B.: Categorical and dimensional affect analysis in continuous input: current trends and future directions. Image Vis. Comput. **31**(2), 120–136 (2013)
31. Haque, A., Reddi, V., Giallanza, T.: Deep learning for suicide and depression identification with unsupervised label correction. In: Farkaš, I., Masulli, P., Otte, S., Wermter, S. (eds.) ICANN 2021. LNCS, vol. 12895, pp. 436–447. Springer, Cham (2021). https://doi.org/10.1007/978-3-030-86383-8_35
32. Harzig, P., Einfalt, M., Lienhart, R.: Synchronized audio-visual frames with fractional positional encoding for transformers in video-to-text translation. In: 2022 IEEE International Conference on Image Processing (ICIP), pp. 2041–2045 (2022)
33. He, K., Zhang, X., Ren, S., Sun, J.: Deep residual learning for image recognition. In: CVPR, pp. 770–778 (2016)
34. He, L., et al.: Deep learning for depression recognition with audiovisual cues: a review. Inf. Fusion **80**, 56–86 (2022)
35. Loshchilov, I., Hutter, F.: Decoupled weight decay regularization. In: ICLR (2019)
36. Jaegle, A., Gimeno, F., Brock, A., Vinyals, O., Zisserman, A., Carreira, J.: Perceiver: general perception with iterative attention. In: International Conference on Machine Learning, pp. 4651–4664. PMLR (2021)
37. Kaushik, R., Gaur, S., Pandit, J.N., Satapathy, S., Behera, C.: Live streaming of suicide on Facebook. Psychiatry Res. Case Rep. **2**(2), 100141 (2023)

38. Lee, J., Lee, Y., Kim, J., Kosiorek, A., Choi, S., Teh, Y.W.: Set transformer: a framework for attention-based permutation-invariant neural networks. In: International Conference on Machine Learning, pp. 3744–3753. PMLR (2019)
39. Leis, A., Ronzano, F., Mayer, M.A., Furlong, L.I., Sanz, F.: Detecting signs of depression in tweets in Spanish: behavioral and linguistic analysis. J. Med. Internet Res. **21**(6), e14199 (2019)
40. Li, B., Xiong, P., Han, C., Guo, T.: Shrinking temporal attention in transformers for video action recognition. In: Proceedings of the AAAI Conference on Artificial Intelligence, vol. 36, pp. 1263–1271 (2022)
41. Lucas, G.M., Gratch, J., Scherer, S., Boberg, J., Stratou, G.: Towards an affective interface for assessment of psychological distress. In: 2015 International Conference on Affective Computing and Intelligent Interaction (ACII), pp. 539–545. IEEE (2015)
42. Ma, X., Yang, H., Chen, Q., Huang, D., Wang, Y.: Depaudionet: an efficient deep model for audio based depression classification. In: Proceedings of the 6th International Workshop on Audio/Visual Emotion Challenge, AVEC 2016, pp. 35–42. Association for Computing Machinery (2016)
43. Mackintosh, J., Kumar, R., Kitamura, T.: Blink rate in psychiatric illness. Br. J. Psychiatry **143**(1), 55–57 (1983)
44. Nguyen, D.K., et al.: Multimodal transformer for automatic depression estimation system. In: The 29th International Workshop on Frontiers of Computer Vision (2023)
45. Oureshi, S.A., Dias, G., Saha, S., Hasanuzzaman, M.: Gender-aware estimation of depression severity level in a multimodal setting. In: 2021 International Joint Conference on Neural Networks (IJCNN), pp. 1–8. IEEE (2021)
46. Pampouchidou, A., et al.: Depression assessment by fusing high and low level features from audio, video, and text. In: Proceedings of the 6th International Workshop on Audio/Visual Emotion Challenge, pp. 27–34 (2016)
47. Pampouchidou, A., et al.: Automatic assessment of depression based on visual cues: a systematic review. IEEE Trans. Affect. Comput. **10**(4), 445–470 (2017)
48. Pascanu, R., Mikolov, T., Bengio, Y.: On the difficulty of training recurrent neural networks. In: International Conference on Machine Learning, pp. 1310–1318. PMLR (2013)
49. Pascual, S., Ravanelli, M., Serrà, J., Bonafonte, A., Bengio, Y.: Learning problem-agnostic speech representations from multiple self-supervised tasks. In: Proceedings of Interspeech, pp. 161–165 (2019)
50. Pavlidou, A., et al.: Hand gesture performance is impaired in major depressive disorder: a matter of working memory performance? J. Affect. Disord. **292**, 81–88 (2021)
51. Pérez, A., Piot-Pérez-Abadín, P., Parapar, J., Barreiro, Á.: Psyprof: a platform for assisted screening of depression in social media. In: Kamps, J., et al. (eds.) ECIR 2023. LNCS, vol. 13982, pp. 300–306. Springer, Cham (2023). https://doi.org/10.1007/978-3-031-28241-6_30
52. Pirina, I., Çöltekin, Ç.: Identifying depression on Reddit: the effect of training data. In: Proceedings of the 2018 EMNLP Workshop SMM4H: The 3rd Social Media Mining for Health Applications Workshop & Shared Task, pp. 9–12 (2018)
53. Ravanelli, M., et al.: Multi-task self-supervised learning for robust speech recognition. In: IEEE International Conference on Acoustics, Speech and Signal Processing (ICASSP), pp. 6989–6993 (2020)
54. Ray, A., Kumar, S., Reddy, R., Mukherjee, P., Garg, R.: Multi-level attention network using text, audio and video for depression prediction. In: Proceedings of the 9th International on Audio/Visual Emotion Challenge and Workshop, pp. 81–88 (2019)
55. Reece, A.G., Danforth, C.M.: Instagram photos reveal predictive markers of depression. EPJ Data Sci. **6**(1), 15 (2017)

56. Renneberg, B., Heyn, K., Gebhard, R., Bachmann, S.: Facial expression of emotions in borderline personality disorder and depression. J. Behav. Ther. Exp. Psychiatry **36**(3), 183–196 (2005)
57. Ringeval, F., et al.: AVEC 2019 workshop and challenge: state-of-mind, detecting depression with AI, and cross-cultural affect recognition. In: Proceedings of the 9th International on Audio/Visual Emotion Challenge and Workshop, pp. 3–12 (2019)
58. Ríssola, E.A., Aliannejadi, M., Crestani, F.: Beyond modelling: understanding mental disorders in online social media. In: Jose, J.M., et al. (eds.) ECIR 2020. LNCS, vol. 12035, pp. 296–310. Springer, Cham (2020). https://doi.org/10.1007/978-3-030-45439-5_20
59. Rottenberg, J., Vaughan, C.: Emotion expression in depression: emerging evidence for emotion context-insensitivity. In: Vingerhoets, A.J., Nyklíček, I., Denollet, J. (eds.) Emotion Regulation: Conceptual and Clinical Issues, pp. 125–139. Springer, Boston (2008). https://doi.org/10.1007/978-0-387-29986-0_8
60. Russell, J.A.: A circumplex model of affect. J. Pers. Soc. Psychol. **39**(6), 1161 (1980)
61. Saggu, G.S., Gupta, K., Arya, K., Rodriguez, C.R.: Depressnet: a multimodal hierarchical attention mechanism approach for depression detection. Int. J. Eng. Sci. **15**(1), 24–32 (2022)
62. Shen, G., et al.: Depression detection via harvesting social media: a multimodal dictionary learning solution. In: Proceedings of the Twenty-Sixth International Joint Conference on Artificial Intelligence. International Joint Conferences on Artificial Intelligence Organization (2017)
63. Sleeman, W.C., IV., Kapoor, R., Ghosh, P.: Multimodal classification: current landscape, taxonomy and future directions. ACM Comput. Surv. **55**(7), 1–31 (2022)
64. Song, S., Shen, L., Valstar, M.: Human behaviour-based automatic depression analysis using hand-crafted statistics and deep learned spectral features. In: 13th IEEE International Conference on Automatic Face & Gesture Recognition (FG), pp. 158–165 (2018)
65. Su, Y., Lan, T., Li, H., Xu, J., Wang, Y., Cai, D.: PandaGPT: one model to instruction-follow them all. arXiv preprint arXiv:2305.16355 (2023)
66. Sundararajan, M., Taly, A., Yan, Q.: Axiomatic attribution for deep networks. In: International Conference on Machine Learning, pp. 3319–3328. PMLR (2017)
67. Tao, Y., Yang, M., Wu, Y., Lee, K., Kline, A., Hu, B.: Depressive semantic awareness from vlog facial and vocal streams via spatio-temporal transformer. Digit. Commun. Netw. (2023)
68. Taylor, B.P., et al.: Psychomotor slowing as a predictor of fluoxetine nonresponse in depressed outpatients. Am. J. Psychiatry **163**(1), 73–78 (2006)
69. Toisoul, A., Kossaifi, J., Bulat, A., Tzimiropoulos, G., Pantic, M.: Estimation of continuous valence and arousal levels from faces in naturalistic conditions. Nat. Mach. Intell. **3**(1), 42–50 (2021)
70. Trifan, A., Antunes, R., Matos, S., Oliveira, J.L.: Understanding depression from psycholinguistic patterns in social media texts. In: Jose, J.M., et al. (eds.) ECIR 2020. LNCS, vol. 12036, pp. 402–409. Springer, Cham (2020). https://doi.org/10.1007/978-3-030-45442-5_50
71. Valstar, M., et al.: AVEC 2013: the continuous audio/visual emotion and depression recognition challenge. In: Proceedings of the 3rd ACM International Workshop on Audio/Visual Emotion Challenge, pp. 3–10 (2013)
72. Vaswani, A., et al.: Attention is all you need. In: NeurIPS, vol. 30, pp. 6000–6010 (2017)
73. Villatoro-Tello, E., Ramírez-de-la Rosa, G., Gática-Pérez, D., Magimai.-Doss, M., Jiménez-Salazar, H.: Approximating the mental lexicon from clinical interviews as a support tool for depression detection. In: Proceedings of the 2021 International Conference on Multimodal Interaction, pp. 557–566 (2021)
74. Wang, J., Zhang, L., Liu, T., Pan, W., Hu, B., Zhu, T.: Acoustic differences between healthy and depressed people: a cross-situation study. BMC Psychiatry **19**, 1–12 (2019)

75. Wang, Y.A., Chen, Y.N.: What do position embeddings learn? An empirical study of pre-trained language model positional encoding. In: Proceedings of the 2020 Conference on Empirical Methods in Natural Language Processing (EMNLP), pp. 6840–6849 (2020)
76. Wei, P.C., Peng, K., Roitberg, A., Yang, K., Zhang, J., Stiefelhagen, R.: Multi-modal depression estimation based on sub-attentional fusion. In: Karlinsky, L., Michaeli, T., Nishino, K. (eds.) ECCV 2022. LNCS, vol. 13806, pp. 623–639. Springer, Cham (2023). https://doi.org/10.1007/978-3-031-25075-0_42
77. Williamson, J.R., et al.: Detecting depression using vocal, facial and semantic communication cues. In: Proceedings of the 6th International Workshop on Audio/Visual Emotion Challenge, AVEC 2016, pp. 11–18. Association for Computing Machinery (2016)
78. Williamson, J.R., Quatieri, T.F., Helfer, B.S., Ciccarelli, G., Mehta, D.D.: Vocal and facial biomarkers of depression based on motor incoordination and timing. In: Proceedings of the 4th International Workshop on Audio/Visual Emotion Challenge, pp. 65–72 (2014)
79. Wolohan, J., Hiraga, M., Mukherjee, A., Sayyed, Z.A., Millard, M.: Detecting linguistic traces of depression in topic-restricted text: attending to self-stigmatized depression with NLP. In: Proceedings of the First International Workshop on Language Cognition and Computational Models, pp. 11–21. Association for Computational Linguistics (2018)
80. Xu, H., Bazavan, E.G., Zanfir, A., Freeman, W.T., Sukthankar, R., Sminchisescu, C.: GHUM & GHUML: generative 3D human shape and articulated pose models. In: IEEE/CVF Conference on Computer Vision and Pattern Recognition (CVPR), pp. 6183–6192 (2020)
81. Xu, X., et al.: Leveraging large language models for mental health prediction via online text data (2023)
82. Yadav, S., Caragea, C., Zhao, C., Kumari, N., Solberg, M., Sharma, T.: Towards identifying fine-grained depression symptoms from memes. In: Proceedings of the 61st Annual Meeting of the Association for Computational Linguistics, pp. 8890–8905 (2023)
83. Yamamoto, M., et al.: Using speech recognition technology to investigate the association between timing-related speech features and depression severity. PLoS ONE **15**(9), e0238726 (2020)
84. Yang, K., Zhang, T., Kuang, Z., Xie, Q., Ananiadou, S.: Mentalllama: interpretable mental health analysis on social media with large language models. arXiv preprint arXiv:2309.13567 (2023)
85. Yates, A., Cohan, A., Goharian, N.: Depression and self-harm risk assessment in online forums. In: Proceedings of the 2017 Conference on Empirical Methods in Natural Language Processing, pp. 2968–2978. Association for Computational Linguistics (2017)
86. Yoon, J., Kang, C., Kim, S., Han, J.: D-vlog: multimodal vlog dataset for depression detection. In: Proceedings of the AAAI Conference on Artificial Intelligence, vol. 36, pp. 12226–12234 (2022)
87. Zanwar, S., Wiechmann, D., Qiao, Y., Kerz, E.: SMHD-GER: a large-scale benchmark dataset for automatic mental health detection from social media in German. In: Findings of the Association for Computational Linguistics: EACL 2023, Dubrovnik, Croatia, pp. 1526–1541. Association for Computational Linguistics (2023)
88. Zeng, W., et al.: Real-time multi-person eyeblink detection in the wild for untrimmed video. In: Proceedings of the IEEE/CVF Conference on Computer Vision and Pattern Recognition (CVPR), pp. 13854–13863 (2023)
89. Zhang, X., Park, S., Beeler, T., Bradley, D., Tang, S., Hilliges, O.: ETH-XGaze: a large scale dataset for gaze estimation under extreme head pose and gaze variation. In: European Conference on Computer Vision (ECCV), pp. 365–381 (2020)
90. Zhang, Y., et al.: Associations between depression symptom severity and daily-life gait characteristics derived from long-term acceleration signals in real-world settings: retrospective analysis. JMIR mHealth uHealth **10**(10), e40667 (2022)

91. Zheng, W., Yan, L., Wang, F.Y.: Two birds with one stone: knowledge-embedded temporal convolutional transformer for depression detection and emotion recognition. IEEE Trans. Affect. Comput. 1–18 (2023)
92. Zhou, L., Liu, Z., Shangguan, Z., Yuan, X., Li, Y., Hu, B.: TAMFN: time-aware attention multimodal fusion network for depression detection. IEEE Trans. Neural Syst. Rehabil. Eng. **31**, 669–679 (2022)
93. Zhou, L., Liu, Z., Yuan, X., Shangguan, Z., Li, Y., Hu, B.: CAIINET: neural network based on contextual attention and information interaction mechanism for depression detection. Digit. Signal Process. **137**, 103986 (2023)
94. Zhu, D., Chen, J., Shen, X., Li, X., Elhoseiny, M.: MiniGPT-4: enhancing vision-language understanding with advanced large language models (2023)

DREQ: Document Re-ranking Using Entity-Based Query Understanding

Shubham Chatterjee[1(✉)], Iain Mackie[2], and Jeff Dalton[1]

[1] University of Edinburgh, Edinburgh, UK
{shubham.chatterjee,jeff.dalton}@ed.ac.uk
[2] University of Glasgow, Glasgow, UK
i.mackie.1@research.gla.ac.uk

Abstract. While entity-oriented neural IR models have advanced significantly, they often overlook a key nuance: the varying degrees of influence individual entities within a document have on its overall relevance. Addressing this gap, we present DREQ, an entity-oriented dense document re-ranking model. Uniquely, we emphasize the query-relevant entities within a document's representation while simultaneously attenuating the less relevant ones, thus obtaining a query-specific entity-centric document representation. We then combine this entity-centric document representation with the text-centric representation of the document to obtain a "hybrid" representation of the document. We learn a relevance score for the document using this hybrid representation. Using four large-scale benchmarks, we show that DREQ outperforms state-of-the-art neural and non-neural re-ranking methods, highlighting the effectiveness of our entity-oriented representation approach.

1 Introduction

During the last decade, the emergence of large-scale Knowledge Graphs (KGs) has motivated the development of entity-oriented search systems. Entities, with their rich semantic information, help bridge the gap between unstructured text and structured knowledge. Prior research [16,36,55,61,62] underscores the significance of entities in feature-based retrieval systems. Their effectiveness within neural IR models has also been demonstrated, with the Entity-Duet Ranking Model (EDRM) from Liu et al. [38] standing out as a pioneering effort. This model synergizes the word-entity duet framework [62] with the strengths of neural networks and KGs. More recently, Tran and Yates [59] introduced an approach that clusters entities within documents, offering multiple "views" or perspectives to enhance the understanding of various document facets.

Yet, amidst these advancements, current models often overlook a crucial aspect: not all entities within a document contribute equally to its relevance. For instance, given the query "Black Bear Attacks" and the document in Fig. 1, while the entity *hand axe* may have peripheral relevance to the given query due to its defensive use during animal encounters, the entity *National Guard* is

N. Goharian et al. (Eds.): ECIR 2024, LNCS 14608, pp. 210–229, 2024.
https://doi.org/10.1007/978-3-031-56027-9_13

likely non-relevant. Conversely, the entity *Parks Highway*, known for bear sightings, may hold greater significance. In the "retrieve-then-rerank" paradigm, a common approach in neural IR, initial retrieval fetches a broad set of candidate documents. For the subsequent re-ranking phase, though, the differential relevance of individual entities becomes particularly critical as the model must sift through the candidates and rank them with high precision. Furthermore, prior work [17,38,59,62] often produce query-agnostic document representations which fail to resonate with the specific nuances and requirements of the query. Our proposition emphasizes a more refined approach: to enhance re-ranking accuracy, entities should be weighted based on their query relevance, ensuring that the document's representation is both influenced by the most relevant entities and tailored to the query's nuances.

Against this backdrop, in this work[1], we introduce ***D****ocument* ***R****e-ranking using* ***E****ntity-based* ***Q****uery Understanding* (`DREQ`), an entity-oriented neural re-ranking model that extends the conventional "retrieve-then-rerank" paradigm by introducing an innovative intermediate step. Given a query Q and a candidate set of documents $\mathcal{D}$ retrieved using an initial retrieval method (e.g., BM25), we want to *re-rank* these candidates to order them by their relevance to the query Q. While prior approaches predominantly utilize entities to gain a fine-grained understanding of documents, our method uniquely perceives entities within candidate documents as overarching concepts essential for a comprehensive understanding of the query. We identify and prioritize entities that align closely with the query. To this end, we emphasize the embeddings of the relevant entities and concurrently attenuate the less relevant ones within the document's representation, thus obtaining a refined, *query-specific* and *entity-centric* document representation. Recognizing that the raw text of a document captures the document's overarching semantic context, we also derive a broader text-centric representation of the document, complementing the focused entity-centric perspective. We meld the entity-centric and text-centric representations of the document, obtaining a "hybrid" document representation that imbibes insights about what the query specifically seeks. Using this hybrid representation, we learn a fine-grained "interaction vector" that imbibes the differences, commonalities, and subtle relationships between the query and document. We then use this vector to learn a relevance score for the document. We term our methodology "retrieve-harness-rerank"[2], emphasizing our process of harnessing the information from entities present in candidate documents to re-rank the candidates.

Contributions. We make the following contributions through this work:

1. We introduce `DREQ`, an entity-oriented re-ranking model that enriches a document's representation with *query-specific* entity knowledge for nuanced relevance matching. We achieve new state-of-the-art results on four major document ranking test collections.

[1] **Code and data:** https://github.com/shubham526/ECIR2024-DREQ.

[2] Term coined by Dr. Laura Dietz at the SIGIR 2023 tutorial (https://github.com/laura-dietz/neurosymbolic-representations-for-IR).

2. We introduce a hybrid representation learning mechanism, blending entity-centric and text-centric representations. This hybrid representation captures a document's broad context and query-specific relevance, enhancing the precision of re-ranking.
3. We show that the effectiveness of re-ranking is significantly enhanced by meticulously selecting and assigning appropriate weights to entities within a document.

2 Related Work

Entity-Oriented Search. Initial attempts at entity-oriented search primarily used entities for query expansion, one notable example being Entity-Query Feature Expansion [11] (EQFE) model which used entity links within documents for query expansion. Entities later became a latent layer [17,36,61] in document and query representations, forming a high-dimensional entity space that improved retrieval by revealing hidden semantics. Research progressed to treat entities as explicit elements in retrieval models, coexisting with term-based approaches. Methods like entity-based language models [55] and semantically-driven models [16] rank documents by their semantic relation to the query. A prominent line of research [61,62,64–66] introduced a dual-layered approach, combining a "bag-of-entities" with the traditional "bag-of-terms" to improve document retrieval.

Neural IR. Recently, deep learning has transformed text ranking, removing the need for handcrafted features and fostering semantic matching. Before BERT [12], models either created vector representations [25,43,44,57] for queries and documents or built similarity matrices to capture term interactions [10,22,26,27, 63]. The advent of BERT and its derivatives [7,24,28,37,71] in a retrieve-then-rerank [49,50] framework ushered in models like Birch [1] and BERT-MaxP [8], emphasizing sentence or passage relevance. Subsequent models, such as CEDR [39] and PARADE [35], utilized BERT's contextual embeddings, with the latter aggregating passage representations. Concurrently, ERNIE [71] enhanced BERT with knowledge graphs and vast text corpora.

Meanwhile, bi-encoders emerged as a foundational mechanism in text ranking, employing distinct encoders to derive query and document vectors. For example, DPR [30] employs BERT's `[CLS]` token for both, gauging similarity through their inner product. The choice of negative examples is crucial: DPR favors BM25-retrieved passages while ANCE [67] leans on ANN techniques. Addressing bi-encoders' challenges with term-level interactions, research has delved into multi-vector text representation, exemplified by ColBERT's [31] "late interaction" using token-level embeddings.

The evolution of pre-trained sequence-to-sequence models like T5 [53] has led to models such as MonoT5 [51] and RankT5 [75]: The former assesses document relevance via probability assignment to "true" labels during decoding whereas the latter provides direct ranking scores for query-document pairs.

Query/Document Expansion. Recent efforts in this direction have been driven by embeddings. Solutions such as DeepCT [9] and docT5query [51] use

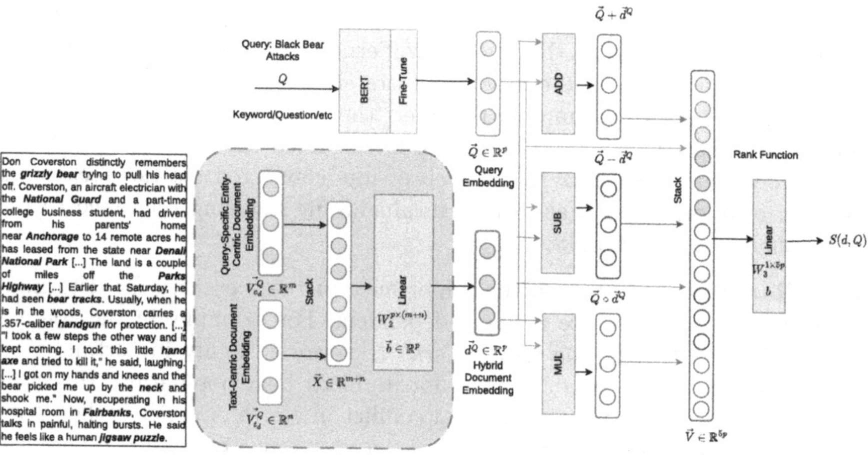

Fig. 1. Our proposed system DREQ uses a hybrid document embedding learnt using (1) the query-specific entity-centric embedding, and (2) text embedding of the document to learn the document score.

transformers to enrich traditional PRF models by highlighting significant terms in documents. Simultaneously, methods like Neural PRF [34] and BERT-QE [72] have adopted neural ranking models to measure document similarity with feedback documents. CEQE [45] has further enhanced this approach by using BERT to create contextualized representations and select expansion terms based on similarity in the embedding space. While ColBERT-PRF [60] directly utilizes BERT embeddings for retrieval without further training, ANCE-PRF [70] requires additional retraining of the query encoder using PRF information

Entity Ranking. Early methods include using MRFs to deal with the joint distribution of entity terms from semi-structured data [23,42,48,54,73], utilizing types [2,18,29] and relations [6,58] from a KG, and using Learning-To-Rank methods [4,13,21,56] to rank entities using a diverse set of features. Recent advancements have been significantly driven by neural models using advanced techniques such as autoregressive entity ranking [3], integrating BERT-based entity rankings with other features [5], and enhancing BERT with Wikipedia2Vec [68] embeddings [20]. Concurrently, the emergence of graph embedding-based models [19,47] has enriched entity ranking approaches by utilizing joint embedding of entities and words within the same vector space.

3 Approach: DREQ

Overarching Idea. Our work is anchored in the intuition that documents contain interconnected entities which provide a distilled understanding of the document's main content, and serve as broad concepts important for understanding the query. For example, in Fig. 1, given the query "Black Bear Attacks", the presence of specific geographical entities such as *Parks Highway* and *Fairbanks* narrows down the context to the Alaskan wilderness, a region known for its bear population. Additionally, the entity *Hand axe* sheds light on human defensive measures during such encounters. Together, these entities reinforce the document's fit for narratives around bear encounters. Even though the central narrative revolves around a grizzly, the web of interconnected entities creates an ecosystem of information that can be invaluable in assessing relevance to the broader theme of bear attacks.

Entity Ranking. Nonetheless, the importance of these entities for determining the document's relevance is query-dependent. For example, in the context of "Black Bear Attacks", the entity *Parks Highway* is highly important as it specifies a geographical context known for its black bear population. However, for a broader query like "Human-animal conflict in North America", the entity is less significant as the theme encompasses a vast range of conflicts, animals, and locations, making the specific location of *Parks Highway* just one of many possible locales of interest and less indicative of the document's relevance.

To address this issue, we pool all entities from all candidate documents $d \in \mathcal{D}$ to obtain a candidate set of entities $\mathcal{E}$ for the query Q. Following previous work [14,15,46], we transfer the relevance labels from documents to entities in the document based on the assumption that a relevant document contains relevant entities. Using this entity ground truth, we train a separate entity ranking model to rank entities $e \in \mathcal{E}$. Specifically, we follow previous work [5,38,41,66,69] and leverage the Knowledge Base (DBpedia [33]) description t_e of entity e to learn an embedding $\mathbf{e} \in \mathbb{R}^k$ of the entity using BERT. The input to BERT is a sequence of query tokens $\tau^q \in Q$ and description tokens $\tau^e \in t_e$, separated by the special token `[SEP]`, and preceded by the special token `[CLS]`. We use the k-dimensional embedding of the `[CLS]` token from the last hidden layer of BERT as the embedding $\mathbf{e}$ of an entity $e \in \mathcal{E}$. To derive a rank score for this entity in relation to the query Q, we learn a linear projection $\mathbb{R}^k \rightarrow \mathbb{R}$ using the embedding $\mathbf{e}$ and a weight matrix $W_1^{1\times k}$. This scoring function $\mathrm{S}(e, Q)$ is formulated as: $\mathrm{S}(e, Q) = W_1 \cdot \mathbf{e} + b$, where b is a scalar bias term.

Document Representation. Entities such as *Parks Highway*, *Neck*, and *Hand axe* in Fig. 1 collectively suggest a bear encounter in Alaska. When related to the query "Black Bear Attacks", the combined narrative from the entities implies the document is probably relevant as it aligns with the query's essence. Based on this idea, and acknowledging that the significance of these entities for determining a document's relevance varies with the query, we learn a ***query-specific entity-centric*** representation $\mathbf{V}_{e_d}^Q \in \mathbb{R}^m$ of a candidate document $d \in \mathcal{D}$. Specifically, we first represent each entity within a document via its embedding from Wikipedia2Vec [68] due to its ability to capture the relationships and deeper contexts between entities. The document representation $\mathbf{V}_{e_d}^Q$ is a weighted sum of embeddings of entities in the document:

$$\mathbf{V}_{e_d}^Q = \sum_{e \in d} w_e \cdot \mathbf{e} \tag{1}$$

where the weight w_e for each entity is query-specific, determined by the rank score of the entity:

$$w_e = \frac{S(e, Q)}{\sum_{e' \in d} S(e', Q)}$$

Note that $S(e, Q)$ are normalized using the `softmax` function. Consequently, the weights w_e can be interpreted as the *probability* of the entity's relevance to the query. Intuitively, using these weights would prioritize query-relevant entities while concurrently down-weighting those of lesser relevance within the document.

While encapsulating key entities helps capture a document's query-specific semantic essence, it's crucial to also consider the broader textual context to capture nuances missed by solely focusing on entities. Hence, we also obtain a text-centric representation $\mathbf{V}_{t_d}^{Q} \in \mathbb{R}^n$ of a candidate document $d \in \mathcal{D}$ using BERT. Following previous work [35,39], we first segment the document into passages using a sliding window of M sentences with a stride of S sentences over the document. We represent each passage in the document via the embedding of the `[CLS]` token from BERT. To obtain the document representation $\mathbf{V}_{t_d}^{Q}$, we average the passage embeddings. We combine the text and entity-centric document representations to learn a single ***hybrid*** representation $\mathbf{d}^Q \in \mathbb{R}^p$ via a linear projection $\mathbb{R}^{m+n} \rightarrow \mathbb{R}^p$ as follows:

$$\mathbf{d}^Q = W_2 \cdot [\mathbf{V}_{t_d}^{Q}; \mathbf{V}_{e_d}^{Q}] + \mathbf{b}$$

where $\mathbf{X} \in \mathbb{R}^{m+n}$ is the concatenated embeddings $[\mathbf{V}_{t_d}^{Q}; \mathbf{V}_{e_d}^{Q}]$, $W_2^{p\times(m+n)}$ is a weight matrix, and $\mathbf{b} \in \mathbb{R}^p$ is a bias vector. Our intuition is that this hybrid embedding $\mathbf{d}^Q$ of a candidate document $d \in \mathcal{D}$ encapsulates both the granular insights from the entities and the overarching narrative provided by the document's text, thereby ensuring a more accurate search.

Document Ranking. To learn the document scoring function $\mathrm{S}(d, Q)$, we first learn several fine-grained interactions between the query embedding $\mathbf{Q} \in \mathbb{R}^p$ (obtained via the embedding of the `[CLS]` token from BERT) and the hybrid document embedding $\mathbf{d}^Q$ as follows: An additive interaction $\mathbf{V}_{\mathbf{add}}^{\mathbf{d},\mathbf{Q}} = \mathbf{Q} + \mathbf{d}^{\mathbf{Q}}$, subtractive interaction $\mathbf{V}_{\mathbf{sub}}^{\mathbf{d},\mathbf{Q}} = \mathbf{Q} - \mathbf{d}^{\mathbf{Q}}$, and multiplicative (Hadamard product) interaction $\mathbf{V}_{\mathbf{mul}}^{\mathbf{d},\mathbf{Q}} = \mathbf{Q} \circ \mathbf{d}^{\mathbf{Q}}$. We then learn the scoring function $\mathrm{S}(d, Q)$ through a linear projection $\mathbb{R}^{5p} \rightarrow \mathbb{R}$ as follows:

$$\mathrm{S}(d, Q) = W_3 \cdot \mathbf{V} + b$$

where $W_3^{1\times 5p}$ is a weight matrix, $\mathbf{V} \in \mathbb{R}^{5p}$ is a vector representing the concatenated embeddings $[\mathbf{Q}; \mathbf{d}^{\mathbf{Q}}; \mathbf{V}_{\mathbf{add}}^{\mathbf{d},\mathbf{Q}}; \mathbf{V}_{\mathbf{sub}}^{\mathbf{d},\mathbf{Q}}; \mathbf{V}_{\mathbf{mul}}^{\mathbf{d},\mathbf{Q}}]$, and b is a scalar bias. We hypothesize that by subtracting, adding, and multiplying the query and hybrid document embeddings, we can explore different aspects of how the query relates to the document. Subtraction might show the differences between them, which can help point out any areas that don't match. On the other hand, addition might reveal what they have in common, highlighting overlapping themes. Multiplying the embeddings goes a step further and might reveal subtle relationships between specific parts of the query and document. By merging these interactions with the original query and document embeddings, our intuition is that the model would garner a rich, composite insight into the document's depth and breadth.

End-To-End Training. We train both our entity and document ranking models using the binary cross-entropy loss below:

$$\mathcal{L} = -\frac{1}{N}\sum_{i=1}^{N}\Big(y_i \cdot \log(p(\hat{y}_i)) + (1 - y_i) \cdot (1 - \log(p(\hat{y}_i)))\Big)$$

where y_i is the label and $p(\hat{y}_i)$ is the predicted probability of the entity/document being relevant. This is analogous to pointwise learning-to-rank where each of N query-entity/query-document pairs is independently classified into relevant or non-relevant. The entire model is optimized end-to-end using back-propagation. The weight matrices, namely W_1 for the entity ranking model and (W_2, W_3) for the document ranking model, are learnt concurrently with the fine-tuning of the initial embeddings. Moreover, the Wikipedia2Vec embeddings undergo end-to-end fine-tuning specifically within the document ranking model, ensuring that these entity embeddings are tailored to enhance document ranking effectiveness.

4 Experimental Methodology

4.1 Datasets

CODEC. CODEC [40] is a benchmark specifically designed for complex research topics in social sciences. The benchmark includes 42 topics, and new entity linked corpus with 729,824 documents with focused content across finance, history, and politics. The corpus contains an average of 159 entities per document. It provides expert judgments on 6,186 documents derived from diverse automatic and manual runs.

TREC Robust 2004. The TREC Robust 2004 track focuses on poorly performing topics. The track provides 250 topics with short "titles" and longer "descriptions"; we report results on both title and description queries. The collection consists of 528,024 documents (containing an average of 116 entities per document) taken from TREC disks 4 and 5 excluding the Congressional Record. The track provides 311,409 graded relevance judgments for evaluation.

TREC News 2021. The TREC News track focuses on search tasks within the news domain. We focus on the background linking task, which involves retrieving news articles that provide relevant context or background information for a given news story. There are 51 topics, each with a title, description, and narrative; in this work, we use all three fields for query formulation. The track uses the TREC Washington Post (v4) collection, encompassing 728,626 documents containing 131 entities per document on average. The track provides 12,908 graded relevance assessments for evaluation.

TREC Core 2018. The TREC Core track offers 50 topics, each consisting of titles, descriptions, and narratives. For this work, we utilize all three components. The track uses the TREC Washington Post (v2) collection, encompassing 595,037 news articles and blog posts and containing 123 entities per document on average. 26,233 graded relevance judgements are available.

4.2 Evaluation Paradigm

Candidate Ranking. We retrieve a candidate set of 1000 documents per query using BM25+RM3 (Pyserini default). The Recall@1000 of this candidate set for each dataset is as follows: (1) CODEC: 0.82, (2) Robust04 (title): 0.77, (3) Robust04 (desc): 0.75, (4) News 2021: 0.94, (5) Core 2018: 0.75. Other metrics shown in Tables 1 and 2.

Evaluation Metrics. (1) Precision at $k = 20$, (2) Normalized Discounted Cumulative Gain (nDCG) at $k = 20$, and (3) Mean Average Precision (MAP). We conduct significance testing using paired-t-tests.

Entity Linking. Our work relies on an entity linked corpus. While any entity linking system may be used, in this work, we use WAT [52].

Train and Test Data. As positive examples during training, we use documents that are assessed as relevant in the ground truth provided with the dataset. Following the standard [30], for negative examples, we use documents from the candidate ranking (BM25+RM3) which are either explicitly annotated as negative or not present in the ground truth. We balance the training data by keeping the number of negative examples the same as the number of positive examples. These examples are then divided into 5-folds for cross-validation. We create these folds at the query level.

Baselines. We compare our proposed re-ranking approach `DREQ` to the following supervised state-of-the-art neural re-rankers: (1) **RoBERTa** [37], (2) **DeBERTa** [24], (3) **ELECTRA** [7], (4) **ConvBERT** [28], (5) **RankT5** [75], (6) **KNRM** [63], (7) **ERNIE** [71], (8) **EDRM** [38]. Furthermore, we also include an unsupervised entity-based baseline (9) **MaxSimCos** which scores documents using the maximum cosine similarity between every pair of query and document entity embedding. On TREC Robust 2004, we include the following additional (full-retrieval) baselines: (1) CEDR [39], (2) EQFE [11], (3) BERT-MaxP [8], and (4) PARADE [35]. All baselines are fine-tuned on the target datasets via 5-fold cross-validation using the binary cross-entropy loss.

4.3 Implementation Details

We use the `bert-base-uncased` model from HuggingFace to obtain the initial query and document/entity representations and fine-tune our model using the `CrossEntropyLoss` function from PyTorch. We use the Adam [32] optimizer with a learning rate of 10^{-5} and batch size of 20. BERT layers were not frozen during fine-tuning. For document segmentation, we apply a 10-sentence sliding

Table 1. Overall results on TREC Robust 2004. Best results in bold. ▲ denotes significant improvement and ▼ denotes significant deterioration compared to ⋆. Paired-t-test at $p < 0.05$. **nDCG and Precision measures at cut-off rank 20**. Unavailable results denoted by "–". Baselines denotes by † are re-ranking the BM25+RM3 candidate set at the top.

		TREC Robust 2004 (Title)			TREC Robust 2004 (Desc)		
		MAP	nDCG	Prec	MAP	nDCG	Prec
	BM25+RM3	0.29⋆	0.44⋆	0.38⋆	0.28⋆	0.42⋆	0.37⋆
Non-entity	RankT5†	0.30	0.50▲	0.43▲	0.33▲	0.54▲	0.46▲
	RoBERTa†	0.29	0.47▲	0.41▲	0.33▲	0.54▲	0.46▲
	DeBERTa†	0.29	0.49▲	0.42▲	0.34▲	0.55▲	0.47▲
	ELECTRA†	0.27▼	0.45	0.39	0.29	0.49▲	0.41▲
	ConvBERT†	0.32▲	0.52▲	0.45▲	0.35▲	0.57▲	0.48▲
	KNRM†	0.11▼	0.18▼	0.16▼	0.08▼	0.14▼	0.12▼
	CEDR	0.37▲	0.55▲	0.48▲	0.40▲	0.60▲	0.52▲
	PARADE	0.30	0.53▲	0.45▲	0.30▲	0.56▲	0.47▲
	BERT-MaxP	0.32▲	0.48▲	0.42▲	0.31▲	0.49▲	0.22▼
Entity-based	ERNIE†	0.29	0.48▲	0.41▲	0.33▲	0.54▲	0.45▲
	EDRM†	0.07▼	0.10▼	0.09▼	0.05▼	0.07▼	0.07▼
	EQFE	0.33▼	0.42▼	0.38▼	–	–	–
	MaxSimCos†	0.17▼	0.26▼	0.24▼	0.13▼	0.20▼	0.18▼
	TREC Best	0.33▲	–	–	0.33▲	–	–
	DREQ	**0.57**▲	**0.75**▲	**0.73**▲	**0.55**▲	**0.78**▲	**0.71**▲

Table 2. Overall results on CODEC, TREC News 2021, and TREC Core 2018. **nDCG and Precision measures at cut-off rank 20**.

	CODEC			TREC News 2021			TREC Core 2018		
	MAP	nDCG	Prec	MAP	nDCG	Prec	MAP	nDCG	Prec
BM25+RM3	0.36⋆	0.38⋆	0.43⋆	0.47⋆	0.48⋆	0.58⋆	0.31⋆	0.45⋆	0.46⋆
RankT5	0.38▲	0.44▲	0.45▲	0.27▼	0.31▼	0.36▼	0.22▼	0.33▼	0.33▼
RoBERTa	0.36	0.38	0.43	0.47	0.52▲	0.58	0.26▼	0.36▼	0.39▼
DeBERTa	0.39▲	0.45▲	0.46▲	0.43▼	0.47	0.56	0.35▲	0.52▲	0.51▲
ELECTRA	0.32▼	0.32▼	0.41▼	0.23▼	0.46▼	0.53▼	0.24▼	0.35▼	0.36▼
ConvBERT	0.38▲	0.44▲	0.46▲	0.44▼	0.47	0.55	0.32	0.50▲	0.49▲
KNRM	0.30▼	0.30▼	0.34▼	0.14▼	0.15▼	0.19▼	0.11▼	0.14▼	0.16▼
ERNIE	0.39▲	0.46▲	0.47▲	0.48	0.53▲	0.60▲	0.34▲	0.52▲	0.51▲
EDRM	0.30▼	0.29▼	0.35▼	0.09▼	0.10▼	0.13▼	0.09▼	0.10▼	0.13▼
MaxSimCos	0.33▼	0.33▼	0.40▼	0.20▼	0.22▼	0.30▼	0.10▼	0.12▼	0.14▼
TREC Best	–	–	–	0.43▼	–	–	0.43▲	–	0.61▲
DREQ	**0.58**▲	**0.50**▲	**0.68**▲	**0.76**▲	**0.56**▲	**0.84**▲	**0.58**▲	**0.66**▲	**0.71**▲

window with a 5-sentence stride using spaCy (v6.3.1). Embeddings (document text, and entity) and entity links are cached off-line for lookup during inference.

5 Results and Discussions

5.1 Effectiveness of DREQ

In this section, we explore the following research question: **(RQ1)** *How does DREQ perform in comparison to state-of-the-art document re-ranking methods? Which type of queries does it help the most?* From Tables 1 and 2, we observe that DREQ outperforms all baselines in terms of all evaluation metrics on all datasets. For example, on Robust04 (title), the best performing baseline is CEDR which obtains nDCG@20 = 0.55; however, DREQ improves performance by 36% over CEDR and achieves nDCG@20 = 0.75. While CEDR improves nDCG@20 of the candidate set (BM25+RM3) by 25% (from 0.44 to 0.55), our model DREQ improves this by 70% (from 0.44 to 0.75).

Query-level Analysis. We further delve into the source of performance improvements by analyzing the results at the query-level. We categorize the queries into different levels of difficulty using (a) the performance of the BM25+RM3 candidate ranking method, and (b) Weighted Information Gain (WIG) [74] Query Performance Prediction (QPP) method. Results shown in Fig. 2.

In Fig. 2a, we place the 5% most difficult queries (according to nDCG@20) for BM25+RM3, the candidate ranking method, on the left, and the 5% easiest ones on the right. Remarkably, DREQ enhances the performance of the most challenging queries (bins 0–25%). For example, there are no relevant documents among the top-20 for queries in bin 0–5% (nDCG@20 = 0.0); however, DREQ improves (helps) the performance of these queries, achieving an nDCG@20 of 0.70 (twice of CEDR which achieves 0.35). Specifically, we find that DREQ improves the performance of 210 of the 250 queries whereas CEDR only helps 160 queries.

In Fig. 2b, we specifically study how the reranking pipeline BM25+RM3 >> DREQ compares to the best performing baseline pipeline BM25+RM3 >> CEDR. For this, we use a well-known QPP method called WIG which provides a measure of the effectiveness of a query in retrieving relevant documents: A higher WIG score means the query is more effective (easier) while a lower score means the query is less effective (harder). Based on this, we divide the query set into three levels of difficulty: easy, medium, and hard. Once again, we observe that compared to CEDR, DREQ improves performance for the most challenging queries.

Example. We further examine some of the challenging queries helped by DREQ from bin 0–5% in Fig. 2a. One such query is "Behavioral Genetics". For this query, a highly relevant document (according to NIST assessments) is placed at rank 434 in the candidate set. However, we find that DREQ promotes this document to rank 1 (CEDR places this document at rank 212). Upon analyzing the entity ranking for this query, we find that entities directly relevant to the

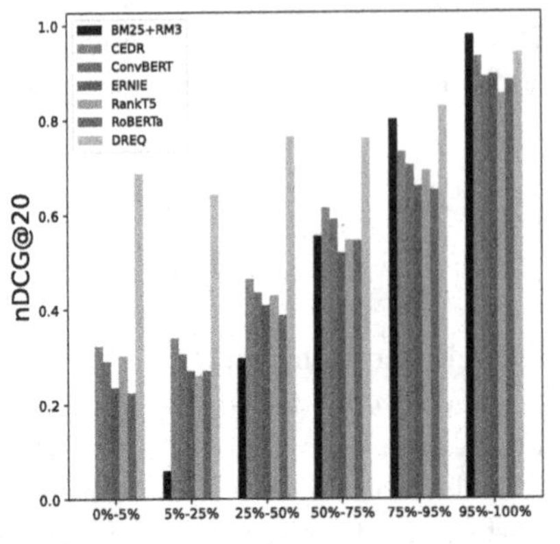

(a) Using performance of BM25+RM3

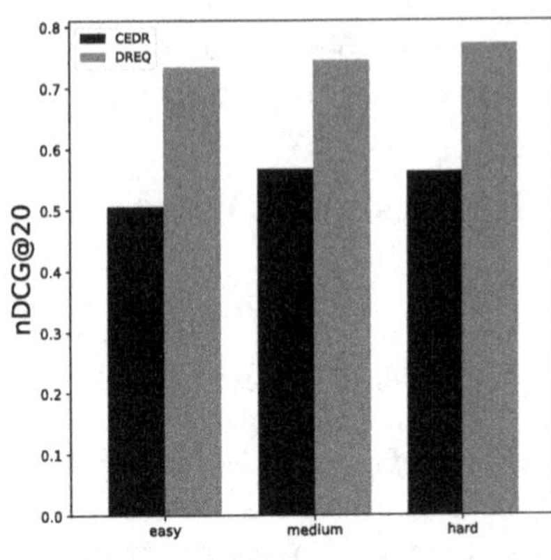

(b) Using WIG QPP Method

Fig. 2. Difficulty test for nDCG@20 on Robust04 (title). DREQ improves performance for the most difficult queries.

query, such as *Sensorineural Hearing Loss*, are assigned higher scores, while non-relevant entities, such as *Home for the Holidays*, receive lower scores. By utilizing these entity scores to weigh the entity embeddings, the model can emphasize the most important entities for the query, thereby enhancing the contribution of these entities to the hybrid document embedding. This, in turn, helps the model understand the nuances of relevance for the query, thereby improving the document's ranking position. We discuss contribution of entities in Sect. 5.2.

Take Away. To answer **RQ1**, DREQ achieves new state-of-the-art results for the document re-ranking task on four document ranking benchmarks across five diverse query sets by promoting relevant documents to the top of the ranking, thereby improving the precision at the top ranks of the candidate set. Our experiments show that DREQ is particularly adept at handling the most challenging queries, a key advantage over competing models. This demonstrates the robustness of DREQ, as it not only improves the performance of "easy" queries but also significantly boosts the performance of the most "difficult" ones. This is crucial for real-world applications where queries may be of varying difficulty.

5.2 Contribution of Entities in DREQ

Entities lie at the heart of our approach. Hence, in this section, we ask: **(RQ2)** *What is the contribution of entities in DREQ? How does the performance change if we change the underlying entity ranking system?* Furthermore, **(RQ3)** *How does changing the entity weighing method alter the performance of DREQ?* In the following, **we discuss results with respect to only Robust04 (title)**.

Ablation Study. To evaluate the contribution of the entity component to our model's performance, we first conduct an ablation study by removing the entity embeddings and retraining the model using only the other components. Our results reveal that, when entities are removed from DREQ, there's a marked performance drop. The nDCG@20 score plunges to 0.26, a decrease from the 0.75 score of the full model, and even below the 0.44 of the BM25+RM3 candidate set. For a clearer context, it is worth noting that the highest nDCG@20 score achieved by the best baseline model, CEDR, is 0.55. Performing the difficulty test described above using the BM25+RM3 candidate ranking, we find that DREQ without entities helps 74 queries; in contrast, the full DREQ model helps 210.

These results underline the pivotal role of entities in DREQ. The marked reduction in performance, when compared to both the full model and the best baseline, indicates that entities are integral to the model's effectiveness in document ranking and relevance interpretation. The comparison with CEDR, the best performing baseline, underscores the significance of incorporating entities, as even CEDR fails to match the nDCG@20 score of the full DREQ model. This suggests that entities are not only essential for DREQ but could also be a valuable addition to other document ranking models to enhance their performance.

Alternative Entity Ranking. As we use the rank scores of entities to weigh the entity embeddings when learning the hybrid document embedding, we study the effect of the entity ranking system on the performance of DREQ. We replace our supervised BERT-based entity ranking system with (1) BM25, a sparse model and, (2) GEEER [19], a recently proposed, state-of-the-art, entity re-ranking method. GEEER first computes the (Wikipedia2Vec [68]) embedding-based score for an entity E in a given candidate set of entities as a weighted sum: $\mathrm{S_{emb}}(E, Q) = \sum_{e \in Q} C(e) \cdot \cos(\mathbf{E}, \mathbf{e})$, where $C(e)$ is the confidence score of entity $e \in Q$ obtained from an entity linker. The final score is obtained by interpolating this embedding-based score with the score of the entity E obtained using a retrieval model. In this work, we use BM25 as the retrieval model in GEEER.

In the results, we find that incorporating the BM25 entity ranking into DREQ leads to a decline in performance, with the nDCG@20 dropping from 0.75 to 0.44. Similarly, substituting our BERT-based entity ranking system with GEEER causes a decrease in the nDCG@20 for DREQ, from 0.75 to 0.42. Our results underscore the importance of carefully selecting an entity ranking system that is well-suited to the specific requirements of the task. It is particularly important to note that merely using an off-the-shelf entity ranking system is insufficient; it is necessary to train the system specifically to predict entities that are likely to be mentioned in documents relevant to the query.

Alternative Entity Weighing. The DREQ model intrinsically uses entity scores, expressed as probabilities, to weight the entity embeddings within documents. Acknowledging the pivotal role these probabilistic weightings have in shaping the model's performance, we perform a series of experiments to examine various alternative weighting schemes.

First, **we remove entity weightings**, treating all entities with equivalent importance within the document's representation ($w_e = 1$ in Eq. 1). This leads to a significant performance decrease in DREQ, from an nDCG@20 of 0.75 to 0.70. The difficulty test, using the BM25+RM3 candidate ranking, shows that DREQ using uniform weights helps 187 queries, whereas the original model helps

210. This underscores the pivotal role of entities in DREQ. Their embeddings, when appropriately weighted by relevance probabilities, significantly amplify the semantic alignment between queries and documents, attesting to the strategic decision to utilize them. Their importance becomes evident when the individual significance is disregarded, leading to a diluted representation that hampers the model's performance. Furthermore, it's clear that the ranking system's ability to discern and prioritize relevant entities is instrumental to DREQ's success.

Additionally, we explore the **reciprocal rank (RR) of entities** as an alternative weighting mechanism for entity embeddings ($w_e = \frac{1}{rank(e)}$ in Eq. 1). The results are telling: nDCG@20 plummets from 0.75 to 0.54. The RR method assumes a precipitous decrease in the importance of entities based on rank. However, this isn't always fitting. Take, for instance, the query "Black Bear Attacks". We find that the entity *George Parks Highway*, with a score of 0.99, is ranked 71st, while *Cantwell, Alaska*, with a nearly identical score of 0.98, stands at 106th rank. Such negligible probability differences juxtaposed against substantial rank disparities highlight the potential limitations of the RR system. It underscores the risk of undervaluing entities that, by probability, are nearly as pertinent as top-tier entities.

Take Away. To answer **RQ2**, the significant role of entities in the performance of DREQ is undeniable, as evident from the ablation study and alternative entity ranking experiments. The experiments indicate that entities are integral to the model's effectiveness, and that the entity ranking system's ability to discern and prioritize query-relevant entities is instrumental to the success of DREQ. However, it is essential to carefully select and train the entity ranking system to predict entities likely to be mentioned in documents relevant to the query.

In addressing **RQ3**, our experiments unveil crucial insights. The inherent design of DREQ, which uses entity scores as probabilities to weight the entity embeddings, stands out as a potent approach, demonstrated by superior performance metrics. When entities are uniformly weighted, stripping them of their probabilistic distinctions, the model's performance takes a tangible hit, underscoring the model's dependence on the nuanced gradation of entity relevance. The experiment with RR as weights further cements this observation. While RR attempts to rank entities based on their ordinal positions, it fails to adequately capture the subtleties in actual relevance, particularly when entities with near-identical probabilities have significantly differing ranks. The experiments not only re-emphasize the instrumental role of entities in DREQ (**RQ2**), but also the imperativeness of a precise entity ranking and weighting mechanism.

6 Conclusion

We introduce DREQ, an entity-oriented document re-ranking model that taps into the semantic richness of entities in candidate documents. While many neural IR models overlook the differential importance of entities within documents, DREQ prioritizes query-relevant entities, crafting a nuanced, query-specific entity-centric representation. This is then melded with the document's text-centric representation to form a *hybrid* embedding, capturing *query-specific insights*. We provide compelling evidence of the effectiveness of `DREQ` in the document re-ranking task across multiple benchmarks and diverse query sets. `DREQ` achieves new state-of-the-art results by enhancing precision at the top ranks of the candidate set. Its proficiency in handling challenging queries demonstrates its robustness and underscores its potential for real-world applications where queries may vary in difficulty. We also demonstrate the pivotal role of entities, and the importance of meticulously selecting and weighing entities in `DREQ`'s performance.

Overall, our work contributes significantly to the ongoing efforts to enhance the effectiveness of information retrieval systems. The `DREQ` model, with its innovative approach to incorporating entities and its proven ability to outperform existing methods, presents a promising avenue for further exploration and development in the field of information retrieval.

References

1. Akkalyoncu Yilmaz, Z., Yang, W., Zhang, H., Lin, J.: Cross-domain modeling of sentence-level evidence for document retrieval. In: Proceedings of the 2019 Conference on Empirical Methods in Natural Language Processing and the 9th International Joint Conference on Natural Language Processing (EMNLP-IJCNLP), Hong Kong, China, pp. 3490–3496. Association for Computational Linguistics, November 2019. https://doi.org/10.18653/v1/D19-1352. https://aclanthology.org/D19-1352
2. Balog, K., Bron, M., De Rijke, M.: Query modeling for entity search based on terms, categories, and examples. ACM Trans. Inf. Syst. **29**(4) (2011). https://doi.org/10.1145/2037661.2037667
3. Cao, N.D., Izacard, G., Riedel, S., Petroni, F.: Autoregressive entity retrieval. In: International Conference on Learning Representations (2021). https://openreview.net/pdf?id=5k8F6UU39V
4. Chatterjee, S., Dietz, L.: Entity retrieval using fine-grained entity aspects. In: Proceedings of the 44th International ACM SIGIR Conference on Research and Development in Information Retrieval, SIGIR 20021, pp. 1662–1666. Association for Computing Machinery, New York (2021). https://doi.org/10.1145/3404835.3463035
5. Chatterjee, S., Dietz, L.: BERT-ER: query-specific BERT entity representations for entity ranking. In: Proceedings of the 45th International ACM SIGIR Conference on Research and Development in Information Retrieval, SIGIR 2022, pp. 1466–1477. Association for Computing Machinery, New York (2022). https://doi.org/10.1145/3477495.3531944
6. Ciglan, M., Nørvåg, K., Hluchý, L.: The SemSets model for ad-hoc semantic list search. In: Proceedings of the 21st International Conference on World Wide Web, WWW 2012, pp. 131–140. Association for Computing Machinery, New York (2012). https://doi.org/10.1145/2187836.2187855
7. Clark, K., Luong, M., Le, Q.V., Manning, C.D.: ELECTRA: pre-training text encoders as discriminators rather than generators. CoRR abs/2003.10555 (2020). https://arxiv.org/abs/2003.10555
8. Dai, Z., Callan, J.: Deeper text understanding for IR with contextual neural language modeling. CoRR abs/1905.09217 (2019). https://arxiv.org/abs/1905.09217

9. Dai, Z., Callan, J.: Context-aware term weighting for first stage passage retrieval. In: Proceedings of the 43rd International ACM SIGIR Conference on Research and Development in Information Retrieval, SIGIR 20020, pp. 1533–1536. Association for Computing Machinery, New York (2020). https://doi.org/10.1145/3397271.3401204
10. Dai, Z., Xiong, C., Callan, J., Liu, Z.: Convolutional neural networks for soft-matching n-grams in ad-hoc search. In: Proceedings of the Eleventh ACM International Conference on Web Search and Data Mining, WSDM 2018, pp. 126–134. Association for Computing Machinery, New York (2018). https://doi.org/10.1145/3159652.3159659
11. Dalton, J., Dietz, L., Allan, J.: Entity query feature expansion using knowledge base links. In: Proceedings of the 37th International ACM SIGIR Conference on Research & Development in Information Retrieval, pp. 365–374. ACM (2014)
12. Devlin, J., Chang, M., Lee, K., Toutanova, K.: BERT: pre-training of deep bidirectional transformers for language understanding. CoRR abs/1810.04805 (2018). https://arxiv.org/abs/1810.04805
13. Dietz, L.: ENT rank: retrieving entities for topical information needs through entity-neighbor-text relations. In: Proceedings of the 42nd International ACM SIGIR Conference on Research and Development in Information Retrieval, SIGIR 2019, pp. 215–224. Association for Computing Machinery, New York (2019). https://doi.org/10.1145/3331184.3331257
14. Dietz, L., Gamari, B., Dalton, J., Craswell, N.: TREC complex answer retrieval overview. In: TREC (2018)
15. Dietz, L., Verma, M., Radlinski, F., Craswell, N.: TREC complex answer retrieval overview. In: Proceedings of Text REtrieval Conference (TREC) (2017)
16. Ensan, F., Bagheri, E.: Document retrieval model through semantic linking. In: Proceedings of the 10th ACM International Conference on Web Search and Data Mining, WSDM 2017, pp. 181–190. Association for Computing Machinery, New York (2017). https://doi.org/10.1145/3018661.3018692
17. Gabrilovich, E., Markovitch, S.: Wikipedia-based semantic interpretation for natural language processing. J. Artif. Intell. Res. **34**, 443–498 (2009)
18. Garigliotti, D., Balog, K.: On type-aware entity retrieval. In: Proceedings of the ACM SIGIR International Conference on Theory of Information Retrieval, ICTIR 2017, pp. 27–34. Association for Computing Machinery, New York (2017). https://doi.org/10.1145/3121050.3121054
19. Gerritse, E.J., Hasibi, F., de Vries, A.P.: Graph-embedding empowered entity retrieval. In: Jose, J.M., et al. (eds.) ECIR 2020. LNCS, vol. 12035, pp. 97–110. Springer, Cham (2020). https://doi.org/10.1007/978-3-030-45439-5_7
20. Gerritse, E.J., Hasibi, F., de Vries, A.P.: Entity-aware transformers for entity search. In: Proceedings of the 45th International ACM SIGIR Conference on Research and Development in Information Retrieval, SIGIR 2022, pp. 1455–1465. Association for Computing Machinery, New York (2022). https://doi.org/10.1145/3477495.3531971
21. Graus, D., Tsagkias, M., Weerkamp, W., Meij, E., de Rijke, M.: Dynamic collective entity representations for entity ranking. In: Proceedings of the Ninth ACM International Conference on Web Search and Data Mining, WSDM 2016, pp. 595–604. Association for Computing Machinery, New York (2016). https://doi.org/10.1145/2835776.2835819

22. Guo, J., Fan, Y., Ai, Q., Croft, W.B.: A deep relevance matching model for ad-hoc retrieval. In: Proceedings of the 25th ACM International on Conference on Information and Knowledge Management, CIKM 2016, pp. 55–64. Association for Computing Machinery, New York (2016). https://doi.org/10.1145/2983323.2983769
23. Hasibi, F., Balog, K., Bratsberg, S.E.: Exploiting entity linking in queries for entity retrieval. In: Proceedings of the 2016 ACM International Conference on the Theory of Information Retrieval, ICTIR 2016, pp. 209–218. Association for Computing Machinery, New York (2016). https://doi.org/10.1145/2970398.2970406
24. He, P., Liu, X., Gao, J., Chen, W.: DeBERta: decoding-enhanced BERT with disentangled attention. CoRR abs/2006.03654 (2020). https://arxiv.org/abs/2006.03654
25. Huang, P.S., He, X., Gao, J., Deng, L., Acero, A., Heck, L.: Learning deep structured semantic models for web search using clickthrough data. In: Proceedings of the 22nd ACM International Conference on Information & Knowledge Management, CIKM 2013, pp. 2333–2338. Association for Computing Machinery, New York (2013). https://doi.org/10.1145/2505515.2505665
26. Hui, K., Yates, A., Berberich, K., de Melo, G.: PACRR: a position-aware neural IR model for relevance matching. In: Proceedings of the 2017 Conference on Empirical Methods in Natural Language Processing, pp. 1049–1058. Association for Computational Linguistics, Copenhagen, September 2017. https://doi.org/10.18653/v1/D17-1110. https://aclanthology.org/D17-1110
27. Hui, K., Yates, A., Berberich, K., de Melo, G.: Co-PACRR: a context-aware neural IR model for ad-hoc retrieval. In: Proceedings of the Eleventh ACM International Conference on Web Search and Data Mining, WSDM 2018, pp. 279–287. Association for Computing Machinery, New York (2018). https://doi.org/10.1145/3159652.3159689
28. Jiang, Z., Yu, W., Zhou, D., Chen, Y., Feng, J., Yan, S.: ConvBERT: improving BERT with span-based dynamic convolution. CoRR abs/2008.02496 (2020). https://arxiv.org/abs/2008.02496
29. Kaptein, R., Serdyukov, P., De Vries, A., Kamps, J.: Entity ranking using wikipedia as a pivot. In: Proceedings of the 19th ACM International Conference on Information and Knowledge Management, CIKM 2010, pp. 69–78. Association for Computing Machinery, New York (2010). https://doi.org/10.1145/1871437.1871451
30. Karpukhin, V., et al.: Dense passage retrieval for open-domain question answering. In: Proceedings of the 2020 Conference on Empirical Methods in Natural Language Processing (EMNLP), pp. 6769–6781. Association for Computational Linguistics, Online, November 2020. https://doi.org/10.18653/v1/2020.emnlp-main.550. https://aclanthology.org/2020.emnlp-main.550
31. Khattab, O., Zaharia, M.: ColBERT: efficient and effective passage search via contextualized late interaction over BERT. In: Proceedings of the 43rd International ACM SIGIR Conference on Research and Development in Information Retrieval, SIGIR 2020, pp. 39–48. Association for Computing Machinery, New York (2020). https://doi.org/10.1145/3397271.3401075
32. Kingma, D.P., Ba, J.: Adam: a method for stochastic optimization. arXiv preprint arXiv:1412.6980 (2014)
33. Lehmann, J., et al.: DBpedia-a large-scale, multilingual knowledge base extracted from Wikipedia. Semant. Web **6**, 167–195 (2015). https://doi.org/10.3233/SW-140134

34. Li, C., et al.: NPRF: a neural pseudo relevance feedback framework for ad-hoc information retrieval. In: Proceedings of the 2018 Conference on Empirical Methods in Natural Language Processing, pp. 4482–4491. Association for Computational Linguistics, Brussels, October–November 2018. https://doi.org/10.18653/v1/D18-1478. https://aclanthology.org/D18-1478
35. Li, C., Yates, A., MacAvaney, S., He, B., Sun, Y.: PARADE: passage representation aggregation for document reranking. CoRR abs/2008.09093 (2020). https://arxiv.org/abs/2008.09093
36. Liu, X., Fang, H.: Latent entity space: a novel retrieval approach for entity-bearing queries. Inf. Retr. J. **18**(6), 473–503 (2015)
37. Liu, Y., et al.: RoBERta: a robustly optimized BERT pretraining approach. CoRR abs/1907.11692 (2019). https://arxiv.org/abs/1907.11692
38. Liu, Z., Xiong, C., Sun, M., Liu, Z.: Entity-duet neural ranking: understanding the role of knowledge graph semantics in neural information retrieval. In: Proceedings of the 56th Annual Meeting of the Association for Computational Linguistics (Volume 1: Long Papers), Melbourne, Australia, pp. 2395–2405. Association for Computational Linguistics, July 2018. https://doi.org/10.18653/v1/P18-1223. https://aclanthology.org/P18-1223
39. MacAvaney, S., Yates, A., Cohan, A., Goharian, N.: CEDR: contextualized embeddings for document ranking. CoRR abs/1904.07094 (2019). https://arxiv.org/abs/1904.07094
40. Mackie, I., Owoicho, P., Gemmell, C., Fischer, S., MacAvaney, S., Dalton, J.: Codec: complex document and entity collection. In: Proceedings of the 45th International ACM SIGIR Conference on Research and Development in Information Retrieval, SIGIR 2022, pp. 3067–3077. Association for Computing Machinery, New York (2022). https://doi.org/10.1145/3477495.3531712
41. Manotumruksa, J., Dalton, J., Meij, E., Yilmaz, E.: CrossBERT: a triplet neural architecture for ranking entity properties. In: Proceedings of the 43rd International ACM SIGIR Conference on Research and Development in Information Retrieval, SIGIR 2020, pp. 2049–2052. Association for Computing Machinery, New York (2020). https://doi.org/10.1145/3397271.3401265
42. Metzler, D., Croft, W.B.: A Markov random field model for term dependencies. In: Proceedings of the 28th Annual International ACM SIGIR Conference on Research and Development in Information Retrieval, SIGIR 2005, pp. 472–479. Association for Computing Machinery (2005). https://doi.org/10.1145/1076034.1076115
43. Mitra, B., Craswell, N.: An updated duet model for passage re-ranking. CoRR abs/1903.07666 (2019). https://arxiv.org/abs/1903.07666
44. Nalisnick, E., Mitra, B., Craswell, N., Caruana, R.: Improving document ranking with dual word embeddings. In: Proceedings of the 25th International Conference Companion on World Wide Web, WWW 2016 Companion, International World Wide Web Conferences Steering Committee, Republic and Canton of Geneva, CHE, pp. 83–84 (2016). https://doi.org/10.1145/2872518.2889361
45. Naseri, S., Dalton, J., Yates, A., Allan, J.: CEQE: contextualized embeddings for query expansion. In: Hiemstra, D., Moens, M.-F., Mothe, J., Perego, R., Potthast, M., Sebastiani, F. (eds.) ECIR 2021. LNCS, vol. 12656, pp. 467–482. Springer, Cham (2021). https://doi.org/10.1007/978-3-030-72113-8_31
46. Nguyen, T., et al.: MS MARCO: a human generated machine reading comprehension dataset. CoRR abs/1611.09268 (2016). https://arxiv.org/abs/1611.09268

47. Nikolaev, F., Kotov, A.: Joint word and entity embeddings for entity retrieval from a knowledge graph. In: Jose, J.M., et al. (eds.) ECIR 2020, Part I. LNCS, vol. 12035, pp. 141–155. Springer, Cham (2020). https://doi.org/10.1007/978-3-030-45439-5_10
48. Nikolaev, F., Kotov, A., Zhiltsov, N.: Parameterized fielded term dependence models for ad-hoc entity retrieval from knowledge graph. In: Proceedings of the 39th International ACM SIGIR Conference on Research and Development in Information Retrieval, SIGIR 2016, pp. 435–444. Association for Computing Machinery, New York (2016). https://doi.org/10.1145/2911451.2911545
49. Nogueira, R.F., Cho, K.: Passage re-ranking with BERT. CoRR abs/1901.04085 (2019). https://arxiv.org/abs/1901.04085
50. Nogueira, R.F., Yang, W., Cho, K., Lin, J.: Multi-stage document ranking with BERT. CoRR abs/1910.14424 (2019). https://arxiv.org/abs/1910.14424
51. Nogueira, R.F., Yang, W., Lin, J., Cho, K.: Document expansion by query prediction. CoRR abs/1904.08375 (2019). https://arxiv.org/abs/1904.08375
52. Piccinno, F., Ferragina, P.: From TagME to wat: a new entity annotator. In: Proceedings of the First International Workshop on Entity Recognition & Disambiguation, ERD 2014, pp. 55–62. Association for Computing Machinery, New York (2014). https://doi.org/10.1145/2633211.2634350
53. Raffel, C., et al.: Exploring the limits of transfer learning with a unified text-to-text transformer. CoRR abs/1910.10683 (2019). https://arxiv.org/abs/1910.10683
54. Raviv, H., Carmel, D., Kurland, O.: A ranking framework for entity oriented search using Markov random fields. In: Proceedings of the 1st Joint International Workshop on Entity-Oriented and Semantic Search, JIWES 2012. Association for Computing Machinery, New York (2012). https://doi.org/10.1145/2379307.2379308
55. Raviv, H., Kurland, O., Carmel, D.: Document retrieval using entity-based language models. In: Proceedings of the 39th International ACM SIGIR Conference on Research and Development in Information Retrieval, SIGIR 2016, pp. 65–74. Association for Computing Machinery, New York (2016). https://doi.org/10.1145/2911451.2911508
56. Schuhmacher, M., Dietz, L., Paolo Ponzetto, S.: Ranking entities for web queries through text and knowledge. In: Proceedings of the 24th ACM International on Conference on Information and Knowledge Management, CIKM 2015, pp. 1461–1470. Association for Computing Machinery, New York (2015). https://doi.org/10.1145/2806416.2806480
57. Shen, W., Wang, J., Han, J.: Entity linking with a knowledge base: issues, techniques, and solutions. IEEE Trans. Knowl. Data Eng. **27**(2), 443–460 (2015). https://doi.org/10.1109/TKDE.2014.2327028
58. Tonon, A., Demartini, G., Cudré-Mauroux, P.: Combining inverted indices and structured search for ad-hoc object retrieval. In: Proceedings of the 35th International ACM SIGIR Conference on Research and Development in Information Retrieval, SIGIR 2012, pp. 125–134. Association for Computing Machinery, New York (2012). https://doi.org/10.1145/2348283.2348304
59. Tran, H.D., Yates, A.: Dense retrieval with entity views. In: Proceedings of the 31st ACM International Conference on Information & Knowledge Management, CIKM 2022, pp. 1955–1964. Association for Computing Machinery, New York (2022). https://doi.org/10.1145/3511808.3557285
60. Wang, X., MacDonald, C., Tonellotto, N., Ounis, I.: ColBERT-PRF: semantic pseudo-relevance feedback for dense passage and document retrieval. ACM Trans. Web **17**(1) (2023). https://doi.org/10.1145/3572405

61. Xiong, C., Callan, J.: EsdRank: connecting query and documents through external semi-structured data. In: Proceedings of the 24th ACM International Conference on Information and Knowledge Management, CIKM 2015, pp. 951–960. ACM, New York (2015). https://doi.org/10.1145/2806416.2806456. https://doi.acm.org/10.1145/2806416.2806456
62. Xiong, C., Callan, J., Liu, T.Y.: Word-entity duet representations for document ranking. In: Proceedings of the 40th International ACM SIGIR Conference on Research and Development in Information Retrieval, SIGIR 2017, pp. 763–772. Association for Computing Machinery, New York (2017). https://doi.org/10.1145/3077136.3080768
63. Xiong, C., Dai, Z., Callan, J., Liu, Z., Power, R.: End-to-end neural ad-hoc ranking with kernel pooling. In: Proceedings of the 40th International ACM SIGIR Conference on Research and Development in Information Retrieval, SIGIR 2017, pp. 55–64. Association for Computing Machinery, New York (2017). https://doi.org/10.1145/3077136.3080809
64. Xiong, C., Liu, Z., Callan, J., Hovy, E.: JointSem: combining query entity linking and entity based document ranking. In: Proceedings of the 2017 ACM SIGIR Conference on Information and Knowledge Management, CIKM 2017, pp. 2391–2394. Association for Computing Machinery, New York (2017). https://doi.org/10.1145/3132847.3133048
65. Xiong, C., Liu, Z., Callan, J., Liu, T.Y.: Towards better text understanding and retrieval through kernel entity salience modeling. In: The 41st International ACM SIGIR Conference on Research and Development in Information Retrieval, SIGIR 2018, pp. 575–584. Association for Computing Machinery, New York (2018). https://doi.org/10.1145/3209978.3209982
66. Xiong, C., Power, R., Callan, J.: Explicit semantic ranking for academic search via knowledge graph embedding. In: Proceedings of the 26th International Conference on World Wide Web, WWW 2017, International World Wide Web Conferences Steering Committee, Republic and Canton of Geneva, CHE, pp. 1271–1279 (2017). https://doi.org/10.1145/3038912.3052558
67. Xiong, L., et al.: Approximate nearest neighbor negative contrastive learning for dense text retrieval. CoRR abs/2007.00808 (2020). https://arxiv.org/abs/2007.00808
68. Yamada, I., et al.: Wikipedia2Vec: an efficient toolkit for learning and visualizing the embeddings of words and entities from Wikipedia. In: Proceedings of the 2020 Conference on Empirical Methods in Natural Language Processing: System Demonstrations, pp. 23–30. Association for Computational Linguistics, Online, October 2020. https://doi.org/10.18653/v1/2020.emnlp-demos.4. https://aclanthology.org/2020.emnlp-demos.4
69. Yamada, I., Shindo, H., Takefuji, Y.: Representation learning of entities and documents from knowledge base descriptions. In: Proceedings of the 27th International Conference on Computational Linguistics, pp. 190–201. Association for Computational Linguistics, Santa Fe, August 2018. https://aclanthology.org/C18-1016
70. Yu, H., Xiong, C., Callan, J.: Improving query representations for dense retrieval with pseudo relevance feedback. In: Proceedings of the 30th ACM International Conference on Information & Knowledge Management, CIKM 2021, pp. 3592–3596. Association for Computing Machinery, New York (2021). https://doi.org/10.1145/3459637.3482124

71. Zhang, Z., Han, X., Liu, Z., Jiang, X., Sun, M., Liu, Q.: ERNIE: enhanced language representation with informative entities. In: Proceedings of the 57th Annual Meeting of the Association for Computational Linguistics, pp. 1441–1451. Association for Computational Linguistics, Florence, July 2019. https://doi.org/10.18653/v1/P19-1139. https://aclanthology.org/P19-1139
72. Zheng, Z., Hui, K., He, B., Han, X., Sun, L., Yates, A.: BERT-QE: contextualized query expansion for document re-ranking. In: Findings of the Association for Computational Linguistics: EMNLP 2020, pp. 4718–4728. Association for Computational Linguistics, Online, November 2020. https://doi.org/10.18653/v1/2020.findings-emnlp.424. https://aclanthology.org/2020.findings-emnlp.424
73. Zhiltsov, N., Kotov, A., Nikolaev, F.: Fielded sequential dependence model for ad-hoc entity retrieval in the web of data. In: Proceedings of the 38th International ACM SIGIR Conference on Research and Development in Information Retrieval, SIGIR 2015, pp. 253–262. Association for Computing Machinery, New York (2015). https://doi.org/10.1145/2766462.2767756
74. Zhou, Y., Croft, W.B.: Query performance prediction in web search environments. In: Proceedings of the 30th Annual International ACM SIGIR Conference on Research and Development in Information Retrieval, SIGIR 2007, pp. 543–550. Association for Computing Machinery, New York (2007). https://doi.org/10.1145/1277741.1277835
75. Zhuang, H., et al.: Rankt5: fine-tuning t5 for text ranking with ranking losses. In: Proceedings of the 46th International ACM SIGIR Conference on Research and Development in Information Retrieval, SIGIR 2023, pp. 2308–2313. Association for Computing Machinery, New York (2023). https://doi.org/10.1145/3539618.3592047

Hypergraphs with Attention on Reviews for Explainable Recommendation

Theis E. Jendal[1(✉)], Trung-Hoang Le[2], Hady W. Lauw[2], Matteo Lissandrini[1], Peter Dolog[1], and Katja Hose[1,3]

[1] Aalborg University, Aalborg, Denmark
{tjendal,matteo,dolog,khose}@cs.aau.dk
[2] Singapore Management University, Singapore, Singapore
{thle.2017,hadywlauw}@smu.edu.sg
[3] Technische Universität Wien,Vienna, Austria
katja.hose@tuwien.ac.at

Abstract. Given a recommender system based on reviews, the challenges are how to effectively represent the review data and how to explain the produced recommendations. We propose a novel review-specific Hypergraph (HG) model, and further introduce a model-agnostic explainability module. The HG model captures high-order connections between users, items, aspects, and opinions while maintaining information about the review. The explainability module can use the HG model to explain a prediction generated by any model. We propose a path-restricted review-selection method biased by the user preference for item reviews and propose a novel explanation method based on a review graph. Experiments on real-world datasets confirm the ability of the HG model to capture appropriate explanations.

1 Introduction

Recommender Systems (RSs) utilize information about users' past interactions for recommendations, commonly referred to as Collaborative Filtering (CF) [9,18,19,22,49,50]. Often, these methods lack scrutability and users may not comprehend the reasons behind the recommendations [23]. Recent methods have exploited reviews for recommendation and explainability [9,27,52], typically by presenting the user with a given recommendation and an accompanying review that indicates what another user said about the item (see Fig. 1). Yet, many review-level recommenders select reviews based solely on the item attention [9,27,35,52], disregarding the target user's preferences in selecting the explanation. Furthermore, some select the complete text, even when a review is verbose and filled with irrelevant information.

In this work, we extract aspects and opinions from reviews and represent them in a Hypergraph (HG) structure, where hyperedges connect sets of nodes representing aspects and opinions occurring together. A review-based HG captures the interdependencies of reviews by different users and items (e.g., *great*

N. Goharian et al. (Eds.): ECIR 2024, LNCS 14608, pp. 230–246, 2024.
https://doi.org/10.1007/978-3-031-56027-9_14

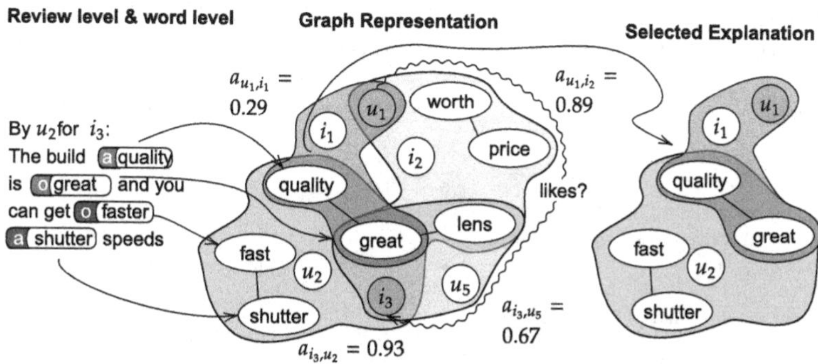

Fig. 1. Review-level information connected to the item i_3 recommended to the user u_1 modelled as a hypergraph. Hyperedges are represented by the colored areas.

quality in Fig. 1), while capturing the intradependencies of aspects and opinions in a review (e.g., *fast shutter* and *great quality*). It also captures n-order correlations providing higher expressivity than binary relations in normal graphs.

Prior work focused on edge-labeled multigraph, which cannot easily model the high-order interdependencies between reviews mentioning different aspects and opinions [6,45,49]. They apply high-order graph convolutions, which can capture interdependencies across, but not within a review. Furthermore, the attention mechanism is often node-specific, as it does not take the user preferences into account when computing the attention of neighboring nodes [44,49,59].

Using a HG to represent reviews, we get a one-to-one mapping between edges and reviews while still capturing high-order interdependencies. Specifically, a review $r_{u,i} \in \mathcal{R}$ is represented as a set of triples consisting of an aspect, an opinion, and a sentiment, (a, o, s), where $a \in \mathcal{A}$, $o \in \mathcal{O}$, and $s \in \{-1, 1\} = \mathcal{S}$, often referred to as phrase-level opinions. The set of aspects $\mathcal{A}$ and opinions $\mathcal{O}$ are extracted from reviews, with the sentiment describing the polarity (e.g. *not* great). We opt for modeling reviews as hyperedges, being sets of nodes, in a graph consisting of users, items, aspects, and opinions. In Fig. 1, we show an example of a review representation where a single edge connects the item to the user and related review's phrase-level opinions; thus, enabling us to capture ternary or higher intradependencies within the reviews. A HG consists of nodes $\mathcal{V}$ and hyperedges $\mathcal{E} \subseteq \mathcal{P}(\mathcal{V}) \setminus \{\emptyset\}$; formally defined as $g = \langle \mathcal{V}, \mathcal{E} \rangle$. We define our HG, containing interactions, aspects, and opinions, as g, creating one edge for each review. Thus, we capture the global connections, i.e., interdependencies between different nodes and reviews, e.g., we can find that the aspect *quality* is important for the collaborative signal, and capture intradependencies within the individual reviews.

Our method, Hypergraph with Attention on Reviews (HypAR), takes into account both the learned attention about the user's reviews as well as the user's historical opinions about aspects when selecting a review as the explanation of a recommendation. Hence, given the predicted user preference for an item, we

Table 1. Overview of methods. Review, Path, Graph Ex., and Word are the levels of the explanations. (*) Could not reproduce methods based on given information.

Methods	Graph	Review	Path	Graph Ex.	Word	Source code
R3* [35]	✗	✓	✗	✗	(✓)	✗
AENAR [61]	✗	✗	✗	✗	(✓)	✗
SGMC [8]	✗	✗	✗	✗	✗	✗
HAGERec* [59], EIUM [20], KPRN [51]	✓	✗	✓	✗	✗	✗
HRDR [27], AHN [12], NARRE [9]	✗	✓	✗	✗	✗	✗
KGCN(-LS) [46, 45]	✓	✗	(✓)	✗	✗	✓
KTUP [6]	✓	✗	✓	(✓)	✗	✓
KGAT [49], RuleRec [31], PGPR [58], RippleNet [44]	✓	✗	✓	✗	✗	✓
RMG [53]	(✓)	✓	✗	✗	✓	✓
HUITA [52], HANN [11]	✗	✓	✗	✗	✓	✓
MTER [48]	✗	✗	✗	✗	✓	✓
TransNets [7]	✗	✓	✗	✗	✗	✓
TriRank [17]	✓	✗	✗	✗	(✓)	✗
HypAR	✓	✓	✓	✓	✓	✓

enable a better-informed attention mechanism that exploits connectivity between users, items, aspects, and opinions to provide an explanation. Furthermore, modeling the reviews in a graph allows us to generate graph-based explanations by selecting relevant aspects and opinions; thus, we will present directly to the user the salient points from the review. For example, based on our review attention assigned to each review in Fig. 1, we can select the most important reviews connecting user u_1 and item i_3 and to extrapolate which parts of the review text are most important, here the great quality of the lens. As such, the graph view captures directly both the path-based reasoning and a succinct and structured review representation as illustrated by the hyperedges.

Contributions. We summarize our contribution as follows: (i) We propose a novel review representation using the HG structure and an accompanying architecture that applies graph convolutions to incorporate sentiment polarity and opinions. (ii) We provide a *dual-view* explanation: review-to-graph, which results in a novel graph explanation taking user preferences w.r.t. items, aspects, and opinions into account. (iii) We construct a framework for explainable recommendations that is agnostic to the preference module. (iv) We define simple quantitative evaluation measures for explainable graph recommendation in the problem formulation. Through studies on four real-world datasets in different domains, we show HypAR's improvements upon baselines. Therefore, HypAR is the first model with the ability to make both review-level and graph-based explanations for its recommendations, providing ad-hoc explanations that are integral to the recommendation process instead of weak post-hoc explanations.

2 Related Work

In recent years, Matrix Factorization (MF) methods [44–46,49,59] and graph-based methods have become popular for CF [17,18,50]. Yet, limited work focus on representing reviews as graphs. In the following, we will introduce recent works using graphs and reviews as explanations, as well as related HG architectures.

Explainability. Multiple types of explanations have been proposed, from identifying areas of interest in product images [10] to finding relevant users or items [28]. Previous works has used the attention mechanisms in multiple research areas [2,10,41], particularly some use it to select the most important review as textual explanations [9,27,52]. In Table 1, we show various explainability options, as reviews written by users, paths selected in a graph, showing a complete graph, and showing individual words. *The first column is graph-based recommenders, clearly showing no graph-based recommendation methods provide review explanations*, with the exception of RMG [53]. However, RMG uses the graph structure as the preference module and not for review-representation.

Reviews contain subjective opinions on different aspects and have been used for explanations [7,9,12,27,52,53]. Specifically, some use review-level attention mechanism [9,27,52] for explanations; however, these are unable to capture high-order relationships between users and items. The attention mechanism used is often item-specific, meaning non-personalized explanations, while those that utilize a user's preference, i.e. personalized explanations [7,11,12], are ill-suited for ranking, requiring distinct unique computations for each user-item pair. *Therefore, instead of increasing the ranking complexity, we propose a novel review selection method utilizing the HG structure based on the non-personalized attention mechanisms.* In the experiment section, we show our selection strategy outperforms the non-personalized methodology.

Knowledge Graphs (KGs) [20,45,49,58,59] can supplement MF methods when ratings are sparse. Information is either propagated inwards using Graph Neural Networks (GNNs) [6,49] or outwards using ripples [44] to capture high-order connectivities. Current KG based methods only exploit factual (instead of opinion-based) explanations, e.g., a path would only describe a product, not any user opinions. To provide an explanation that would match the subjective judgment the user may have about a product, methods have explicitly extracted aspects and opinions from reviews to generate opinionated explanations [48] or aspect-level explanations [4,17,56,63], but they cannot capture high-order connections.

Hypergraph. The Hypergraph Neural Network (HGNN) [14] uses a GNN on HGs; however as a HG's edges are sets of nodes, the HGNN first aggregates nodes occurring in the edges and then the aggregates the edges a node occurs in. HGNNs variants have been used for multiple tasks [16,47,54,55], often differing in the HG construction methodology [16,42,60]. However, none of these methods are explainable [24,38], or, if explainably, use a naïve explanation like producing k-most similar items [8]. Yet, such 'explanations' are still opaque, as they not explain why items are similar. *Instead, we set out to select reviews (hyperedges, not nodes) as explanations, which provide relationships between items and aspects.*

3 Methodology

Our method HypAR (Fig. 2) consists of four modules: (i) review representation, which computes the embeddings for each review; (ii) review aggregation, which aggregates the reviews' embedding generating an opinion vector; (iii) preference computation, computing the user and item preference vectors; and finally (iv) combine and predict module, which combines the vectors for ranking. In the following, we define our problem and then expand on each module.

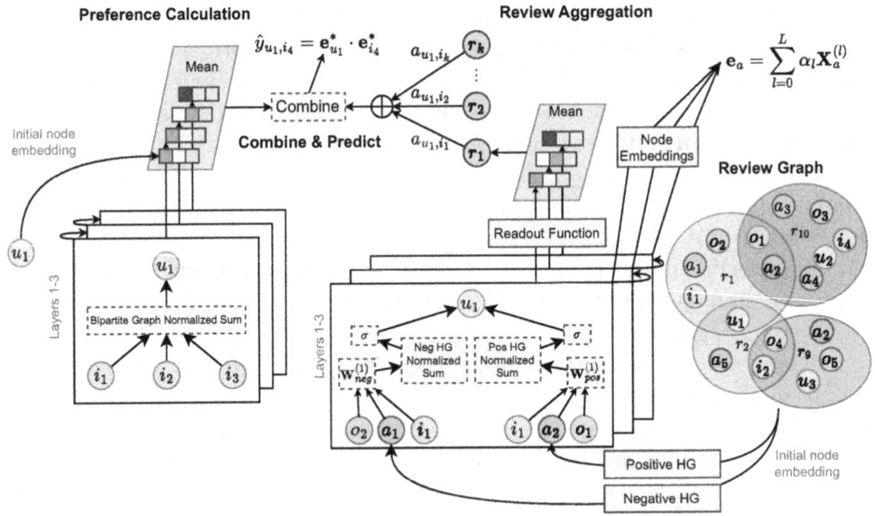

Fig. 2. Illustration of the embedding process of user u_1.

Problem Formulation. We define the interaction matrix, given a set of users $\mathcal{U}$ and items $\mathcal{I}$, as $\mathbf{I} \in \{0,1\}^{|\mathcal{U}| \times |\mathcal{I}|}$, s.t. $\mathbf{I}_{u,i} = 1$ if the user u has interacted with the item i; otherwise 0. Similarly, for all interactions, we have a corresponding review $r_{u,i}$ if $\mathbf{I}_{u,i} = 1$. For each interaction, we extract aspects mentioned in the reviews and the given opinion, such as *fast* or *worth*, along with the sentiment.

The objective of our method is: (i) to rank items according to a user's preference and (ii) to ensure that the ranking of items is explainable. Regarding the recommendation objective, we model our task as a top-k recommendation problem, s.t., given a learned model Θ, we are able to rank the items $\mathcal{I}$ based on their likelihood of being liked by the user. For explainability, we predict both whether the user will like/buy the product as well as the reason behind the choice. In our model, we assume reviews to be concrete manifestations of the reasoning behind the user preference. Thus, here we assume an explanation to be either a given review or to be comparable to a review, i.e., a set $\varepsilon \subset \mathcal{A} \times \mathcal{O} \times \mathcal{S}$ of aspects, opinion, sentiment triples that justify the (predicted) user choice.

Based on this, we can identify *two new forms of explainability metrics* describing a good explanation ε: (i) ε is the set of aspect, opinion, sentiment triples that constitute the actual review the user will write; or (ii) ε can be used to deterministically separate items that will be ranked higher by the user. The first case can intuitively be understood as a strong correlation between a given set of aspects (e.g., *fast shutter*) and if a user would generally prefer items described with those aspects to items selected based on another disjoint set of random aspects (e.g., *long cable*). For the second objective, we are interested in knowing if the generated explanation matches the method's ranking, i.e., if the user prefers the item due to the explanation (*fast shutter*), we assume items matching the explanation would be ranked higher than another random ranking.

Review Representation. Given an initial representation of nodes $\mathbf{X}^0 \in \mathbb{R}^{|\mathcal{V}|\times d}$ and our HG g, we define the incidence matrix $\mathbf{H} \in \{0,1\}^{|\mathcal{V}|\times|\mathcal{E}|}$, indicating whether $v \in \mathcal{V}$ occurs in $e \in \mathcal{E}$, the diagonal matrices of edge degrees $\mathbf{D}_e$ and nodes $\mathbf{D}_v$, where the node degree is defined as $d(v) = \sum_{e\in\mathcal{E}} \mathbf{H}_{v,e}$, and the edge degree as $d(e) = \sum_{v\in\mathcal{V}} \mathbf{H}_{v,e}$ [14]. To capture the semantics of each word we use Word2Vec [33], as in [9,35], to learn the word embeddings and apply 2 layers of Multi-Layer Perceptrons (MLPs) with `tanh` activation to transform the initial aspect and opinion vectors from the word-embedding space into $\mathbf{X}^0$.

We employ sentiment-specific linear transformation matrices and split our HG into two HGs: one containing only positive phrase-level opinions and one containing only negative phrase-level opinions (see Fig. 2), allowing our model to differentiate between positive and negative phrases. As such, we are able to capture both the sentiment and the high-order connections, using:

$$\mathbf{X}^{(l)} = \tfrac{1}{2}\textstyle\sum_{s\in\mathcal{S}} \mathbf{D}_{vs}^{-\frac{1}{2}}\mathbf{H}_s\mathbf{D}_{es}^{-1}\mathbf{H}_s^\top\mathbf{D}_{vs}^{-\frac{1}{2}}\mathbf{X}^{(l-1)}\mathbf{W}_s^{(l)} \quad (1)$$

where H_s, D_{vs}, and D_{es} are the sentiment-specific incidence, node degree, and edge degree matrices, respectively, and $\mathbf{W}_s^l \in \mathbb{R}^{d\times d}$ is the sentiment specific transformation. The HG convolution is very similar to Graph Convolutional Network (GCN) layers with an extra normalization using edge degrees. Yet, we are interested in capturing review-specific occurrences; therefore, we define an edge-wise readout function using the mean aggregator. Thus, the review representation at layer l is $\mathbf{r}_{u,i}^{(l)} = \frac{1}{|r_{u,i}|}\sum_{v\in r_{u,i}} \mathbf{X}_v^{(l)}$. To better capture the connectivities in the HG, we propagate the initial embeddings through the convolutional layer L times and aggregate the layers using the mean, leaving the study of other aggregators as future work, and define the vector as $\mathbf{r}_{u,i} = \sum_{l=0}^{L} \alpha_l\, \mathbf{r}_{u,i}^{(l)}$, where $\alpha_l = \frac{1}{L+1}$.

Attention-Based Review Aggregation. We employ an attention-specific user and item representation vector to capture the quality of the reviews written by the user and about the item [9], defined as:

$$a_{i,u}^* = \mathbf{h}^\top \mathrm{ReLU}(\mathbf{W}_a(\mathbf{r}_{u,i}\|\mathbf{q}_u) + \mathbf{b}_1) + b_2 \quad (2)$$

where $\mathbf{W}_a \in \mathbb{R}^{d'\times 2d}$ transforms the review representation into the attention space taking the quality of the user into account, $\mathbf{h} \in \mathbb{R}^{d'}$ is the attention vector, $\mathbf{b}_1$ and b_2 are learned biases, ReLU [34] is a non-linear activation function, and $\|$ is the concatenation operation. We can calculate the user-specific review attention $a_{u,i}$ by substituting $\mathbf{q}_u$ with $\mathbf{q}_i$. The aggregation for a user is illustrated in Fig. 2 under review aggregation. The current attention score is unbounded; we normalize the attention weight w.r.t. all other reviews of the item using softmax $a_{i,u} = \frac{\exp(a^*_{i,u})}{\sum_{(u',i)\in R}\exp(a^*_{i,u'})}$. Given the attention mechanism, we calculate the weighted mean of the item-specific review embeddings. Our method selects the more "important" reviews for the item as $\mathbf{r}_i = \sum_{(u,i)\in R} a_{i,u}\mathbf{r}_{u,i}$. As a single attention kernel may not be sufficient to capture complex explanations, we learn multiple kernels, taking the average embedding over all kernels' final output.

Preference Computation. We note that our review aggregation module is agnostic to the preference computation module (see Fig. 2), i.e., our architecture that learns to select explanations does not impose any restriction on the module that predicts which item to recommend to the user. Therefore, our architecture learns in parallel both to provide recommendations as well as to explain them. Aggregating information across high-order connectivity has been found to greatly increase performance for top-n recommendation in the preference computation module [18,45,46,49]. To show the effectiveness of representing the extracted aspects and opinions in a HG, we adopt two preference computation modules: (i) using MF, as in NARRE [9] and HRDR [27], allowing for direct comparison of performance, simply representing users and items in a latent space $X_p^{(0)} \in \mathbb{R}^{|\mathcal{V}|\times d}$, s.t. a user u and item i each have a unique row in $X_p^{(0)}$, defined as $\mathbf{e}_u$ and $\mathbf{e}_i$. (ii) using the well-performing LightGCN [18], where users and items are represented as the average embedding of neighbors, through multiple layers of GNN convolutions, as: $\mathbf{e}_u^{(l)} = \sum_{i\in\mathcal{N}_u}\frac{1}{\sqrt{|\mathcal{N}_u||\mathcal{N}_i|}}\mathbf{e}_i^{(l-1)}$, $\mathbf{e}_i^{(l)} = \sum_{u\in\mathcal{N}_i}\frac{1}{\sqrt{|\mathcal{N}_u||\mathcal{N}_i|}}\mathbf{e}_u^{(l-1)}$, where $e_i^{(0)}$ is the item i's row in the learned preference matrix $\mathbf{X}_p^{(0)}$.

Combine and Predict. There are multiple ways of combining the review and aggregation modules, such as adding them [9,27]. Yet, simply combining the embeddings using addition may lead to subpar performance. We, therefore, investigate three different combination methods: (i) **addition** combines the preference and review embeddings using element-wise addition as $\mathcal{C}_{add}(u) = \mathbf{e}_u + \mathbf{r}_u$; (ii) **multiplication** uses element-wise multiplication, ($\odot$), thereby capturing the affinity between the two module representations as $\mathcal{C}_{mul}(u) = \mathbf{e}_u \odot \mathbf{r}_u$; and (iii) **concatenation** creates a vector of new dimension as it concatenates the two embeddings as $\mathcal{C}_{cat}(u) = \mathbf{e}_u\|\mathbf{r}_u$. In this case, if the output embeddings are of unequal size, then the larger embedding would have a higher weight in the final prediction step. Yet, this method is also the only one proposed here that handles the case where preference and review modules produce output embeddings of unequal size. As such, using any of the combination functions $\mathcal{C}$, we can compute the embeddings of the users and items as: $\mathbf{e}^*_u = \mathcal{C}(u), \quad \mathbf{e}^*_i = \mathcal{C}(i)$.

Prediction. Our framework is agnostic to the prediction method. We here describe to possibilities. The learned similarity, $f_{NARRE}(u,i) = \mathbf{W}_p(\mathbf{e}^*_u \odot \mathbf{e}^*_i) + b_u + b_i + \mu$ [9], is a linear transformation of the affinity between the users and items representation, where $\mathbf{W}_p \in \mathbb{R}^{1\times d'}$ and b_u, b_i, and μ denotes user, item and global biases, respectively. Here, the weight matrix $\mathbf{W}_p$ is able to select which features are most important for ranking user and item affinities. The other proposed prediction method is the inner product $(\cdot)$ between a user and an item, which in certain settings outperform the learned [37], as $f_{dot}(u,i) = \mathbf{e}_u^* \cdot \mathbf{e}_i^*$.

Optimization. We use Bayesian Personalized Ranking (BPR) as the collaborative loss function [36] as $L_{CF} = \sum_{(u,i,j)\in\{(u,i,j)|\mathbf{I}_{u,i}=1,\mathbf{I}_{u,j}=0\}} -\ln\sigma\left(\hat{y}_{u,i} - \hat{y}_{u,j}\right)$, item i is preferred over item j; σ is the sigmoid function; and $\hat{y}_{u,i}$ is the output of either f_{NARRE} or f_{dot}. To accommodate the quantitative explainability reasoning, we develop an explainability-specific loss function. Intuitively, if a user mentions some aspects and opinions about an item, we assume them to be important. We, therefore, propose a TransR [26]-like similarity function augmented for AOS triple as $f_{TR}(u,i,a,o,s) = \left(\mathbf{W}_s^1(\mathbf{e}_u^* \| \mathbf{e}_i^*) + \mathbf{s}\right)^\top \mathbf{W}_s^2(\mathbf{e}_a \| \mathbf{e}_o)$, where $\mathbf{W}_s^1, \mathbf{W}_s^2 \in \mathcal{R}^{d'' \times d'}$ are sentiment-specific weight matrices, such that we can rank aspect and opinion of both positive and negative sentiment. Here $\mathbf{e}_a$ and $\mathbf{e}_o$ are the average node representations of the aspect and opinion, computed using the HG convolutions calculated (see Fig. 2); and $\mathbf{s} \in \mathcal{R}^{d''}$ is a sentiment-specific relation vector. We maximizes the similarity using BPR, minimizing:

$$L_{AOS} = \sum_{(u,i,a,o,s,\bar{a},\bar{o},\bar{s})\in\mathcal{B}} -\ln\sigma\left(f_{TR}(u,i,a,o,s) - f_{TR}(u,i,\bar{a},\bar{o},\bar{s})\right) \qquad (3)$$

where $\mathcal{B} = \{(u,i,a,o,s,\bar{a},\bar{o},\bar{s})|\ \mathbf{I}_{u,i} = 1, (a,o,s) \in r_{u,i}, (\bar{a},\bar{o},\bar{s}) \notin r_{u,i}\}$.

The final loss function includes Θ, the set of all learnable parameters, including $\mathbf{X}^{(0)}$ and $\mathbf{X}_p^{(0)}$ as $L = L_{CF} + \gamma L_{AOS} + \lambda\|\Theta\|_2^2$, where γ and λ are parameters for tuning the L_{AOS} loss and L_2 regularization, respectively. When training, we exclude the interactions we are ranking from the preference module and the accompanying reviews from the review aggregation module, ensuring the method does not have a bias towards already seen items. For each excluded review, we sample an AOS triple for the explainability loss such that we can optimize both the preference and review aggregation modules, concurrently. In practice, we optimize using AdamW with decoupled weight decay. Weight decay is equivalent to L_2 when using SGD, but L_2 does not scale properly with adaptive gradients [29].

Generating Explanations. Our method produces explanations that can then be employed at three different levels (see Fig. 1): (i) at review level, selecting a review that can provide information that is *user specific*; (ii) at word level, highlighting important aspects and opinions; and (iii) at graph level, explanations are paths connecting the user to the item via aspects-opinion pairs and items in other reviews. The graph review level, allows a user to quickly understand why an item is recommended, by giving an easy-to-understand connection between previous purchases and the item. However, this view lacks the context of the extracted phrases, which is often necessary for purchase decisions. Thus, the underlying review text can be used for context, and the extracted phrase-level words are highlighted to draw attention to important areas. Yet, the reviews selected in previous works [9,27] are not user-specific, meaning the same review is given to all users when selecting an item. Such review selection is suboptimal, as it does not consider the user's preferences. Instead, we propose a path-based restriction of the reviews we can use as explanations for an item towards a user.

In Fig. 1, there are multiple paths from the user to the item. However, connecting on the opinion word is most likely uninformative, as reviews may use *great* in very different contexts. Instead, we define the matching criterion as matching on aspects and opinion pairs, creating a path from r_{u_1,i_1} to r_{u_2,i_3}, via the pair (*quality, great*). While in Fig. 1, there is a direct link between a user review and the selected item review, longer paths are possible. In such cases, paths such as $i{\rightarrow}u$ or $i{\rightarrow}v{\rightarrow}u$ are possible, where v is the connecting user and $\rightarrow$ indicates a connection through an aspect-opinion pair. We limit to a max of one intermediary user for reduced graph sizes. Intuitively, if a user writes about similar aspects and we have learned that these are important for the respective user, their reviews about a possible recommendation might also be important. The undirected labeled graph g_m used to compute the explanation contains only weighted paths from user to item satisfying the connectivity constraints. We define the graph as $g_m = \{\{u, r_{u,i}, \{a, o\}\} | r_{u,i} \in \mathcal{R}, (a, o, s) \in r_{u,i}\}$.

We are interested in limiting the possible reviews to a single review for review explanations and to a set of reviews for graph-recommendation. Based on the graph constructed either directly between a user and an item or through multiple hops, we will use the user's attention on reviews along the path, except for the last hop, where we use the item's attention, to select the best path. Intuitively, we are interested in knowing how important users find the reviews, and by extension, the aspects and opinions, but also which reviews are important for the item we are recommending. For example, in Fig. 1, we could select either the review $r_{u_1,i_2}{\rightarrow}r_{u_5,i_3}$ or $r_{u_1,i_1}{\rightarrow}r_{u_2,i_3}$. If we greedily select starting from the user, we would select the path $r_{u_1,i_2}{\rightarrow}r_{u_5,i_3}$, leading to the review level explanation of r_{u_5,i_3}, while greedily selecting, starting from the item would lead to the same selection as shown in Fig. 1. Starting from the user, would create more diverse, user-specific explanations but could disregard the attention score on the item side, while greedily selecting reviews from the item side, could lead to less diversity. Based on these limitations, we propose three selection methods which we will study in the experimental section: (i) greedily selecting reviews, starting from the user until we find the item; (ii) greedily selecting reviews, starting from the item until we find the user; and (iii) finding the path with maximum weight.

Complexity Analysis. HypAR is bounded by the HG convolutions. Thus, to estimate the time complexity of our method, it is sufficient to study the complexity of the HGNN. The convolutions can be described as four operations applied sequentially: (i) feature transformation, with complexity $O(|\mathcal{V}|d^2)$; (ii) two node degree normalizations, of $O(2|\mathcal{V}|^3)$; (iii) transformations from nodes to edges and back again, with $O(2|\mathcal{V}||\mathcal{R}|^2)$; and (iv) neighborhood aggregation, being $O(|\mathcal{V}|^2 d)$. Of these, the node to edge transformations are of highest complexity, as $|\mathcal{R}| \gg |\mathcal{V}| \gg d$. The complexity of the HGNN is $O(|\mathcal{V}||\mathcal{R}|^2)$; however, this naïvely assumes that we utilize dense Matrix Multiplication (MM) instead of Sparse MM (SPMM). Using SPMM we reduce the complexity of (iii) to $O(\|\mathbf{H}\|_1|\mathcal{R}|)$, which

Table 2. Data statistics

Dataset	#User	#Product	#Aspect	#Opinion	#Review
Computer	19,818	8,431	5,046	4,017	92,761
Camera	4,770	2,612	2,182	2,218	21,122
Toy	2,672	1,919	780	1,186	16,070
Cellphone	2,340	1,350	817	1,196	11,134

can be rewritten using the average number of edges $\mathcal{E}_a$, as $O(|\mathcal{V}|\mathcal{E}_a|\mathcal{R}|)$. In most cases $\mathcal{E}_a \ll |\mathcal{V}|$, greatly reducing the complexity. Furthermore, as operations (ii) and (iii) are computed once they have no influence on time complexity during forward propagation, and HypAR is thus bounded by operation (iv).

4 Experiments

The experimental objectives revolve around how our HG-based model compare to state-of-the-art models at providing high-quality recommendations; explanation quality, whether our model is better suited for providing high-quality explanations; and how well the model upholds our explanation objectives.

Datasets. We utilize data based on the four public datasets of 2014 Amazon review dataset [32] (Table 2): *Computer and Accessories* (Computer), *Camera and Photo* (Camera), *Toys and Games* (Toy), and *Cell Phones and Accessories* (Cellphone), for which aspects, opinions and sentiments have been extracted [23]. However, due to space restrictions, we only show our explainability experiments on the smallest and largest datasets, being Cellphone and Computer, respectively, having similar results on the other datasets withheld. We filter users and items with fewer than five ratings and split the datasets with the ratio 0.6:0.2:0.2, for each user based on time. Furthermore, we use a porter-stemmed version of all aspects and opinions by applying Gensim's text preprocessor[1].

Baselines. As shown in Table 1, there are two major categories of explanations, either based on reviews or paths. We have selected NARRE [9] and HRDR [27] for the first group of explainable recommenders. Both use Convolutional Neural Networks on word vectors to represent a review, with attention mechanisms for review selection as explanation. The most prominent method using GNN for recommendation that also produces some form of explanation exploiting connections in the graphs is KGAT [49]. KGAT utilizes TransR [26] to learn weights between entities in a KG. For word level, we use TriRank [17], using a tripartite graph of user, items, and aspects for ranking smoothing. We further compare to MF [36] and LightGCN [18], as we model our preference module after them. We have implemented all methods in the Cornac [40] framework as it supports multi-modal information, such as aspects and reviews[2].

[1] https://radimrehurek.com/gensim/parsing/preprocessing.html#gensim.parsing.preprocessing.stem_text.

[2] All methods are available at https://github.com/PreferredAI/cornac.

Evaluation Metrics. For each user in the test set, we rank all items not interacted with in the train and validation sets [18,49,50]. To evaluate the ranking quality, we measure AUC [15], MAP [5], and NDCG [21]. To evaluate the methods' explainability, we opt for three different methodologies: (i) compare the selected review with the review written by the user; (ii) select a graph (or path) as the explanation and compare it to the ground truth graph constructed from the user's review; and (iii) given a set of aspects and opinions assumed to describe the user's preferences, we study the quality of the approximate ranking obtained by ranking higher items matching them (described in Sect. 3).

To compare a selected review with a ground truth review, we adopt five different sentence similarity metrics for evaluation: BERTScore [62], being based on textual embedding similarities; and BLEU [57], METEOR with alpha = 0.9 [3], and ROUGE [25] which are based on n-gram overlaps. For graph overlap, we use Precision [43], Recall [43], F1-measure, overlap-coefficient [30], and Diversity [1]; measuring the methods' ability to retrieve correct and unique aspects and opinions. For fairness, we allow KGAT to sample an unlimited number of paths until it has chosen (close to) as many nodes as HypAR to study KGAT's ability to select a diverse and relevant subset of nodes.

Recommendation Performance. The results are shown in Table 3, with the default HypAR using LightGCN as the preference module, concatenation as the combiner, and dot product for prediction. Our experiments show concatenation to outperform addition and multiplication (not detailed here for brevity). HypAR$_e$ refers to the method with the explainability loss, HypAR$_{NARRE}$ to experiments with f_{NARRE}, and HypAR-MF with MF as the preference module.

We outperform both review-level recommenders, NARRE and HRDR, with both MF and LightGCN as preference modules, having up to 255% increase in performance using MAP. Using high-order connections is crucial for recommendation performance. While TriRank outperforms all non-graph-based methods, yet HypAR, NGCF, KGAT, and LightGCN outperform this method on all metrics. Thus, as the number of users and items increases, so does the running time and memory footprint, contrary to all other graph-based methods chosen here.

Of all methods, both LightGCN and HypAR consistently perform well across datasets. However, LigthGCN cannot explain its recommendations, motivating our model-agnostic explainability module, i.e., *we aim to maintain the prediction power of LigthGCN while also learning to explain its recommendations.* We see HypAR performing consistently better or similarly to LightGCN.

Explanation Quality. In Table 5, we have HypAR where gi, gu, w, i being the selection strategies greedy user, greedy item, weighted, and naïve selection using only item attention. Furthermore, γ is the AOS loss weight. The experiments on the quality of the explanations (Table 4) show that HypAR selects most often reviews of higher quality than both NARRE and HRDR. We often see a statistically significant increase in performance when using t-test over users (p-value of 0.05). Using explainability loss (Eq. 3), we have a general increase in explainability performance but also recommendation. Given that our method outperforms the two baselines without the path restriction methodology, indi-

Table 3. Recommendation performance on different datasets. **Bold** is the best performing and underline is the second best.

Method	Cellphone			Toy			Camera			Computer		
	AUC	MAP	NDCG	AUC	MAP	NDCG	AUC	MAP	NDCG	AUC	MAP	NDCG
MostPop	0.5370	0.0314	0.1709	0.4412	0.0047	0.1338	0.5916	0.0171	0.1528	0.6604	0.0148	**0.1390**
BPR-MF [36]	0.5662	0.0317	0.1711	0.4729	0.0050	0.1337	0.6190	0.0174	0.1527	0.6785	**0.0150**	**0.1390**
NARRE [9]	0.5123	0.0077	0.1385	0.5135	0.0053	0.1357	0.4983	0.0030	0.1217	0.5020	0.0011	0.1036
HRDR [27]	0.5317	0.0206	0.1541	0.4964	0.0053	0.1346	0.5088	0.0060	0.1272	0.4522	0.0059	0.1120
HypAR-MF	**0.6440**	**0.0334**	**0.1789**	**0.6462**	**0.0128**	**0.1544**	**0.6635**	**0.0177**	**0.1549**	**0.6890**	0.0124	0.1355
HypAR-MF_{NARRE}	0.6422	0.0272	0.1732	0.6302	0.0116	0.1524	0.6314	0.0145	0.1484	0.6684	0.0110	0.1330
TriRank [17]	0.6965	0.0248	0.1769	0.6666	0.0136	0.1585	0.6887	0.0117	0.1507	0.7054	0.0048	0.1247
NGCF [50]	0.7430	0.0365	0.1900	0.7138	0.0158	0.1622	0.7122	0.0216	0.1631	0.6978	0.0130	0.1358
KGAT [49]	0.7295	0.0500	0.2017	0.6830	0.0155	0.1599	0.6928	0.0202	0.1602	0.7105	0.0113	0.1352
LightGCN [18]	0.7448	0.0507	0.2037	0.7129	0.0184	0.1648	0.7294	0.0293	0.1741	0.7181	0.0187	0.1458
HypAR	**0.7533**	**0.0517**	**0.2054**	0.7169	0.0199	0.1663	0.7325	0.0286	0.1734	0.7278	0.0194	**0.1473**
HypAR ($\gamma = .1$)	0.7515	0.0507	0.2045	**0.7172**	0.0200	**0.1665**	**0.7348**	**0.0297**	**0.1747**	0.7280	**0.0196**	0.1471
HypAR_{NARRE}	0.7293	0.0501	0.2029	0.6826	**0.0201**	0.1656	0.7207	0.0278	0.1718	**0.7308**	0.0191	0.1472

Table 4. The results of the review selection.

Method	Cellphone				Computer			
	BERTScore	BLEU	METEOR	ROUGEL	BERTScore	BLEU	METEOR	ROUGEL
HRDR	0.8390	0.0143	0.0680	0.0943	0.8341	0.0106	0.0561	0.0793
NARRE	**0.8408**	0.0220	0.0920	0.1049	**0.8341**	0.0070	0.0429	0.0727
HypAR_{i}	0.8389	0.0291	0.1090	0.1082	0.8328	0.0250	0.0993	0.0968
HypAR_{gi}	0.8392	0.0360	0.1141	0.1141	0.8321	0.0282	0.0990	**0.0993**
HypAR_{gu}	0.8389	0.0295	0.1114	0.1085	0.8324	0.0258	0.1015	0.0971
HypAR_{w}	0.8389	0.0295	0.1106	0.1088	0.8323	0.0259	**0.1019**	0.0969
$\text{HypAR}_{gi}(\gamma = .05)$	0.8396	**0.0374**	**0.1152**	**0.1154**	0.8322	**0.0283**	0.0991	**0.0993**
$\text{HypAR}_{gi}(\gamma = .1)$	0.8391	0.0362	0.1143	0.1138	0.8321	**0.0283**	0.0993	0.0991
Improv %	−0.15*	70.23*	25.19*	10.05*	−0.16*	165.87*	81.62*	25.24*

Table 5. The results of the graph selection.

Method	Cellphone					Computer				
	Div	F1	OC	Prec.	Recall	Div	F1	OC	Prec.	Recall
KGAT	0.4851	0.1863	0.3610	0.1346	0.3598	0.4874	0.1612	0.3513	0.1141	0.3500
HypAR_{i}	0.3167	0.2627	0.3420	**0.2595**	0.3050	0.3163	0.2349	0.3183	**0.2325**	0.2829
HypAR_{gi}	0.5573	0.2893	0.5487	0.2110	0.5470	0.5274	0.2417	0.5169	0.1733	0.5145
HypAR_{gu}	0.5672	0.2892	0.5469	0.2111	0.5452	0.5634	0.2448	0.5081	0.1768	0.5056
HypAR_{w}	**0.5702**	0.2900	0.5483	0.2116	0.5465	**0.5707**	**0.2451**	0.5106	0.1767	0.5081
HypAR_{gi} ($\gamma = .05$)	0.5644	**0.2916**	0.5495	0.2136	0.5472	0.5271	0.2417	**0.5171**	0.1730	**0.5149**
HypAR_{gi} ($\gamma = .1$)	0.5568	0.2911	**0.5527**	0.2122	**0.5508**	0.5258	0.2421	0.5169	0.1735	0.5146
Improv %	17.53	56.50*	53.08*	92.77*	53.08*	17.10	52.11*	47.21*	103.75*	47.10*

cates that the HG structure is essential for the review explainability. We perform slightly worse with the BERTScore metric; however, BERTScore suffers from the antonym embedding problem, where antonyms have similar embeddings due to similar contextual information [13,39].

We are outperforming KGAT (Table 5) on all metrics with statistical significance. KGAT selects a smaller subset of nodes, as seen with the diversity metric, even when selecting as many nodes as HypAR. The nodes selected are of less importance for the user, as KGAT only selects half as many relevant nodes as HypAR, which can be extrapolated from its recall. We see the path selection algorithm affect the performance; particularly, the performance between diversity, precision, and recall. Which to use therefore depends on the specific explainability scenario. When selecting a single review, we reduce the number of selected nodes and thereby increasing the precision at the cost of diversity and recall. Furthermore, in both explainability settings, using our path restriction methodology increases performance over the naïve attention mechanism.

Explainability Criteria. We test our method using the two explainability criteria defined in Sect. 3. Specifically, based on the top-5 recommended items for each user, we find the explanation ε of each item and all other items matching the explanation ε. We then conduct two studies comparing to: (i) a random ordering of items and (ii) items ranked using a different explanation of similar size. This evaluation works as a *litmus test* where we inspect the ability of the explanation to carry some information about the user preference beyond the specific item for which it was generated. For example, if we recommend item i_3 and we provide *great quality* instead of *fast shutter* as explanation, we would expect then that items with *great quality* are, on average, preferred to items with *fast shutter*. Any method not passing these criteria would not be explainable as the explanations would be indistinguishable from random explanation. This test is designed specifically to validate the explanation informativeness for set-based techniques as ours. Thus, these metrics and the results are not comparable to methods that select explanations based on other criteria, e.g., MostPop. On all datasets, we see a statistically significant increase over both the random ordering and the random explanation when using the average rank of the selected items. For example, for Cellphone, we have an average rank (lower is better) of 543.8, with random ordering having 674.5 and random explanation having 628.0 leading to a p-value less than < 0.01. As such, our method is, firstly, able to select a set of nodes that correlate to the actual ordering of items by our method, and secondly, the set of selected items correlates with our method's understanding of the user's preferences. Since our method outperforms the baselines, w.r.t. the evaluation metrics in Table 4 and Table 5, and the explanations selected correlate with the HypAR's learned user preferences; we have strong evidence that the graph selection strategy is sound.

Case Study. We present here an illustrative case study (Fig. 3) by randomly selecting a user (`A143LJ4G20PP7T`) and providing an explanation for the highest-ranked item (`B00D64PN36`). The extrapolated explanation would be that both products are dual-chargers based on the intersection; however, `B00D...` has a smaller size. This is due to the user's displeasure of `B007...`'s larger size, which makes `B00D...` likely preferable. As such, our method selects the relevant information for the user, and this can then be adopted for word-level highlighting of the parts of text in review of the recommended item.

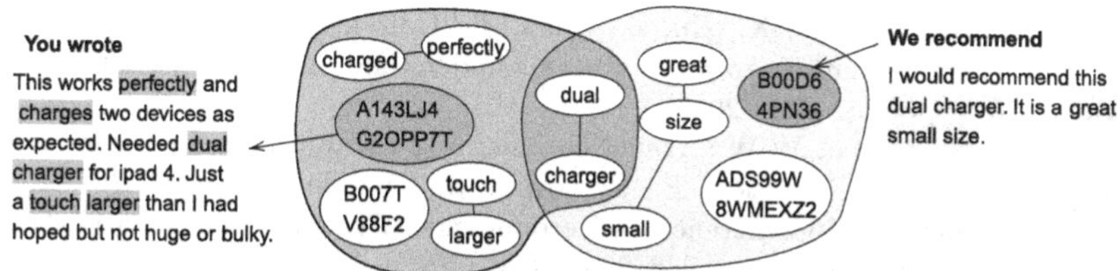

Fig. 3. Real example from the Cellphone dataset.

5 Conclusion

In this work, we propose a novel model-agnostic review-based HG architecture for explainable recommendation. Our new graph model, HypAR, is based on review-induced hyperedges and illustrate its possible use cases. We demonstrate both the recommendation abilities of our method and the power of its explanations compared to existing review-based explanation methods. We show that HypAR either improves or maintains the performance of the underlying recommendation method we provide explanations for while improving the explanation quality compared to state-of-the-art methods; concluding that more attention is required for graph-based review explanations, as existing methods underperform. However, if the number of phrase-level sentences explodes, the graph view may become indigestible. Future work could therefore focus on pruning and highlighting the graph view for easy understanding and digestibility.

Acknowledgements. Katja Hose and Theis Jendal are supported by the Poul Due Jensen Foundation and the Independent Research Fund Denmark (DFF) under grant agreement no. DFF-8048-00051B. Hady W. Lauw and Trung-Hoang Le acknowledge that this research/project is supported by the National Research Foundation, Singapore under its AI Singapore Programme (AISG Award No: AISG2-RP-2021-020).

Disclosure of Interests. The authors have no competing interests to declare.

References

1. Adomavicius, G., Kwon, Y.: Improving aggregate recommendation diversity using ranking-based techniques. TKDE **24**(5), 896–911 (2012)
2. Al-Taie, M.Z., Kadry, S.: Visualization of explanations in recommender systems. J. Adv. Manag. Sci. **2**(2), 140–144 (2014)
3. Banerjee, S., Lavie, A.: METEOR: an automatic metric for MT evaluation with improved correlation with human judgments. In: WIEEMMTS 2005, pp. 65–72 (2005)
4. Bauman, K., Liu, B., Tuzhilin, A.: Aspect based recommendations: recommending items with the most valuable aspects based on user reviews. In: KDD 2017 (2017)
5. Beitzel, S.M., Jensen, E.C., Frieder, O.: MAP. In: Encyclopedia of Database Systems, 2nd edn. (2018)

6. Cao, Y., Wang, X., He, X., Hu, Z., Chua, T.: Unifying knowledge graph learning and recommendation: towards a better understanding of user preferences. In: WWW 2019, pp. 151–161 (2019)
7. Catherine, R., Cohen, W.W.: TransNets: learning to transform for recommendation. In: Cremonesi, P., Ricci, F., Berkovsky, S., Tuzhilin, A. (eds.) Proceedings of the Eleventh ACM Conference on Recommender Systems, RecSys 2017, Como, Italy, 27–31 August 2017, pp. 288–296. ACM (2017)
8. Chen, C., Li, D., Yan, J., Huang, H., Yang, X.: Scalable and explainable 1-bit matrix completion via graph signal learning. In: AAAI 2021, pp. 7011–7019 (2021)
9. Chen, C., Zhang, M., Liu, Y., Ma, S.: Neural attentional rating regression with review-level explanations. In: WWW 2018, pp. 1583–1592 (2018)
10. Chen, X., et al.: Personalized fashion recommendation with visual explanations based on multimodal attention network: towards visually explainable recommendation. In: SIGIR 2019, pp. 765–774 (2019)
11. Cong, D., et al.: Hierarchical attention based neural network for explainable recommendation. In: ICMR 2019 (2019)
12. Dong, X., et al.: Asymmetrical hierarchical networks with attentive interactions for interpretable review-based recommendation. In: AAAI 2020, pp. 7667–7674 (2020)
13. Etcheverry, M., Wonsever, D.: Unraveling antonym's word vectors through a Siamese-like network. In: Korhonen, A., Traum, D.R., Màrquez, L. (eds.) Proceedings of the 57th Conference of the Association for Computational Linguistics, ACL 2019, Florence, Italy, July 28–August 2 2019, Volume 1: Long Papers, pp. 3297–3307. Association for Computational Linguistics (2019). https://doi.org/10.18653/V1/P19-1319
14. Feng, Y., You, H., Zhang, Z., Ji, R., Gao, Y.: Hypergraph neural networks. In: AAAI 2019, pp. 3558–3565 (2019)
15. Flach, P.A.: ROC analysis. In: Encyclopedia of Machine Learning and Data Mining, pp. 1109–1116 (2017)
16. Gao, Y., Feng, Y., Ji, S., Ji, R.: HGNN$^+$: general hypergraph neural networks. IEEE Trans. Pattern Anal. Mach. Intell. **45**(3), 3181–3199 (2023)
17. He, X., Chen, T., Kan, M., Chen, X.: TriRank: review-aware explainable recommendation by modeling aspects. In: CIKM 2015, pp. 1661–1670 (2015)
18. He, X., Deng, K., Wang, X., Li, Y., Zhang, Y., Wang, M.: LightGCN: simplifying and powering graph convolution network for recommendation. In: SIGIR 2020, pp. 639–648 (2020)
19. He, X., Liao, L., Zhang, H., Nie, L., Hu, X., Chua, T.: Neural collaborative filtering. In: WWW 2017, pp. 173–182 (2017)
20. Huang, X., Fang, Q., Qian, S., Sang, J., Li, Y., Xu, C.: Explainable interaction-driven user modeling over knowledge graph for sequential recommendation. In: MM 2019, pp. 548–556 (2019)
21. Järvelin, K., Kekäläinen, J.: Discounted cumulated gain. In: Encyclopedia of Database Systems, 2nd edn. (2018)
22. Koren, Y., Bell, R.M., Volinsky, C.: Matrix factorization techniques for recommender systems. Computer **42**(8), 30–37 (2009)
23. Le, T., Lauw, H.W.: Synthesizing aspect-driven recommendation explanations from reviews. In: IJCAI 2020, pp. 2427–2434 (2020)
24. Li, Y., et al.: Hyperbolic hypergraphs for sequential recommendation. In: CIKM 2021, pp. 988–997 (2021)
25. Lin, C.Y.: ROUGE: a package for automatic evaluation of summaries. In: Text Summarization Branches Out, pp. 74–81, July 2004

26. Lin, Y., Liu, Z., Sun, M., Liu, Y., Zhu, X.: Learning entity and relation embeddings for knowledge graph completion. In: AAAI 2015, pp. 2181–2187 (2015)
27. Liu, H., et al.: Hybrid neural recommendation with joint deep representation learning of ratings and reviews. Neurocomputing **374**, 77–85 (2020)
28. Liu, H., Wen, J., Jing, L., Yu, J., Zhang, X., Zhang, M.: In2Rec: influence-based interpretable recommendation. In: CIKM 2019, pp. 1803–1812 (2019)
29. Loshchilov, I., Hutter, F.: Decoupled weight decay regularization. In: ICLR 2019 (2019)
30. Vijaymeena, M.K., Kavitha, K.: A survey on similarity measures in text mining. Mach. Learn. Appl. Int. J. **3**, 19–28 (2016)
31. Ma, W., et al.: Jointly learning explainable rules for recommendation with knowledge graph. In: WWW 2019, pp. 1210–1221 (2019)
32. McAuley, J.J., Targett, C., Shi, Q., van den Hengel, A.: Image-based recommendations on styles and substitutes. In: SIGIR 2015, pp. 43–52 (2015)
33. Mikolov, T., Chen, K., Corrado, G., Dean, J.: Efficient estimation of word representations in vector space. In: ICLR 2013 (2013)
34. Nair, V., Hinton, G.E.: Rectified linear units improve restricted Boltzmann machines. In: ICML 2010, pp. 807–814 (2010)
35. Pan, S., Li, D., Gu, H., Lu, T., Luo, X., Gu, N.: Accurate and explainable recommendation via review rationalization. In: WWW 2022 (2022)
36. Rendle, S., Freudenthaler, C., Gantner, Z., Schmidt-Thieme, L.: BPR: Bayesian personalized ranking from implicit feedback. In: UAI 2009, pp. 452–461 (2009)
37. Rendle, S., Krichene, W., Zhang, L., Anderson, J.R.: Neural collaborative filtering vs. matrix factorization revisited. In: RecSys 2020, pp. 240–248 (2020)
38. Rong, G., Zhang, Y., Yang, L., Zhang, F., Kuang, H., Zhang, H.: Modeling review history for reviewer recommendation: a hypergraph approach. In: ICSE 2022 (2022)
39. Saadany, H., Orăsan, C.: BLEU, METEOR, BERTscore: evaluation of metrics performance in assessing critical translation errors in sentiment-oriented text. In: TRITON 2021 (2021)
40. Salah, A., Truong, Q., Lauw, H.W.: Cornac: a comparative framework for multimodal recommender systems. J. Mach. Learn. Res. **21**, 95:1–95:5 (2020)
41. Sánchez, L.Q., Sauer, C., Recio-García, J.A., Díaz-Agudo, B.: Make it personal: a social explanation system applied to group recommendations. Expert Syst. Appl. **76**, 36–48 (2017)
42. Sun, X., et al.: Heterogeneous hypergraph embedding for graph classification. In: WSDM 2021, pp. 725–733 (2021)
43. Ting, K.M.: Precision and recall. In: Encyclopedia of Machine Learning and Data Mining, pp. 990–991 (2017)
44. Wang, H., et al.: RippleNet: propagating user preferences on the knowledge graph for recommender systems. In: CIKM 2018, pp. 417–426 (2018)
45. Wang, H., et al.: Knowledge-aware graph neural networks with label smoothness regularization for recommender systems. In: SIGKDD 2019, pp. 968–977 (2019)
46. Wang, H., Zhao, M., Xie, X., Li, W., Guo, M.: Knowledge graph convolutional networks for recommender systems. In: WWW 2019, pp. 3307–3313 (2019)
47. Wang, J., Zhang, Y., Wang, L., Hu, Y., Piao, X., Yin, B.: Multitask hypergraph convolutional networks: a heterogeneous traffic prediction framework. IEEE Trans. Intell. Transp. Syst. **23**(10), 18557–18567 (2022)
48. Wang, N., Wang, H., Jia, Y., Yin, Y.: Explainable recommendation via multi-task learning in opinionated text data. In: SIGIR 2018, pp. 165–174 (2018)
49. Wang, X., He, X., Cao, Y., Liu, M., Chua, T.: KGAT: knowledge graph attention network for recommendation. In: SIGKDD 2019, pp. 950–958 (2019)

50. Wang, X., He, X., Wang, M., Feng, F., Chua, T.: Neural graph collaborative filtering. In: SIGIR 2019, pp. 165–174 (2019)
51. Wang, X., Wang, D., Xu, C., He, X., Cao, Y., Chua, T.: Explainable reasoning over knowledge graphs for recommendation. In: AAAI 2019, pp. 5329–5336 (2019)
52. Wu, C., Wu, F., Liu, J., Huang, Y.: Hierarchical user and item representation with three-tier attention for recommendation. In: NAACL-HLT 2019, pp. 1818–1826 (2019)
53. Wu, C., Wu, F., Qi, T., Ge, S., Huang, Y., Xie, X.: Reviews meet graphs: enhancing user and item representations for recommendation with hierarchical attentive graph neural network. In: EMNLP-IJCNLP 2019 (2019)
54. Wu, L., Wang, D., Song, K., Feng, S., Zhang, Y., Yu, G.: Dual-view hypergraph neural networks for attributed graph learning. Knowl. Based Syst. **227**, 107185 (2021)
55. Wu, X., Chen, Q., Li, W., Xiao, Y., Hu, B.: AdaHGNN: adaptive hypergraph neural networks for multi-label image classification. In: MM 2020, pp. 284–293 (2020)
56. Wu, Y., Ester, M.: FLAME: a probabilistic model combining aspect based opinion mining and collaborative filtering. In: WSDM 2015, pp. 199–208 (2015)
57. Wu, Y., et al.: Google's neural machine translation system: bridging the gap between human and machine translation. CoRR (2016)
58. Xian, Y., Fu, Z., Muthukrishnan, S., de Melo, G., Zhang, Y.: Reinforcement knowledge graph reasoning for explainable recommendation. In: SIGIR 2019 (2019)
59. Yang, Z., Dong, S.: HAGERec: hierarchical attention graph convolutional network incorporating knowledge graph for explainable recommendation. Knowl. Based Syst. **204**, 106194 (2020)
60. Yu, J., Yin, H., Li, J., Wang, Q., Hung, N.Q.V., Zhang, X.: Self-supervised multi-channel hypergraph convolutional network for social recommendation. In: WWW 2021, pp. 413–424 (2021)
61. Zhang, T., Sun, C., Cheng, Z., Dong, X.: AENAR: an aspect-aware explainable neural attentional recommender model for rating predication. Expert Syst. Appl. **198**, 116717 (2022)
62. Zhang, T., Kishore, V., Wu, F., Weinberger, K.Q., Artzi, Y.: BERTScore: evaluating text generation with BERT. In: ICLR (2020)
63. Zhang, Y., Lai, G., Zhang, M., Zhang, Y., Liu, Y., Ma, S.: Explicit factor models for explainable recommendation based on phrase-level sentiment analysis. In: SIGIR 2014, pp. 83–92 (2014)

Investigating the Usage of Formulae in Mathematical Answer Retrieval

Anja Reusch(✉), Julius Gonsior, Claudio Hartmann, and Wolfgang Lehner

Dresden Database Research Group, Technische Universität Dresden, Dresden, Germany
{anja.reusch,julius.gonsior,claudio.hartmann, wolfgang.lehner}@tu-dresden.de

Abstract. This work focuses on the task of Mathematical Answer Retrieval and studies the factors a recent Transformer-Encoder-based Language Model (LM) uses to assess the relevance of an answer for a given mathematical question. Mainly, we investigate three factors: (1) the general influence of mathematical formulae, (2) the usage of structural information of those formulae, (3) the overlap of variable names in answers and questions. The findings of the investigation indicate that the LM for Mathematical Answer Retrieval mainly relies on shallow features such as the overlap of variables between question and answers. Furthermore, we identified a malicious shortcut in the training data that hinders the usage of structural information and by removing this shortcut improved the overall accuracy. We want to foster future research on how LMs are trained for Mathematical Answer Retrieval and provide a basic evaluation set up (Link to repository: https://github.com/AnReu/math_analysis) for existing models.

Keywords: Mathematical Information Retrieval · Transformer-Encoders

1 Motivation

Mathematical Answer Retrieval (Math AR) deals with the task of ranking a set of answers for their relevance to a given mathematical question. As in general Information Retrieval, Transformer-Encoder-based Language Models (LMs) are part of the most successful approaches for Math AR [7,21,32], but are usually applied as black box models. Recently, research has found that LMs adapted to mathematics encode mathematical parse trees, namely Operator Trees, in their contextualized embeddings [20]. However, when the model is probed for these tree structures after it was fine-tuned on Math AR, the performance degrades. In Fig. 1, we visualize how much structural information can be extracted from the embeddings of each of the Transformer layers following the methodology of the original authors. After the model was fine-tuned for Math AR (blue line), the performance on the structural probing task was significantly lower in all layers greater than 2. This finding demonstrates that the probe was less successful in

N. Goharian et al. (Eds.): ECIR 2024, LNCS 14608, pp. 247–261, 2024.
https://doi.org/10.1007/978-3-031-56027-9_15

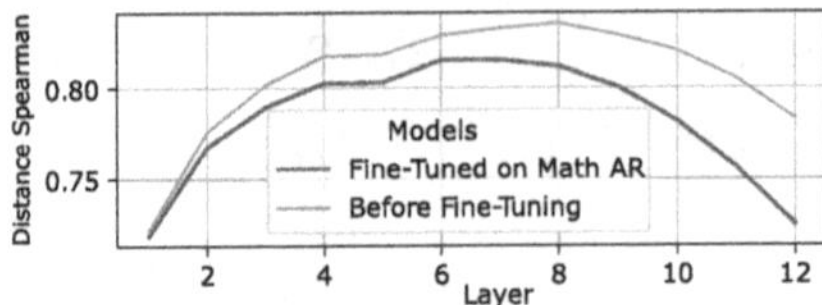

Fig. 1. Layerwise results of an LM before and after fine-tuning on Math AR.

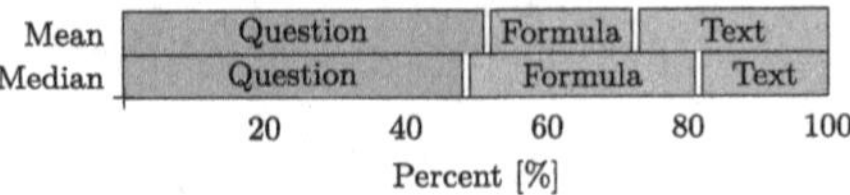

Fig. 2. IG scores aggregated on question tokens (red), and answer tokens (blue). (Color figure online)

extracting structural information from the model's embeddings after fine-tuning, which indicates that the structural information got lost during the fine-tuning process, since the model "overwrote" it with other information about formulae. Van Aken et al. [27] showed in a similar experiment that factual knowledge that is not required for solving a fine-tuning task is lost after fine-tuning, while relevant information is more extractable by their probe. Therefore, our question here is which information on mathematical formulae instead got reinforced by the model during fine-tuning on Math AR. Hence, the goal of this work is to study how the LM that demonstrated the best ability for extracting structural information of mathematical formulae, `AnReu/math_pretrained_bert`, is using formulae when assessing the relevance of a mathematical answer. These insights are then used to identify and verify shortcomings in the training and optimize the model's retrieval performance.

Before we begin, we motivate our methodology by performing an initial experiment where we investigate whether the model considers formula token at all during the retrieval process. Hereby, we utilize the Integrated Gradient (IG) attribution method [25], which measures how much of the model's prediction can be attributed to which token. We use `AnReu/math_pretrained_bert` after it was fine-tuned in a Cross-Encoder setup for Math AR using the ARQMath data set [11] following the methodology of Reusch et al. [21]. Since cross-encoders simply perform a classification on top of the [CLS] embedding of the LM, IG calculates the importance of each token with regard to the class denoting that the answer is relevant for the mathematical question. Similarly to prior work [31], we break up the scores by query (here: question) and document (answer), but further differentiate between token types of the answer to study whether the respective formula parts are used for the relevance assessment. Transformer-interpret's implementation[1] of IG is applied and we aggregated (mean and median) the top 50 answers for each topic for tokens that belong to the mathematical question, the text part of the answer, and the mathematical tokens of the answer. The results can be found in Fig. 2 where the proportion of IG scores for each part of the input is plotted. It is visible that the model's decision for the relevance score is in part based on the answer and in part also based on the question, which is as expected. The most important observation can be made when looking at the IG scores on the formula tokens of the answer. Here, we can see that the model

[1] https://pypi.org/project/transformers-interpret/.

indeed considers the formulae along with the textual part of the answer. Thus, we can conclude that the model uses information from formulae when assessing the relevance of an answer, but it is still an open question how the formulae are used and to which extent the model is relying on this information.

Therefore, this work begins by further analyzing what influence formulae in the answer have when the model assesses the relevance of an answer. Here, we investigate how much the retrieval performance degrades when the model has no or limited access to formula information in the answers. Afterwards, we compare the ability of the LM before and after fine-tuning on Math AR on the task of predicting variables that occur in the token overlap of a question and an answer.

This task is motivated by the observation that several results, which had the same structure as the formula in the question but no common variables, were ranked low by the model, even though they were considered relevant by experts. In contrast, non-relevant results, which shared the same variable names with a different structure, were ranked high by the model. Hence, the model seems to be biased towards answers that have a higher number of common variables with the question. During training, it could have learned a malicious shortcut of only considering the variable overlap and not the actual semantics of a formula. To verify the insights from the first experiments described here, we construct two additional training sets for fine-tuning. The first data set artificially adds the mentioned shortcut, while the second completely removes it such that the model has to focus on other features than the shortcut. By comparing the results of the model on these two fine-tuning sets to the baseline data set, we can conclude that the model indeed relied on this shortcut.

To summarize, this paper aims to contribute to the following: Our evaluation shows that the model `AnReu/math_pretrained_bert` uses information from the formulae when assessing the relevance of an answer given a question (Sect. 3), but probing revealed that during fine-tuning, the importance of overlapping variables between question and answers is reinforced (Sect. 4). We further demonstrate that substituting variables in the fine-tuning set leads to an overall improved performance in the Math AR task (Sect. 5).

2 Preliminaries

2.1 Related Work

To interpret the knowledge and skills Transformer-Encoder models learn during training, several *post-hoc* techniques have been introduced. One of these techniques is the probing classifier or the probe where a simple classifier is trained in order to predict a certain property from the model's learned contextualized embeddings. If the classifier successfully predicts the property, sufficient information about this property has to be encoded in the contextualized embeddings. Therefore, one concludes that the model encoded or learned this property during the initial training. For more details on probing, we refer the reader to this survey [2]. The application of these probing classifiers for knowledge of retrieval models was studied in previous research: Fan et al. [6] evaluated 16 different tasks for natural language understanding while Van Aken et al. [27] probed for layer-wise information in question answering. Also, Zhan et al. [31] applied a probing classifier for a model's ranking performance on different layers. Wallat et al. [29] studied which information can be extracted by a probing classifier designed for factual knowledge before and after it was fine-tuned on a fine-tuning task and found that the fine-tuning task forces the model to forget and reinforce certain information.

In the context of mathematical models, [20] trained a structural probe to evaluate how much information on mathematical parse trees can be extracted from the contextualized embeddings of math-adapted Transformer-Encoder models. While several studies cover new data sets to inspect how well Transformer models solve mathematical questions [5,8,18,24], no research was specifically conducted to analyze the layer-wise information of Transformer-Encoder models for Math AR. In a similar way, previous research analyzed the IG attribution and the attention patterns of BERT models for document ranking [19,31]. However, since the interpretability of attention is disputed [28], we refrain from using it. The removal and scrambling of words have been used to study their effects on LMs [16,17]. However, we apply these ideas to formulae in a mathematical context.

2.2 Fine-Tuning Setup

Model Choice. The recent series of ARQMath Labs [11–13] has moved more attention to the development of Transformer-Encoder models that are adapted for Math AR. During the course of the lab series, several teams adapted exisiting Transformer-Encoder models that were originally pre-trained on natural language, to the domain of mathematical documents [15,21–23]. These models use LaTeX as the form of representing formulae. In contrast, other works make use of (linearized) tree structures of mathematical parse trees to represent formulae in their models' input [14,32]. Even though several Transformer-Encoder models for mathematical documents have been proposed, this work analyzes the model `AnReu/math_pretrained_bert`. We chose this model because the goal of this work is to study which mathematical information is used when a model assesses the relevance of an answer given a question, and this particular model displayed the best results when extracting syntactic information from its contextualized embeddings [20]. In addition, this is also the BERT-based version of the successful `AnReu/math_albert`, which was the best performing cross-encoder model in the ARQMath Lab 3 [21].

Training Details. In order to perform Math AR, we train the models in a cross-encoder setup [9], where a binary classification is used to assess if a given answer addresses a mathematical question. Question and answer tokens are concatenated and provided as an input to the LM up to a limit of 512 tokens. A classification head (linear classifier) on top of the [SEP] token embedding is trained to decide for either "relevant answer" or "non-relevant answer". The question-answer pairs we use are based on the official ARQMath 2020 corpus [11]. Relevant answers are answers that a user posted in response to a mathematical question. For each question, we use up to ten relevant answers or as many as the questions had if less were present. Non-relevant answers are sampled by chance from the set of answers with at least one similar topic. Both classes, "relevant" and "non-relevant", are equally distributed among the 2.7 M pairs, of which 90% are used for training and 10% for validation. The training and evaluation was performed using Huggingface's transformers library [30]. We followed Reusch et al. [21] for pre-processing and hyperparameter settings, because this set up demonstrated the most promising results in a cross-encoder set up.

The evaluation of our models is performed in the same way as for the ARQ-Math Lab that provides 77 questions from 2019 with relevance judgment for several answers annotated using pooling. Each of these questions is paired with every answer from the corpus, then each pair is provided as input to the model. The classification score for label "relevant answer" is used to rank the answers. We follow ARQMath and evaluate the top 1,000 ranking answers for each topic using nDCG and the top 10 using precision. For each question from the evaluation corpus, the organizers of the ARQMath Lab 2020 also divided the questions into the three categories *Formulae*, *Text*, and *Both* indicating whether answering a question depended on understanding mainly the question's formulae, the natural language text or both, respectively. During evaluation we will break down the performance of our models by these categories.

Judged Evaluation. The ARQMath Lab does not provide relevance judgments for every answer regarding each question, but only for a small subset that was annotated during the pooling procedure. Hence, there exists a large number of answers that do not contain scores. When evaluating the performance of a model, non-judged answers are removed and the nDCG and Precision are calculated only on basis of the judged answers. However, evaluating the models on the entire set of answers is costly. Therefore, in most cases evaluations on the Math AR test set were performed only using judged answers meaning that for each question in the test set only answers whose relevance was judged during the ARQMath Lab were considered as candidates. When only comparing the approaches and models in this paper, this is a valid approach since only using the judged answers does not influence the ranking of the answers nor the relations between the scores and the models. But we cannot compare these model results to the results of the original ARQMath Lab because their ranking contains non-judged results which were removed but still influence the ranks of relevant answers when computing the metrics. Therefore, we provide an evaluation of the final models on the non-judged test set using the original metrics as used in the ARQMath Lab (nDCG′ and p′@10) at the end of the work.

Significance Testing. In order to compare models, we use the Almost Stochastic Order (ASO) test [3,4] as implemented by Ulmer et al. [26]. We compared all pairs of models based on five random seeds each using ASO with a confidence level of $\alpha = 0.05$ (before adjusting for all pair-wise comparisons using the

Bonferroni correction). We report almost stochastic dominance ($\epsilon_{min} < \tau$ with $\tau = 0.2$) in all results sections.

3 Usage of Mathematical Formulae for Answer Retrieval

First of all, we study if models fine-tuned for Mathematical Answer Retrieval incorporate mathematical information at all, when determining the relevance of an answer given a question. We start with three simple experiments where we evaluate the results on the ARQMath 2020 test set when removing all mathematical formulae (1) or when replacing each formula with a dummy expression (2). Thereby, we identify if the models use the formulae at all when differentiating between answers. If the results on the modified two test sets are lower than on the original test set, the models relies on the formula information to judge the relevance. If the scores are higher than on the original test set, the formula information might confuse the model or lead it in the wrong direction. Additionally, each formula in the test set is replaced by a string containing the same tokens, but sorted (3), evaluating the usage of structural information. Thus, we end up with three variants (1)–(3) of the test set, each enabling us to show a specific aspect of the trained knowledge.

3.1 Experiment Setup

In these three experiments, we use the models trained by applying our baseline setup as detailed in Sect. 2.2 and evaluate them in the same way. The only difference compared to experiment (1), (2), and (3) is that we evaluate them on a slightly modified test set as we will detail in the following section.

Each test set contains the same questions from the ARQMath 2020 test set along with the same set of answers. Each question is paired with each answer from the set of judged answers, which are provided by the organizer of the ARQMath Lab. The unprocessed data is in XML format while each post is formatted in HTML. The formulae are present in ⟨math⟩ containers and therefore easily removable. For test set (1), we remove each mathematical formula entirely. Since this could cause a break in the coherence of a post, we create test set (2) where each formula is replaced by $a + b$. Because each formula contains now the same information, the model cannot differentiate between two answers only on basis of these formulae. For test set (3), we tokenized each formula using a LaTeX tokenizer and sorted the tokens. This way structural information can not be identified by the models. An illustration of the applied test sets can be seen in Table 1.

3.2 Results

The results of the experiments on the three modified test sets as well as for the default test set are displayed in Table 1. There is a visible and significant drop on both metrics when comparing the default test set with each of the modified test sets. The differences between test sets (1) and (2) are not significant. In addition, we break the score down regarding the three dependency categories *Formulae*, *Text* and *Both* (not in the table) for a more detailed analysis: The largest drop in performance is as one would expect in the *Formulae* category (on average –0.23 on nDCG@1,000 on *Formulae*, –0.11 on *Both* and –0.04 in *Text*). This indicates that the model focuses more on mathematical formulae for questions where these formulae are also more important to retrieve a relevant answer.

Table 1. Results for all applied test sets including an example.

Test Set	Example	nDCG@1,000	p@10
Default	How to simplify $\int_a^b f(x)dx+...$	0.6990	0.3733
No Math (1)	How to simplify	0.5438	0.1349
Dummy Math (2)	How to simplify $a+b$	0.5309	0.1434
Sorted formulae (3)	How to simplify $?(())+\int ab...$	0.5917	0.2221

From these results, it can be inferred that formula information is used by the models to determine the relevance of an answer since on both test sets the metrics deteriorate in comparison to the default test set that contained formulae. Because the formula information was not available in the two modified test sets, the models had to rely on the textual information, which in some cases was not sufficient to rate the answer accordingly. Since we now know that formulae are actually beneficial to answer a question and are not random by-standing tokens, we can investigate how they are used by the models.

Here, we can look at the scores for test set (3), which are between the ones of the default test set and test set (1) and (2). This indicates that the model does - in contrast to what Fig. 1 suggests - use some information on the structure of formulae as the scores drop when the structure is removed. However, after sorting the tokens of the formulae some information is still recovered and used to assess the relevance of the answers given the questions. This can be inferred because the model is able to better rank the test set (3) in contrast to test set (1) and (2). A possible reason could be that the sorting destroys also simple relationships such as operator-argument relationships, which could be used by a model to identify common patterns such as $f(x)$ or $sin(\pi)$. Here, further investigations are necessary to evaluate cases where the model uses structural information. We have seen that some structural information is used, but the results suggest that the model is also relying on other information, which is visible by the gap in the scores between test set (3) and (2). Therefore, the next section will evaluate another candidate for information the models might rely on: Variable Overlap.

4 Variable Overlap Prediction

The observation that question and answers, which are retrieved by the model, share more common tokens than relevant answers from the test set in general led us to investigate the variable overlap in answers that were ranked high by the model in comparison to the overlap of variables in actual relevant answers. The overlap of variables in the top 1,000 answers for ARQMath 2020 ranked by `AnReu/math_pretrained_bert` was 50.0% while when only considering relevant answers it is only 34.5%. The models seem to be biased towards answers that have a higher number of common variables with the question which it could have learned during the fine-tuning. Hence, it is reasonable to investigate the model's ability of capturing which variables occur in both the question and the answer. We train a probing classifier to detect common tokens between the two input segments of the models and call this task *Variable Overlap Prediction*. As Sect. 1, we evaluate the model's performance before and after fine-tuning and compare them to see if it learned to reinforce this ability or replace it by something else. In the following sections, we will provide details on the data set and the training of the probing classifier and analyze the results of our experiment.

4.1 Training Details

Our data set comprises pairs of formulae as input data and the labels indicating the variables they have in common. The formulae are taken from the train and test split of the MATH data set [8]. We randomly paired two formulae and intersected their tokens. These intersection was then filtered for variables. We chose the set of single letters in the Latin alphabet as variables, since Greek letters or other variable names occurred only in less than 20 examples. Each variable denotes a single class. If the variable is included in the overlap of the two formulae, its class gets the label 1; if it is not present in the overlap, the label for the class is set to 0. If the overlap of the two formulae contains another token except of the variables, it was added as an instance of the fallback class "other". This class is needed when no variable overlap between the formulae is present, but other tokens still overlap, e.g. $a + 1$ and $b + 1$. Therefore, each sample from the data set consists of two formulae in LaTeX and one Boolean vector with 27 dimensions denoting the labels of the 26 variables and the fallback class. The training examples were selected from the train split of the MATH data set, while the examples in the test set were selected from its test split and deduplicated since formulae can occur in both splits. In total, the data set consists of 5k formula pairs with their labels for training, 2k for validation and 10k for the test set. As before, training and evaluation were performed using the transformers library [30]. We conducted hyperparameter tuning using Optuna [1].

In order to probe the LMs for their information on variable overlap, we input a concatenation of the two formulae in the model. On top of the [CLS] token embedding of the respective layer, we train a classification head which consists of a single linear layer. Identifying the variable overlap between the two formulae is a multi class, multi label classification with 27 classes. The output vector $o \in [0, 1]^{27}$ of the model is trained to contain a value between 0 and 1 for each of the 27 classes. $o_i = 0$ if the ith variable is not present in the overlap, and $o_i = 1$ if it is present.

4.2 Results

Figure 3 shows the results for the [CLS] embedding of each layer, when evaluated on the Variable Overlap Prediction task (the results for this section are shown in solid lines, dashed are the results for Sect. 5). For each layer except the first two, the difference between the fine-tuned model and the model before fine-tuning is significant. It was possible to extract the information about the variable overlap with the highest F_1 score in Layer 6 and 11. The lowest scores receive Layer 3 for both setups.

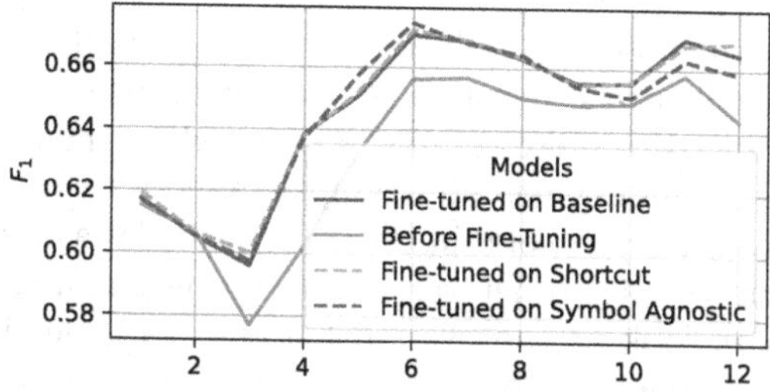

Fig. 3. F_1 Scores on Variable Overlap Prediction for each layer before and after fine-tuning on the different Math AR data sets.

This significant gap in almost all layers demonstrates that the information on variable overlap between both segments is more easily extractable from the [CLS] embedding. That indicates that fine-tuning reinforced this knowledge as it was more useful or important for the model. It could be the case that the relevance of a formula in an answer is determined only by judging the overlapping variables between question and answer. This way the model would not learn to rely on the structure of the formula, but only on the shallow feature of the names of the variables. A probable reason for the model to learn such a shortcut, which would make it easier for the model to receive a low loss during training, is that the fine-tuning data is constructed in such a way. The data contains the same amount of correct and wrong answers for each question. The correct answers are the answers to that question provided by users of StackExchange. These answers are likely to contain the same variables as the question since they refer to what was written in the question. Wrong answers are sampled from the corpus. The user who wrote these answers referred to other questions which in most cases contained other variables. Therefore, the model only needs to learn to look at the variables in question and answer. If the overlap is high, the answer is probably one of the correct ones, while it is a wrong answer when the variables do not match. However, we base this claim only on the observation of the gap in the Variable Overlap results. In the next section, we will train two more models that model the variable overlap between question and answer explicitly or not at all to verify our claim.

Table 2. Example data pairs for the three fine-tuning data sets.

	Baseline	Shortcut	Symbol Agnostic
Question	I have a question regarding this calculation: How can I expand $(x+1) \cdot y$?		
Correct	The solution is $x \cdot y + y$	The solution is $x \cdot y + y$	The solution is $c \cdot r + r$
Incorrect	Let the probability $P(x)$	Let the probability $P(a)$	Let the probability $P(a)$

5 Symbol-Agnostic Training

To verify whether our models heavily rely on the shortcut of only considering overlapping variables between question and answer when judging an answer's relevance, we construct two more training sets based on the initial training set. In the first modified training set, all examples adhere to the shortcut, which means that all correct answers contain the same variables as the question, while all incorrect answers contain other variables. During training, the model simply needs to learn to look at the overlap between variables in questions and answers to classify the relevance of an answer. We therefore call this fine-tuning set "shortcut data set". By evaluating models trained on this data set, we can compare it against models trained on the baseline data set and analyze if the renaming leads to significant drops in performance. Furthermore, when rerunning the evaluation of the variable overlap prediction we can see if the models trained on the shortcut data set receive even higher scores. In addition to this first set, we construct a training set where the models cannot learn the shortcut to look only at the variable overlap, but instead - hopefully - capture more structural information of the formulae. In this data set no overlapping variables exist between questions and answers and the model needs to focus on other features. We compare the performance of models trained on this second data set, which we call "symbol agnostic data set", to models trained on the baseline data set and the shortcut data set. If no significant differences can be found, the models do not rely on the shortcut.

In the following, we will explain in detail how the data sets are constructed. The models are trained in the same way as the baseline model in Sect. 2.2, but with the modified fine-tuning data set. For each data set we train five models with different random seeds. The evaluation of each sub-task is carried out as outlined in the previous two sections.

5.1 Renamed Fine-Tuning Data Sets

The base for the two additional data sets is the baseline data set, which consists of pairs of questions and answers, and their labels ("relevant" or "non-relevant" answer). For each question, we tokenized[2] the content of the math containers in the underlying HTML and stored its variables. Each answer first remained split into text and math container parts. We left the text parts as they were,

[2] We use a custom LaTeX tokenizer, e.g., `\sum_ix_i` is tokenized as `\sum _ i x _ i`.

but tokenized the math container content. For the shortcut data set, we only further processed the "non-relevant" answers. Here, we determined the overlapping variable names with the variables from the question. The variables in both posts were renamed by other variables from the set of single letter variables from a to z and Greek letters, both upper and lower case. From this set of variables, we sampled randomly a variable name that was not contained in the question and did not already exist in the post. All instances of overlapping variables were renamed using this procedure. We kept the renaming consistent among all formulae of an answer post. After the renaming was conducted, text and formula parts were concatenated as in the baseline data set. The text and formula parts of the correct answers were concatenated in the same way, but without renaming. For the symbol-agnostic fine-tuning set, we follow the same procedure, but rename all overlapping variables in "relevant" and "non-relevant" answers. It should be noted, that the three sets, the baseline set, the shortcut set, and the symbol agnostic set, contain the same question-answer pairs, only the variable names differ. This way we minimize the influence of the concrete pairing of questions and answers and the influence of how the negative examples were selected. An example of the fine-tuning sets is displayed in Table 2.

5.2 Results

Results of the models on Math AR on each of the test sets of Sect. 3 can be seen in Table 3. Overall, the scores of the models trained on the three data sets are relatively close in all three metrics and on all test sets. All models show the same losses as the baseline model when removing, replacing, or sorting mathematical formulae in the answers. However, on each data set and each metric the symbol agnostic model demonstrates the highest relative and absolute losses in comparison to the default test data, except for p@10 on the sorted formula test set. For example, the relative loss on nDCG@1,000 when comparing the default data set to the sorted data set is 15.35% for the baseline model, 15.55% for the shortcut model, and 16.39% for the symbol agnostic model. We can therefore conclude that the symbol agnostic model relied the most on formulae in the answers when comparing it to the other models. Another possible explanation for this behavior could have been that the symbol agnostic model learns to not focus on formulae at all. But this would have led to a lower loss on the test sets without formula information (*No Math*). Since this is not the case, we conclude that the symbol agnostic models actually focus more on formulae. In addition, eliminating the shortcut also led to the (significantly) highest results of this model in both nDCG and p@10.

The F_1 scores on the Variable Overlap task follow the same trend for the models on all three data sets (dashed lines in Fig. 3). Nevertheless, it is noteworthy that significant differences between the models trained on the symbol agnostic data and the baseline and shortcut models exist in the layers 5 and 12. This could indicate that in the later layers, where the information is needed by the shortcut model, the symbol agnostic model does not rely on this information and therefore use its capacities to encode other information in its embeddings.

Table 3. Results of the additional models on all test sets for Math AR, rows indicate different test sets, columns indicate different fine-tuning data, *All Answers* denotes the default test set including judged and non-judged answers.

Data Set	nDCG@1,000			p@10		
	Baseline	Shortcut	Symbol Agn.	Baseline	Shortcut	Symbol Agn.
All Answers	0.4487	0.4437	0.4559	0.3733	0.3656	0.3794
Default	0.6990	0.6943	0.7019	0.3733	0.3656	0.3795
No Math	0.5438	0.5440	0.5422	0.1349	0.1410	0.1360
Dummy Math	0.5309	0.5334	0.5302	0.1434	0.1374	0.1397
Sorted Formulae	0.5917	0.5864	0.5868	0.2221	0.2183	0.2259

Overall, our evaluation also showed that the differences between models trained on the baseline set and the shortcut set are small and mostly not significant. Especially on the Variable Overlap Prediction, there was not a single layer in which significant differences between the models were visible. We can therefore conclude that the shortcut of only looking at the overlap of the variables in question and answer, which we artificially introduced with the shortcut data set, was already there and the baseline models learned to rely on it. That means that training sets for Math AR that only consider correct answers from their corresponding question will always contain the shortcut as long as the variables are not renamed.

6 Summary

The goal of this paper was to investigate which factors Transformer-Encoder-based Language Models rely on when they are employed for Mathematical Answer Retrieval. We developed an evaluation set up which includes a probing classifier and three evaluation data sets for the retrieval task to study the usage of formulae and their structural features as well as the overlap of variables between question and answer. In this study, we applied this evaluation set up to the model `AnReu/math_pretrained_bert`.

Our analysis demonstrated that the model considers formula information when assessing the relevance of an answer given a mathematical question. We further showed that the model loses a significant part of its modeling capacities of structural relationships of formulae after being fine-tuned on Mathematical Answer Retrieval. We attribute this to the fact that the fine-tuning data forces the model to consider mostly the overlap of variables between question and answer formulae, and not to use structural features. We verified this claim by demonstrating that there are no significant differences between a model that is fine-tuned on a data set that explicitly contains this malicious shortcut and the baseline data set. Furthermore, by removing the shortcut from the data, we improved the retrieval performance of the model. Our evaluation set up can easily be applied to further models for Mathematical Answer Retrieval and other fine-tuning tasks. Even though we only studied a model trained in a cross-encoder set up, the presented set up can also be used with ColBERT-based models [10] such as the one presented by Zhong et al. [33].

However, certain limitations still hold for our set up. All presented data sets on ARQMath data only considered formulae which were originally enclosed in math containers in the original HTML source, but users can also write formulae without LaTeX environments in their posts. These formulae would be unnoticed, since we considered everything outside of math containers as text. However, the proportion of unnoticed formulae is rather small, because the MathStack-Exchange community moderates and corrects posts, which leads to a high data quality in this regard. Nevertheless, it could still influence the results, since formulae whose variables were not renamed could still be present in the symbol agnostic fine-tuning set.

Future work should include further possible features that the models could use, for example, the matching of topic categories (such as matrix calculus, probability theory, etc.) of question and answers or the interplay between formulae and text. In this work, we only looked at overlapping variables between question and answer. In addition, one could also investigate the degree to which the tree distance between formulae in questions and answers is captured. The identified shortcut of only looking at the overlapping strings between questions and answers is most likely not limited to the formulas. The model might also learn to look at reoccurring natural language tokens, because answers will probably pick up terms and phrases from the question text. Because this hinders the model from learning to actually capturing the relevance between question and answer, a similar counteraction as renaming the variables during training should be conducted for the natural language text of the answers. Furthermore, future research should identify the components in the model that correspond to structural modeling, variable overlap or other features in order to gain more insights about where in the model knowledge is stored. Overall, these insights should eventually be used optimize to the fine-tuning procedure to train models for Math AR which make use of formulas in a more meaningful way.

Acknowledgements. The authors would like to thank the anonymous reviewers for their helpful feedback and comments. This work was supported by the DFG under Germany's Excellence Strategy, Grant No. EXC-2068-390729961, Cluster of Excellence "Physics of Life" of TU Dresden. Furthermore, the authors are grateful for the GWK support for funding this project by providing computing time through the Center for Information Services and HPC (ZIH) at TU Dresden.

References

1. Akiba, T., Sano, S., Yanase, T., Ohta, T., Koyama, M.: Optuna: a next-generation hyperparameter optimization framework. In: Proceedings of the 25rd ACM SIGKDD International Conference on Knowledge Discovery and Data Mining (2019)

2. Belinkov, Y.: Probing classifiers: promises, shortcomings, and advances. Comput. Linguist. **48**(1), 207–219 (2022)
3. del Barrio, E., Cuesta-Albertos, J.A., Matrán, C.: An optimal transportation approach for assessing almost stochastic order. In: Gil, E., Gil, E., Gil, J., Gil, M.Á. (eds.) The Mathematics of the Uncertain. SSDC, vol. 142, pp. 33–44. Springer, Cham (2018). https://doi.org/10.1007/978-3-319-73848-2_3
4. Dror, R., Shlomov, S., Reichart, R.: Deep dominance - how to properly compare deep neural models. In: Korhonen, A., Traum, D.R., Màrquez, L. (eds.) Proceedings of the 57th Conference of the Association for Computational Linguistics, ACL 2019, Florence, Italy, 28 July–2 August 2019, vol. 1: Long Papers, pp. 2773–2785. Association for Computational Linguistics (2019). https://doi.org/10.18653/v1/p19-1266
5. Dua, D., Wang, Y., Dasigi, P., Stanovsky, G., Singh, S., Gardner, M.: Drop: a reading comprehension benchmark requiring discrete reasoning over paragraphs. In: Proceedings of NAACL-HLT, pp. 2368–2378 (2019)
6. Fan, Y., Guo, J., Ma, X., Zhang, R., Lan, Y., Cheng, X.: A linguistic study on relevance modeling in information retrieval. In: Proceedings of the Web Conference 2021, pp. 1053–1064 (2021)
7. Geletka, M., Kalivoda, V., Štefánik, M., Toma, M., Sojka, P.: Diverse semantics representation is king. In: Proceedings of the Working Notes of CLEF 2022 (2022)
8. Hendrycks, D., et al.: Measuring mathematical problem solving with the math dataset. In: NeurIPS (2021)
9. Humeau, S., Shuster, K., Lachaux, M.A., Weston, J.: Poly-encoders: architectures and pre-training strategies for fast and accurate multi-sentence scoring. In: International Conference on Learning Representations (2019)
10. Khattab, O., Zaharia, M.: Colbert: efficient and effective passage search via contextualized late interaction over bert. In: Proceedings of the 43rd International ACM SIGIR Conference on Research and Development in Information Retrieval, pp. 39–48 (2020)
11. Mansouri, B., Agarwal, A., Oard, D., Zanibbi, R.: Finding old answers to new math questions: the ARQMath lab at CLEF 2020. In: Jose, J.M., et al. (eds.) ECIR 2020. LNCS, vol. 12036, pp. 564–571. Springer, Cham (2020). https://doi.org/10.1007/978-3-030-45442-5_73
12. Mansouri, B., Agarwal, A., Oard, D., Zanibbi, R.: Advancing math-aware search: the arqmath-2 lab at clef 2021, pp. 631–638 (2021)
13. Mansouri, B., Novotný, V., Agarwal, A., Oard, D.W., Zanibbi, R.: Overview of arqmath-3 (2022): third clef lab on answer retrieval for questions on math (working notes version). In: Proceedings of the Working Notes of CLEF 2022 (2022)
14. Mansouri, B., Oard, D.W., Zanibbi, R.: DPRL systems in the clef 2021 arqmath lab: sentence-bert for answer retrieval, learning-to-rank for formula retrieval (2021)
15. Novotný, V., Štefánik, M.: Combining sparse and dense information retrieval. In: Proceedings of the Working Notes of CLEF (2022)
16. O'Connor, J., Andreas, J.: What context features can transformer language models use? In: Proceedings of the 59th Annual Meeting of the Association for Computational Linguistics and the 11th International Joint Conference on Natural Language Processing, vol. 1: Long Papers, pp. 851–864 (2021)
17. Pham, T., Bui, T., Mai, L., Nguyen, A.: Out of order: how important is the sequential order of words in a sentence in natural language understanding tasks? In: Findings of the Association for Computational Linguistics: ACL-IJCNLP 2021, pp. 1145–1160 (2021)

18. Polu, S., Sutskever, I.: Generative language modeling for automated theorem proving. arXiv preprint arXiv:2009.03393 (2020)
19. Qiao, Y., Xiong, C., Liu, Z., Liu, Z.: Understanding the behaviors of bert in ranking. arXiv preprint arXiv:1904.07531 (2019)
20. Reusch, A., Lehner, W.: Extracting operator trees from model embeddings. In: Proceedings of the 1st MathNLP Workshop (2022)
21. Reusch, A., Thiele, M., Lehner, W.: Transformer-encoder and decoder models for questions on math. In: Proceedings of the Working Notes of CLEF 2022, pp. 5–8 (2022)
22. Reusch, A., Thiele, M., Lehner, W.: Transformer-encoder-based mathematical information retrieval. In: International Conference of the Cross-Language Evaluation Forum for European Languages, pp. 175–189. Springer, Heidelberg (2022). https://doi.org/10.1007/978-3-031-13643-6_14
23. Rohatgi, S., Wu, J., Giles, C.L.: Psu at clef-2020 arqmath track: unsupervised reranking using pretraining. In: CEUR Workshop Proceedings. Thessaloniki, Greece (2020)
24. Saxton, D., Grefenstette, E., Hill, F., Kohli, P.: Analysing mathematical reasoning abilities of neural models. In: International Conference on Learning Representations (2019)
25. Sundararajan, M., Taly, A., Yan, Q.: Axiomatic attribution for deep networks. In: International Conference on Machine Learning, pp. 3319–3328. PMLR (2017)
26. Ulmer, D., Hardmeier, C., Frellsen, J.: deep-significance: easy and meaningful signifcance testing in the age of neural networks. In: ML Evaluation Standards Workshop at the Tenth International Conference on Learning Representations (2022)
27. Van Aken, B., Winter, B., Löser, A., Gers, F.A.: How does bert answer questions? a layer-wise analysis of transformer representations. In: Proceedings of the 28th ACM International Conference on Information and Knowledge Management, pp. 1823–1832 (2019)
28. Vashishth, S., Upadhyay, S., Tomar, G.S., Faruqui, M.: Attention interpretability across NLP tasks. arXiv preprint arXiv:1909.11218 (2019)
29. Wallat, J., Singh, J., Anand, A.: Bertnesia: investigating the capture and forgetting of knowledge in bert. CoRR abs/2106.02902 (2021). https://arxiv.org/abs/2106.02902
30. Wolf, T., et al.: Transformers: state-of-the-art natural language processing, pp. 38–45. Association for Computational Linguistics (2020). https://www.aclweb.org/anthology/2020.emnlp-demos.6
31. Zhan, J., Mao, J., Liu, Y., Zhang, M., Ma, S.: An analysis of bert in document ranking. In: Proceedings of the 43rd International ACM SIGIR Conference on Research and Development in Information Retrieval, pp. 1941–1944 (2020)
32. Zhong, W., Lin, S.C., Yang, J.H., Lin, J.: One blade for one purpose: advancing math information retrieval using hybrid search. In: Proceedings of the 46th International ACM SIGIR Conference on Research and Development in Information Retrieval, pp. 141–151 (2023)
33. Zhong, W., Yang, J.H., Lin, J.: Evaluating token-level and passage-level dense retrieval models for math information retrieval. arXiv preprint arXiv:2203.11163 (2022)

Beyond Topicality: Including Multidimensional Relevance in Cross-encoder Re-ranking
The Health Misinformation Case Study

Rishabh Upadhyay[1], Arian Askari[2], Gabriella Pasi[1], and Marco Viviani[1](✉)

[1] Università degli Studi di Milano-Bicocca, Department of Informatics, Systems, and Communication, Viale Sarca 336, 20126 Milan, Italy
{rishabh.upadhyay,gabriella.pasi,marco.viviani}@unimib.it
[2] Universiteit Leiden, Leiden Institute of Advanced Computer Science, Snellius Gebouw, Niels Bohrweg 1, 2333 CA Leiden, The Netherlands
a.askari@liacs.leidenuniv.nl

Abstract. In this paper, we propose a novel approach to consider multiple dimensions of relevance in cross-encoder re-ranking. On the one hand, cross-encoders constitute an effective solution for re-ranking when considering a single relevance dimension such as topicality, but are not designed to straightforwardly account for additional relevance dimensions. On the other hand, the majority of re-ranking models accounting for multdimensional relevance are often based on the aggregation of multiple relevance scores at the re-ranking stage, leading to potential compensatory effects. To address these issues, in the proposed solution we enhance the candidate documents retrieved by a first-stage lexical retrieval model with suitable relevance statements related to distinct relevance dimensions, and then perform a re-ranking on them with cross-encoders. In this work we focus, in particular, on an extra dimension of relevance beyond topicality, namely, credibility, to address health misinformation in the Consumer Health Search task. Experimental evaluations are performed by considering publicly available datasets; our results show that the proposed approach statistically outperforms state-of-the-art aggregation-based and cross-encoder re-rankers.

Keywords: Cross-encoders · Re-ranking · Health misinformation · Consumer Health Search

1 Introduction

In recent years, there has been an increasing interest in addressing the problem of implementing effective retrieval models that consider different dimensions of relevance across various domains and tasks in the field of Information Retrieval [11,18,27,51]. Today, document ranking is often achieved by performing a *first-stage retrieval*, usually focused on topical relevance, to efficiently identify a subset

N. Goharian et al. (Eds.): ECIR 2024, LNCS 14608, pp. 262–277, 2024.
https://doi.org/10.1007/978-3-031-56027-9_16

of relevant documents from the entire collection; on this subset, a *re-ranking* stage is performed, where topicality and/or additional dimensions of relevance may be considered [21,42,57,58].

BM25 [47] is often used as a first-stage retrieval model because of its effectiveness and efficiency. Concerning re-ranking models, some are based on the *aggregation* of the topicality score obtained from the first-stage retrieval model and other relevance scores related to additional relevance dimensions [16,17]. In doing so, simple and compensatory aggregation techniques are used in most cases [23]. Other re-rankers exploit the potential of *cross-encoders* [20,40], but consider one-dimensional relevance. A recent study, in particular, has improved the effectiveness of such kind of re-ranker by injecting the BM25 score obtained in the first-stage retrieval as an input token to the cross-encoder re-ranker [5]. However, despite the effectiveness of this type of approach, the limitation of considering only topical relevance remains.

Therefore, in this paper, we aim to explore the impact of incorporating other dimensions of relevance into a cross-encoder for document re-ranking. In particular, instead of manipulating the input sequence of the cross-encoder with an additional *relevance score* for an additional relevance dimension, we integrate a so-called *relevance statement* into the document. This statement is constituted by a text related to the relevance dimension under consideration and its associated relevance score. This "enhanced" document is provided, along with the query, as input of a cross-encoder to obtain the final relevance score. To illustrate and assess the proposed model we consider, as a case study, the *health misinformation* problem in the *Consumer Health Search* (CHS) task [29]. Hence, as an additional dimension of relevance beyond topicality, we take into account information *credibility* [52]. Hence, the contributions of our work are the following:

- We propose the $CE_{rel.stat}$ model, whose purpose is to exploit the potential of cross-encoders in the re-ranking phase by exploiting the multidimensional nature of relevance via document enhancement; specifically, health-related documents are enhanced with credibility-related relevance statements;
- We analyze the effectiveness of the proposed model compared to state-of-the-art baselines – i.e., multidimensional aggregation-based approaches and cross-encoder re-rankers – by conducting extensive experiments on the TREC-2020 Health Misinformation and CLEF-2020 eHealth datasets [11,29];
- We provide a qualitative analysis of the model for explainability, showing the potential impact of including relevance statements into documents also to this aim.

2 Related Work

Prevailing approaches that consider multiple relevance dimensions in re-ranking are based on the *aggregation* of the topicality score with other relevance dimension scores [14,19,38,44]. In particular, when considering the health misinformation issue, the TREC Health Misinformation tracks spanning 2019 to 2022 [2,11–13] and CLEF eHealth track in 2017 [26], 2020 [28], and 2021 [27] have been established to assess the robustness of IR systems against this issue. In the context of such initiatives, several models have been proposed, considering multiple relevance dimensions such as *credibility*, *correctness*, and *understandability* [1,7,22,42,43,49,50,56]. While these methods employ varied techniques to calculate the relevance scores, a common thread among many is their reliance on simple *linear* or *non-linear aggregation* [1,7,22,43,50,58]. Others [1,21,30,35,49,56] leverage *rank fusion* methods, mainly based on *Reciprocal Rank Fusion* [15], *CombSUM* [24], and *Borda count* [6]. An intrinsic challenge with these methods lies in their compensatory effect, which poses difficulties in elucidating the interactions between various relevance dimensions.

Approaches that have proven effective for re-ranking are today based on the use of *cross-encoders*, but have so far been used with respect to a single dimension of relevance, namely topicality [32,46,59]. In these approaches, two so-called *sequences* – i.e., the *query* q and a candidate *document* d – are concatenated and fed into a *Transformer* model like BERT [20]. Thus, Transformer *attention heads* can directly model which elements of one sequence are correlated with elements of the other, allowing a (topical) relevance score σ to be calculated. Formally:

$$\sigma(q, d) = \mathrm{CE}([\mathrm{CLS}]\ q\ [\mathrm{SEP}]\ d\ [\mathrm{SEP}]) \cdot W \tag{1}$$

where CE is the cross-encoder, CLS and SEP are special tokens to represent the classifier token and the separator token, and W is a learned matrix that represents the relationship between the query and document representations. In this context, the recent CE_{BM25CAT} model [5] has been proposed to improve the effectiveness of BERT-based re-rankers by injecting the topicality score obtained by a first-stage BM25 model as a token (BM25) into the input of the cross-encoder. Formally:

$$\sigma(q, d) = \mathrm{CE}([\mathrm{CLS}]\ q\ [\mathrm{SEP}]\ \mathrm{BM25}\ [\mathrm{SEP}]\ d\ [\mathrm{SEP}]) \cdot W \tag{2}$$

However, this approach does not account for multidimensional relevance. Other approaches that modify the input of cross-encoders to more effective retrieval, are those described in [8,9], which are based on highlighting exact matching signals by marking the start and the end of each occurrence of the query terms by adding markers to the input. In [3], the authors experimented with the inclusion of various supervised signals into the input of the cross-encoder to emphasize target words in context, while in [34] the authors injected boundary markers between contiguous words for Chinese named entity identification. Recently, the model proposed in [39] for Question-Answering modified the cross-encoder input by integrating it with the primary query, answer passages, and neighboring context passages. However, also these studies do not account for multidimensional relevance. In this paper, we fill this research gap by considering multiple dimensions of relevance in cross-encoder re-ranking – specifically the topicality and information credibility dimensions – in the manner that is explained in detail in the next section.

3 The $CE_{rel.stat}$ Model for Re-ranking

The cross-encoder-based model proposed in this paper to perform re-ranking, namely $CE_{rel.stat}$, is based on performing four steps, i.e., (*i*) an initial *retrieval* phase by using BM25, (*ii*) the *computation* of a *relevance score* for an relevance dimension, (*iii*) the *enhancement* of the retrieved documents with a text related to the additional relevance dimension in the form of a *relevance statement*, and (*iv*) the actual *re-ranking* that occurs by feeding the cross-encoder with the *query* and the related *enhanced documents*. Figure 1 illustrates these steps in the context of the CHS task, considering *credibility* as an additional dimension of relevance.

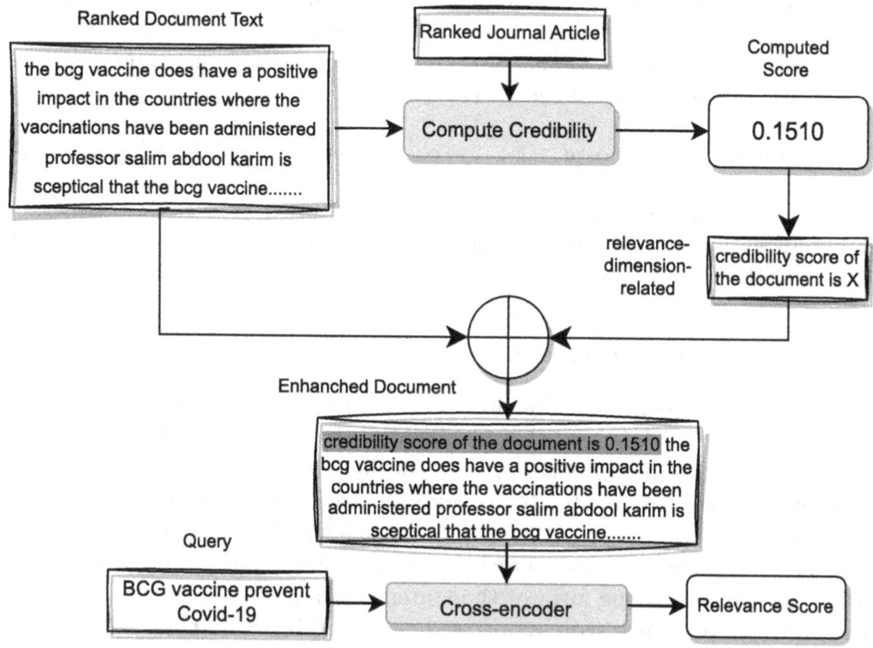

Fig. 1. The $CE_{rel.stat}$ re-ranker.

The steps shown in Fig. 1 are detailed below, also instantiated with respect to CHS and credibility.

3.1 First-Stage Retrieval: BM25

The first-stage retrieval in our framework is based on the BM25 model [48]. BM25 calculates a *topicality score* based on word frequency and distribution in the query and document, providing a list of the most relevant documents. It is effective and efficient, making it a popular choice as the first-stage ranker in Information Retrieval Systems [4, 25, 31].

3.2 Calculation of the Relevance Score

Depending on which dimension(s) of relevance to consider in addition to topicality, different methods can be taken into account to obtain a relevance score for each dimension, based on the nature of the relevance dimension(s). In our case, to calculate the *credibility score*, we employed a recent state-of-the-art approach [50]. It is an unsupervised solution that overcomes the problem of dealing with labeled datasets and obtains high effectiveness on the CHS datasets that are used in our work for evaluation purposes. This approach involves comparing the content of retrieved documents, given a query, with scientific articles,[1] which are considered reliable sources of evidence for the same query. Both the documents and scientific articles are represented using BioBERT [33]; to calculate the credibility score of each retrieved document d with respect to a query q, denoted as $cred(d, q)$, a linear combination of the cosine similarity scores between d and the top-k scientific articles j_j that were deemed relevant to the same query for which d was retrieved, is performed. Formally:

$$cred(d, q) = \omega_1 \cdot \cos(d, j_1) + \omega_2 \cdot \cos(d, j_2) + \ldots + \omega_k \cdot \cos(d, j_k) \tag{3}$$

where $\omega_1, \omega_2, \ldots, \omega_k$, such that $\sum \omega_i = 1$, and $\omega_i \geq \omega_{i+1}$ $(1 \leq i \leq k-1)$. These weights allow assigning greater emphasis to the similarity scores according to the *rank* of the retrieved articles j_j, as illustrated in detail in [50].

3.3 Document Enhancement

In this phase, we enhance each document retrieved in the first-stage phase with a so-called *relevance statement* that is related to the additional relevance dimension(s) considered. This statement consists of the *relevance score* obtained in the second phase plus a text related to the additional relevance dimension associated with the score. Of course, this textual information should vary depending on the characteristics of the relevance dimension being considered. For example, when considering *credibility*, the form of the statement can be: "*credibility score X*", where X is the relevance score associated with credibility. In this case, being d = "*the bcg vaccine does have a positive …*" the original document, its enhanced version becomes: $\tilde{d}$ = "*STAT the bcg vaccine does have a positive …*", where *STAT* represents the considered statement. The statements actually tested in this work to enhance the documents will be illustrated in the section devoted to experimental evaluations.

3.4 Cross-encoder Re-ranking

With respect to Eq. (1) illustrated in Sect. 2, to go beyond topicality and consider an extra relevance dimension, here we modify the basic input of the cross-encoder

[1] I.e., open-source scientific articles extracted from reputed and trustworthy medical journals such as the *Journal of the American Medical Association* (JAMA) and *eLife*.

to perform re-ranking. Specifically, we replace the original document with the enhanced document representation, i.e., $\tilde{d}$. Formally:

$$\sigma(q, \tilde{d}) = \text{CE}([\text{CLS}]\ q\ [\text{SEP}]\ \tilde{d}\ [\text{SEP}]) \cdot W \tag{4}$$

It is important to underline that we use enhanced documents in both the *fine-tuning* and *inference* phases of the $CE_{rel.stat}$ model. Further details on model training and validation are provided in Sect. 4.1.

4 Experimental Evaluations

For evaluation purposes, we focused on the *ad-hoc retrieval* task within the TREC-2020 Health Misinformation Track [11] and CLEF-2020 eHealth Track [29]. Both tracks relate to *Consumer Health Search* and consider *credibility* an important relevance dimension in addition to topicality. A subset of 1 million documents from each track was used, with the TREC-2020 Track covering 46 topics related to Coronavirus and the CLEF-2020 Track covering 50 medical condition topics. The TREC-2020 Health Misinformation Track provides binary labels for documents, with those meeting the criteria of being "topical and credible" labeled as "1" and the rest labeled as "0". As for the CLEF-2020 eHealth Track, the same binary labeling applies for both topicality and credibility.

4.1 Implementation Details

We employed PyTerrier [37] for indexing and implementing the BM25 model. We created two indexes, one for TREC-2020 and another for CLEF-2020. As the considered document set is health-related, we used BioBERT ($dmis-lab/biobert-v1.1$[2]) [33] along with the base version of the BERT model for cross-encoder re-ranking training and inference. It should be noted that the tokenizers of both BERT and BioBERT were modified to recognize and handle decimal numbers up to four digits. This technical modification ensures the accurate and optimal processing of such fine-grained decimal values. In effect, the tokenizer's capability to recognize and correctly process these decimals was a determinant factor, significantly influencing the model's interpretative processing and effective ranking outcomes.

In the absence of predefined *training* and *test* splits for the TREC-2020 and CLEF-2020 datasets, we adopted a *cross-dataset evaluation* approach [10]. This choice allowed us to assess our model's adaptability across similar yet distinct contexts, a preliminary test of the method's generalizability. We trained the cross-encoder on 80% of the queries from one dataset (e.g. TREC-2020) and used the other query set as the validation set, and selected all queries and documents from the other dataset (CLEF-2020) as the evaluation set and vice versa. We fine-tuned both the BERT and BioBERT models with a batch size of 4 and 512 tokens as the maximum sequence length for 10 epochs using the Adam optimizer with

[2] https://huggingface.co/dmis-lab/biobert-v1.1.

an initial learning rate set to 2×10^{-5}. We employed the Huggingface library [55], the Cross-encoder package of Sentence-transformers library [45], and PyTorch [41] for training and inferencing.

4.2 Baselines

The baseline models considered to evaluate the proposed approach comparatively include:

- BM25: the BM25 retrieval model as implemented by PyTerrier;
- *WAM*: the current state-of-the-art aggregation-based multidimensional relevance model presented in [50], which is based on the *weighted average* of distinct relevance scores. Specifically, weights associated with topicality and credibility are set as in the best model described in [50];
- *CE*: the original cross-encoder model for re-ranking as presented in [20], based on Eq. (1);
- $CE_{BM25CAT}$: the cross-encoder re-ranker defined in [5], where the normalized BM25 score is injected into the input sequence of the cross-encoder, based on Eq. (2);
- $CE_{CredCAT}$: a cross-encoder re-ranker, where a credibility score is injected into the input sequence of the cross-encoder instead of the BM25 score Formally:

$$\sigma(q,d) = \text{CE}([\text{CLS}]\ q\ [\text{SEP}]\ \mathit{cred}\ [\text{SEP}]\ d\ [\text{SEP}]) \cdot W \tag{5}$$

  where *cred* is the credibility score;
- $CE_{BM25CredCAT}$: a cross-encoder re-ranker, where both BM25 and credibility scores are injected into the input sequence of the cross-encoder. Formally:

$$\sigma(q,d) = \text{CE}([\text{CLS}]\ q\ [\text{SEP}]\ \text{BM25}\ [\text{SEP}]\ \mathit{cred}\ [\text{SEP}]\ d\ [\text{SEP}]) \cdot W \tag{6}$$

4.3 Results

This section provides the experimental results of the proposed $CE_{rel.stat}$ model compared with the baselines, using *Normalized Discounted Cumulative Gain* at 10 (NDCG@10), *Precision* at 10 (P@10), *Mean Reciprocal Rank* at 10 (MRR@10) and *Mean Average Precision* (MAP) as evaluation metrics. All results are statistically significant according to a *paired t-test* ($p < 0.05$) with Bonferroni correction for multiple testing [54]. Specifically, before providing the final results, research questions had to be answered to optimize the parameters to be used in the proposed model.

RQ1: What is the best representation of the credibility score for cross-encoder re-ranking? Existing research has highlighted the sensitivity of BERT and other NLP models to the representation of numerical values [53]. To explore this aspect, we evaluated three types of numerical representations for the credibility score: *decimal* (dec.), *integer* (int), and *segmented* (seg.), where each digit of the score

Table 1. Evaluation of the effectiveness of numerical representation for ranking. BioBERT models fine-tuned on TREC 2020 and evaluated on CLEF 2020.

Model	Represent.	CLEF 2020			
		NDCG@10	P@10	MRR@10	MAP
CE (BioBERT)		0.2743	0.2811	0.4801	0.1474
$CE_{rel.score}$ (dec.)	1 Decimal	0.2563	0.2461	0.3622	0.1164
	2 Decimals	0.2612	0.2578	0.3861	0.1194
	3 Decimals	0.2796	0.2703	0.4612	0.1374
	4 Decimals	**0.2867**	**0.2861**	**0.4859**	**0.1457**
$CE_{rel.score}$ (int.)	X100	0.1897	0.1974	0.2453	0.0742
	X1000	0.2013	0.2243	0.3012	0.0985
$CE_{rel.score}$ (seg.)		0.2196	0.2231	0.3564	0.1001

is considered individually. For these experiments, we employed the BioBERT-based *CE* model as a baseline and an ad hoc model, $CE_{rel.score}$, which only enhances documents with *credibility scores* and not with textual statements.

The results, presented in Table 1, illustrate that the decimal representation outperforms both the integer and segmented representations across all evaluation metrics. The optimal performance is observed when using four decimal places, suggesting that higher precision in the input helps the model to better capture the nuance of the credibility dimension. This potentially reduces noise in the input, allowing for more effective re-ranking.

In contrast, the integer and segmented representations underperform compared to the decimal representation. When using an integer representation by multiplying the original scores by 100 or 1000, the model's effectiveness drops substantially. This could be because the integer representation may miss nuances captured by decimal points, thereby introducing noise.

As for the segmented representation, although it seemingly provides more information than the integer representation by breaking down each digit of the score, this approach does not translate into better performance. It appears that the segmented method might dilute the model's focus, making it harder to understand the relative importance of the scores. In summary, our results indicate that the decimal representation yields the best performance.

RQ2: To what extent the effectiveness of the model is affected by the textual information in the statement in addition to the score? In this case, we evaluated document enrichment with respect to the injection of a credibility-related statement, a topicality-related statement, and both. We experimented with the following statements for *credibility*:

c1. *Credibility score is X*
c2. *Credibility score of the document is X*

where X is the credibility score (represented as a 4-decimal floating-point number). We also considered the further addition of *topicality:*

*t*1. *Topicality score is Y*
*t*2. *Topicality score of the document is Y*

where Y is the normalized BM25 score. For the *combination* of topicality and credibility, we considered the following statement:

tc. *c*2 and *t*2

Table 2. Evaluation of the effectiveness of statement representation for ranking. BioBERT models fine-tuned on TREC 2020 and evaluated on CLEF 2020.

Model	Statement	CLEF 2020			
		NDCG@10	P@10	MRR@10	MAP
$CE_{rel.score}$ (dec.)		0.2867	0.2861	0.4859	0.1457
$CE_{rel.stat}$ (cred., dec.)	*c*1	0.3487	0.3412	0.5664	0.1664
	*c*2	**0.3762**	**0.3669**	**0.6187**	**0.1964**
$CE_{rel.stat}$ (top., BM25)	*t*1	0.2765	0.2721	0.4838	0.1154
	*t*2	0.2876	0.2878	0.5002	0.1191
$CE_{rel.stat}$ (cred., dec. – top., BM25)	*tc*	0.3324	0.331	0.5576	0.1745

Results are illustrated in Table 2, where $CE_{rel.score}$ (dec.) denotes the best baseline model from Table 1 (decimal representation, 4 decimals), $CE_{rel.stat}$ (cred., dec.) denotes the here proposed model where documents are enhanced with just a credibility-related statement, $CE_{rel.stat}$ (top., BM25) denotes our model where documents are enhanced with just a topicality-related statement (based on BM25), and $CE_{rel.stat}$ (cred., dec. – top., BM25) denotes our model where both the credibility- and topicality-related statements are considered. First, these results show that, for specific statement formats, adding a relevance statement in the documents (including the relevance score), instead of the score alone, improves the overall results. Second, enhancing documents with just a topicality-related statement does not lead to a significant increase in effectiveness. Third, the use of the *c*2 statement related to credibility alone outperforms all models, also the one in which topicality- and credibility-related statements are considered together. This suggests that the specificity of the statement to the relevance dimension and the document is able to provide valuable information to the model.

RQ3: What is the effectiveness of the $CE_{rel.stat}$ *model compared to state-of-the-art baselines?* At this point, having selected the best numerical representation for the scores (i.e., decimal with four digits) and the best form for the statement (i.e., c2), it is possible to answer this research question, which is related to the

Table 3. Evaluation of the effectiveness of the $CE_{rel.stat}$ model w.r.t. to baselines. Re-ranking over the first 500 retrieved documents. Results are referred to TREC and CLEF 2020.

TREC 2020

Represent.	Model	TREC 2020			
		NDCG@10	P@10	MRR@10	MAP
Lexical	BM25	0.4166	0.4177	0.5107	0.2142
	WAM	0.5065	0.4976	0.5546	0.2453
BERT	$CE_{rel.stat}$	**0.6157**	**0.5977**	**0.7101**	**0.3208**
	$CE_{BM25CredCAT}$	0.5784	0.5671	0.6823	0.2875
	$CE_{CredCAT}$	0.5587	0.5581	0.6622	0.2652
	$CE_{BM25CAT}$	0.5374	0.5398	0.6341	0.2499
	CE	0.5589	0.5501	0.6619	0.2664
BioBERT	$CE_{rel.stat}$	**0.6704**	**0.6622**	**0.7961**	**0.3865**
	$CE_{BM25CredCAT}$	0.6219	0.6245	0.7512	0.3324
	$CE_{CredCAT}$	0.6111	0.6001	0.7061	0.3015
	$CE_{BM25CAT}$	0.5875	0.5812	0.6801	0.2765
	CE	0.6055	0.6059	0.6997	0.2986

CLEF 2020

Represent.	Model	CLEF 2020			
		NDCG@10	P@10	MRR@10	MAP
Lexical	BM25	0.1054	0.1081	0.1578	0.1064
	WAM	0.0865	0.1002	0.1232	0.1102
BERT	$CE_{rel.stat}$	**0.3327**	**0.3401**	**0.5403**	**0.1601**
	$CE_{BM25CredCAT}$	0.3098	0.3141	0.5173	0.1356
	$CE_{CredCAT}$	0.2633	0.2703	0.4543	0.1198
	$CE_{BM25CAT}$	0.2288	0.2301	0.4147	0.0964
	CE	0.2579	0.2601	0.4456	0.1165
BioBERT	$CE_{rel.stat}$	**0.3762**	**0.3669**	**0.6187**	**0.1964**
	$CE_{BM25CredCAT}$	0.3221	0.3221	0.5731	0.1642
	$CE_{CredCAT}$	0.2805	0.2824	0.4812	0.1437
	$CE_{BM25CAT}$	0.2414	0.2522	0.4702	0.1274
	CE	0.2743	0.2811	0.4801	0.1474

final evaluation of the proposed model. The results illustrated in Table 3 show that the proposed model outperforms other baselines across all four evaluation metrics on both datasets, when leveraging both the BERT and BioBERT textual representations. This result, in particular, suggests how enhancing the document with statements related to the considered relevance dimension is more effective than methods that modify the structure of the cross-encoder input by relevance score injection alone. This confirms, compared with literature baselines, what was also revealed by the results shown in Table 2 with respect to the $CE_{rel.score}$ model.

RQ4: Is it possible to explain the contribution that document enhancement makes with respect to ranking effectiveness? To provide a reply to this last research question, we computed SHAP (*Shapley Additive Explanations*) [36] values. The SHAP method allows us to estimate the importance of each input token, enabling us to understand the model's decision-making process. By applying SHAP to the $CE_{rel.stat}$ model, we were able to gain insights into the contribution of the relevance statement and its impact on the model's effectiveness as shown in Fig. 2. The darker (red) highlighted parts, corresponding to higher SHAP values, are those that contribute most to relevance, both with respect to topicality (the query is represented by the text before [SEP]) and with respect to credibility. We can also see, by analyzing the "Model Rank" column, how our model causes a document deemed irrelevant to fall in the ranking and how it instead places a relevant document at the top.

Doc Text	Rel.	Model Rank
can 5g antennas cause covid 19 [SEP] credibility score of the document is 0.3023 the red dots represent 5g towers while the green dots represent the spread of covid 19 now people are theorizing that 5g cell phone towers are the leading cause of the spread of coronavirus the first is that 5g is the actual cause of covid 19	0	$CE_{rel.stat}$ - 28 $CE_{BM25credCAT}$ - 13 $CE_{CredCAT}$ - 10 $CE_{BM25CAT}$ - 5 CE - 7 WAM - 3 BM25- 1
can 5g antennas cause covid 19 [SEP] credibility score of the document is 0.5009 during an on screen slot about the farcical notion that 5g causes covid 19 holmes said progressive society has been since the days of ancient greece but the argument about disease carrying 5g equipment is so dumb it makes	1	$CE_{rel.stat}$ - 1 $CE_{BM25credCAT}$ - 15 $CE_{CredCAT}$ - 23 $CE_{BM25CAT}$ - 36 CE - 31 WAM - 40 BM25- 45

Fig. 2. Example query and two passages in the input of $CE_{rel.stat}$ highlighted by Shapley values.

5 Conclusion and Further Research

In this paper, we introduced the $CE_{rel.stat}$ model, which integrates a relevance score and associated textual information in the form of a relevance statement to documents in the re-ranking step performed through cross-encoders. This model was implemented and evaluated against the health misinformation problem within the Consumer Health Search task. Experimental results over publicly available datasets (TREC and CLEF 2020) show that the proposed model statistically significantly outperforms several baselines in the considered domain, either based on the aggregation of multiple relevance scores or cross-encoders that only take topicality into account. The SHAP method used to provide explainability to the results highlighted the contribution of adding relevance statements to documents to increase re-ranking effectiveness.

Possible future work involves: investigating the impact of repositioning the relevance statements in the document to discern any positional effects on re-ranking; evaluating the effectiveness of the $CE_{rel.stat}$ model across diverse domains and considering the integration of other relevance dimensions such as correctness and readability; exploring the application of active learning and other semi-supervised learning techniques to further augment the model's efficiency and effectiveness.

Acknowledgements. This work was supported by the EU Horizon 2020 ITN/ETN on Domain Specific Systems for Information Extraction and Retrieval (DoSSIER), H2020-EU.1.3.1., ID: 860721, https://dossier-project.eu/.

Data and Code Availability. The data used in this article are publicly accessible at: https://github.com/ikr3-lab/TREC-CLEF-HealthMisinfoSubdatasets. The code for the implementation and evaluation of the proposed model is publicly accessible at: https://github.com/ikr3-lab/Multidimensional-Cross-Encoder-Reranking.

References

1. Abualsaud, M., et al.: Uwaterloomds at the TREC 2021 health misinformation track. In: Proceedings of the Thirtieth Retrieval Conference Proceedings (TREC 2021), pp. 1–18. National Institute of Standards and Technology (NIST), Special Publication (2021)
2. Abualsaud, M., Lioma, C., Maistro, M., Smucker, M.D., Guido, Z.: Overview of the TREC 2019 decision track (2020). https://api.semanticscholar.org/CorpusID:221857114
3. Al-Hajj, M., Jarrar, M.: Arabglossbert: fine-tuning bert on context-gloss pairs for WSD. arXiv preprint arXiv:2205.09685 (2022)
4. Anand, M., Zhang, J., Ding, S., Xin, J., Lin, J.: Serverless bm25 search and bert reranking. In: DESIRES, pp. 3–9 (2021)
5. Askari, A., Abolghasemi, A., Pasi, G., Kraaij, W., Verberne, S.: Injecting the bm25 score as text improves bert-based re-rankers. arXiv preprint arXiv:2301.09728 (2023)
6. Aslam, J.A., Montague, M.: Models for metasearch. In: Proceedings of the 24th Annual International ACM SIGIR Conference on Research and Development in Information Retrieval, pp. 276–284 (2001)
7. Bondarenko, A., et al.: Webis at TREC 2021: deep learning, health misinformation, and podcasts tracks. In: The Thirtieth Retrieval Conference Proceedings (TREC 2021), pp. 500, 335 (2021)
8. Boualili, L., Moreno, J.G., Boughanem, M.: Markedbert: integrating traditional IR cues in pre-trained language models for passage retrieval. In: Proceedings of the 43rd International ACM SIGIR Conference on Research and Development in Information Retrieval (SIGIR 2020), pp. 1977–1980. Association for Computing Machinery, New York (2020). https://doi.org/10.1145/3397271.3401194

9. Boualili, L., Moreno, J.G., Boughanem, M.: Highlighting exact matching via marking strategies for ad hoc document ranking with pretrained contextualized language models. Inf. Retriev. J. **25**(4), 414–460 (2022). https://doi.org/10.1007/s10791-022-09414-x
10. Chen, Y., et al.: Cdevalsumm: an empirical study of cross-dataset evaluation for neural summarization systems. arXiv preprint arXiv:2010.05139 (2020)
11. Clarke, C.L.A., Maistro, M., Rizvi, S., Smucker, M.D., Zuccon, G.: Overview of the TREC 2020 health misinformation track (2020). https://trec.nist.gov/pubs/trec29/papers/OVERVIEW.HM.pdf
12. Clarke, C.L.A., Maistro, M., Seifikar, M., Smucker, M.D.: Overview of the TREC 2022 health misinformation track. In: 30th Retrieval Conference, TREC 2021, vol. 500, 338, pp. 15–19. Gaithersburg, Maryland (2021)
13. Clarke, C.L.A., Rizvi, S., Smucker, M.D., Maistro, M., Zuccon, G.: Overview of the TREC 2021 health misinformation track. In: Text Retrieval Conference (2021). https://api.semanticscholar.org/CorpusID:235600234
14. Cong, G., Jensen, C.S., Wu, D.: Efficient retrieval of the top-k most relevant spatial web objects. Proc. VLDB Endowm. **2**(1), 337–348 (2009)
15. Cormack, G.V., Clarke, C.L., Buettcher, S.: Reciprocal rank fusion outperforms condorcet and individual rank learning methods. In: Proceedings of the 32nd International ACM SIGIR Conference on Research and Development in Information Retrieval, pp. 758–759 (2009)
16. da Costa Pereira, C., Dragoni, M., Pasi, G.: Multidimensional relevance: a new aggregation criterion. In: Boughanem, M., Berrut, C., Mothe, J., Soule-Dupuy, C. (eds.) Advances in Information Retrieval. ECIR 2009. LNCS, vol. 5478, pp. 264–275. Springer, Heidelberg (2009). https://doi.org/10.1007/978-3-642-00958-7_25
17. da Costa Pereira, C., Dragoni, M., Pasi, G.: A prioritized "and" aggregation operator for multidimensional relevance assessment. In: Serra, R., Cucchiara, R. (eds.) AI*IA 2009: Emergent Perspectives in Artificial Intelligence. AI*IA 2009. LNCS, vol. 5883, pp. 72–81. Springer, Heidelberg (2009). https://doi.org/10.1007/978-3-642-10291-2_8
18. da Costa Pereira, C., Dragoni, M., Pasi, G.: Multidimensional relevance: prioritized aggregation in a personalized information retrieval setting. Inf. Process. Manag. **48**(2), 340–357 (2012). https://doi.org/10.1016/j.ipm.2011.07.001
19. Daoud, M., Tamine, L., Boughanem, M.: A personalized graph-based document ranking model using a semantic user profile. In: De Bra, P., Kobsa, A., Chin, D. (eds.) User Modeling, Adaptation, and Personalization. UMAP 2010. LNCS, vol. 6075, pp. 171–182. Springer, Heidelberg (2010). https://doi.org/10.1007/978-3-642-13470-8_17
20. Devlin, J., Chang, M.W., Lee, K., Toutanova, K.: BERT: pre-training of deep bidirectional transformers for language understanding. In: Proceedings of the 2019 Conference of the North American Chapter of the Association for Computational Linguistics: Human Language Technologies, Volume 1 (Long and Short Papers), pp. 4171–4186. Association for Computational Linguistics, Minneapolis (2019). https://doi.org/10.18653/v1/N19-1423
21. Fernández-Pichel, M., Losada, D.E., Pichel, J.C.: A multistage retrieval system for health-related misinformation detection. Eng. Appl. Artif. Intell. **115**, 105211 (2022), https://api.semanticscholar.org/CorpusID:250932569
22. Fernández-Pichel, M., Losada, D.E., Pichel, J.C., Elsweiler, D.: Citius at the trec 2020 health misinformation track. In: TREC (2020)
23. Fox, E.A.: Combination of multiple searches. In: Proceedings of the Second Text Retrieval Conference, August/September 1993 (1993)

24. Fox, E.A., Koushik, M.P., Shaw, J., Modlin, R., Rao, D., et al.: Combining evidence from multiple searches. In: The First Text Retrieval Conference (TREC-1), pp. 319–328 (1993)
25. Gao, L., Dai, Z., Chen, T., Fan, Z., Van Durme, B., Callan, J.:. Complement lexical retrieval model with semantic residual embeddings. In: Hiemstra, D., Moens, MF., Mothe, J., Perego, R., Potthast, M., Sebastiani, F. (eds) Advances in Information Retrieval. ECIR 2021. LNCS, vol. 12656, pp. 146–160. Springer, Cham (2021). https://doi.org/10.1007/978-3-030-72113-8_10
26. Goeuriot, L., et al.: Clef 2017 ehealth evaluation lab overview. In: Conference and Labs of the Evaluation Forum (2017). https://api.semanticscholar.org/CorpusID:206705118
27. Goeuriot, L., et al.: CLEF eHealth evaluation lab 2021. In: Hiemstra, D., Moens, M.-F., Mothe, J., Perego, R., Potthast, M., Sebastiani, F. (eds.) Advances in Information Retrieval, ECIR 2021. LNCS, vol. 12657, pp. 593–600. Springer, Cham (2021). https://doi.org/10.1007/978-3-030-72240-1_69
28. Goeuriot, L., et al.: Overview of the clef ehealth 2020 task 2: consumer health search with ad hoc and spoken queries. In: Conference and Labs of the Evaluation Forum (2020). https://api.semanticscholar.org/CorpusID:225073918
29. Goeuriot, L., et al.: Overview of the clef ehealth evaluation lab 2020. In: Arampatzis, A., et al. (eds.) Experimental IR Meets Multilinguality, Multimodality, and Interaction. LNCS, vol. 12260, pp. 255–271. Springer, Cham (2020)
30. Huang, Y., Xu, Q., Wu, S., Nugent, C., Moore, A.: Fight against covid-19 misinformation via clustering-based subset selection fusion methods. In: ROMCIR 2022 CEUR Workshop Proceedings, vol. 3138, pp. 11–26 (2022)
31. Kamphuis, C., de Vries, A.P., Boytsov, L., Lin, J.: Which bm25 do you mean? a large-scale reproducibility study of scoring variants. In: Jose, J.M., et al. (eds.) Advances in Information Retrieval. LNCS, vol. 12036, pp. 28–34. Springer, Cham (2020). https://doi.org/10.1007/978-3-030-45442-5_4
32. Khattab, O., Zaharia, M.: Colbert: efficient and effective passage search via contextualized late interaction over bert. In: Proceedings of the 43rd International ACM SIGIR Conference on Research and Development in Information Retrieval, pp. 39–48 (2020)
33. Lee, J., et al.: Biobert: a pre-trained biomedical language representation model for biomedical text mining. Bioinformatics **36**(4), 1234–1240 (2020)
34. Li, L., et al.: Markbert: marking word boundaries improves chinese bert. arXiv preprint arXiv:2203.06378 (2022)
35. Lima, L.C., Wright, D.B., Augenstein, I., Maistro, M.: University of copenhagen participation in TREC health misinformation track 2020. arXiv preprint arXiv:2103.02462 (2021)
36. Lundberg, S.M., Lee, S.I.: A unified approach to interpreting model predictions. Adv. Neural Inf. Process. Syst. **30** (2017)
37. Macdonald, C., Tonellotto, N., MacAvaney, S., Ounis, I.: Pyterrier: declarative experimentation in python from bm25 to dense retrieval. In: Proceedings of the 30th ACM International Conference on Information Knowledge Management (CIKM 2021), pp. 4526–4533. Association for Computing Machinery, New York (2021). https://doi.org/10.1145/3459637.3482013
38. Moulahi, B., Tamine, L., Yahia, S.B.: i a ggregator: multidimensional relevance aggregation based on a fuzzy operator. J. Am. Soc. Inf. Sci. **65**(10), 2062–2083 (2014)

39. Nguyen, M., Kishan, K., Nguyen, T., Chadha, A., Vu, T.: Efficient fine-tuning large language models for knowledge-aware response planning. In: Koutra, D., Plant, C., Gomez Rodriguez, M., Baralis, E., Bonchi, F. (eds.) Joint European Conference on Machine Learning and Knowledge Discovery in Databases. LNCS, vol. 14170, pp. 593–611. Springer, Cham (2023). https://doi.org/10.1007/978-3-031-43415-0_35
40. Nogueira, R., Cho, K.: Passage re-ranking with bert. arXiv preprint arXiv:1901.04085 (2019)
41. Paszke, A., et al.: PyTorch: An Imperative Style, High-Performance Deep Learning Library, p. 12. Curran Associates Inc., Red Hook (2019)
42. Pradeep, R., Ma, X., Nogueira, R., Lin, J.J., Cheriton, D.R.: Vera: prediction techniques for reducing harmful misinformation in consumer health search. In: Proceedings of the 44th International ACM SIGIR Conference on Research and Development in Information Retrieval (2021). https://api.semanticscholar.org/CorpusID:235477259
43. Pradeep, R., et al.: H2oloo at TREC 2020: when all you got is a hammer... deep learning, health misinformation, and precision medicine. Corpus **5**(d3), d2 (2020)
44. Putri, D.G.P., Viviani, M., Pasi, G.: Social search and task-related relevance dimensions in microblogging sites. In: Aref, S., et al. (eds.) Social Informatics (SocInfo 2020). LNCS, vol. 12467, pp. 297–311. Springer, Cham (2020). https://doi.org/10.1007/978-3-030-60975-7_22
45. Reimers, N., Gurevych, I.: Sentence-bert: sentence embeddings using siamese bert-networks. In: Proceedings of the 2019 Conference on Empirical Methods in Natural Language Processing. Association for Computational Linguistics (2019). https://arxiv.org/abs/1908.10084
46. Ren, R., et al.: Rocketqav2: a joint training method for dense passage retrieval and passage re-ranking. arXiv preprint arXiv:2110.07367 (2021)
47. Robertson, S.E., Walker, S.: Some simple effective approximations to the 2-Poisson model for probabilistic weighted retrieval. In: Croft, B.W., van Rijsbergen, C.J. (eds.) SIGIR 1994, pp. 232–241. Springer, Heidelberg (1994). https://doi.org/10.1007/978-1-4471-2099-5_24
48. Robertson, S., Zaragoza, H., et al.: The probabilistic relevance framework: Bm25 and beyond. Found. Trends® Inf. Retriev. **3**(4), 333–389 (2009)
49. Schlicht, I.B., de Paula, A.F.M., Rosso, P.: UPV at TREC health misinformation track 2021 ranking with SBERT and quality estimators. arXiv preprint arXiv:2112.06080 (2021)
50. Upadhyay, R., Pasi, G., Viviani, M. (2022). An unsupervised approach to genuine health information retrieval based on scientific evidence. In: Chbeir, R., Huang, H., Silvestri, F., Manolopoulos, Y., Zhang, Y. (eds.) Web Information Systems Engineering (WISE 2022). LNCS, vol. 13724, pp. 119–135. Springer, Cham (2022). https://doi.org/10.1007/978-3-031-20891-1_10
51. Van Opijnen, M., Santos, C.: On the concept of relevance in legal information retrieval. Artif. Intell. Law **25**, 65–87 (2017)
52. Viviani, M., Pasi, G.: Credibility in social media: opinions, news, and health information-a survey. Wiley Interdiscip. Rev.: Data Mining Knowl. Discov. **7**(5), e1209 (2017)
53. Wallace, E., Wang, Y., Li, S., Singh, S., Gardner, M.: Do NLP models know numbers? probing numeracy in embeddings. arXiv preprint arXiv:1909.07940 (2019)
54. Weisstein, E.W.: Bonferroni correction (2004). https://mathworld.wolfram.com/
55. Wolf, T., et al.: Huggingface's transformers: state-of-the-art natural language processing. arXiv preprint arXiv:1910.03771 (2019)

56. Zhang, B., Naderi, N., Jaume-Santero, F., Teodoro, D.: Ds4dh at TREC health misinformation 2021: multi-dimensional ranking models with transfer learning and rank fusion. arXiv preprint arXiv:2202.06771 (2022)
57. Zhang, B., Naderi, N., Mishra, R., Teodoro, D.: Improving online health search via multi-dimensional information quality models based on deep learning. medRxiv, pp. 2023–04 (2023)
58. Zhang, D., Vakili Tahami, A., Abualsaud, M., Smucker, M.D.: Learning trustworthy web sources to derive correct answers and reduce health misinformation in search. In: Proceedings of the 45th International ACM SIGIR Conference on Research and Development in Information Retrieval, pp. 2099–2104 (2022)
59. Zhuang, S., Zuccon, G.: Tilde: term independent likelihood model for passage re-ranking. In: Proceedings of the 44th International ACM SIGIR Conference on Research and Development in Information Retrieval, pp. 1483–1492 (2021)

Query Obfuscation for Information Retrieval Through Differential Privacy

Guglielmo Faggioli(✉) and Nicola Ferro

University of Padua, Padua, Italy
guglielmo.faggioli@unipd.it

Abstract. Protecting the privacy of a user querying an Information Retrieval (IR) system is of utmost importance. The problem is exacerbated when the IR system is not cooperative in satisfying the user's privacy requirements. To address this, obfuscation techniques split the user's sensitive query into multiple non-sensitive ones that can be safely transmitted to the IR system. To generate such queries, current approaches rely on lexical databases, such as WordNet, or heuristics of word co-occurrences. At the same time, advances in Natural Language Processing (NLP) have shown the power of Differential Privacy (DP) in releasing privacy-preserving text for completely different purposes, such as spam detection and sentiment analysis. We investigate for the first time whether DP mechanisms, originally designed for specific NLP tasks, can effectively be used in IR to obfuscate queries. We also assess their performance compared to state-of-the-art techniques in IR. Our empirical evaluation shows that the Vickrey DP mechanism based on the Mahalanobis norm with privacy budget $\epsilon \in [10, 12.5]$ achieves state-of-the-art privacy protection and improved effectiveness. Furthermore, differently from previous approaches that are substantially on/off, by changing the privacy budget ϵ, DP allows users to adjust their desired level of privacy protection, offering a trade-off between effectiveness and privacy.

1 Introduction

Information Retrieval (IR) systems are a commodity used for many tasks, including searching for personal information, such as symptoms and diseases [7], political opinions, or egosurfing, i.e., searching the own name or social profile, among others. Such searches can be used to profile the user and can put at risk their privacy [6]. For example, an insurance company might try to access the user's queries to determine if they have any disease, or a malicious employee of a search engine might access the query log to blackmail them. To alleviate this, proxy obfuscation approaches hide the sensitive information need, by breaking it down into multiple non-sensitive queries that are less exposing and can be safely transmitted to the IR system. To this end, some approaches rely on replacing words with generalizations, i.e., hypernyms [2,4]. Other strategies use a local corpus to determine which words, by co-occurring in the documents with those in the query, induce the same ranked list [3,5,19]. We investigate for the first time whether Differential Privacy (DP) mechanisms, originally designed for specific Natural Language Processing

N. Goharian et al. (Eds.): ECIR 2024, LNCS 14608, pp. 278–294, 2024.
https://doi.org/10.1007/978-3-031-56027-9_17

(NLP) tasks, can effectively be used in IR to obfuscate queries. DP [15] is a state-of-the-art framework meant to release privately sensitive information. The general idea is to use a randomized mechanism that introduces noise into the computation. Thanks to this, the user can "plausibly deny" the output: it is impossible to prove that the output corresponds to the input of the user and is not due to the randomness of the mechanism. DP is particularly effective in the NLP domain. A line of research [18,37,45,46] operationalizes DP to release text by obfuscating each word individually. Such mechanisms work as follows: i) each word in the text is mapped to a non-contextual embedding space; ii) the embeddings are perturbed with noise drawn from a specific distribution; iii) each word is replaced with the word closest to the noisy embedding. A major advantage of DP is that it allows setting the privacy budget based on the needs of the user. This is different from current obfuscation mechanisms in IR, which are either active or not and cannot be tuned based on the user's needs.

In this work, we focus on three of such mechanisms: the Calibrated Multivariate Perturbation (CMP) [18], the Mahalanobis [45] and the Vickrey [46]. These approaches were originally devised and tested for NLP tasks that include text classification and sentiment analysis. When it comes to the NLP scenario, the model can be trained directly on the obfuscated documents and this allows the model itself to learn how to account for the noise within the documents. However, this is not the case in IR: we assume the IR system to not preserve user privacy, and to possibly be malicious. In our use case, users are the ones concerned about their privacy. They don't want to reveal their real information needs and prefer to transmit obfuscated queries to the IR system while still retrieving relevant documents. Thus, in our use case, the IR system cannot be trained on obfuscated queries or documents, nor should be aware that an obfuscation mechanism has been used. Therefore, to operationalize our mechanism, we assume each user to locally obfuscate their query and transmit the obfuscated query, or possibly multiple queries, to the IR system instead of their real query.

Our goal is to determine if the DP mechanisms introduced above can successfully obfuscate users' information needs while still retrieving relevant documents. More in detail, our research questions are as follows:

RQ1. *Privacy Guarantees*: How much the original query leaks within queries obfuscated using DP obfuscation approaches, originally developed for NLP?

RQ2. *Relevant Documents Retrieved*: Is it feasible to exploit such DP mechanisms to retrieve relevant documents?

RQ3. *Comparing DP and non-DP Approaches*: What is the equivalent DP level for the scrambling approaches, which are the current state-of-the-art in IR?

The paper is organized as follows: Sect. 2 reports the main related works while Sect. 3 introduces the approaches to text obfuscation developed for IR and those developed for NLP. Section 4 delineates our experimental methodology and Sect. 5 details our findings. Finally, Sect. 6 draws our conclusion and outlines the future work.

2 Related Work

In this work, we assume the user to be interested in querying a search engine, without disclosing their real interest. Furthermore, the IR system is not cooperative and does not operate toward protecting the privacy of the user. Therefore, the query needs to be obfuscated on the user side and transmitted to the search engine so that the latter cannot understand the real user's interest. Three main approaches to obfuscate the query from the IR system exist: i) Approaches based on dummy queries; ii) Approaches based on unlinkability; iii) Approaches based on proxy queries. Approaches based on dummy queries [14,16,44,47] transmit to the IR system, along the user's query, a set of unrelated queries, but syntactically similar to the user's query. Approaches based on unlinkability [8,12,13,38] rely on cryptographic and Private Information Retrieval (PIR) primitives to allow a federation of users to exchange the query with each other so that each user submits the query of someone else and the system cannot profile the user. Finally, approaches based on proxy queries [2–5,19] rely on breaking down the query into multiple non-sensitive queries, whose combined results might provide an answer to the user's query. For example, the query "throat cancer" could be transformed into "neck", and "tumour", reducing the information disclosed by each query. The disadvantage of dummy queries and unlinkability approaches is that the original query is sent to the IR system. Therefore, approaches based on dummy queries and unlinkability are vulnerable to machine learning attacks that aim to identify automatically generated queries: as shown by Khan et al. [24,31,32] this is relatively easy, with access to a small number of real user queries. Additionally approaches based on unlinkability move the problem from the IR system to a different user within the federation. This vulnerability does not occur with approaches based on proxy queries. The downside of approaches based on proxy queries is that there is a physiological decrease in effectiveness, which does not occur for dummy queries and unlinkability approaches. Nevertheless, we argue that a user wanting to achieve strong privacy guarantees should be able to do so, even though this might mean renouncing part of the effectiveness in favour of privacy. Therefore we focus on the obfuscation approaches that rely on proxy queries.

The usage of DP in IR involves some efforts to use it to release the local updates in the Federated Learning scenario [25,41] and to release privacy-preserving query logs [20,26,36,48]. We are not aware of any work employing it as the framework to obfuscate queries to be sent to the central server. Therefore, this can be considered the first work using DP for query obfuscation.

3 Approaches

We describe here two approaches designed for IR obfuscation query generation. We also introduce the DP framework. Finally, we present the DP mechanisms, designed for privately releasing text, considered in this work.

3.1 Native IR Approaches

We describe here the two major proxy query obfuscation approaches designed for IR tasks. Arampatzis et al. [2] propose a obfuscation method based on WordNet [29]. For each query term, Arampatzis et al. use WordNet to extract the set of synonyms, hypernyms, and holonyms. The approach considers sets of terms that are two steps away on the WordNet hierarchy. The candidate obfuscation queries are the cartesian product of the term sets. To filter out exposing queries, a similarity measure between each candidate query and the original query is considered. The similarity is the average wup similarity [43] between each term of the obfuscation query with all the terms of the original query. Arampatzis et al. empirically select obfuscated queries within the range (1., 0.7] of wup similarity.

Fröbe et al. [19] extended the earlier work of Arampatzis et al. [3,5] and develop a statistical query obfuscation method. The approach consists of using a local corpus to select and filter candidate obfuscation queries. Using the user's query on the local corpus, Fröbe et al. consider all the possible combinations of n terms from a sliding window of terms within the top-k documents. To enhance privacy, all the candidates containing a query term or a synonym, hypernym, or hyponym, of it are dropped. To decide which candidate queries to submit, the top-k documents retrieved from the local corpus using the original query are considered pseudo-relevant. The candidate queries are ranked according to nDCG achieved in retrieving the pseudo-relevant documents from the local corpus.

3.2 Differential Privacy

Differential Privacy (DP) is considered a state-of-the-art approach for data release. Assume we have a private dataset D and we wish to compute and release some statistics $f(D)$, e.g., documents' scores in response to a query. A "randomized mechanism" $\mathcal{M}_f$ is a function that takes in input a dataset D and outputs the privacy-preserving result of $f(D)$, by introducing some noise. Two datasets D and D' are defined as "neighbouring" if differ by at most one record. A randomized mechanism $\mathcal{M}_f$ satisfies ϵ-DP iff, given a privacy budget $\epsilon \in \mathbb{R}^+$, for any pair of neighbouring datasets D and D':

$$\frac{Pr[\mathcal{M}_f(D) \in \mathcal{S}]}{Pr[\mathcal{M}_f(D') \in \mathcal{S}]} \leq \exp(\epsilon), \forall \mathcal{S} \subseteq \text{Image}(\mathcal{M}_f)$$

The smaller ϵ is, the larger the privacy guarantees, but also the larger the noise introduced in the data. For a DP mechanism, thanks to the introduced noise, the output of the mechanism on the two neighbouring datasets is likely similar.

Metric DP. To achieve DP in a metric space, an obfuscation mechanism should have an equal probability of obfuscating any pair of points as the same point, irrespective of their distance. While this grants the highest level of privacy, it also requires high levels of noise, decreasing the utility of the data. In the case of metric spaces, it is often sufficient if the probability of obfuscating two points with the same one is proportional to the distance between the two points. Or, alternatively, the proportion of sampling a certain noise is inversely proportional to the norm of the noise itself. To this end, a relaxation of DP, called Metric-DP, has been introduced. Metric-DP [1,9,27] is defined as follows: given a privacy budget ϵ and a distance measure $d : \mathbb{R}^p \times \mathbb{R}^p \rightarrow [0, \infty)$, a randomized mechanism $\mathcal{M} : \mathbb{R}^p \rightarrow \mathbb{R}^p$ defined over a geometric space is Metric-DP iff, for any three points in the space $w, w', \hat{w} \in \mathbb{R}^p$, the following holds:

$$\frac{Pr\{\mathcal{M}(w) = \hat{w}\}}{Pr\{\mathcal{M}(w') = \hat{w}\}} \leq \exp(\epsilon d\{w, w'\})$$

If the $d\{w, w'\}$ is small, w and w' are more likely to be obfuscated with the same point. Vice-versa, far apart points might be obfuscated with different points, without violating privacy constraints.

3.3 DP Mechanisms Designed for NLP Tasks

In this work, we evaluate if it is possible to adapt to the IR scenario three DP mechanisms that obfuscate text on a per-word basis, originally developed in the NLP context. More in detail, these approaches take as input a sequence of words. Each word is mapped into a non-contextual embedding, such as GloVe [33]. Then, the embedding is obfuscated by adding some appositely sampled noise to it. To ensure that Metric-DP is achieved, the noise vector z is expected to be sampled from a distribution f such that the probability of observing z is $f(z) \propto \exp(-\epsilon||z||)$, i.e., the probability of sampling a noise with norm $||z||$ is inversely proportional to $||z||$. Finally, the closest word to the noisy embedding is used to obfuscate the corresponding word in the original text. We propose to use these approaches in the IR scenario to perturb the queries instead of the documents, as done for NLP tasks. We chose to use approaches that work on a per-word basis, and not directly on the actual representation vector used by the IR system. Indeed, this would require the user to have access to the encoding procedure of the query (which can be computationally expensive). Moreover, knowing how such representation vectors are computed, would mean that the IR system cooperates to protect the user privacy, which is the opposite of our case. Methods based on encoding the words locally, i.e., Local DP, ensure that the system will not be aware of the privacy mechanism being in place, as it will receive just a query composed of terms, as usual. Finally, being transparent to the IR system, this approach is applicable to any IR system and portable, avoiding the need to develop ad-hoc solutions for a specific system.

Calibrated Multivariate Perturbation (CMP) Mechanism. The obfuscation of a word according to the CMP mechanism, defined by Feyisetan et al. [18], is based on sampling a noise vector following an n-dimensional Laplace distribution. Such sampling works in two phases: i) an n-dimensional unitary vector $p \in \mathbb{R}^n$ is sampled uniformly. This vector represents the direction of the perturbation. ii) the radius of the perturbation $r \in \mathbb{R}^+$ is sampled from a Gamma distribution. To sample p, a vector $N \in \mathbb{R}^n$ is sampled from a multivariate normal distribution, with location 0 and identity covariance matrix $\mathbf{I}_n$: $N \sim \mathcal{N}(\mathbf{0}, \mathbf{I}_n)$. Then $p = N/||N||_2$. The radius of the noise is sampled from a Gamma distribution with shape n and scale $\frac{1}{\epsilon}$ as $r \sim Gam(n, \frac{1}{\epsilon})$. It is possible to observe that, the larger the privacy requirement, i.e., the smaller the ϵ, the bigger the noise. The noise vector z is such that $z = p{\cdot}r$ and, as proven by Fernandes et al. [17], z corresponds to a random vector sampled from a multivariate Laplace distribution with scale $1/\epsilon$, thus $f(z) \propto \exp(-\epsilon||z||_2)$. To perturb a word w, the noise vector z, is added to the original word embedding $\phi(w) \in \mathbb{R}^n$, and the word closest to the noisy word embedding is used as obfuscation. Feyisetan et al. [18] demonstrate that for any word sequence $\mathcal{W}^l$ of length $l \geq 1$ and any $\epsilon > 0$, the mechanism CMP: $\mathcal{W}^l \to \mathcal{W}^l$ satisfies ϵd-privacy with respect to d, where d is the Euclidean distance. The CMP mechanism [18] to obfuscate a single word is presented in Algorithm 1.

Algorithm 1: The CMP mechanism [18]

Data: a string w, privacy parameter ϵ
$v = \phi(w)$;
sample z such that $f(z) \propto \exp(-\epsilon||z||_2)$ as described above;
$\hat{v} = v + z$;
$w = \text{argmin}\,||\phi(u) - \hat{v}||$;
return w;

Feyisetan et al. [18] originally employed this mechanism to release documents for NLP use cases. In the IR scenario, we employ it to perturb the query by applying Algorithm 1 to each query term.

Mahalanobis Mechanism. Xu et al. [45] noticed how the perturbation induced by CMP mechanism tends to be weak, especially for high ϵ. In particular, they consider this to be caused by how the direction of the noise is chosen. Their hypothesis is that sampling the direction of the perturbation on a circumference ($||p||_2 = 1$) increases the risk of sampling a point on an empty region. If this occurs, the embedding for the original word remains the closest to the noisy vector and the word is obfuscated with itself. Therefore, Xu et al. adapt the CMP mechanism to increase the likelihood that the sampled noise will be toward the direction where most of the embeddings are. Practically, this corresponds to transforming the direction of the noise from a circumference to an ellipsis whose orientation can be set to be towards the other embeddings. To do so, it is necessary to modify the sampling mechanism, so that, instead of sampling p such that $||p||_2 = 1$, p is sampled so that $||p||_M = 1$ where $||\cdot||_M$ is the Mahalanobis norm [28]. The Mahalanobis norm $||\cdot||_M$ is defined as follows: for a positive definite matrix Σ, the Mahalanobis norm of a vector $x \in \mathbb{R}^n$ is $||x||_M = \sqrt{x^T\Sigma^{-1}x}$. By properly setting the matrix Σ and normalizing x by its Mahalanobis norm, we can change the orientation and eccentricity of the ellipse around x. Xu et al. [45] define $\Sigma \in \mathbb{R}^{n\times n}$ as the covariance matrix of all the word embeddings, divided by the mean of the variances of each embedding so that the trace of Σ is equal to n. Using Σ defined as above ensures that the noise is stretched toward the direction that corresponds to the largest variability in the embedding space: where it is more likely to find other embeddings. Notice that, the procedure provides a single matrix Σ for all the word embeddings, regardless of the words we are trying to obfuscate. To ensure that the noise z is sampled such that its probability distribution is $f(z) \propto \exp(-e||z||_M)$ a vector N is sampled from the multivariate normal distribution $N \sim \mathcal{N}(\mathbf{0}, I_n)$. Then, p is such that $p = \Sigma^{1/2} \cdot (N/||N||_2)$. The sampling of the norm of the noise r remains the same as for CMP. The Mahalanobis Mechanism, which we refer to as Mhl, obfuscates every single word as in Algorithm 1, with the only difference that, in step 2, z is sampled so that $f(z) \propto \exp(-e||z||_M)$. To obfuscate a query composed of multiple words, Mhl is applied independently to each word. Xu et al. [45] demonstrate that, for any $\epsilon > 0$ and for any sequence of words $\mathcal{W}^l$ of length l, Mhl satisfies ϵd-privacy with respect to the Mahalanobis distance.

Vickrey Mechanism. The Mhl still tends to obfuscate a word with itself for large ϵ. To reduce the probability of masking a token with itself, Xu et al. [46] defines the Vickrey [39] DP mechanism (we refer to it as Vkr). The Vickrey mechanism draws upon the Vickrey auction, a type of auction in which the highest bidder wins but the price paid is the second-highest bid. Vkr is based on two steps: in the first step, a noisy vector is sampled using any of the mechanisms described above – Xu et al. [46] illustrate their approach using Mhl as the instantiating mechanism (we indicate it with Vkr_{Mhl}), but the results hold also for CMP (Vkr_{CMP}). In the second step, with probability Pr the word corresponding to the closest embedding to the noisy vector is used as the obfuscation word. Vice versa, with probability $1 - Pr$ the word corresponding to the second closest embedding is used as obfuscation. The probability Pr is defined as follows. We call $\phi(u_1)$ and $\phi(u_2)$ respectively the closest and second closest word embeddings to $\hat{v}$, the perturbed embedding of w, and t an additional free parameter. $Pr(t, \hat{v}) = \frac{(1-t)||\phi(u_2)-\hat{v}||_2}{t||\phi(u_1)-\hat{v}||_2+(1-t)||\phi(u_2)-\hat{v}||_2}$. If $t = 0$, then the Vkr mechanism falls back to the instantiating mechanism. Intermediate values of t allow selecting either the first or the second nearest neighbour depending on the size of t, but also on the distance of the second closest neighbour to the noisy vector. We set $t = 0.75$, being the most performing according to Xu et al. [46].

4 Evaluation Methodology

We would like to point out that, interpreting the results when it comes to privacy requires weighing the risk of information leakage and the effectiveness. Full privacy protection is achieved only by transmitting white noise to the IR system while to preserve entirely the effectiveness it is necessary to destroy the privacy. A privacy-preserving approach is as good as it is capable of reducing privacy leakage, while still obtaining satisfactory effectiveness, but also according to how easy it is for the user to tune such tradeoff, based on their needs.

Answering RQ1. To investigate how effective DP approaches are in protecting the user information need, we compute the average similarity of the obfuscation queries with respect to the original query. High similarity indicates that the approach does not protect the user's privacy, as the information need can be inferred also by looking at the obfuscation queries. We propose to adopt two similarity measures: the Jaccard similarity between the terms of the queries and a neural sentence similarity approach relying on MiniLM [42]. The former similarity allows us to verify to what extent the terms of the two queries overlap, while the latter allows us to verify that query terms have not simply been replaced with synonyms. This second strategy consists of encoding each obfuscation query and the original query using MiniLM and obtaining an embedding for each query. then, the average cosine similarity is computed between the embeddings for each obfuscation query and the one for the original query.

Answering RQ2. We investigate whether DP approaches produce obfuscation queries that retrieve relevant documents. Using each DP approach, we produce a set of obfuscation queries and use it to retrieve the documents, we then evaluate the number of relevant documents for the original user's query retrieved by at least one obfuscation query. If this holds, in a real-life scenario, the user interested in retrieving documents while protecting their privacy can issue the obfuscation queries in place of the real one, obtain the results and rerank or reindex them locally to improve the precision, as proposed by Arampatzis et al. [2].

Answering RQ3. We consider obfuscation approaches originally devised explicitly for the IR task and measure to what level of ϵ they can be considered equivalent. We take into consideration the seminal work by Arampatzis et al. [2], labelled AED, and the recent state-of-the-art solution by Fröbe et al. [19], labelled FSH. We will compare these approaches with the DP mechanisms, based on three axes: i) the *obfuscation measure*, which we define as 1 minus the sentence similarity computed using the MiniLM representations; ii) the pooled recall; iii) the nDCG@10 observed if we re-rank the documents pooled by the obfuscation queries. Upon receiving 100 documents for each obfuscation query, we rerank them using TAS-B and evaluate the quality of this ranked list. This goes beyond the current state-of-the-art [2,19], which does not evaluate the final rank that the user observes. For each approach, these measures are reported on a radar plot where, as a rule of thumb, a larger area corresponds to more desirable results.

5 Experimental Results

5.1 Experimental Setup

We considered two different collections TREC Robust '04 [40] and TREC Deep Learning (DL '19) [10]. The former relies on a corpus of documents: disks 4 and 5 of the TIPSTER collection, minus the congressional records. The latter is based on the MS MARCO [30] passages corpus. As word embeddings, we used GloVe [33] with 300 dimensions trained on the Common Crawl. We also

Table 1. Average Jaccard similarity and MiniLM-based sentence similarity between the original query and 20 obfuscation queries generated with different approaches.

	Robust '04									DL '19								
ϵ	1	5	10	12.5	15	17.5	20	50	No DP	1	5	10	12.5	15	17.5	20	50	No DP
	Jaccard Similarity																	
CMP	0.000	0.006	0.225	0.512	0.772	0.915	0.965	0.988		0.000	0.002	0.109	0.299	0.537	0.731	0.855	0.976	
Mhl	0.000	0.005	0.101	0.259	0.470	0.679	0.841	0.988		0.000	0.004	0.051	0.145	0.291	0.475	0.648	0.975	
Vkr_{CMP}	0.000	0.005	0.096	0.159	0.188	0.196	0.195	0.239		0.000	0.002	0.054	0.099	0.139	0.171	0.167	0.200	
Vkr_{Mhl}	0.000	0.005	0.049	0.096	0.147	0.179	0.186	0.231		0.000	0.002	0.030	0.068	0.103	0.135	0.157	0.194	
AED									0.200									0.338
FSH									0.000									0.000
scrambling	MiniLM Sentence Similarity																	
CMP	0.074	0.100	0.396	0.672	0.871	0.961	0.987	0.996		0.024	0.032	0.214	0.458	0.681	0.824	0.903	0.952	
Mhl	0.077	0.095	0.244	0.427	0.627	0.794	0.907	0.996		0.020	0.034	0.119	0.241	0.427	0.610	0.750	0.951	
Vkr_{CMP}	0.077	0.100	0.278	0.412	0.511	0.578	0.622	0.760		0.028	0.032	0.137	0.211	0.308	0.372	0.413	0.565	
Vkr_{Mhl}	0.076	0.096	0.188	0.282	0.382	0.472	0.533	0.746		0.023	0.026	0.084	0.149	0.215	0.284	0.333	0.553	
AED									0.487									0.509
FSH									0.203									0.077

experiment with other sizes of vectors, obtaining substantially identical findings, not reported for space reasons. In terms of retrieval models, we consider two sparse bag-of-word models, BM25 [34] and Vector Space Model (TF-IDF) [35], and two dense bi-encoders, TAS-B [21] and Contriever [23]. The choice of using these retrieval models stems from the fact that BM25 is a widely adopted lexical method, mostly based on exact matching: by changing the terms in the query we might end up losing specific terms that allow us to retrieve relevant documents. Vice versa, both TAS-B and Contriever are dense IR models that project and compute the similarity of queries and documents in a dense space and thus do not rely explicitly on the exact matching of terms. Nevertheless, by obfuscating the query terms, we lose the semantics of the query, and this might impair the retrieval phase. By using these IR systems, we can observe how the obfuscation approaches interact with IR systems based on different rationales.

In AED, to avoid bias, i.e., too similar or different queries from the original one, we select for each query as obfuscation queries the 10 queries above median wup similarity [43] and the 10 queries below. For FSH we employ the sliding window candidate generator with a window size of 16 as it is the best performing as originally observed by Fröbe et al. [19]. We use the parametrization reported by Fröbe et al. [19], considering the first 10 documents retrieved by TF-IDF from the local corpus as target documents, obfuscation queries of at most three terms, and remove queries retrieving less than 100 documents. We use 50,000 documents randomly sampled from the MS MARCO collection and from the TIPSTER disks 4 and 5, as local corpora for the Robust '04 and DL '19 respectively. For each query, we generate 20 obfuscation queries. The code is publicly available at: https://github.com/guglielmof/24-ECIR-FF.

5.2 RQ1: Privacy Guarantees

Table 1 shows, as a proxy of the privacy achieved by the mechanisms, the similarity between the original query and the obfuscation queries generated to hide it. As expected from a differential privacy mechanism, the higher the ϵ the higher the similarity between the queries – with $\epsilon = 50$ for both Robust '04 and DL '19, CMP and Mhl achieve a Jaccard similarity higher than 97.5%. This indicates that overall the generated queries are almost identical to the original ones and there is no substantial privacy protection. Depending on the collection, CMP and Mhl obtain Jaccard similarity which falls in the range 20%–30%, with ϵ in the range $[10, 12.5]$. This indicates that, on average, 1 in 3 to 5 words remain equal to the original query. Similar results can also be observed for AED. Interestingly, when it comes to Vkr-based mechanisms, they tend to be much safer, as they obtain, with $\epsilon = 50$ less than 0.24 of Jaccard similarity on the Robust '04 collection, and 0.20 on the DL '19, 40.8% less than AED. Notice that, according to Jaccard similarity, FSH achieves perfect privacy, i.e., zero similarity, thanks to the fact that the words of the query are removed from the vocabulary of words that can be used to generate obfuscation queries. However, the approach based on the Jaccard similarity fails in assessing privacy leakage when synonyms are used to obfuscate the words of the query. Therefore, we also measure the similarity between the obfuscation queries and original queries using a more semantic-oriented approach, as described in Sect. 4. All the approaches have a much higher semantic similarity than what was observed for the Jaccard: in most of the cases, words are replaced with synonyms or highly correlated words. As for the Jaccard similarity, FSH, which explicitly removes synonyms and hypernyms from the queries, is particularly safe and corresponds to a DP Vkr_{CMP} mechanism with $\epsilon \in [5, 10]$ or a Vkr_{Mhl} with $\epsilon \in [10, 12.5]$ for the Robust '04, and DP Vkr_{CMP} and a Vkr_{Mhl} mechanism with $\epsilon \in [5, 10]$ for the DL '19. As observed for the Jaccard similarity, the privacy achieved by AED can be achieved with ϵ in the range $[10; 12.5]$ by CMP and Mhl on both collections. ϵ values that grant a comparable level of privacy are much higher for Vkr-based mechanisms, especially Vkr_{Mhl}, on both collections. As a general observation, privacy is a trade-off between noise and performance. It is reasonable that privacy is almost negligible for high values of ϵ and maximized for low values of ϵ. Furthermore, it is also intuitive that there are ϵ levels for which the DP mechanisms perform better, at least in terms of privacy, than any other baseline. We argue that, besides granting lower similarity at specific ϵ levels, the major advantage of DP is that it allows to meet the privacy requirements of the user, who can specify the privacy they would like to obtain and adapt the obfuscation mechanism consequently.

5.3 RQ2: Relevant Documents Retrieved

To assess the effectiveness of approaches based on DP, we measure the number of relevant documents retrieved, by pooling 100 documents from the 20 obfuscation queries representing the same information need. Not retrieving enough relevant documents would render the obfuscation approach unusable. As a reference point, we report the recall observed for the top 100 documents retrieved with the original query. Notice that, the number of used obfuscation queries is the same for DP-based approaches and for both AED and FSH. Using multiple obfuscation queries is generally a widely adopted procedure [2,19]. Since the

Table 2. Mean recall achieved by pooling 100 documents retrieved for each obfuscation query.

		Robust '04 ϵ									DL '19 ϵ								
model	mechanism	1	5	10	12.5	15	17.5	20	50	No DP	1	5	10	12.5	15	17.5	20	50	No DP
	CMP	0.020	0.146	0.483	**0.510**	0.489	0.442	0.421	0.407		0.011	0.135	0.384	**0.534**	0.517	0.514	0.480	0.444	
	Mhl	0.032	0.089	0.398	0.500	**0.512**	0.501	0.466	0.407		0.000	0.118	0.294	0.411	0.515	**0.529**	**0.529**	0.444	
	Vkr_{CMP}	0.041	0.152	0.407	0.506	0.548	0.554	**0.561**	0.518		0.000	0.073	0.282	0.363	0.498	**0.539**	0.498	0.533	
BM25	Vkr_{Mhl}	0.021	0.133	0.304	0.409	0.493	0.544	**0.556**	0.530		0.016	0.039	0.263	0.286	0.419	0.479	**0.514**	0.506	
	AED									0.420									0.445
	FSH									0.140									0.231
	Original									0.410									0.454
	CMP	0.020	0.146	0.487	**0.512**	0.491	0.444	0.423	0.408		0.011	0.135	0.386	**0.535**	0.515	0.515	0.478	0.442	
	Mhl	0.032	0.089	0.398	0.504	**0.516**	0.504	0.468	0.408		0.000	0.118	0.295	0.412	0.515	**0.530**	0.527	0.442	
	Vkr_{CMP}	0.039	0.151	0.407	0.506	0.551	0.557	**0.563**	0.521		0.000	0.070	0.274	0.363	0.497	0.534	0.499	**0.533**	
TF-IDF	Vkr_{Mhl}	0.021	0.132	0.305	0.411	0.494	0.547	**0.559**	0.532		0.016	0.039	0.263	0.285	0.419	0.479	**0.512**	0.505	
	AED									0.420									0.443
	FSH									0.139									0.231
	Original									0.411									0.451
	CMP	0.034	0.106	0.469	**0.507**	0.481	0.433	0.406	0.392		0.000	0.077	0.446	0.615	**0.644**	0.628	0.576	0.512	
	Mhl	0.026	0.088	0.345	0.473	**0.503**	0.497	0.460	0.392		0.000	0.057	0.264	0.476	0.601	0.641	**0.650**	0.512	
	Vkr_{CMP}	0.038	0.125	0.397	0.486	0.510	0.518	**0.520**	0.475		0.000	0.010	0.280	0.430	0.537	0.598	0.579	**0.608**	
Contriever	Vkr_{Mhl}	0.024	0.094	0.269	0.392	0.462	0.504	**0.518**	0.481		0.000	0.027	0.254	0.321	0.418	0.522	0.580	**0.604**	
	AED									0.419									0.497
	FSH									0.204									0.204
	Original									0.392									0.528
	CMP	0.027	0.080	0.434	**0.477**	0.444	0.398	0.369	0.356		0.000	0.063	0.392	0.615	**0.636**	0.622	0.575	0.498	
	Mhl	0.028	0.064	0.310	0.438	**0.464**	0.460	0.423	0.356		0.000	0.042	0.275	0.455	0.584	**0.645**	0.638	0.499	
	Vkr_{CMP}	0.025	0.086	0.355	0.448	0.483	0.490	**0.495**	0.446		0.000	0.025	0.245	0.398	0.534	0.585	0.579	**0.600**	
TAS-B	Vkr_{Mhl}	0.023	0.073	0.237	0.355	0.423	0.476	**0.482**	0.452		0.000	0.019	0.206	0.267	0.375	0.503	0.553	**0.603**	
	AED									0.387									0.491
	FSH									0.161									0.238
	Original									0.358									0.518

queries are noisy, the adversarial is not able to recognize what was the topic of interest for the user. In turn, we expect each obfuscation query to return some relevant documents, most likely in low positions of the ranking, as it is not directly linkable to the original query. Once the results are available on a secure machine (e.g., the user's client) they can be reordered using the original query.

Following what was observed for the similarity, as observable in Table 2, the effectiveness varies widely over different mechanisms, with CMP and Mhl achieving higher recall for lower ϵ compared to Vkr-based mechanisms. Interestingly, the best-pooled recall is seldom achieved with $\epsilon = 50$ – exclusively using Vkr mechanisms and on the DL '19 collection. Moreover, for all the mechanisms, there is at least one value of ϵ for which the recall is higher than the one observed using the original queries. This is due to the fact that DP approaches with intermediate levels of ϵ automatically implement query rewriting. The usage of query variations has a strong impact on the performance of a system [11]: by automatically changing the words within a query with terms that are correlated but not identical, we can pool relevant documents that are lost when we use queries that are almost identical to the original ones, i.e., with $\epsilon = 50$ and CMP or Mhl mechanisms. By selecting either the closest or the second closest term, the Vkr-based mechanisms still apply implicit query rewriting also for higher levels of ϵ. Regardless of the collection considered, we notice that for ϵ ranging between 10 and 17.5, almost all mechanisms are able to overcome the IR state-of-the-art

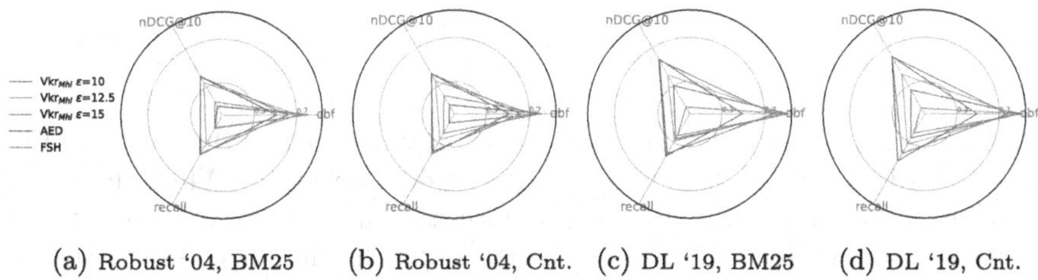

(a) **Robust '04, BM25** (b) **Robust '04, Cnt.** (c) **DL '19, BM25** (d) **DL '19, Cnt.**

Fig. 1. Performance of different obfuscation mechanisms over three axes: pooled recall, nDCG@10 of the reranked documents, obfuscation (obf), measured as 1-similarity. Cnt. stands for "Contriever". (Color figure online)

approaches in terms of recall. In particular, the DP approaches overcome FSH already with low ϵ, while $\epsilon \geq 15$ allows for overcoming AED as well. This suggests that using DP mechanisms that were originally thought to be used in NLP scenarios, and with additional training of the models, successfully allows us to obtain satisfactory performance in retrieval.

5.4 RQ3: Comparing DP and Non-DP Approaches

As a final analysis, we compare the most promising DP approach, the Vickrey mechanism based on the Mahalanobis norm, with the current state-of-the-art approaches in IR. To avoid cluttering, we focus only on BM25 and Contriever as IR systems. Figure 1 reports the radar plots, showing the performance of different obfuscation approaches over the three axes mentioned above. We notice that the area corresponding to AED approach (in red) is encompassed within the area corresponding to Vkr_{Mhl} with $\epsilon = 15$ (green). In fact, on the Robust '04 collection, AED achieves nDCG@10 of 0.410 and 0.424 for BM25 and Contriever respectively, recall of 0.420 and 0.419, and obfuscation of 0.513. Vice versa Vkr_{Mhl} with $\epsilon = 15$ obtains nDCG@10 of 0.416 and 0.431, recall of 0.493 and 0.462, and obfuscation of 0.618. The exception is DL '19 with Contriever as the IR system, where AED has higher recall than Vkr_{Mhl} (0.497 against 0.418). Nevertheless, this larger recall does not correspond to much larger nDCG@10, indicating that Vkr_{Mhl} is preferable over AED, as it has comparable nDCG@10 (0.604 for Vkr_{Mhl} against 0.607 for AED), with improved obfuscation (0.785 against 0.491). When it comes to FSH (purple), the behaviour depends on the collection. In the DL '19, using Vkr_{Mhl} with $\epsilon = 10$ (blue) provides an edge over FSH: they have comparable obfuscation (0.916 the former, 0.923 the latter), but Vkr_{Mhl} has much larger nDCG@10 (0.254 compared to 0.064). On the Robust '04 collection, to observe an improvement in terms of nDCG@10, it is necessary to use Vkr_{Mhl} with $\epsilon = 12.5$ (nDCG@10 of 0.349 and 0.355 for BM25 and Contriever respectively) to overcome FSH in terms of nDCG@10 (0.140 and 0.194). Nevertheless, while Vkr_{Mhl} with $\epsilon = 12.5$ exhibits nDCG@10 performance slightly lower than AED, it also has obfuscation (0.719) relatively close to FSH, which has obfuscation of 0.797, much closer than AED, with obfuscation

0.513. As a general guideline, our proposal is to use Vkr_{Mhl} as the obfuscation mechanism, with ϵ chosen in the interval [10, 15], depending on the optimal trade-off between privacy and effectiveness, as chosen by the user.

An important aspect that should be discussed, is whether the user could be profiled or identified by looking at the documents retrieved. While the list of documents returned in response to an obfuscated query contains some relevant documents (see Subsect. 5.3), it is also true that, as the query is obfuscated, not all the documents retrieved will be strictly related to the topic of interest. Therefore, each obfuscated query will contain some "speck of gold", the relevant documents, which will be filtered on the user side, so that the adversarial cannot reconstruct the information need of the user by looking at the retrieved documents. Similarly, in more search engine-oriented scenarios, we could consider that the system might profile the user, based on which documents they click. We argue that, while present, this is an orthogonal problem to the task investigated here. In fact, there exist approaches, such as TrackMeNot [22], which simulate user clicks on random or non-relevant documents, to prevent the adversarial from using the clicks done by the user to profile them.

6 Conclusion and Future Work

In this work, we analyzed for the first time the performance of three DP mechanisms, originally designed for NLP, in the proxy query obfuscation IR task. These mechanisms are the Calibrated Multivariate Perturbation, the Mahalanobis, and the Vickrey mechanisms. We evaluated these mechanisms on the IR setting by considering three aspects: their obfuscation capabilities, their effectiveness in terms of recall, and their ability in allowing to retrieve highly relevant documents. To measure the obfuscation, we considered the dissimilarity between the original query and the obfuscation queries produced by different approaches. To measure their recall and effectiveness, we generated 20 obfuscation queries and used them to retrieve documents from Robust '04 and DL '19. Our findings highlight that the Vickrey mechanism with $\epsilon \in [10, 12.5]$ achieves higher privacy guarantees, with improved effectiveness, than current state-of-the-art approaches. Furthermore, lower or higher levels of ϵ allow for better satisfy the user, either in terms of privacy or accuracy, depending on their inclinations. As a future work, we plan to investigate how to perturb dense representations of the queries and combine them with generative language models to produce obfuscation queries with the same dense representation, but different terms.

Acknowledgments. This work has received support from CAMEO, PRIN 2022 n. 2022ZLL7MW.

References

1. Andrés, M.E., Bordenabe, N., Chatzikokolakis, K., Palamidessi, C.: Geo-indistinguishability: differential privacy for location-based systems. In: Sadeghi, A., Gligor, V.D., Yung, M. (eds.) 2013 ACM SIGSAC Conference on Computer and Communications Security, CCS 2013, Berlin, Germany, 4–8 November 2013, pp. 901–914. ACM (2013). https://doi.org/10.1145/2508859.2516735
2. Arampatzis, A., Efraimidis, P.S., Drosatos, G.: Enhancing deniability against query-logs. In: Clough, P.D., et al. (eds.) Advances in Information Retrieval - 33rd European Conference on IR Research, ECIR 2011, Dublin, Ireland, 18–21 April 2011. Proceedings. LNCS, vol. 6611, pp. 117–128. Springer, Cham (2011). https://doi.org/10.1007/978-3-642-20161-5_13
3. Arampatzis, A., Drosatos, G., Efraimidis, P.: A versatile tool for privacy-enhanced web search. In: Serdyukov, P., et al. (eds.) Advances in Information Retrieval - 35th European Conference on IR Research, ECIR 2013, Moscow, Russia, 24–27 March 2013. Proceedings. LNCS, vol. 7814, pp. 368–379. Springer, Cham (2013). https://doi.org/10.1007/978-3-642-36973-5_31
4. Arampatzis, A., Efraimidis, P.S., Drosatos, G.: A query scrambler for search privacy on the internet. Inf. Retr. **16**(6), 657–679 (2013). https://doi.org/10.1007/s10791-012-9212-1
5. Arampatzis, A., Drosatos, G., Efraimidis, P.S.: Versatile query scrambling for private web search. Inf. Retr. J. **18**(4), 331–358 (2015). https://doi.org/10.1007/s10791-015-9256-0
6. Barbaro, M., Zeller, T.: A Face is Exposed for AoL Searcher No. 4417749. New York Times (2006)
7. Bavadekar, S., et al.: Google COVID-19 search trends symptoms dataset: anonymization process description (version 1.0). CoRR, abs/2009.01265 (2020). https://arxiv.org/abs/2009.01265
8. Castellà-Roca, J., Viejo, A., Herrera-Joancomartí, J.: Preserving user's privacy in web search engines. Comput. Commun. **32**(13–14), 1541–1551 (2009). https://doi.org/10.1016/j.comcom.2009.05.009
9. Chatzikokolakis, K., Andrés, M., Bordenabe, N., Palamidessi, C.: Broadening the scope of differential privacy using metrics. In: Cristofaro, E.D., Wright, M.K. (eds.) Privacy Enhancing Technologies - 13th International Symposium, PETS 2013, Bloomington, IN, USA, 10–12 July 2013. Proceedings. LNCS, vol. 7981, pp. 82–102. Springer, Cham (2013). https://doi.org/10.1007/978-3-642-39077-7_5
10. Craswell, N., Mitra, B., Yilmaz, E., Campos, D., Voorhees, E.M.: Overview of the TREC 2019 deep learning track. CoRR, abs/2003.07820 (2020). https://arxiv.org/abs/2003.07820
11. Culpepper, J.S., Faggioli, G., Ferro, N., Kurland, O.: Topic difficulty: collection and query formulation effects. ACM Trans. Inf. Syst. **40**(1), 19:1–19:36 (2022). https://doi.org/10.1145/3470563
12. Domingo-Ferrer, J., González-Nicolás, Ú.: Rational behavior in peer-to-peer profile obfuscation for anonymous keyword search. Inf. Sci. **185**(1), 191–204 (2012). https://doi.org/10.1016/j.ins.2011.09.010
13. Domingo-Ferrer, J., Bras-Amorós, M., Wu, Q., Manjón, J.A.: User-private information retrieval based on a peer-to-peer community. Data Knowl. Eng. **68**(11), 1237–1252 (2009). https://doi.org/10.1016/j.datak.2009.06.004
14. Domingo-Ferrer, J., Solanas, A., Castellà-Roca, J.: H(k)-private information retrieval from privacy-uncooperative queryable databases. Online Inf. Rev. **33**(4), 720–744 (2009). https://doi.org/10.1108/14684520910985693

15. Dwork, C., Roth, A.: The algorithmic foundations of differential privacy. Found. Trends Theor. Comput. Sci. **9**(3–4), 211–407 (2014). https://doi.org/10.1561/0400000042
16. Elovici, Y., Shapira, B., Maschiach, A.: A new privacy model for web surfing. In: Halevy, A.Y., Gal, A. (eds.) Next Generation Information Technologies and Systems, 5th International Workshop, NGITS 2002, Caesarea, Israel, 24–25 June 2002, Proceedings. LNCS, vol. 2382, pp. 45–57. Springer, Cham (2002). https://doi.org/10.1007/3-540-45431-4_5
17. Fernandes, N., Dras, M., McIver, A.: Generalised differential privacy for text document processing. In: Nielson, F., Sands, D. (eds.) Principles of Security and Trust - 8th International Conference, POST 2019, Held as Part of the European Joint Conferences on Theory and Practice of Software, ETAPS 2019, Prague, Czech Republic, 6–11 April 2019, Proceedings. LNCS, vol. 11426, pp. 123–148. Springer, Cham (2019). https://doi.org/10.1007/978-3-030-17138-4_6
18. Feyisetan, O., Balle, B., Drake, T., Diethe, T.: Privacy- and utility-preserving textual analysis via calibrated multivariate perturbations. In: Caverlee, J., Hu, X.B., Lalmas, M., Wang, W. (eds.) Proceedings of the 13th International Conference on Web Search and Data Mining, pp. 178–186. ACM, January 2020. https://doi.org/10.1145/3336191.3371856
19. Fröbe, M., Schmidt, E.O., Hagen, M.: Efficient query obfuscation with keyqueries. In: He, J., et al. (eds.) WI-IAT 2021: IEEE/WIC/ACM International Conference on Web Intelligence, Melbourne VIC Australia, 14–17 December 2021, pp. 154–161. ACM (2021). https://doi.org/10.1145/3486622.3493950
20. Götz, M., Machanavajjhala, A., Wang, G., Xiao, X., Gehrke, J.: Publishing search logs - a comparative study of privacy guarantees. IEEE Trans. Knowl. Data Eng. **24**(3), 520–532 (2012). https://doi.org/10.1109/TKDE.2011.26
21. Hofstätter, S., Lin, S., Yang, J., Lin, J., Hanbury, A.: Efficiently teaching an effective dense retriever with balanced topic aware sampling. In: Diaz, F., Shah, C., Suel, T., Castells, P., Jones, R., Sakai, T. (eds.) SIGIR 2021: The 44th International ACM SIGIR Conference on Research and Development in Information Retrieval, Virtual Event, Canada, 11–15 July 2021, pp. 113–122. ACM (2021). https://doi.org/10.1145/3404835.3462891
22. Howe, D., Nissenbaum, H.: TrackMeNot: resisting surveillance in web search. Technical report queries (2009)
23. Izacard, G., et al.: Unsupervised dense information retrieval with contrastive learning. Trans. Mach. Learn. Res. **2022** (2022). https://openreview.net/forum?id=jKN1pXi7b0
24. Khan, R., Ullah, M., Khan, A., Uddin, M.I., Al-Yahya, M.: NN-QuPiD attack: neural network-based privacy quantification model for private information retrieval protocols. Complexity **2021**, 6651662:1–6651662:8 (2021). https://doi.org/10.1155/2021/6651662
25. Kharitonov, E.: Federated online learning to rank with evolution strategies. In: Culpepper, J.S., Moffat, A., Bennett, P.N., Lerman, K. (eds.) Proceedings of the Twelfth ACM International Conference on Web Search and Data Mining, WSDM 2019, Melbourne, VIC, Australia, 11–15 February 2019, pp. 249–257. ACM (2019). https://doi.org/10.1145/3289600.3290968
26. Korolova, A., Kenthapadi, K., Mishra, N., Ntoulas, A.: Releasing search queries and clicks privately. In: Quemada, J., León, G., Maarek, Y.S., Nejdl, W. (eds.) Proceedings of the 18th International Conference on World Wide Web, WWW 2009, Madrid, Spain, 20–24 April 2009, pp. 171–180. ACM (2009). https://doi.org/10.1145/1526709.1526733

27. Laud, P., Pankova, A., Pettai, M.: A framework of metrics for differential privacy from local sensitivity. Proc. Priv. Enhancing Technol. **2020**(2), 175–208 (2020). https://doi.org/10.2478/popets-2020-0023
28. Mahalanobis, P.C.: On the generalized distance in statistics. Sankhyā: Indian J. Stat. Ser. A (2008-) **80**, S1–S7 (2018). ISSN 0976836X, 09768378. https://www.jstor.org/stable/48723335
29. Miller, G.A.: WordNet: a lexical database for English. Commun. ACM **38**(11), 39–41 (1995). https://doi.org/10.1145/219717.219748
30. Nguyen, T., et al.: MS MARCO: a human generated machine reading comprehension dataset. In: Besold, T.R., Bordes, A., d'Avila Garcez, A.S., Wayne, G. (eds.) Proceedings of the Workshop on Cognitive Computation: Integrating Neural and Symbolic Approaches 2016 Co-located with the 30th Annual Conference on Neural Information Processing Systems (NIPS 2016), Barcelona, Spain, 9 December 2016, vol. 1773 of CEUR Workshop Proceedings. CEUR-WS.org (2016). https://ceur-ws.org/Vol-1773/CoCoNIPS_2016_paper9.pdf
31. Peddinti, S.T., Saxena, N.: On the effectiveness of anonymizing networks for web search privacy. In: Cheung, B.S.N., Hui, L.C.K., Sandhu, R.S., Wong, D.S. (eds.) Proceedings of the 6th ACM Symposium on Information, Computer and Communications Security, ASIACCS 2011, Hong Kong, China, 22–24 March 2011, pp. 483–489. ACM (2011). https://doi.org/10.1145/1966913.1966984
32. Peddinti, S.T., Saxena, N.: Web search query privacy: evaluating query obfuscation and anonymizing networks. J. Comput. Secur. **22**(1), 155–199 (2014). https://doi.org/10.3233/JCS-130491
33. Pennington, J., Socher, R., Manning, C.D.: GloVe: global vectors for word representation. In: Moschitti, A., Pang, B., Daelemans, W. (eds.) Proceedings of the 2014 Conference on Empirical Methods in Natural Language Processing, EMNLP 2014, 25–29 October 2014, Doha, Qatar, A Meeting of SIGDAT, A Special Interest Group of the ACL, pp. 1532–1543. ACL (2014). https://doi.org/10.3115/v1/d14-1162
34. Robertson, S.E., Walker, S., Jones, S., Hancock-Beaulieu, M., Gatford, M.: Okapi at TREC-3. In: Harman, D.K. (ed.) Proceedings of The Third Text REtrieval Conference, TREC 1994, Gaithersburg, Maryland, USA, 2–4 November 1994, vol. 500–225 of NIST Special Publication, pp. 109–126. National Institute of Standards and Technology (NIST) (1994). http://trec.nist.gov/pubs/trec3/papers/city.ps.gz
35. Salton, G., Wong, A., Yang, C.: A vector space model for automatic indexing. Commun. ACM **18**(11), 613–620 (1975). https://doi.org/10.1145/361219.361220
36. Sánchez, D., Batet, M., Viejo, A., Rodriguez-Garcia, M., Castellà-Roca, J.: A semantic-preserving differentially private method for releasing query logs. Inf. Sci. **460–461**, 223–237 (2018). https://doi.org/10.1016/j.ins.2018.05.046
37. Tang, J., Zhu, T., Xiong, P., Wang, Y., Ren, W.: Privacy and utility trade-off for textual analysis via calibrated multivariate perturbations. In: Kutylowski, M., Zhang, J., Chen, C. (eds.) Network and System Security - 14th International Conference, NSS 2020, Melbourne, VIC, Australia, 25–27 November 2020, Proceedings. LNCS, vol. 12570, pp. 342–353. Springer, Cham (2020). https://doi.org/10.1007/978-3-030-65745-1_20
38. Ullah, M., Islam, M.A., Khan, R., Aleem, M., Iqbal, M.A.: ObSecure Logging (OSLo): a framework to protect and evaluate the web search privacy in health care domain. J. Med. Imaging Health Inform. **9**(6), 1181–1190 (2019). https://doi.org/10.1166/jmihi.2019.2708
39. Vickrey, W.: Counterspeculation, auctions, and competitive sealed tenders (1961)

40. Voorhees, E.M.: Overview of the TREC 2004 robust track. In: Voorhees, E.M., Buckland, L.P. (eds.) Proceedings of the Thirteenth Text REtrieval Conference, TREC 2004, Gaithersburg, Maryland, USA, 16–19 November 2004, vol. 500-261 of NIST Special Publication. National Institute of Standards and Technology (NIST) (2004). http://trec.nist.gov/pubs/trec13/papers/ROBUST.OVERVIEW.pdf
41. Wang, S., Liu, B., Zhuang, S., Zuccon, G.: Effective and privacy-preserving federated online learning to rank. In: Hasibi, F., Fang, Y., Aizawa, A. (eds.) ICTIR 2021: The 2021 ACM SIGIR International Conference on the Theory of Information Retrieval, Virtual Event, Canada, 11 July 2021, pp. 3–12. ACM (2021). https://doi.org/10.1145/3471158.3472236
42. Wang, W., Wei, F., Dong, L., Bao, H., Yang, N., Zhou, M.: MiniLM: deep self-attention distillation for task-agnostic compression of pre-trained transformers. In: Larochelle, H., Ranzato, M., Hadsell, R., Balcan, M., Lin, H. (eds.) Advances in Neural Information Processing Systems: Annual Conference on Neural Information Processing Systems 2020, NeurIPS 2020, 6–12 December 2020, vol. 33, Virtual (2020)
43. Wu, Z., Palmer, M.S.: Verb semantics and lexical selection. In: Pustejovsky, J. (ed.) 32nd Annual Meeting of the Association for Computational Linguistics, 27–30 June 1994, New Mexico State University, Las Cruces, New Mexico, USA, Proceedings, pp. 133–138. Morgan Kaufmann Publishers/ACL (1994). https://doi.org/10.3115/981732.981751, https://aclanthology.org/P94-1019/
44. Wu, Z., Shen, S., Lian, X., Su, X., Chen, E.: A dummy-based user privacy protection approach for text information retrieval. Knowl. Based Syst. **195**, 105679 (2020). https://doi.org/10.1016/j.knosys.2020.105679
45. Xu, Z., Aggarwal, A., Feyisetan, O., Teissier, N.: A differentially private text perturbation method using regularized Mahalanobis metric. In: Proceedings of the Second Workshop on Privacy in NLP. Association for Computational Linguistics (2020). https://doi.org/10.18653/v1/2020.privatenlp-1.2
46. Xu, Z., Aggarwal, A., Feyisetan, O., Teissier, N.: On a utilitarian approach to privacy preserving text generation. CoRR, abs/2104.11838, April 2021. https://doi.org/10.48550/ARXIV.2104.11838
47. Yu, P., Ahmad, W., Wang, H.: Hide-n-Seek: an intent-aware privacy protection plugin for personalized web search. In: Collins-Thompson, K., Mei, Q., Davison, B.D., Liu, Y., Yilmaz, E. (eds.) The 41st International ACM SIGIR Conference on Research & Development in Information Retrieval, SIGIR 2018, Ann Arbor, MI, USA, 8–12 July 2018, pp. 1333–1336. ACM (2018). https://doi.org/10.1145/3209978.3210180
48. Zhang, S., Yang, G.H., Singh, L.: Anonymizing query logs by differential privacy. In: Perego, R., Sebastiani, F., Aslam, J.A., Ruthven, I., Zobel, J. (eds.) Proceedings of the 39th International ACM SIGIR conference on Research and Development in Information Retrieval, SIGIR 2016, Pisa, Italy, 17–21 July 2016, pp. 753–756. ACM (2016). https://doi.org/10.1145/2911451.2914732

On-Device Query Auto-completion for Email Search

Yifan Qiao[1(✉)], Otto Godwin[2], and Hua Ouyang[2]

[1] University of California, Santa Barbara, CA 93106, USA
yifanqiao@cs.ucsb.edu
[2] Apple Inc., Cupertino, CA 95014, USA
{otto_g,hua_ouyang}@apple.com

Abstract. Traditional query auto-completion (QAC) relies heavily on search logs collected over many users. However, in on-device email search, the scarcity of logs and the governing privacy constraints make QAC a challenging task. In this work, we propose an on-device QAC method that runs directly on users' devices, where users' sensitive data and interaction logs are not collected, shared, or aggregated through web services. This method retrieves candidates using pseudo relevance feedback, and ranks them based on relevance signals that explore the textual and structural information from users' emails. We also propose a private corpora based evaluation method, and empirically demonstrate the effectiveness of our proposed method.

Keywords: Email search · Query auto-completion · On-device search

1 Introduction and Motivations

Email is one of the most popular tools for personal, business, and organizational communications. With billions of emails circulating every day [14], email inboxes are expanding at a staggering rate, making the task of searching for specific emails quite challenging for users. Sometimes, users vaguely remember the existence of an email but fail to recall the exact words in the content to use as search queries. Other times, their queries may be too narrow to retrieve the desired target, or too broad, resulting in an overwhelming number of mostly irrelevant emails. Query auto-completion (QAC) [5–7] is a key feature designed to address these issues, assisting users in formulating queries by suggesting a list of plausible candidates with each keystroke. A well-designed QAC system should be able to significantly cut down users' typing, memory and browsing efforts, enhancing the overall email search experience [18].

QAC is a ubiquitous feature in search systems. In a typical web-based QAC system, candidates are firstly retrieved from an index (e.g. a trie or inverted index), then the top-k suggestions are sorted by ranking models. During this process, web-based QAC presumes that data and logs can be collected from a vast number of users, and are constantly accessible for aggregation, feature

N. Goharian et al. (Eds.): ECIR 2024, LNCS 14608, pp. 295–309, 2024.
https://doi.org/10.1007/978-3-031-56027-9_18

generation, and model training. However, in personal search, this assumption is often not valid due to privacy constraints. In response to these restrictions, *we propose an on-device QAC setting for email search.* In on-device QAC, users can only access their own personal data from their own devices (e.g. mobile phones or laptops). Search and QAC are purely powered by on-device indices and algorithms, without relying on the availability of web-based search services. Users' interaction logs are generated on their devices, and will not be collected, shared or aggregated through web services.

On-device QAC for email search has distinct characteristics when compared to web-based QAC. First, due to corpora differences in email search, on-device QAC is inherently personalized. Given the sensitivity of the information in emails, collecting a global email dataset from numerous users is also prohibitive. Second, email search logs are generated on device. They will not be collected by centralized services. Third, an on-device email index will change much more frequently than a web index. For example, email users may receive new emails or delete old ones at any time. An on-device email search system should be able to reflect these changes and users' engagements in real-time. Lastly, there is a lot of structural information in emails that are different from web pages. These structural differences can be leveraged to improve the quality of QAC algorithms. All of these characteristic differences make traditional log-based web search QAC algorithms not directly applicable, and motivate the design of our on-device QAC methods.

Our proposed on-device QAC method comprises two stages. In the retrieval stage, we use pseudo relevance feedback (PRF) to generate potential completion and suggestion candidates. During this stage, emails that are most relevant to the users' queries are retrieved, from which QAC candidates are extracted. In the ranking stage, relevance signals based on the structural and textual information derived from the users' personal corpora are employed for ranking candidates and post-processing. To evaluate our method, we propose a novel grader-based offline evaluation pipeline. Extensive experiments show that graders are more satisfied with the quality of our QAC results, and our method outperforms strong baselines.

2 Related Work

Web-based QAC systems collect user interaction data centrally across many users. For example, Most Popular Completion (MPC) [6] utilizes search logs and generates candidates from the most frequently issued queries. Contextual, time sensitive and engagement signals collected from user activities are proposed for QAC candidates ranking [10,28–30]. In order to improve personal search quality, [8,17] aggregate non-private query-document associations from user interactions. In case query logs are absent, generative QAC methods are used [15,26]. Mitra et al. [20] generates candidates for rare prefixes using popular n-gram suffixes. Park et al. [21] further propose a character-level neural language model trained on query logs to generate QAC candidates. Dehghani et al. [12] then utilize the attention mechanism to capture the structure of the session context.

Emails, different from web pages, have a lot of unique characteristics. Wang et al. [27] classify enterprise email intents into 4 categories, showing the topics of emails are less diversified than web pages. Alrashed et al. [4] show there is a positive correlation between user interaction and significance of emails. Ai et al. [3] study large-scale behavior logs in the email search. They found that compared to web search, email queries tend to be more specific, shorter, and less repetitive. This difference makes it hard to directly apply web search techniques to the email domain. Some existing works exploit these specific characteristics to develop more effective search systems for emails. For example, Meng et al. [19] combine the token-level sparse features and email-level dense features to better capture users' intents. Carmel et al. [11] explore the importance of using freshness signals in email search. Horovitz et al. [13] propose to use freshness signals along with the structural information to enhance the query completion results. In our experiments, it turns out that these additional features can significantly boost the performance.

In personalized search, there is typically a shared corpus, and users have their own individual profiles. Different users may expect different results with the same query. Teevan et al. [24] form users' profiles according to their historical search behaviors, and use such profiles to personalize their search results. Zhou et al. [30] encode users' search history using transformers [25] to build users' profile. In order to solve words ambiguity across different users, Yao et al. [29] build a personalized model based on personal embeddings. In the industry, Shokouhi et al. [23] propose a labelling strategy for generating offline training labels and train a personalized QAC model based on users' search history. Aberdeen et al. [2] first train a global model, then apply transfer learning to adapt it to individual users.

In our on-device QAC setting, we keep users' personal data on-device. Under this constraint, existing massive log-based training techniques (e.g. [8,17]) cannot be applied. [9,13] have the most similar setup to our work, hence serve as the baselines for our experiments. They propose to first construct the candidates set by selecting all the n-grams in the mailbox containing the users' input as prefix, then rank based on the term-frequency scores. Horovitz et al. [13] use multiple ranking signals, while Bhatia et al. [9] extend TF-IDF by taking into account the term-to-phrase probability. However, these works assume the best QAC result is a consecutive n-gram in the corpus, which is not always the case.

3 On-Device QAC

3.1 System and Settings

Our email search system mainly consists of a trie-based inverted index, a key-value store for metadata storage, and components for tiered retrieval, ranking, and QAC. All of these components are on-device. Users' emails are indexed instantly upon receipt. Figure 1 illustrates our user interface for email search and QAC. When a user taps on the search bar and begins typing, the system receives an *original query* upon each keystroke. For each original query, a list of emails are retrieved and ranked. QAC then fetches metadata from the top ranked emails and generates QAC results from them. After the user views the QAC list and clicks a QAC result, a list of emails are retrieved and presented back to the user. There are interesting challenges in each of these components. The focus of this paper is on the QAC method. In cases where the original query is misspelled or includes synonyms that do not appear in the email contents, this will be left for future work.

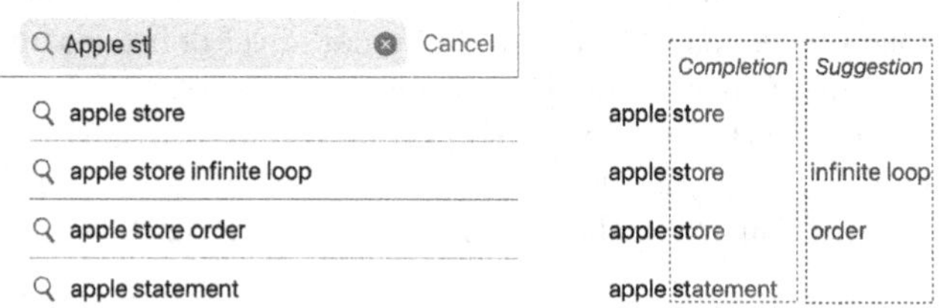

Fig. 1. Left: Search and QAC user interface. Right: *Completion* and *suggestion* examples. *Completion* part completes the last prefix of the user's original query to a word. *Suggestion* part adds relevant word(s). Final QAC result is a composition of original query, completion, and suggestion parts.

3.2 Candidates Generation

We propose to generate QAC candidates from PRF [1,16,22]. The classic PRF paradigm assumes that the set of top-k ranked documents (emails) are relevant to the query. This assumption holds naturally in our system which includes an on-device email ranker. This email ranker leverages a few important signals such as textual match between query and various zones of emails, freshness of activities, and users' historical engagements.

After the top ranked emails are returned from the ranker, their metadata such as subjects, body texts, senders and recipients are obtained from a key-value store. QAC then identifies matches against the original query and extracts candidates from these texts on-the-fly.

Compared with existing work which explores terms and phrases from *all* emails (e.g. [9,13]), retrieving candidates via PRF brings many benefits. We can avoid aggregating candidates from irrelevant emails, hence QAC results are potentially more relevant to the query. When users engage with a QAC result, their chance of getting relevant emails is also higher. This idea is motivated by the fact that users' goal is to find target emails that are relevant to them, while QAC serves as an intermediate interface that helps users formulate their queries. As described in Sect. 3.3, QAC can take advantage of the ranking signals that come along with the email results. Moreover, there is no need to maintain a phrase dictionary that could be costly and non-trivial to be kept up-to-date.

We propose to formulate a QAC candidate as the combination of the following parts: 1) original query, excluding the last term; 2) n-gram completion of the last query term, defined as the ***completion***; 3) optional suggested n-gram that can be anywhere in the text, defined as the ***suggestion***. These concepts are illustrated in Fig. 1. For example, imagine the intent of the original query "apple st" is to retrieve an order confirmation email for a recent order placed on the Apple online store. The QAC candidate "apple store order" is constructed from the non-tailing part of the *original query* "apple", the *completion* "store", and the *suggestion* "order". An important difference from previous work [9,13] is that, the QAC candidates do not have to be contiguous n-grams extracted from emails. For example, in candidate "apple store order", "store" and "order" may not consecutively occur in the email.

From the top ranked emails, QAC first extracts all the n-grams as suggestion candidates. These n-grams do not contain punctuation marks and do not cross sentences, paragraphs or zones. Following [9], when counting n-grams, common stop-words will be "jumped over" and retained, so that the resulting phrases do not start or end with stop-words, preventing unintelligible candidates from being generated.

For each suggestion candidate within an email, QAC finds the nearest token that matches the trailing term of the original query. This matched token could be part of an n-gram. If we only use this token as the completion, a QAC result could be incomplete. For example, imagine the intent of the original query "americ" is to retrieve a recent annual statement email from American Express, and the suggestion is "statement". A uni-gram completion could generate an unintelligible result "american statement". If the completion is extended to n-grams, we can complete "americ" to "american express" and formulate a better result "american express statement". Because of this, completions are generated as n-grams around the matched token. For each suggestion, completions could also come from multiple emails. For example, "american airlines statement" is another QAC result for "americ". In this case, suggestion "statement" corresponds to two completions: "american express" and "american airlines".

Our QAC system responds to user's keystrokes instantly, hence the size of the candidates set has to be controlled for ranking and computational feasibility. We consider the following parameters: 1) k, the maximum number of top ranked emails we use to generate candidates; 2) N_c and N_s, the maximum size of the completion and the suggestion n-grams. For example, when the original query is "apple st" and $N_c = N_s = 1$, "apple store" is a valid candidate because the completion size $n_c = 1 = N_c$ and the suggestion size $n_s = 0 < N_s$; "apple store order" is also valid since $n_c = 1 = N_c$ and $n_s = 1 = N_s$; however, "apple store order shipment" is not a valid candidate since the suggestion size $n_s = 2 > N_s$.

3.3 Candidates Ranking and Post-processing

In this section, we describe how the suggestion and completion candidates are ranked and combined. First, all the suggestion candidates are ranked. Then for each top ranked suggestion candidate, its corresponding completion candidates are ranked and the final QAC candidates are formulated.

To rank all the suggestion candidates generated from top-k emails, we utilize the following relevance features listed below. These features are efficient, explainable, and have been widely adopted in ranking systems of modern search engines.

- Term frequency (TF) and inverse document frequency (IDF). TF measures the popularity of suggestion candidates in the top-k documents, and IDF measures their popularity among all emails in the user's mailbox.
- Proximity: $e^{1/d}$, where d is the distance between a suggestion s_w and a completion c_w. If a suggestion and a completion overlap, then $d = 0$. In this scenario, candidates are more conservative and likely to be relevant, and we set a large value to the proximity. If there are multiple completion choices, the most adjacent one to the suggestion s_w will be used.
- Zone weights. Intuitively, tokens from sender or subject zones are likely to be more relevant than those from body zone. Therefore, candidates from the "sender" or "subject" zones have higher zone weights than those from "body".
- Document score. It captures the importance of the document where candidates are generated from. Candidates generated from more relevant emails are likely more valuable. Inappropriate candidates from unwanted emails are demoted due to low document scores. We set

$$DocScore = e^{1/r} \cdot p(q|D_i) \tag{1}$$

where r is the rank of the email provided by the email ranker. Note the freshness of the email, which is an important signal [18], is incorporated in r. Then,

$$p(q|D_i) = \Pi_{tok \in q} \frac{tf(tok, D_i)}{|D_i|} \tag{2}$$

measures the similarity between the original query q and email D_i following relevance models [16].
- Completion cost. It can be too aggressive if the suggestion is very long while the original query is short. In order to relieve this issue, candidates with shorter lengths receive more scores: $CompCost = 1 + (N_s - n_s)$.

The ranking score for a suggestion s_w and corresponding completion c_w is a weighted TF-IDF that aggregates all these features:

$$\begin{aligned} Score(s_w, c_w, q) \\ &= \sum_{D_i \in D} \sum_{s_w \in D_i \wedge c_w \in D_i} \left(\frac{1}{|D_i|} \cdot IDF(s_w) \right. \\ &\cdot Proximity(s_w, c_w, D_i) \cdot ZoneWeight(s_w, D_i) \\ &\left. \cdot DocScore(q, D_i) \cdot CompCost(s_w) \right), \end{aligned} \tag{3}$$

where D is the list of top-k relevant emails. The final ranking score for suggestion s_w is the sum of the scores over all completions c_w:

$$Score(s_w, q) = \sum_{c_w} Score(s_w, c_w, q). \tag{4}$$

For each suggestion s_w, the best $\hat{c_w}$ is used to construct the final QAC result:

$$\hat{c_w} = \arg\max_{c_w} Score(s_w, c_w, q). \tag{5}$$

To construct the final QAC candidate for s_w, the last step is to stitch the three parts together: the original query excluding the last term $\tilde{q}$, $\hat{c_w}$ and s_w. In most cases, the candidate can be formulated as the concatenation of $\langle\tilde{q}, \hat{c_w}, s_w\rangle$. However, if in the original email, completion $\hat{c_w}$ appears consecutively after suggestion s_w, the order of $\hat{c_w}$ and s_w should be switched. For example, imagine a user is looking for their reservation confirmation email from the Hyatt Hotel. Given the original query "hyatt conf", the suggestion s_w "reservation" has the best corresponding completion $\hat{c_w}$ "confirmation". The expected candidate "hyatt reservation confirmation" is actually the concatenation of $\langle\tilde{q}, s_w, \hat{c_w}\rangle$, since s_w ="reservation" and $\hat{c_w}$ ="confirmation" are consecutive in the email.

Lastly, it's possible that some of the final QAC candidates are similar to each other. To avoid duplication, we go through the ranked list of candidates. If a candidate does not contain any new tokens compared with all the other candidates that are ranked higher, it will be eliminated from the list.

4 Evaluation Method and Experiments

Evaluating personalized on-device search is non-trivial. Offline experiments must remain sensitive and insightful without breaking privacy promises whilst online A/B experiments are challenged by on-device model deployment and minimal instrumentation under privacy preserving conditions. To evaluate the efficacy of our QAC method, we propose a novel, grader-based offline evaluation that enables direct measurement of QAC quality without compromising grader privacy. This method also affords greater control over unintentional factors that can impact QAC quality including: on-device indexing status, query sampling, display position bias, and results scraping consistency.

4.1 Experiment Setup

To evaluate our method, we set up four, double-blind, offline experiments using 47 US-based graders. Each experiment evaluated the quality of a different QAC method using the same graders, the same email accounts, and the same query test sets. QAC results generated by each method were scraped on-device in quick succession in order to fix the state of each grader's email account.

To thoroughly protect grader privacy and ensure that no sensitive information ever left a grader's device, all QAC result scraping and grading happened

Table 1. Fixed test set distribution by scenario.

Email Scenario	Queries (%)
online order's shipment status	21
online order's cancellation or refund status	18
booking confirmation for a restaurant, flight, hotel, or car rental	18
bank or insurance statement	17
upcoming appointment or event	16
travel tip, recipe, or recommendation	10

solely on-device. Only quality grades, grader comments, and unpersonalized meta information collected during the experiment were sent to our server, along with optionally donated queries and QAC results.

The 47 US-based graders selected for these experiments were specialized in English annotation tasks and represented a diverse demographic pool with 47% identifying as women and 53% as men. Experiment eligibility required graders to be active email users with at least basic general technical skills. Advanced technical knowledge and skills were neither required nor sought. To prevent grader fatigue, the experiments were completed over a three-week period and each experiment restricted grading to no more than forty distinct query prefixes, which each generated no more than eight QAC results.

Before beginning these experiments, graders synced their primary personal email accounts to their evaluation devices. These accounts each contained between approximately 5,000 and 50,000 emails. Sufficient time was left prior to beginning the experiments to allow these email accounts to completely index.

Evaluating personalized search requires personalized queries that are relevant to each grader's email account. To collect personalized queries we presented six different email-search scenarios that prompted graders to think of an email in their inbox. If graders could think of a relevant email, they were then asked to provide search queries they would use to retrieve that email. The scenarios were selected from a query-traffic analysis that identified the most common use cases for email search.

QAC systems suggest QAC results upon each typed keystroke therefore to replicate these keystrokes, each grader's query was deconstructed into prefixes which were then weighted by length and randomly sampled to create one test set per grader. These test sets were fixed such that each grader used the same prefixes to evaluate QAC results in all four experiments. The final test set across all graders included 1,854 prefixes and the query distribution by scenario is included in Table 1.

Table 2. QAC String Quality guidelines and examples. The intent of the original query "americ" is to retrieve a recent annual statement email from American Express.

Rating	Meaning	Example: "americ"	Example Explanation
Helpfulness			
Helpful	QAC result fully matches the intent of the original query	american express statement	Fully matches intent of original query. Clicking this QAC result will likely return the desired email
Slightly Helpful	QAC result partially matches the intent of the original query; but, is either too general or too specific to fully match the intent	american express	Partially matches intent of original query but is too general to fully match. Clicking this QAC result will likely return the desired email amongst other undesired emails from American Express
Unhelpful	QAC result is unrelated to the intent of the original query	american airlines	Unrelated to the intent of original query. Clicking this QAC result will likely fail to return the desired email
Serious defect flags			
Unintelligible	QAC result is very difficult to understand (e.g. the QAC result is gibberish, incomplete, or in a language that you do not understand)	america by	QAC result is grammatically incomplete
Inappropriate	QAC result contains or seeks pornographic, disturbing, or offensive content	american viagra	QAC result is potentially offensive and was likely generated from a spam email.

4.2 Offline Evaluation and Metrics

QAC results are helpful shortcuts for users to arrive at their desired email with less effort. There are two dimensions that must be considered when evaluating the end-to-end quality of a QAC result: 1) *QAC String Quality*, the QAC result is intelligible and aligns with the search intent of the query; 2) *QAC-to-Email Quality*, clicking the QAC result returns the desired email. Both our evaluation and metrics are designed to capture these two dimensions.

Quality Evaluation. Each QAC result was graded independent of display position but dependent on the prefix and search intent of the original query.

To evaluate *QAC String Quality*, graders recorded a Helpfulness score on a 3-point scale based on the extent to which the QAC result aligned with the grader's search intent. To summarize the grading criteria, score "0" denotes an "Unhelpful" result unaligned with user intent, score "1" denotes a "Slightly Helpful" result somewhat aligned with user intent, and score "2" denotes a "Helpful" result fully aligned with user intent. For QAC results with a "0" score, graders were also able to select one or more defect flags to help categorize the "Unhelpful" result. The flags, "Unintelligible" and "Inappropriate", were further tagged as *serious defect* flags that represent the most critical product issues. See Table 2 for further guidelines and examples to evaluating *QAC String Quality.*

To evaluate *QAC-to-Email Quality*, graders were also asked to confirm whether the top six emails returned after selecting the QAC result contained their desired email(s). This step was completed independently of the string quality evaluation to prevent data peeking that might have influenced Helpfulness scores.

Metrics. To measure *QAC String Quality* we first adopt the widely used Normalized Discounted Cumulative Gain to compute NDCG Helpfulness@k using the 3-point Helpfulness score. Additionally, to control for highly defective QAC results, we also compute a binary, secondary metric, unweighted Defect Rate@k which measures the percentage of test set prefixes which generated one or more top-k QAC results with at least one *serious defect.* To measure *QAC-to-Email Quality* we use unweighted Email Recall@k which computes the percentage of top-k QAC results which successfully returned the desired email within the top 6 emails.

Metrics are measured @3 and @5 and for each metric we calculate the mean value across all test set prefixes. These prefix-averaged metrics most closely capture the end-user's experience who will see a new list of QAC results for each newly typed prefix.

4.3 Baseline Methods

Most existing QAC methods utilize query logs, and learn from public or shared corpora. To our best knowledge, [9,13] are the only methods that do not solely rely on query logs or shared corpora, hence serving as our baselines. These methods generate n-gram candidates which match the original query as a prefix. They are described as follows.

- **TF-IDF** [13] used term frequency (TF) and inverse document frequency (IDF) features to calculate ranking scores for all n-gram candidates extracted from the email corpus.
- **TF-IDF+** [9] used TF-IDF to capture the word completion probability (e.g., the probability of completing "appl" to "apple"). TF-IDF is then multiplied with a word-to-phrase probability (e.g., the probability of completing "apple" to "apple store").
- **Multiple** ranking features [13] used document scores and zone weights as additional features on top of TF-IDF. Feature weights are from a model where labels are derived from centrally collected logs. We have no access to massive logs due to the on-device setting, hence weights are manually tuned with a labeled set.

4.4 Results

Table 3. Evaluation results.

Methods	NDCG Helpfulness		Defect Rate		Email Recall	
	@3	@5	@3	@5	@3	@5
TF-IDF	43.17%	39.83%	3.22%	4.06%	41.05%	38.04%
TF-IDF+	46.69%	47.83%	4.68%	6.57%	45.59%	44.50%
Mul	44.67%	42.92%	4.19%	4.88%	44.20%	42.58%
Our Method	**53.79%**	**53.29%**	**1.72%**	**1.93%**	**46.76%**	**46.49%**

In total 19,584 QAC results were graded and experimental results are summarized in Table 3. It can be observed that our proposed method consistently outperformed all of the baselines. It achieved the highest NDCG Helpfulness, while maintaining the lowest Defect Rate. The gains show that the graders were generally more satisfied with the quality of our QAC results. Our proposed method also achieved the highest Email Recall, indicating that with our method it was much easier for the graders to locate their desired emails. The metrics gap between **TF-IDF+** and **TF-IDF** suggests the effectiveness of separating completion and suggestion parts during candidates ranking. **Mul** takes advantage of the additional document features and exhibited slightly higher NDCG and Email Recall, compared with **TF-IDF**.

The significant gains of our proposed method can be attributed to the following factors. First, our method is not limited to contiguous n-grams during candidates retrieval. By separating the completion and suggestion parts and progressively calculating the rank score, our method incorporates a larger candidates set, and produces higher NDCG Helpfulness and Email Recall at the same time. Second, our utilization of the structural ranking signals such as zone weight and proximity also helps boost NDCG Helpfulness and lower the Defect Rate. Third, the document score feature connects the email relevance and the QAC candidates ranking, which ensures that the generated candidates favor those top ranked documents that are supposed to be more relevant to the user, hence producing higher Email Recall. Fourth, the completion cost feature prevents a result from being too aggressive, hence lowers the Defect Rate.

In addition to the metrics gains, we also carried out voluntary case studies with our graders to make sure these gains are indeed intended. Two examples are presented in Table 4. In the first example, the grader's query is "bm", and the intent is to find a recent event email with subject "Registration is open for the BMW Ultimate Driving Experience". "bmw" is ranked 1st for all methods. **Our method** is the only one that can retrieve and rank result "bmw registration" to the top of the list. The token, "registration", is not consecutive to "bmw" in any zones of the matched emails, hence the baseline methods do not even have a chance to retrieve it as a candidate. **Mul** takes the zone weights into consideration, hence it promotes the result "bmw santa monica" which comes

Table 4. Case study examples. For "bm", the 2nd results are listed. For "cha", the 2nd and 3rd results are listed.

	"bm"	"cha"	
TF-IDF	bmw santa	chargers	chargers holders
TF-IDF+	bmw santa	chargers	change
Mul	bmw santa monica	chase credit journey	chargers holders
Our Method	bmw registration	chase credit journey	charles schwab

from the sender zone. **TF-IDF** and **TF-IDF+** both surface "bmw santa", which was graded as "Unintelligible" because it is incomplete and therefore the grader found it very difficult to understand.

In the second example, the grader's query is "cha", and the intent is to find emails from Chase Bank. All four methods rank "chase" to the 1st, hence it is not listed. TF-IDF's 2nd and 3rd results are "chargers" and "chargers holders", which come from a large number of promotional emails. These two results were graded as "Unintelligible" because the grader had never read these promotional emails and therefore found the results very difficult to understand. **Our method** and **Mul** were able to find a "Helpful" result, "chase credit journey", which comes from a sender, and surfaced due to higher zone weights. Our method's 3rd result, "charles schwab", has a low TF in the grader's mailbox, but comes from a recent email with a high document score. Although it is not aligned with the grader's main intent, it is better than "change" which comes from an unintended sender "Change.org". It is also better than "chargers holders" since it is potentially more useful and not defective.

Apart from quality evaluations, we also carried out latency measurements on the performance of our system. Our implementation has a low latency overhead of 50ms on average, which is fast enough for instant search, and is not perceivable by the end users.

5 Conclusions and Future Work

In this paper, we propose an on-device QAC method for email search. Our QAC system is seamlessly integrated with the on-device email retrieval and ranking systems, hence it is personalized and adaptive to a user's ever-changing email corpus. QAC candidates are generated from top ranked emails. Their retrieval is not limited to consecutive n-grams. Candidates ranking features are efficient and easy to implement. We also propose a novel, private corpora based offline evaluation method to measure on-device QAC quality. Experiments show that our method outperforms strong baseline algorithms.

There are promising directions for future work. Current ranking signals used in our method are mainly from lexical information, while rich semantic information can be extracted from email texts, entities, connections and user engagements. On-device personalized modeling or model fine-tuning based on a user's personal data is a very interesting and challenging direction, where federated learning and transfer learning can be employed to protect user privacy. Another potential direction is to study how the QAC results can be diversified to cover multiple possible query intents, and to design evaluation methods to measure the diversity of the generated QAC results.

References

1. Abdul-Jaleel, N., et al.: UMass at TREC 2004: novelty and hard. Computer Science Department Faculty Publication Series, p. 189 (2004)
2. Aberdeen, D., Pacovsky, O., Slater, A.: The learning behind gmail priority inbox (2010)
3. Ai, Q., Dumais, S.T., Craswell, N., Liebling, D.: Characterizing email search using large-scale behavioral logs and surveys. In: Proceedings of the 26th International Conference on World Wide Web, pp. 1511–1520 (2017)
4. Alrashed, T., Lee, C.J., Bailey, P., Lin, C., Shokouhi, M., Dumais, S.: Evaluating user actions as a proxy for email significance. In: The World Wide Web Conference, pp. 26–36 (2019)
5. Baeza-Yates, R., Hurtado, C., Mendoza, M.: Query recommendation using query logs in search engines. In: Lindner, W., Mesiti, M., Türker, C., Tzitzikas, Y., Vakali, A.I. (eds.) EDBT 2004. LNCS, vol. 3268, pp. 588–596. Springer, Heidelberg (2004). https://doi.org/10.1007/978-3-540-30192-9_58
6. Bar-Yossef, Z., Kraus, N.: Context-sensitive query auto-completion. In: Proceedings of the 20th International Conference on World Wide Web, pp. 107–116 (2011)
7. Barouni-Ebarhimi, M., Ghorbani, A.A.: A novel approach for frequent phrase mining in web search engine query streams. In: Fifth Annual Conference on Communication Networks and Services Research (CNSR 2007), pp. 125–132. IEEE (2007)
8. Bendersky, M., Wang, X., Metzler, D., Najork, M.: Learning from user interactions in personal search via attribute parameterization. In: Proceedings of the Tenth ACM International Conference on Web Search and Data Mining, pp. 791–799 (2017)
9. Bhatia, S., Majumdar, D., Mitra, P.: Query suggestions in the absence of query logs. In: Proceedings of the 34th International ACM SIGIR Conference on Research and Development in Information Retrieval, pp. 795–804 (2011)
10. Cai, F., Liang, S., De Rijke, M.: Time-sensitive personalized query auto-completion. In: Proceedings of the 23rd ACM International Conference on Conference on Information and Knowledge Management, pp. 1599–1608 (2014)
11. Carmel, D., Halawi, G., Lewin-Eytan, L., Maarek, Y., Raviv, A.: Rank by time or by relevance? Revisiting email search. In: Proceedings of the 24th ACM International on Conference on Information and Knowledge Management, pp. 283–292 (2015)
12. Dehghani, M., Rothe, S., Alfonseca, E., Fleury, P.: Learning to attend, copy, and generate for session-based query suggestion. In: Proceedings of the 2017 ACM on Conference on Information and Knowledge Management, pp. 1747–1756 (2017)

13. Horovitz, M., Lewin-Eytan, L., Libov, A., Maarek, Y., Raviv, A.: Mailbox-based vs. log-based query completion for mail search. In: Proceedings of the 40th International ACM SIGIR Conference on Research and Development in Information Retrieval, pp. 937–940 (2017)
14. INC., T.R.G.: Email Statistics Report, 2021–2025 (2021). https://www.radicati.com/wp/wp-content/uploads/2020/12/Email-Statistics-Report-2021-2025-Executive-Summary.pdf
15. Kang, Y.M., Liu, W., Zhou, Y.: QueryBlazer: efficient query autocompletion framework. In: Proceedings of the 14th ACM International Conference on Web Search and Data Mining, pp. 1020–1028 (2021)
16. Lavrenko, V., Croft, W.B.: Relevance-based language models. In: ACM SIGIR Forum, vol. 51, pp. 260–267. ACM New York, NY, USA (2017)
17. Li, C., Zhang, M., Bendersky, M., Deng, H., Metzler, D., Najork, M.: Multi-view embedding-based synonyms for email search. In: Proceedings of the 42nd International ACM SIGIR Conference on Research and Development in Information Retrieval, pp. 575–584 (2019)
18. Mackenzie, J., Gupta, K., Qiao, F., Awadallah, A.H., Shokouhi, M.: Exploring user behavior in email re-finding tasks. In: The World Wide Web Conference, pp. 1245–1255 (2019)
19. Meng, Y., Karimzadehgan, M., Zhuang, H., Metzler, D.: Separate and attend in personal email search. In: Proceedings of the 13th International Conference on Web Search and Data Mining, pp. 429–437 (2020)
20. Mitra, B., Craswell, N.: Query auto-completion for rare prefixes. In: Proceedings of the 24th ACM International on Conference on Information and Knowledge Management, pp. 1755–1758 (2015)
21. Park, D.H., Chiba, R.: A neural language model for query auto-completion. In: Proceedings of the 40th International ACM SIGIR Conference on Research and Development in Information Retrieval, pp. 1189–1192 (2017)
22. Rocchio, J.: Relevance feedback in information retrieval. In: The Smart Retrieval System-experiments in Automatic Document Processing, pp. 313–323 (1971)
23. Shokouhi, M.: Learning to personalize query auto-completion. In: Proceedings of the 36th International ACM SIGIR Conference on Research and Development in Information Retrieval, pp. 103–112 (2013)
24. Teevan, J., Dumais, S.T., Horvitz, E.: Personalizing search via automated analysis of interests and activities. In: Proceedings of the 28th Annual International ACM SIGIR Conference on Research and Development in Information Retrieval, pp. 449–456 (2005)
25. Vaswani, A., et al.: Attention is all you need. In: Advances in Neural Information Processing Systems, pp. 5998–6008 (2017)
26. Wang, P.W., Zhang, H., Mohan, V., Dhillon, I.S., Kolter, J.Z.: Realtime query completion via deep language models. In: eCOM@ SIGIR (2018)
27. Wang, W., Hosseini, S., Awadallah, A.H., Bennett, P.N., Quirk, C.: Context-aware intent identification in email conversations. In: Proceedings of the 42nd International ACM SIGIR Conference on Research and Development in Information Retrieval, pp. 585–594 (2019)
28. Yadav, N., Sen, R., Hill, D.N., Mazumdar, A., Dhillon, I.S.: Session-aware query auto-completion using extreme multi-label ranking. In: Proceedings of the 27th ACM SIGKDD Conference on Knowledge Discovery & Data Mining, pp. 3835–3844 (2021)

29. Yao, J., Dou, Z., Wen, J.R.: Employing personal word embeddings for personalized search. In: Proceedings of the 43rd International ACM SIGIR Conference on Research and Development in Information Retrieval, pp. 1359–1368 (2020)
30. Zhou, Y., Dou, Z., Wen, J.R.: Encoding history with context-aware representation learning for personalized search. In: Proceedings of the 43rd International ACM SIGIR Conference on Research and Development in Information Retrieval, pp. 1111–1120 (2020)

Empowering Legal Citation Recommendation via Efficient Instruction-Tuning of Pre-trained Language Models

Jie Wang[1](✉), Kanha Bansal[1], Ioannis Arapakis[2], Xuri Ge[1], and Joemon M. Jose[1]

[1] University of Glasgow, Glasgow, Scotland
j.wang.9@research.gla.ac.uk
[2] Telefonica Research, Barcelona, Spain

Abstract. The escalating volume of cases in legal adjudication has amplified the complexity of citing relevant regulations and authoritative cases, posing an increasing challenge for legal professionals. Current legal citation prediction methods, which are predominantly reliant on keyword or interest-based retrieval, are proving insufficient. In particular, Collaborative Filtering (CF) based legal recommendation methods exhibited low accuracy. In response to these challenges, we propose the Instruction GPT with Low-Rank Adaptation architecture (IGPT-LoRA), aiming to enhance the performance of legal citation recommendations and reduce computational demands by tuning Pre-trained Language Models (PLMs). IGPT-LoRA leverages prompting and efficient tuning strategies, thus offering a significant improvement over previous context-aware legal citation prediction methods. We design effective domain-specific instruction templates to guide the adaptation of PLMs for recommendation purposes, shedding light on the potential of prompt-based learning in the legal domain. Furthermore, we optimize the learning process with an efficient tuning layer - the Low-Rank Adaptation (LoRA) architecture - to bolster applicability. Experimental results on a real-world legal data set (BVA) demonstrate that IGPT-LoRA outperforms state-of-the-art methods, delivering substantial improvements in accuracy and also in training time and computational efficiency.

Keywords: Legal Citation Recommendation · Prompt · Tuning PLMs

1 Introduction

In the realm of legal adjudication, lawyers and judges bear the responsibility of drafting case documents or judgments, which frequently necessitates citing statutes and regulations set forth by legislative bodies and law enforcement agencies or referencing pertinent prior judgments. These citations form the bedrock of the analysis and decision-making process. However, given the constraints of

N. Goharian et al. (Eds.): ECIR 2024, LNCS 14608, pp. 310–324, 2024.
https://doi.org/10.1007/978-3-031-56027-9_19

available resources and the ever-expanding cases, identifying the relevant citations is becoming an increasingly formidable and time-consuming task. The prolific accumulation of cases can also precipitate slower decision-making and elevate the potential for errors. Therefore, tools that can assist legal professionals in efficiently and accurately locating relevant citations are indispensable for enhancing the efficiency of legal adjudication.

Legal citation prediction methods [49] typically rely on retrieving citations from a database, guided by keywords or attributes furnished by the user. However, this approach is generally not effective, as it necessitates decision-makers to extract essential attributes from lengthy cases—a task often intricate and time-consuming. Thankfully, recent Recommender Systems (RSs) [46], which quickly provide users with candidate choices in massive data, have broad prospects in the legal field [6,20,51]. However, the Collaborative Filtering (CF) based legal recommendation method exhibited exceedingly low accuracy. This result significantly deviates from that of other recommendation systems [20,48,53], potentially because legal citation recommendations rely more heavily on the intrinsic content of the cases rather than the relationships between citing behaviors.

Researchers have shifted their focus toward employing deep learning for legal citation recommendations, owing to its impressive performance in general RSs [15,47,52]. Recent RS methods [9,14,42] calculating context similarity through a bag-of-words approach have significantly outperformed CF and achieved higher accuracy. This strengthens the assertion that the cited object can be represented by the local context of words surrounding it, thus enhancing the effectiveness of retrieval. [20] further demonstrated improved performance by encoding case context through the RoBERTa [30], a pre-trained language model improves multiple downstream Natural Language Process (NLP) tasks, e.g., question answering [41], reading comprehension [23]. These findings suggest that domain-specific fine-tuning Pre-trained Language Models (PLMs) have considerable potential to yield superior results for legal citation recommendations. However, PLMs that perform better in downstream tasks often possess larger model sizes, e.g., GPT3 (175 B) [3], PaLM (540 B) [1] and Llama2 (70 B) [45]. Fine-tuning such models, with their extensive number of parameters, can be computationally heavy. Moreover, when applied to limited downstream data, these models often result in overfitting, which can adversely affect stability and scalability.

To further efficiently tune PLMS, Prompting [29,33] is proposed to incorporate extra text generally with inputs to encourage the performance adapted to specific domains. The existing strategies, e.g., in-context learning [3] and instruction learning [34], aim to reduce computational effort by not updating PLMs parameters. The former requires a context containing multiple related examples to help the model understand and perform the current task. For example, we first give the model a series of English-to-French translation examples and then provide a new English sentence for the model to translate. However, the inclusion of additional samples proportionally increased the demand for computational resources. While the instruction learning approach guides the model to learn how to generate appropriate responses based on the given instruction.

For instance, we might present the PLM with a prompt like '*Translate the following English text to French:* Hello, how are you?'. The model is expected to understand and follow this directive, translating the English text into French. Motivated by this,

In addition, due to constraints in computational resources, numerous PLMs endeavor to minimize computational overhead. It lies in two main approaches: (i) adapters [13,19] for PLMs that need to add some additional learnable layer in original layers in models, and (ii) prefix-tuning [27] of PLMs that adds some prefixes for specific parameters of trained networks. The former causes inference delays due to the introduction of additional model layers, while the latter occupies the input sequence space of downstream tasks due to the addition of prefixes, affecting model performance and making it difficult for the model to converge. To address the above issues, we employ Low-Rank Adaptation (LoRA) [18] technique to reduce the computational effort and improve the training efficiency of PLM adaptation to the legal domain.

In light of the aforementioned challenges and motivations, we summarize that the objective of our paper is to achieve comparable or superior performance in legal citation recommendation while reducing computation to improve efficiency. We propose to empower legal citation recommendation with PLMs (e.g., GPT family [3,38,39]) by prompting and efficient-tuning framework: Instruction learning GPT with Low-Rank Adaptation architecture (IGPT-LoRA). The main contributions of this paper are summarized as follows:

- We introduce a novel framework - *Instruction GPT with LoRA* (called *IGPT-LoRA*) - for modeling the task of legal citation recommendation. The accuracy of IGPT-LoRA's legal citation recommendations significantly outperforms state-of-the-art methods, while also achieving considerable improvements in training speed and computational efficiency.
- We formulate effective prompt templates to facilitate the adaptation of PLMs for legal citation recommendations. This exploration uncovers the potential of prompt learning in the legal citation domain and offers valuable insights.
- We further optimize the learning procedure by employing the efficient tuning layer for PLMs - *Low-Rank Adaptation (LoRA) architecture* - to improve applicability. Experimental results on the real-world legal dataset from BVA[1] corpus demonstrates the practical applicability of IGPT-LoRA.

2 Related Work

2.1 Legal Citation Recommendation

Legal citation prediction is an essential aspect of legal research, enabling legal professionals to efficiently identify relevant legal authorities [3,6,8,20,51]. Context-aware [20] method has been proposed directly automate the recommendation of legal citation. In this literature review, we provide an overview of the existing research and highlight the key advancements made in this domain.

[1] https://reglab.stanford.edu/data/bva-case-citation-dataset/.

Collaborative Filtering (CF). Early works [4,28,32] in legal citation recommendation focused on collaborative filtering techniques, where the recommendation of legal authorities was based on the patterns of citation behaviour observed in legal documents. These approaches utilized matrix factorization and user-item interaction models to make citation recommendations. While collaborative filtering showed promise in some cases, it often struggled with the sparsity of citation data, limiting its effectiveness in real-world applications.

Text Similarity-Based Models. Another set of methods leveraged text similarity [2,11,14,20,21] to predict legal citations. These models measure the semantic similarity between legal documents and recommend relevant citations. While text similarity-based models showed improvements in accuracy compared to collaborative filtering, they still faced challenges in handling long-contextual dependencies and capturing complex legal language nuances.

Context-aware Legal Citation Recommendation. Decision text of cases is considered to predict relevant citations. The introduction of BiLSTM [16,26,31] networks revolutionized legal citation prediction. BiLSTM models allowed for context-based citation prediction by capturing the sequential dependencies in legal text. The pre-trained RoBERTa model is fine-tuned to realize better performance. These models demonstrated impressive results in accurately predicting citations based on the surrounding text context. However, the computational complexity limited their scalability for large-scale legal datasets.

2.2 Pre-trained Language Models

In this section, we first introduce Pre-trained Language Models (PLMs). Then, we discuss how downstream tasks can learn from these PLMs and how they can be tuned effectively.

Pre-trained Language Models (PLMs). Transformer-based Language Model contains three classifications: decoder-only (e.g., GPT [38]), encoder-only (e.g., BERT [7]), or encoder-decoder (e.g., BART [25], T5 [40]). Typically, decoder-only models are trained autoregressively, while encoder-only and encoder-decoder models are trained using masked language modeling (MLM). Generative Pre-trained Transformer (GPT) models, e.g., GPT, GPT2 [39], and GPT3 [3], have positioned themselves as powerful tools in the domain of language generation tasks. Despite the impressive performance, their model size also increases, making them known as Large Language Models (LLMs). GPT3 is about 175 billion parameters and uses a huge amount of raw data for pre-training. There are also many other LLMs, such as PaLM (540 B) [5], T5 (11 B) and the recently proposed Llama2 (70 B) [45]. In our study of enhancing legal citation prediction, considering both computation resource limitation and dataset, we select GPT2 to explore the effectiveness of PLMs for generating legal

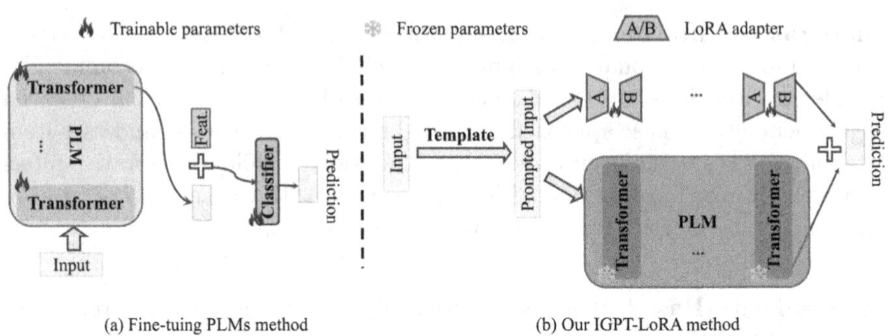

Fig. 1. (a) Fine-tuning full parameters of PLMs for legal citation recommendation. The input is case text. Features are concatenated with the last hidden state of the sequential outputs. (b) Our IGPT-LoRA is tuned based on efficient adapters. The input is prompted by domain-specific instructions.

citations since it has several versions (Small[2] and XL[3]) and relatively suitable model size.

Learning Paradigms. There are primarily two paradigms of leveraging PLMs, namely, fine-tuning and prompting the PLMs for specific tasks. The fine-tuning approach involves adjusting some or all layers of the PLM, and subsequently adding prediction heads [50], which are jointly trained in an end-to-end manner, with the majority of the computation focused on fine-tuning the language model to yield the desired input representation. Prompting [33], on the other hand, entails appending natural language text to the input or output to guide PLMs in performing specific tasks. Compared with fine-tuning, prompting may not necessitate updates to the PLM's parameters, thereby reducing computational demands. Prompts also facilitate a better alignment between the new task formulation and the pre-training objective. It's important to note that encountering tuning methods that employ both paradigms concurrently is common in recent research.

Efficient Tuning. While PLMs have been extremely successful, they are becoming larger and fine-tuning is more computationally expensive. This can make them difficult to use in practice, especially in resource-constrained environments. There are several strategies for addressing this issue. One approach is to use smaller, more efficient models. This might involve using a smaller version of a PLM (such as DistilBERT [43], a smaller version of BERT), or fine-tuning a model that has been specifically designed for efficiency (such as ALBERT [24]). Another approach is to apply techniques such as quantization [22] or pruning [37] to reduce the size of the model or the computational requirements for inference.

[2] https://huggingface.co/gpt2.
[3] https://huggingface.co/gpt2-xl.

Table 1. Prompting PLMs for recommending legal citations by instructions.

Instruction	Input ([X])	Output ([Y])	Template
Given a legal case, to infer the citations accordingly.	Case text	Citation label	**Instruction**. The context is [X]. The legal citation for the case is [Y].
You are a legal professional working on the case of the Board of Veterans. Your task is to predict the citation that a lawyer or judgment assigns based on the given context. Review the context below and provide the anticipated legal citation for the case.	Case text	Citation label	**Instruction.** The context is [X]. The anticipated legal citation for this case is [Y].

While these techniques can often produce loss on performance. Yet another approach is to use techniques such as Parameter-Efficient Fine-Tuning [17] (PEFT) by adapters [18,35,36] to reduce the number of parameters that need to be updated during fine-tuning. This can make fine-tuning more efficient and less prone to overfitting, especially when the amount of task-specific data is limited.

3 Framework of IGTP-LoRA

In this section, we first introduce the technology for prompting and describe our domain-specific prompt strategy for the legal citation prediction task. Subsequently, we elaborate on the integration of the adapter with GPT and elucidate the efficient tuning process employed in our experimental setup.

3.1 Prompt for Legal Citation Recommendation

We present two prompt templates for the legal citation prediction task, as detailed in Table 1. In these templates, the input denoted as X represents the text extracted from legal cases, and Y represents the label corresponding to the correct citation. Together with the accompanying instructions, these templates constitute the framework for the model's input. Here's a refined version of your statement: For cases with metadata, our approach differs from methods that concatenate feature embeddings with the output state and then make recommendations through a classifier. Alternatively, we incorporate the feature content into the case context, thereby generating the new input X. The first prompt serves the purpose of informing the Pre-trained Language Model (PLM) about the task at hand and instructs it to predict the appropriate citation. In contrast, the second prompt not only conveys the task but also provides guidance to the model regarding the underlying motivation and rationale for its predictions. This includes essential background information and contextual factors. These two distinct prompts are employed to assess the influence of varying levels of instructional granularity on the experimental outcomes.

3.2 Prompt-Tuning via LoRA

Figure 1 illustrates two learning paradigms for harnessing the capabilities of PLMs in the context of legal citation recommendation task: (a) the pre-train then fine-tune approach, and (b) the pre-train and prompt by LoRA strategy - our IGPT-LoRA -. In this paper, we have adopted the former as a comparative methodology. In this subsection, we detail the proposed IGPT-LoRA for legal citation recommendation by efficiently adapting PLMs through prompt tuning. The overall process is shown in Fig. 1(b). LoRA adapter is incorporated into each Transformer block, which consists of a small number of parameters enclosed within low-rank matrices A and B. Our choice to integrate LoRA techniques is motivated by their potential to enhance the feasibility of deploying large-scale PLMs in practical applications. After configuring PLM with LoRA, the case input is prompted through the instruction template 1. Specifically, for each legal case, designated as input X, along with its corresponding citation denoted as Y, our approach involves several steps. Initially, we generate a novel input utilizing a predefined template, which subsequently serves as the training data. In order to alleviate computational demands, we restrict the citation generation to the final position, indicated as $\tilde{Y}$, which aligns with the conventional practice of causal language models predicting the subsequent input sequence. The process is formulated as:

$$T = Template(X), \tag{1}$$

$$[Y_1, Y_2, ..., Y_n] = P_{\Phi+\Theta}(T), \tag{2}$$

where n is the number of candidate legal citations, T is the prompted input generated by an instruction template. Θ is the LoRA parameters that are updated during the training process.

Finally, the citations are predicted through the utilization of a cross-entropy loss function defined as follows:

$$\mathcal{L}_S = -\sum_{i=1}^{n} Y_i \log\left(\tilde{Y}_i\right). \tag{3}$$

$Y_i = 1$ if the truth label is the i-th citation within the recommendation candidates, otherwise, $Y_i = 0$.

4 Experiments

We examine the performance of IGPT-LoRA and baselines via experiments to answer the following research questions:

RQ1: How does the proposed IGPT-LoRA compare in terms of legal citation recommendation performance with the baselines?

Table 2. Metadata statistics.

Feature	year	issue area	VLG
#class	19	17	289

RQ2: How do the variations in tuning strategies of PLMs influence the performance? Can the models with reduced parameters maintain performance while being more computationally efficient?
RQ3: How does our proposed model reduce computation measured by the number of training parameters?
RQ4: How does our model improve training efficiency in terms of training time?

4.1 Experimental Setup

Dataset and Metrics. One of the prominent benchmark models in this context is the context-aware legal citation prediction model, as introduced in [20]. To ensure a fair comparison, we conducted our experiments employing the same dataset, comprising 324,309 legal cases sourced from the BVA corpus, and adhering to the original data pre-processing procedures. A 'legal case' refers to an appellate decision extracted from the BVA corpus, and citations typically pertain to instances where a decision references prior cases, legal statutes, or established legal precedents in support of its argumentation. In our study, we employ the decision text as the primary representation of the cases and incorporate three metadata features, namely the decision year, issue area, and Veterans' Law Judge (VLJ). Detailed metadata information is available in Table 2. We maintain an identical data split as outlined in [20], with 72% allocated to training, 18% to validation, and 10% to the test set. For evaluating the performance of legal citation recommendations, as in [20], we utilize the Recall metric (Recall@k) [10,12], where $k \in 1, 5, 20$. Recall@k quantifies the proportion of cases in which the correct citations are present among the top k predictions.

Baselines. In their study, [20] established BiLSTM and RoBERTa as the leading-edge methods for legal citation. In line with this, we primarily compare our proposed IGPT-LoRA with these models as detailed in [20]: **(i)** BiLSTM [20]: The Bi-directional Long Short-Term Memory (BiLSTM) layer, is employed to encode input sequences in both forward and backward directions. **(ii)** RoBERTa [20]: pre-trained using the Transformer Encoder and Mask Language Modelling (MLM), represents an advancement over BERT through prolonged sequence training and fine-tuning of hyperparameters. Given that we follow the same experimental setup, we report the results of these models as presented in [20]. Additionally, we include results from our own implementations of these models to investigate the reproducibility of the approach:

Table 3. Recommendation performance. Boldface denotes the highest scores and the second-best scores are marked with __. R is short for Recall.

Models	No Metadata			All Metadata		
	R@1	R@5	R@20	R@1	R@5	R@20
BiLSTM	65.2%	81.8%	91.1%	65.8%	82.4%	91.3%
BiLSTM-REP	64.9%	81.5%	90.8%	65.0%	81.8%	91.1%
RoBERTa	<u>65.6%</u>	82.8%	91.7%	<u>66.2%</u>	83.2%	92.1%
RoBERTa-REP	63.5%	81.1%	91.2%	64.2%	82.0%	91.5%
IGPT-LoRA$_{p1}$	65.2%	<u>84.4%</u>	<u>92.1%</u>	65.5%	<u>84.6%</u>	<u>92.4%</u>
IGPT-LoRA$_{p2}$	**65.8%**	**84.8%**	**93.2%**	**66.3%**	**84.9%**	**93.5%**

Table 4. Variants of GPT2.

Models	R@5	R@20
-Small	82.5%	91.6%
-XL	84.6%	92.6%
-XL$_{ft}$	82.8%	91.6%
-LoRA	84.3%	92.3%

(iii) BiLSTM-REP and **(iv)** RoBERTa-REP. For baselines with all metadata, the features are concatenated with the last hidden state and utilized as inputs for the classification layers, as illustrated in Fig. 1(a). Note that in the 'All Metadata' experiments, our IGPT-LoRA incorporates the textual content of the features directly into the case to represent the input context.

Variations. To assess the performance and computational aspects of IGPT-LoRA, we conducted experiments using three additional configurations. These are: **(v)** GPT2-Small and XL: In this setup, all parameters of the selected GPT2 model were fine-tuned; **(vi)**GPT2-XL$_{ft}$: This approach involves the freezing of upper-layer parameters and exclusively fine-tuning the final layers, matching the same number of trainable parameters as IGPT-LoRA. In addition to these, we introduced a straightforward adaptation of LoRA for GPT2, referred to as: **(vii)** GPT-LoRA, whose input is the original case context, to demonstrate the impact of instructions/prompts on model performance. All the baseline models and their variants are provided with input in the form of word sequences extracted from legal cases. Their task is to predict the most probable citation within the given context. Note that we introduce two types of prompts, resulting in two variations of our IGPT-LoRA model: IGPT-LoRA$p1$ and IGPT-LoRA$p2$. Both versions have been evaluated and their results are included in our study.

Implementation Details. The parameter settings of BiLSTM and RoBERTa are directly taken from [20]. We use the pre-trained RoBERTa[4] and GPT2 released by the huggingface community[5]. All the experiments are performed under one RTX A6000 48 GB GPU. For our IGPT-LoRA and variants, we set batch size to 64 and use Adam [44] optimizer with a learning rate of 1e-4.

[4] https://huggingface.co/roberta-base.
[5] https://huggingface.co/.

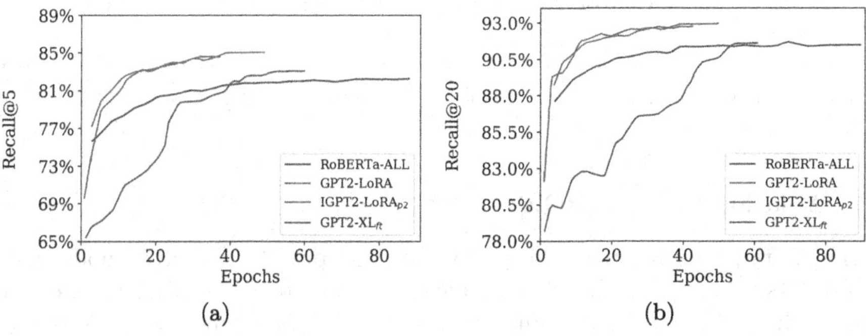

Fig. 2. Fluctuations in Recall@5 and Recall@20 with incremental epochs.

4.2 Performance Comparison (RQ1)

As presented in Table 3, we provide a comprehensive comparison of legal citation recommendation results with baselines. To begin with, the results of BiLSTM-REP and RoBERTa-REP are largely consistent with the results in the original paper[6]. For example, The BiLSTM models, irrespective of whether metadata is incorporated or not, demonstrate relatively consistent performance across all metrics, achieving their peak Recall@20 at 91.1%.

Our proposed model, IGPT-LoRA, excels across all evaluation metrics. Notably, it attains the highest performance in Recall@1, reaching 65.8% and 66.3% with all and with no metadata, respectively. Specifically, it outperforms the best-performing baseline model by relative improvements of 3.1%. Moreover, a slight improvement is also observed in the Recall@5 and Recall@20 metrics. These outcomes indicate that the IGPT-LoRA model not only excels at suggesting the most pertinent legal citations (as evident in Recall@1) but also shines in recommending a broader selection of relevant citations (as demonstrated by Recall@5 and Recall@20), surpassing the performance of the other models.

In conclusion, the IGPT-LoRA model outperforms other models in the task of legal citation recommendation. Particularly in recommending a wider array of pertinent citations, it offers significant advantages for legal research and practice by aiding legal professionals in efficiently identifying relevant citations. Figure 2 illustrates the evolution of performance as the number of training epochs increases. It is evident that models incorporating LoRA consistently enhance performance and converge fast.

[6] Overall, we observed a slight decrease in the range of 0.2% to 0.4%.

Table 5. Number of trainable parameters at training stage for all models.

Model	RoBERTa	GPT2-Small	GPT2-XL	GPT2-XL$_{ft}$	IGPT-LoRA
Training params (Million)	335	259	12,462	250	**247**

4.3 Ablation Studies (RQ2)

In this section, we compare the performance of our IGPT-LoRA model with its variants, results are shown in Table 4, to evaluate the contribution of our model. The GPT-XL model, which fine-tunes all parameters, significantly outperforms the GPT2-Small model. This outcome underscores the positive correlation between the size of the PLMs and their performance in legal citation recommendation tasks. However, when we align the number of trainable parameters with those in the LoRA model, GPT-XL$_{ft}$ model exhibits lower performance and lags behind the metrics of the LoRA architecture. From Fig. 2, we can see that GPT2-XL$_{ft}$, which involves updates to a subset of parameters (similar to LoRA), exhibits substantial variability. This observation spotlights the adaptability and effectiveness of the LoRA adapters. Further, a comparison between LoRA and IGPT-LoRA reveals that the latter model advances improvements across all evaluated metrics. This comparison underscores the significant impact of the employed prompts on the model's performance. Analyzing the results between prompts (IGPT-LoRA$_{pt1}$ and IGPT-LoRA$_{pt2}$), we find that IGPT-LoRA$_{pt2}$, which provides the PLM with more detailed background and instructions on how to perform recommendations, outperforms IGPT-LoRA$_{pt1}$, which simply narrates the task. This suggests that more explicit and informative prompts can enhance the performance of PLMs in domain-specific recommendation tasks. These insights are instrumental in understanding the superior performance of our IGPT-LoRA model. The IGPT-LoRA model exhibits outstanding adaptability to customized prompts, which renders it a promising solution for legal citation prediction, aligning perfectly with the overarching objective of enhancing precision and reliability in this field.

4.4 Parameter Comparison (RQ3)

As illustrated in Table 5, while the GPT2-XL model can achieve desirable performance, it requires tuning of more than 50 times the number of parameters compared to LoRA architecture. When the number of parameters is similar, our IGPT-LoRA model outperforms others in legal citation recommendation. This emphasizes the efficiency of the IGPT-LoRA model - it maintains high performance while significantly reducing the number of parameters, thereby increasing the model's efficiency.

Compared to the GPT2-Small and GPT2-XL$_{ft}$ models, the number of parameters in the IGPT-LoRA model is slightly reduced, yet the performance is improved. This underscores the optimization in the design and training process of the IGPT-LoRA model and the effectiveness of the LoRA adapters.

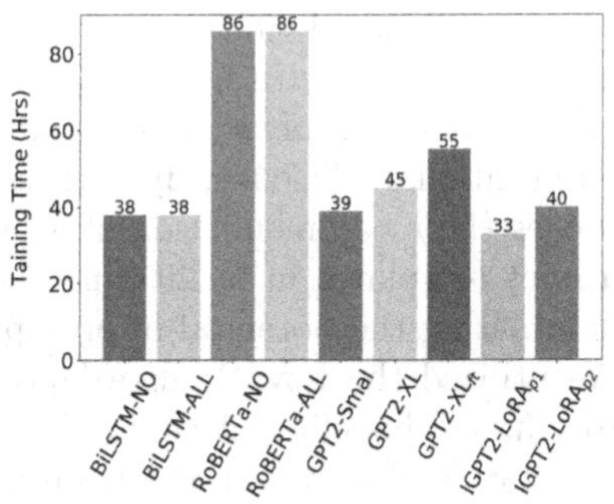

(a) Comparison of training time.

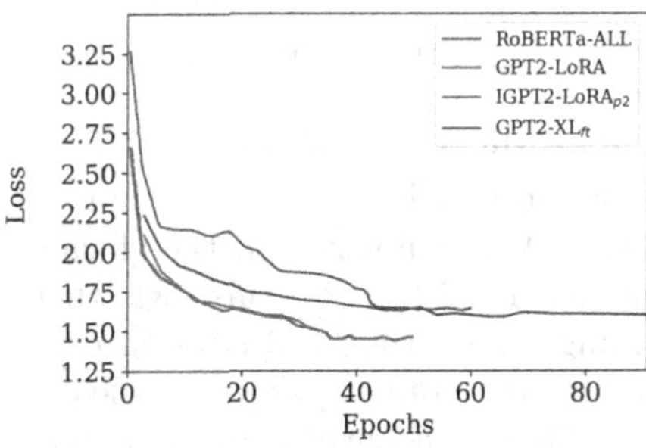

(b) The loss curves during training.

Fig. 3. (a) The total hours required for each method to complete a single training session. (b) The reduction of loss during the training stage varies among different methods, and the number of epochs required for each to stabilize on the validation set determines this trend.

In summary, this parameter comparison highlights that the superior performance of the IGPT-LoRA model does not solely rely on a large number of training parameters. This gives the IGPT-LoRA model a significant advantage in practical applications, particularly in scenarios where a balance between computational resources and performance is required.

4.5 Training Time Comparison (RQ4)

In this section, we compare the training time and the number of epochs required by each model. From Fig. 3, it can be observed that the GPT2-LoRA model requires the least amount of training time while maintaining a reasonable number of epochs, indicating its efficiency. When compared to the BiLSTM, RoBERTa, GPT2-XL, IGPT-LoRA requires fewer hours of training and fewer epochs, yet still maintains or even surpasses the performance of the other models. This demonstrates the efficiency of the IGPT-LoRA model in terms of training time and the number of epochs.

In summary, this training time comparison highlights that the desirable efficiency of the IGPT-LoRA model surpasses that of other models not only in terms of the number of training parameters but also in terms of training time. This gives the GPT2-LoRA model a significant advantage in practical legal citation prediction scenarios where both time efficiency and performance are critical.

5 Conclusion

The objective of this research was to evaluate the potential of Pre-trained Language Models (PLMs) in handling the complex task of legal citation prediction. Our journey was guided by clear objectives: not only to improve the precision of citation predictions but also to optimize the computational efficiency of these models. This paper proposed a framework, Instruction GPT with LoRA (IGPT-LoRA), which surpasses state-of-the-art methods in terms of accuracy, while concurrently improving training speed and computational efficiency. By harnessing the potential of GPT, our approach improves the recommendation of legal citation. We demonstrated the efficacy of prompt templates in facilitating the adaptation of PLMs for this task, thereby illuminating the potential of prompt learning within the legal citation domain. We utilized the Low-Rank Adaptation architecture to optimize the learning procedure. This efficient tuning layer for PLMs significantly reduces computational effort and enhances the training efficiency of PLM adaptation to the legal domain.

References

1. Ames, D., Handan-Nader, C., Ho, D.E., Marcus, D.: Due process and mass adjudication: crisis and reform. Stan. L. Rev. **72**, 1 (2020)
2. Brin, S.: The PageRank citation ranking: bringing order to the web. In: Proceedings of ASIS, vol. 98, pp. 161–172 (1998)
3. Brown, T., et al.: Language models are few-shot learners. Adv. Neural. Inf. Process. Syst. **33**, 1877–1901 (2020)
4. Caragea, C., Silvescu, A., Mitra, P., Giles, C.L.: Can't see the forest for the trees? A citation recommendation system. In: Proceedings of the 13th ACM/IEEE-CS Joint Conference on Digital Libraries, pp. 111–114 (2013)
5. Chowdhery, A., et al.: PaLM: scaling language modeling with pathways. arXiv preprint arXiv:2204.02311 (2022)
6. Dadgostari, F., Guim, M., Beling, P.A., Livermore, M.A., Rockmore, D.N.: Modeling law search as prediction. Artif. Intell. Law **29**, 3–34 (2021)
7. Devlin, J., Chang, M.W., Lee, K., Toutanova, K.: BERT: pre-training of deep bidirectional transformers for language understanding. arXiv preprint arXiv:1810.04805 (2018)
8. Fowler, J.H., Johnson, T.R., Spriggs, J.F., Jeon, S., Wahlbeck, P.J.: Network analysis and the law: measuring the legal importance of precedents at the us supreme court. Polit. Anal. **15**(3), 324–346 (2007)
9. Ge, X., Chen, F., Jose, J.M., Ji, Z., Wu, Z., Liu, X.: Structured multi-modal feature embedding and alignment for image-sentence retrieval. In: Proceedings of the 29th ACM International Conference on Multimedia, pp. 5185–5193 (2021)
10. Ge, X., Chen, F., Xu, S., Tao, F., Jose, J.M.: Cross-modal semantic enhanced interaction for image-sentence retrieval. In: Proceedings of the IEEE Winter Conference on Applications of Computer Vision, pp. 1022–1031 (2023)
11. Gori, M., Pucci, A.: Research paper recommender systems: a random-walk based approach. In: 2006 IEEE/WIC/ACM International Conference on Web Intelligence (WI 2006 Main Conference Proceedings)(WI 2006), pp. 778–781. IEEE (2006)
12. Gunawardana, A., Shani, G.: Evaluating recommender systems. In: Ricci, F., Rokach, L., Shapira, B. (eds.) Recommender Systems Handbook, pp. 265–308. Springer, Boston, MA (2015). https://doi.org/10.1007/978-1-4899-7637-6_8
13. He, J., Zhou, C., Ma, X., Berg-Kirkpatrick, T., Neubig, G.: Towards a unified view of parameter-efficient transfer learning. arXiv preprint arXiv:2110.04366 (2021)

14. He, Q., Pei, J., Kifer, D., Mitra, P., Giles, L.: Context-aware citation recommendation. In: Proceedings of the 19th International Conference on World Wide Web, pp. 421–430 (2010)
15. He, X., Liao, L., Zhang, H., Nie, L., Hu, X., Chua, T.S.: Neural collaborative filtering. In: Proceedings of the 26th International Conference on World Wide Web, pp. 173–182 (2017)
16. Hochreiter, S., Schmidhuber, J.: Long short-term memory. Neural Comput. **9**(8), 1735–1780 (1997)
17. Houlsby, N., et al.: Parameter-efficient transfer learning for NLP. In: International Conference on Machine Learning, pp. 2790–2799. PMLR (2019)
18. Hu, E.J., et al.: LoRA: low-rank adaptation of large language models. arXiv preprint arXiv:2106.09685 (2021)
19. Hu, Z., et al.: LLM-adapters: an adapter family for parameter-efficient fine-tuning of large language models. arXiv preprint arXiv:2304.01933 (2023)
20. Huang, Z., et al.: Context-aware legal citation recommendation using deep learning. In: Proceedings of the Eighteenth International Conference on Artificial Intelligence and Law, pp. 79–88 (2021)
21. Jeh, G., Widom, J.: SimRank: a measure of structural-context similarity. In: Proceedings of the Eighth ACM SIGKDD International Conference on Knowledge dIscovery and Data Mining, pp. 538–543 (2002)
22. Kim, S., Gholami, A., Yao, Z., Mahoney, M.W., Keutzer, K.: I-BERT: integer-only BERT quantization. In: International Conference on Machine Learning, pp. 5506–5518. PMLR (2021)
23. Lai, G., Xie, Q., Liu, H., Yang, Y., Hovy, E.: RACE: large-scale reading comprehension dataset from examinations. arXiv preprint arXiv:1704.04683 (2017)
24. Lan, Z., Chen, M., Goodman, S., Gimpel, K., Sharma, P., Soricut, R.: ALBERT: a lite BERT for self-supervised learning of language representations. arXiv preprint arXiv:1909.11942 (2019)
25. Lewis, M., et al.: BART: denoising sequence-to-sequence pre-training for natural language generation, translation, and comprehension. arXiv preprint arXiv:1910.13461 (2019)
26. Li, P.H., Fu, T.J., Ma, W.Y.: Why attention? Analyze BiLSTM deficiency and its remedies in the case of NER. In: Proceedings of the AAAI Conference on Artificial Intelligence, vol. 34, pp. 8236–8244 (2020)
27. Li, X.L., Liang, P.: Prefix-tuning: Optimizing continuous prompts for generation. arXiv preprint arXiv:2101.00190 (2021)
28. Liben-Nowell, D., Kleinberg, J.: The link prediction problem for social networks. In: Proceedings of the Twelfth International Conference on Information and Knowledge Management, pp. 556–559 (2003)
29. Liu, X., et al.: P-Tuning v2: prompt tuning can be comparable to fine-tuning universally across scales and tasks. arXiv preprint arXiv:2110.07602 (2021)
30. Liu, Y., et al.: RoBERTa: a robustly optimized BERT pretraining approach. arXiv preprint arXiv:1907.11692 (2019)
31. Ma, J., Ganchev, K., Weiss, D.: State-of-the-art Chinese word segmentation with Bi-LSTMs. arXiv preprint arXiv:1808.06511 (2018)
32. McNee, S.M., et al.: On the recommending of citations for research papers. In: Proceedings of the 2002 ACM Conference on Computer Supported Cooperative Work, pp. 116–125 (2002)
33. Min, B., et al.: Recent advances in natural language processing via large pre-trained language models: a survey. ACM Comput. Surv. **56**(2), 1–40 (2023)

34. Ouyang, L., et al.: Training language models to follow instructions with human feedback. Adv. Neural. Inf. Process. Syst. **35**, 27730–27744 (2022)
35. Pfeiffer, J., et al.: AdapterHub: a framework for adapting transformers. arXiv preprint arXiv:2007.07779 (2020)
36. Pfeiffer, J., Vulić, I., Gurevych, I., Ruder, S.: MAD-X: an adapter-based framework for multi-task cross-lingual transfer. arXiv preprint arXiv:2005.00052 (2020)
37. Prasanna, S., Rogers, A., Rumshisky, A.: When BERT plays the lottery, all tickets are winning. arXiv preprint arXiv:2005.00561 (2020)
38. Radford, A., Narasimhan, K., Salimans, T., Sutskever, I., et al.: Improving language understanding by generative pre-training (2018)
39. Radford, A., Wu, J., Child, R., Luan, D., Amodei, D., Sutskever, I., et al.: Language models are unsupervised multitask learners. OpenAI blog **1**(8), 9 (2019)
40. Raffel, C., et al.: Exploring the limits of transfer learning with a unified text-to-text transformer. J. Mach. Learn. Res. **21**(1), 5485–5551 (2020)
41. Rajpurkar, P., Jia, R., Liang, P.: Know what you don't know: unanswerable questions for squad. arXiv preprint arXiv:1806.03822 (2018)
42. Ritchie, A., Robertson, S., Teufel, S.: Comparing citation contexts for information retrieval. In: Proceedings of the 17th ACM Conference on Information and Knowledge Management, pp. 213–222 (2008)
43. Sanh, V., Debut, L., Chaumond, J., Wolf, T.: DistilBERT, a distilled version of BERT: smaller, faster, cheaper and lighter. arXiv preprint arXiv:1910.01108 (2019)
44. Stamenkovic, D., Karatzoglou, A., Arapakis, I., Xin, X., Katevas, K.: Choosing the best of both worlds: diverse and novel recommendations through multi-objective reinforcement learning. In: Proceedings of the Fifteenth ACM International Conference on Web Search and Data Mining, pp. 957–965 (2022)
45. Touvron, H., et al.: Llama 2: open foundation and fine-tuned chat models. arXiv preprint arXiv:2307.09288 (2023)
46. Wang, J., et al.: TransRec: learning transferable recommendation from mixture-of-modality feedback. arXiv preprint arXiv:2206.06190 (2022)
47. Wang, J., Zhu, L., Dai, T., Wang, Y.: Deep memory network with Bi-LSTM for personalized context-aware citation recommendation. Neurocomputing **410**, 103–113 (2020)
48. Wang, J., Zhu, L., Dai, T., Xu, Q., Gao, T.: Low-rank and sparse matrix factorization with prior relations for recommender systems. Appl. Intell. **51**, 3435–3449 (2021)
49. Winkels, R., Boer, A., Vredebregt, B., Van Someren, A.: Towards a legal recommender system. In: JURIX, pp. 169–178 (2014)
50. Wolf, T., et al.: Transformers: state-of-the-art natural language processing. In: Proceedings of the 2020 Conference on Empirical Methods in Natural Language Processing: System Demonstrations, pp. 38–45 (2020)
51. Zhang, P., Koppaka, L.: Semantics-based legal citation network. In: Proceedings of the 11th International Conference on Artificial Intelligence and Law, pp. 123–130 (2007)
52. Zhang, S., Yao, L., Sun, A., Tay, Y.: Deep learning based recommender system: a survey and new perspectives. ACM Comput. surv. (CSUR) **52**(1), 1–38 (2019)
53. Zhang, T., Zhu, L., Wang, J.: Neighborhood constraints based bayesian personalized ranking for explainable recommendation. In: Li, B., Yue, L., Tao, C., Han, X., Calvanese, D., Amagasa, T. (eds.) Asia-Pacific Web (APWeb) and Web-Age Information Management (WAIM) Joint International Conference on Web and Big Data, pp. 166–173. Springer, Cham (2022). https://doi.org/10.1007/978-3-031-25201-3_12

A Streaming Approach to Neural Team Formation Training

Hossein Fani[(✉)], Reza Barzegar, Arman Dashti, and Mahdis Saeedi

University of Windsor, Windsor, ON, Canada
{hfani,barzegar,vaghehd,msaeedi}@uwindsor.ca

Abstract. Predicting *future* successful teams of experts who can effectively collaborate is challenging due to the experts' temporality of skill sets, levels of expertise, and collaboration ties, which is overlooked by prior work. Specifically, state-of-the-art neural-based methods learn vector representations of experts and skills in a *static* latent space, falling short of incorporating the possible drift and variability of experts' skills and collaboration ties in time. In this paper, we propose (1) a streaming-based training strategy for neural models to capture the evolution of experts' skills and collaboration ties over time and (2) to consume time information as an additional signal to the model for predicting future successful teams. We empirically benchmark our proposed method against state-of-the-art neural team formation methods and a strong temporal recommender system on datasets from varying domains with distinct distributions of skills and experts in teams. The results demonstrate neural models that utilize our proposed training strategy excel at efficacy in terms of classification and information retrieval metrics. The codebase is available at https://github.com/fani-lab/OpeNTF/tree/ecir24.

Keywords: Neural Team Formation · Training Strategy · OpeNTF

1 Introduction

Teamwork has shown to be crucial in today's interdisciplinary environment, like in academia [15,28,46], industry [2,6,18], law [17,42], freelancing [4], and the healthcare system [8,40]. Team formation problem aims to automate forming teams of experts whose combined skills, applied in coordinated ways, can solve difficult tasks such as science projects whose success can be measured by publications, or the next blockbuster '*thriller*' with a touch of '*sci-fi*' in the movie industry. Team formation can also be seen as social information retrieval (Social IR) where the *right* group of experts is required to solve the task at hand. Forming teams is challenging due to the large number of candidates from various cultural backgrounds and personality traits as well as unknown synergistic balance among them. More importantly, experts' interests, skills, and levels of expertise change due to society's demands, novel technologies, and working experience. For instance, with the growth of automation, more and more experts are

N. Goharian et al. (Eds.): ECIR 2024, LNCS 14608, pp. 325–340, 2024.
https://doi.org/10.1007/978-3-031-56027-9_20

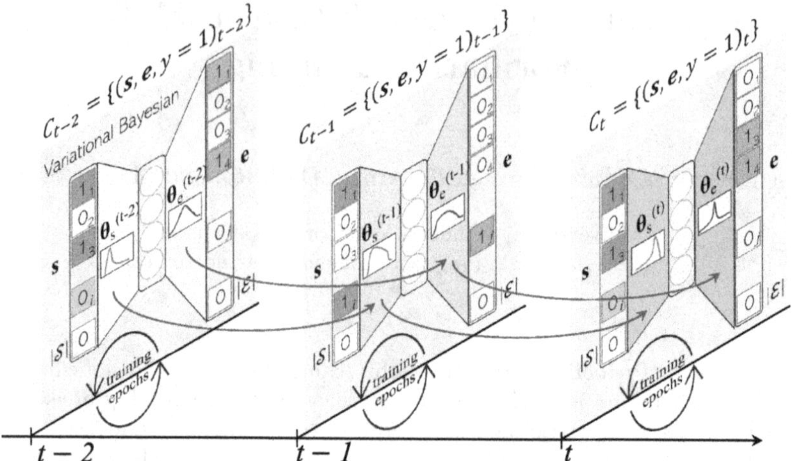

Fig. 1. Streaming training strategy for future team prediction. A neural model learns from the collaborations of experts C_t at time t to kick-start learning the collaborations at the next time interval $t + 1$. Best viewed in color.

acquiring skills related to computer science, as seen in social science, biology, and linguistics, among other sciences [14,31]. Therefore, a successful collaboration of experts *years ago* would not tailor a successful one in the *future*.

Despite a large body of computational methods to address the team formation problem for an overwhelming number of experts, the positive impacts of considering temporality are yet to be studied. Operations research (OR)-based methods, wherein multiple objective functions are optimized with respect to constraints such as planned budget and timeline via integer programming, forego the temporality of experts' skills and collaboration ties [3,11,39,45]. Graph-based team formation methods represent the expert network as a *static* graph and overlook the evolution of the expert network in time and the emergence of new collaborations [19,22,27,29]. State-of-the-art neural-based methods learn experts and skills vector representations in a *static* latent space, and hence, fall short of incorporating the possible drift of experts' skills in time and its impact on the prediction of future successful teams [9,10,33,34,36–38,41]. Little work considers time but as a *constraint* to model the projects' deadlines, the availability of experts, or uncertainty about the duration of the projects [3,39].

In this paper, we propose a streaming training strategy to encode temporal aspects in neural-based team formation methods. Specifically, given the stream of experts' collaborations in each time interval, a neural model learns the collaborations of experts at time interval t to kick-start learning the collaborations of the next time interval $t + 1; 1 \leq t \leq \mathrm{T}$, as shown in Fig. 1. Our proposed training strategy, when employed by neural models, allows experts to change their vector positions in latent space as their skills and collaboration ties evolve over time, and captures the change trajectories up until time interval T to accurately predict experts' vector positions in *future* time interval T+1. Contrary to non-temporal methods that assume the independent and identically distributed (i.i.d) instances of teams (bag of teams) [9,19,27,36,41], our approach incorporates temporality by streaming the teams within time intervals in its training step. In contrast to considering time as a constraint, we study the horizontal nature of time to learn the evolution of experts' skills and collaboration ties in time. We perform experiments on datasets that enjoy distinct distributions of skills and experts in teams, namely `dblp`[1] [26,27], `imdb`[2] [19,21], `uspt`[3] [24], `gith`[4] [25] to demonstrate the domain-free effectiveness of our proposed method. Comparing our work with the state-of-the-art, our results show that incorporating the temporal evolution of experts' skills and collaboration ties exhibits superior performance in predicting future successful teams of experts.

2 Related Works

Since Zakarian and Kusiak's work [47], there has been a surge of research in the team formation problem that can be differentiated based on their optimization method: (1) search-based, where the task of searching for the best team is executed over *all* the subgraphs of the expert network or via integer programming, and (2) learning-based, where a machine learning algorithm, a neural network in particular, is utilized to form teams of experts by learning the distributions of experts and skills in the context of successful teams in the past. Nonetheless, literature related to the team formation problem has ignored the impact of experts' temporal behavior, by and large, despite widespread successful incorporation of temporality in other domains such as temporal information retrieval, temporal knowledge graphs, and temporal recommender systems [12,30], to name a few. There is little work in team formation [3,11,39,45] that studied time but as a constraint such as the projects' deadlines in the optimization functions. In this section, we review some of the prominent works in the team formation literature.

2.1 Non-temporal Methods

Search-based Methods. The foremost method of team formation was conceived in operations research (OR), where multiple objectives must be optimized simultaneously via integer or real programming to find the *optimum* team, given constraints for human and non-human factors and scheduling preferences. Based on the engineering characteristics of a product and the importance of customer requirements, Zakarian *et al.* [47] used the integer linear programming approach to form multi-functional teams. They imposed integer constraints on experts, such as a cap on the number of projects each expert as a team member may take on. More recently, Neshati *et al.* [32] translate team formation to facility location analysis to form groups of experts to perform a multi-aspect task that

[1] https://aminer.org/citation.
[2] https://imdb.com/interfaces/.
[3] https://uspto.gov/ip-policy/economic-research/research-datasets.
[4] https://codelabs.developers.google.com/codelabs/bigquery-github.

requires a diverse set of skills. Therein, teams were defined as locations, a set of skills as facilities, and experts' membership in teams as customers' needs and the optimization happens for optimal locations of facilities while simultaneously satisfying customers' needs. Such works, however, were premised on the mutually independent selection of experts and overlooked the collaborative and social ties among experts.

Although Chen and Lin [7] were among the first to consider experts' ties for team formation, they were Lappas *et al.* [27] who employed social network analysis to fill the gap by incorporating social ties and interpersonal collaboration features. They represented the experts' social network with a graph where nodes are experts with their set of skills, and edges represent the previous collaboration between them. The optimum team hence can be found by a search on *all* possible subgraphs. They proposed two algorithms based on the diameter of the graph and the cost of the minimum spanning tree (MST) to find a subgraph in which experts collectively hold the required skills and can collaborate effectively with minimum communication cost. However, the diameter of a subgraph or the minimum spanning tree is *in*accurate estimators of the true communication costs in a team, and also sensitive to slight changes in the graph that yield a radical change in the solution. To overcome these issues, Kargar and An [19] proposed two novel communication cost functions that minimize the sum of distances function for teams with a leader and lack thereof. Later, Kargar *et al.* [22] further proposed to consider additional budget constraints (expert salary) on top of communication costs as in real-world scenarios. They proposed a *bi*-objective approximation algorithm to optimize communication cost and salary in tandem.

Methods of efficient keyword search on attributed graphs have also been employed for team formation [25,26]. For instance, given a set of query keywords as skills and the desired size of the subgraph as the team size, Khan *et al.* [25] aimed to find closely connected subgraphs with the specific number of nodes wherein nodes contain as many query keywords as possible. Since the total number of answers is exponential in the number of query keywords and the size of the group, they proposed a method to find the approximate top-k groups with polynomial delay. Nonetheless, OR or graph-based optimization models for the task of team formation are computationally intractable and have to be followed by polynomial heuristic solutions such as multichoice [1] for subgraph identification with shortest diameter [27], or simulated annealing [3], branch-and-cut, genetic algorithms [43], and balanced placement [13] for those based on integer programming (IP). Indeed, IP is NP-hard, and subgraph optimization can be reduced to the decision version of the Steiner-tree problem, which is also proved to be NP-hard [23].

Learning-based Methods. Recently, a paradigm shift to machine learning has been observed in team formation literature, opening doors to the analysis of massive collections of experts from different domains. Machine learning approaches efficiently learn relationships between experts and their skills in the context of successful (positive samples) and unsuccessful teams (negative samples) from all past instances [9,10,33,34,36–38,41]. Sapienza *et al.* [41] employed a deep neural autoencoder to form teams and to capture which teammates foster the growth of their peers. However, when training data suffers from the popularity bias, such as in the team formation problem where a few experts have participated in the majority of teams for a small subset of skills while many experts have participated sparingly, autoencoder neural networks are prone to overfitting [5]. Rad *et al.* [36] proposed a variational Bayesian neural architecture to employ uncertainty in learnable parameters and overcome popularity bias. However, they only utilized past successful teams to train their neural model. Dashti *et al.* [9] proposed to utilize negative sampling heuristics to incorporate both successful and *virtually* unsuccessful teams in their training, which resulted in more efficient and effective neural models during training and inference, respectively. Nonetheless, existing learning-based methods neglected the *temporal* nature of experts' skills and collaborative ties.

2.2 Time as Constraint

There has been little work that used time as a constraint to model experts' availabilities or predefined start and due dates of projects. Durfee *et al.* [11] take into account scheduling constraints or preferences in a two-step team formation process. First, teams are built in the matchmaking optimization stage using integer linear programming, taking into account the required skills as well as the ability to be more readily (re)scheduled with respect to the timing requirement. Next, in the scheduling optimization stage, time slots are allotted to the team for completing the task using integer *non*linear programming optimization in a way that minimizes the total delay of the starting times of all the members while satisfying sequential and concurrent ordering constraints. Rahmanniyay *et al.* [39] studied the impact of various factors like weather conditions that can change the duration of a project or delay the delivery of material to a manufacturing company. Yang *et al.* [45] apply integer programming to determine the optimum team of experts *available* at a certain point in time. Contrary to considering time as an optimization constraint, we propose to treat time as an *aspect* through which experts' skills and collaboration ties evolve.

3 Problem Definition

We aim to incorporate the evolution of experts' skills and collaborative ties over time in order to predict *future* teams of experts who collectively hold a set of required skills and can effectively cooperate toward a shared goal based on their gained experience through time. Let $\mathcal{S}$ and $\mathcal{E}$ be the sets of skills and experts, and $\mathcal{C}_t = \{(s, e, y)_t | s \subseteq \mathcal{S}, e \subseteq \mathcal{E}, y \in \{0, 1\}\}$ be the set of collaborations at time t where (s, e) is a team whose members are a subset of experts, e, that collectively hold the subset of skills, s, and has been either successful $y = 1$ or a failure $y = 0$, and t is a discrete entity showing the time intervals. Intuitively, $\mathcal{C}_t$ is a snapshot of all teams of experts over skills during the time interval t and $[\mathcal{C}_1..\mathcal{C}_t..\mathcal{C}_\mathrm{T}]$ streams the dynamic distribution of experts over skills within T consecutive time intervals in the context of teams. Examples of teams include research groups where researchers are the experts and fields of study are the skills, movies consisting of casts and crews such as actors and directors as the experts and the genres as the skills, patents consisting of inventors as the experts and categories (classes) as the skills, or software projects where software developers and programming languages are the experts and the skills, respectively. Figure 2 demonstrates the *non*-uniform and temporal distribution of movies over genres (skills) and casts and crews within time in `imdb` dataset. As seen, although the set of genres remains the same over 100 years, the number of movies that adopt each genre varies over time. Also, an actor (expert) adopts various genres (skills) during his career. In real world, a similar trend can also be observed in research (`dblp`), patents (`uspt`), and computer software (`gith`) domains.

Strangely, the basic question of *"what it means for a team to be successful"* has gone underexamined and has remained controversial in the literature. Finding experts who collectively cover the required skills for a team is *in*sufficient and error-prone for a successful team since skillful experts enjoy various cultural backgrounds and personality traits that result in an unknown synergistic balance among them. Recently, little learning-based work (Sect. 2.1) has defined success (failure) based on the tangible outcomes of a team, like the number of publications for a research group, or the number of issued patents for a team of inventors. In some domains, however, what constitutes success remains controversial. For example, in the movie industry, a movie's success can be measured based on its immediate reception by the people (box office) or critical reviews (ratings) within a long span of time. Nonetheless, a team's label of success y can be redefined without loss of generality in our proposed method. For instance, success can be redefined based on the number of citations for a research paper, critical acclaim for a movie, and commercialization for a patent. In the absence of unlabeled unsuccessful teams, state-of-the-art learning-based methods follow the closed-world assumption; they presume existing instances of teams in the training dataset as successful ($y = 1$) and subsets of experts who have *not* collaborated yet for the input skills as unsuccessful teams (*virtually* negative samples).

Given the stream of collaboration sets $[\mathcal{C}_1..\mathcal{C}_t..\mathcal{C}_\mathrm{T}]$ in the past, we aim to recommend a *new* team of experts e' for a given subset of skills s at a yet-to-be-seen time interval T+1 whose collaboration has a high chance of success, i.e., $(s, e', 1)_{\mathrm{T}+1}$, also referred to as an *optimum* team. More formally, we aim to estimate a mapping function f of parameters θ from the stream of collaboration sets and a subset of skills to a subset of experts whose collaboration in a team is almost surely successful for the one-step-ahead *future* time interval T+1; that is, $f([\mathcal{C}_1..\mathcal{C}_t..\mathcal{C}_\mathrm{T}], s; \theta) = e'$ such that $(s, e', y = 1)_{\mathrm{T}+1}$.

4 Proposed Method

The main contribution of this paper is not a novel machine learning model but a training strategy for such models to take into account the temporal nature of

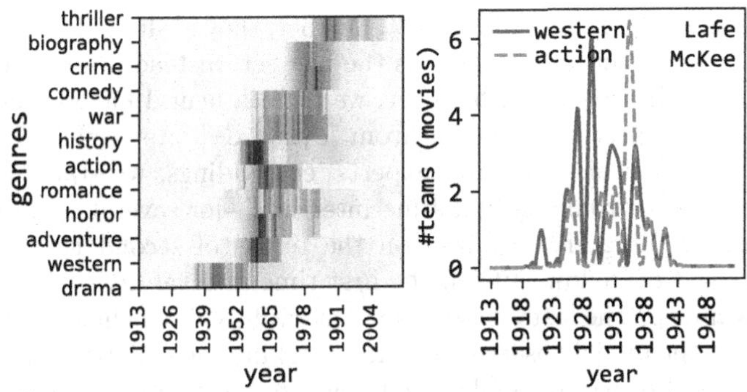

Fig. 2. Temporal distribution of movies over genres (left), and temporal inclination of an actor toward two genres (right). Best viewed in color.

the data in team formation. Let $[\mathcal{C}_1..\mathcal{C}_t..\mathcal{C}_\mathrm{T}]$ be the ordered list of all previous collaborations at each time interval t until T in which experts' collaborations and their skills in teams are evolving over time. We estimate f using a neural model that maximizes the average log probability of successful subsets of experts:

$$\frac{1}{|\mathcal{C}_{\mathrm{T}+1}|} \sum_{(s,e,y)\in\mathcal{C}_{\mathrm{T}+1}} \log \mathrm{p}(y|(s,e):\mathrm{T}+1) \tag{1}$$

where $\mathcal{C}_{\mathrm{T}+1}$ is the collection of yet-to-be-formed unseen (un)successful teams (s, e, y) in the *future* time interval T+1. Since $\mathcal{C}_{\mathrm{T}+1}$ is unseen, we optimized Eq. 1 through observed teams of (s, e, y) in the past:

$$\sum_{t=1}^{\mathrm{T}} \frac{1}{|\mathcal{C}_t|} \sum_{(s,e,y)\in\mathcal{C}_t} \log \mathrm{p}(y|(s,e):t) \tag{2}$$

The same team (s, e) may experience success and/or failure in different time intervals. Therefore, $p(y|(s,e):t)$ depends on the time interval information. To maximize Eq. 2, we map each subset of skills s and each subset of expert e to a low-rank d-dimensional vector in the same latent space, denoted by v_s and v_e, whose positions up until time interval T depend on the preceding movements in the latent space since the first time interval via observation of $[\mathcal{C}_1..\mathcal{C}_t..\mathcal{C}_\mathrm{T}]$ while imposing the following assumptions: (i) skills and experts change their latent representations over time, (ii) subsets of experts who collaborated in teams over similar subsets of skills within $[\mathcal{C}_1..\mathcal{C}_t..\mathcal{C}_\mathrm{T}]$ remain close in latent space, (iii) subsets of experts and skills who are close in latent space at their final positions in the latent space are presumably the optimum teams whose successes are almost surely guaranteed in the *future* time interval T+1.

4.1 Streaming Learning

Previous work in team formation assumed teams are independent and identically distributed and followed the bag of teams approach during model training on a shuffled dataset [3,9,19,20,27,36,39,41]. Further, they evaluated their models on a randomly selected subset of teams as the test set, instead of predicting future successful teams. In this work, however, we train a neural model incrementally over an ordered collection of teams from $[\mathcal{C}_1..\mathcal{C}_t..\mathcal{C}_\mathrm{T}]$. As seen in Fig. 1, after random initialization of skills' and experts' embeddings, we start training the model on the teams from the first time interval $\mathcal{C}_1$ for several epochs, then we continue with training (fine-tuning) on the teams of second time interval $\mathcal{C}_2$ using the learned embeddings from the first time interval and so forth until we finish the training on the last training time interval $\mathcal{C}_\mathrm{T}$. We believe that using this approach helps the model observe how experts' skills and collaborative ties evolve through time, and hence the final embeddings are their optimum representations in the latent space to predict future successful teams.

At each time interval t, we estimate $p(y|(s,e):t)$ through pairwise cosine similarities of embeddings for the subset of experts e and the subset of skills s through all (un)successful teams at time interval t in $\mathcal{C}_t$. More specifically, we estimate $p(y=1|(s,e):t)$ by learning v_e and v_s that are close (high cosine similarity) in the latent space if the subset of experts e has successful collaborations in $\mathcal{C}_t$ with the subset of skills s during the time interval t and estimate $p(y=0|(s,e):t)$ by learning v_e and v_s that are distant (low cosine similarity) otherwise. Hence, $p(y|(s,e):t)$ can be formulated with the sigmoid function σ:

$$p(y|(s,e):t) = \sigma(v_e^\top \cdot v_s) \tag{3}$$

Like Dashti *et al.* [9], when no unsuccessful team is available in the training set, we follow the closed-world assumption to generate *virtually* unsuccessful teams (negative samples), that is, if no successful team for the subset of skills s is known for a randomly selected subset of experts e'' at time interval t, i.e., $(s,e'') \notin \mathcal{C}_t$, the team is considered to be unsuccessful $(s,e'',y=0)$. To this end, we employ an optimization function that discriminates successful and unsuccessful teams through negative sampling from a distribution over the subsets of experts:

$$\sum_{(s,e)\in\mathcal{C}_t \leftrightarrow (s,e,y=1)} [\log \sigma(v_e^\top \cdot v_s) + \sum_{(s,e'')\sim\mathbb{P}:(s,e'')\notin\mathcal{C}_t \leftrightarrow (s,e'',y=0)}^{k} \log \sigma(-v_{e''}^\top \cdot v_s)] \tag{4}$$

where $\mathbb{P}$ is the probability distribution from which we randomly draw k subsets of experts e'' as negative samples for a given subset of skills s. The input layer of the neural model is either (i) sparse occurrence vector representations for skills of size $|\mathcal{S}|$, (ii) pre-trained dense vector representations (`emb`) for the subsets of skills as suggested by Rad *et al.* [36], or (iii) temporal dense skill vector representations (`dt2v`) using temporal word embedding method by Hamilton *et al.* [16] to directly incorporate temporal evolution of skills into the underlying neural model in addition to our proposed streaming strategy. The output layer of the model is sparse occurrence vector representations for experts of size $|\mathcal{E}|$.

Table 1. Statistics of the raw and preprocessed datasets.

	dblp		uspt		imdb		gith	
	raw	filtered	raw	filtered	raw	filtered	raw	filtered
#teams	4,877,383	99,375	7,068,508	152,317	507,034	32,059	132,851	11,312
#unique experts	5,022,955	14,214	3,508,807	12,914	876,981	2,011	452,606	2,686
#unique skills	89,504	29,661	241,961	67,315	28	23	20	19
avg #expert per team	3.06	3.29	2.51	3.79	1.88	3.98	5.52	7.53
avg #skill per team	8.57	9.71	6.29	9.97	1.54	1.76	1.37	1.57
avg #team per expert	2.97	23.02	5.05	44.69	1.09	62.45	1.62	31.72
avg #skill per expert	16.73	96.72	19.49	102.53	1.59	10.85	2.03	5.18
#team w/ single expert	768,956	0	2,578,898	0	322,918	0	0	0
#team w/ single skill	5,569	56	939,955	8,110	315,503	15,180	69,131	6014

5 Experiments

In this section, we lay out the details of our experiments and findings toward answering the following research questions:

RQ1: Does moving embeddings of experts and skill through time in the latent space improve the performance of neural models for the prediction of *future* successful teams? To this end, we benchmark state-of-the-art variational Bayesian neural network [9] (`bnn-*`) that utilizes negative sampling heuristics with our proposed streaming scenario training approach (t`bnn-*`) and lack thereof.

RQ2: Does adding time explicitly to the input embeddings of skills boost neural models performance? We compare the performance of neural models with utilizing temporal skills in the input t`bnn_dt2v_emb` and lack thereof t`bnn_emb`.

RQ3: Is the impact of our proposed training strategy consistent across datasets from various domains with distinct statistical distributions? We benchmark our proposed training approach on `dblp`, `imdb`, `uspt`, and `gith` datasets.

5.1 Setup

Dataset. We evaluate our proposed method on four well-known benchmark datasets in team formation literature, namely `dblp` [26,27], `imdb` [19,21], `uspt` [24], `gith` [25]. In `dblp`, each instance is a publication in computer science consisting of authors, the fields of study (fos), and the year it was published including papers from 1979 to 2018. We map each publication to a team whose authors are the experts and fields of studies are the set of skills. In `imdb`, each instance is a movie consisting of its cast and crew such as director, producer, actors, genre and the year the movie was released, spanning from 1914 to 2020. We consider each movie as a team whose members are the cast and crew, and the movies' genres are the teams' skills. The choice of `imdb` in team formation literature is not to be confused with its use cases in recommender systems or review analysis research; herein, the goal is to form a team of casts and

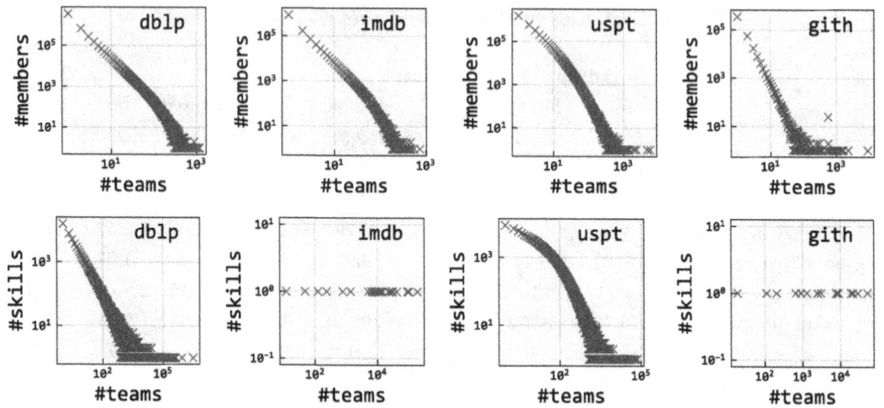

Fig. 3. Distribution of teams over members and skills for all datasets *before* preprocessing.

crews for a movie production as opposed to a movie recommendation [19,21]. In `uspt`, each instance is a patent invention in the United States Patents and Trademarks consisting of inventors (experts) and subcategories (skills) and the time the patent is issued, consisting of patents from 1976 to 2019. In `gith`, each instance is a GitHub repository consisting of the contributors of the repository (experts), the title and programming languages of the project (skills), and the time of the project's release, consisting of repositories from 2008 to 2022.

In all datasets, we can observe the long tail problem in the distributions of teams over experts. As shown in Fig. 3, many experts (researchers in `dblp`, cast and crew in `imdb`, inventors in `uspt`, and developers in `gith`) have participated in very few teams (papers in `dblp`, movies in `imdb`, inventions in `uspt`, and repositories in `gith`). For instance, 10^6 researchers have participated in 1 team only while few researchers have co-authored more than 10^3 papers in `dblp`. With respect to the set of skills, `dblp` and `uspt` are clearly following different distributions compared to `imdb` and `gith`. While `dblp` and `uspt` suffer further from the long-tailed distribution of skills in teams, `imdb` and `gith` follow a more fair distribution, as shown in Fig. 3. Specifically, `imdb` and `gith` have a limited variety of skills (genres and programming languages) which are, by and large, employed by many movies and repositories, respectively.

We filter out singleton and sparse teams with less than `3` members as well as experts who relatively participated in very few teams, as suggested by [9,36]. The latter also reduced the computational complexity of the neural models in their last layer where the size equals the number of experts. We filter out experts who participated in less than `75` teams for `dblp`, `imdb`, and `uspt`, and less than `10` teams for `gith`. From Fig. 4, we ensured that the preprocessing step made no major change to the statistical distributions of the datasets. Also, Table 1 reports additional point-wise statistics on the datasets.

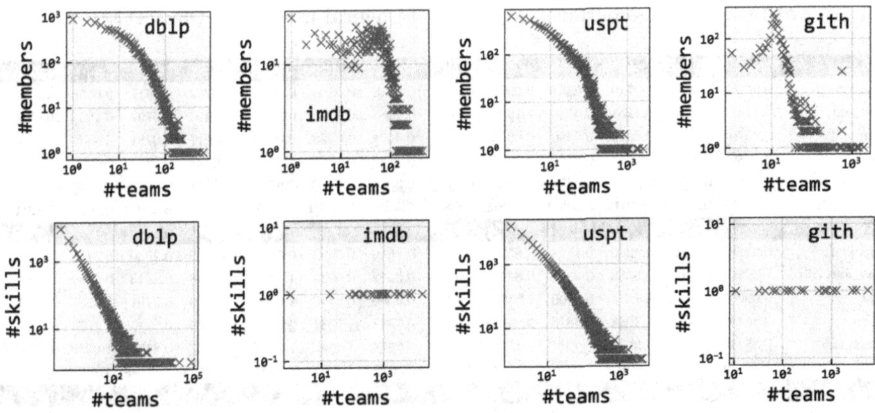

Fig. 4. Distribution of teams over members and skills for all datasets *after* preprocessing.

Baselines. We compare our temporal neural models using streaming training strategy ($t$bnn-*) with (1) non-temporal Bayesian (variational) neural network [9] (bnn-*) and (2) recurrent recommender networks [44] (rrn), where we recommend experts as items to input skills as users. In contrast to conventional recommender systems that assume users' profiles and items' attributes are static, rrn captures their temporal dynamics to predict future behavioral trajectories using a long short-term memory (lstm) autoregressive model and to excel at prediction accuracy. Both temporal and non-temporal Bayesian neural networks utilize the negative sampling objective function (Eq. 4) and include a single hidden layer of size `d = 128`, and `relu` and `sigmoid` are the activation functions for the hidden layer and the output layer, respectively. We used the smoothed unigram distribution in each training mini-batch [9] to generate the negative samples in Eq. 4. We train the model at each time interval t with a learning rate of `0.1` over `20` epochs with mini-batches of size `128` and use `Adam` as the optimizer. We used the same hyper-parameters for rrn.

Evaluation. To test the impact of the streaming training strategy and incorporation of time information to the input embeddings in the prediction of *future* successful teams, we take the last year of each dataset for the test set. To ensure the effectiveness of our approach, we perform `5`-fold cross-validation on the teams in each year for model training and validation. Given a team $(s, e)_{T+1}$ from the test set, we compare the ranked list of predicted experts e' by the model of each fold with the observed subset of experts e and report the average performance of models trained on each fold by information retrieval metrics including normalized discounted cumulative gain (`ndcg`), and mean average precision (`map`) at top-{2,5,10} as well as classification metrics including precision (`pr`) and recall (`rec`) at top-{2,5,10} and area under the receiver operating characteristic curve (`aucroc`).

Table 2. Average performance of 5-fold neural models on the test set.

dblp	%pr2	%pr5	%pr10	%rec2	%rec5	%rec10	%ndcg2	%ndcg5	%ndcg10	%map2	%map5	%map10	%aucroc
bnn [36]	0.0570	0.0663	0.0710	0.0351	0.0993	0.2118	0.0538	0.0806	0.1330	0.0242	0.0411	0.0558	63.52
bnn_emb [35]	0.1124	0.1290	0.1251	0.0668	0.1909	0.3699	0.1083	0.1555	0.2397	0.0474	0.0792	0.1033	66.81
rrn [44]	0.0570	0.0391	0.0472	0.0380	0.0630	0.1552	0.0478	0.0523	0.0959	0.0217	0.0281	0.0446	50.73
*t*bnn	0.1189	0.1413	0.1664	0.0706	0.2090	0.4984	0.1126	0.1689	0.3031	0.0484	0.0845	0.1223	73.08
*t*bnn_emb	0.2996	0.2938	0.2811	0.1816	0.4433	0.8431	0.3048	0.3860	0.5721	0.1411	0.2095	0.2635	74.83
*t*bnn_dt2v_emb	**0.4299**	**0.3973**	**0.3612**	**0.2601**	**0.5963**	**1.0801**	**0.4284**	**0.5221**	**0.7465**	**0.1947**	**0.2864**	**0.3520**	**77.01**
gith													
bnn [36]	0.2128	0.5106	0.4255	0.1418	0.8511	1.3050	0.1646	0.5699	0.7848	0.0709	0.2600	0.3148	51.16
bnn_emb [35]	0.4255	0.5106	0.6383	0.2837	0.8511	1.9574	0.3292	0.5923	1.1358	0.1418	0.2813	0.4389	51.82
rrn [44]	0.0000	0.8511	0.8511	0.0000	1.4184	2.8369	0.0000	0.8163	1.4606	0.0000	0.3191	0.6265	52.22
*t*bnn	0.8511	**1.5319**	1.4043	0.5319	**2.4610**	4.4965	0.7548	1.7381	2.6829	0.3369	0.8215	1.1674	63.46
*t*bnn_emb	0.8511	1.1064	1.0638	0.5674	1.7518	1.3262	0.9474	1.4848	2.2007	0.4965	0.8138	1.0099	**66.87**
*t*bnn_dt2v_emb	**1.9149**	1.1915	**1.4468**	**1.2411**	1.9504	**4.5532**	**1.8667**	**1.8703**	**3.0303**	**0.9043**	**1.1099**	**1.4293**	66.56
uspt													
bnn [36]	0.0657	0.0769	0.0910	0.0353	0.0976	0.2212	0.0655	0.0883	0.1481	0.0266	0.0433	0.0592	64.54
bnn_emb [35]	0.3663	0.4123	0.3748	0.1608	0.4509	0.8141	0.3652	0.4531	0.6094	0.1212	0.2027	0.2583	69.85
rrn [44]	0.0239	0.0383	0.0654	0.0140	0.0500	0.1370	0.0221	0.0408	0.0868	0.0096	0.0186	0.0340	51.60
*t*bnn	0.1843	0.1841	0.2029	0.0933	0.2321	0.5158	0.1794	0.2152	0.3481	0.0681	0.1056	0.1429	75.44
*t*bnn_emb	0.8272	0.7539	0.7042	0.3970	0.9021	1.6933	0.8457	0.9057	1.2657	0.3104	0.4533	0.5679	83.59
*t*bnn_dt2v_emb	**1.2268**	**1.0583**	**0.9324**	**0.6037**	**1.2928**	**2.2518**	**1.2322**	**1.2960**	**1.7348**	**0.4626**	**0.6659**	**0.8118**	**85.34**
gith													
bnn [36]	3.0693	2.8515	2.6931	1.2164	2.8846	5.1174	3.1365	3.2893	4.2340	1.0104	1.5706	2.1633	56.18
bnn_emb [35]	**7.3267**	**4.7129**	3.3861	**3.5441**	**5.1580**	**6.1885**	**6.4753**	**5.8418**	**6.2665**	**2.3424**	**3.0822**	**3.3837**	62.65
rrn [44]	0.0000	0.1980	0.0990	0.0000	0.0619	0.0619	0.0000	0.1679	0.1090	0.0000	0.0206	0.0206	52.26
*t*bnn	3.8614	2.8515	2.3564	1.8801	3.1525	4.5754	4.3319	3.9721	4.5031	1.8025	2.3978	2.8768	56.65
*t*bnn_emb	4.9505	3.5248	3.1287	1.9434	3.0770	4.3718	5.0849	4.4715	4.9844	1.6957	2.1431	2.5949	62.20
*t*bnn_dt2v_emb	5.7426	4.5941	**3.8020**	2.1874	3.8474	4.7855	5.6081	5.3287	5.6670	1.7131	2.4258	2.7858	**64.89**

5.2 Results

Foremost, we acknowledge that baselines achieve low values of evaluation metrics for practical application of team formation, which is primarily due to the simplicity of the neural model architectures and the small number of training epochs; metric values are reported in % for ease of readability and comparison. Our main goal is *not* to report state-of-the-art results for a novel model *but* to showcase the synergistic effects of our proposed training strategy for such models.

In response to **RQ1**, i.e., whether the streaming training strategy improves the predictive power of state-of-the-art neural models, from Table 2, comparing bnn and bnn_emb with *t*bnn and *t*bnn_emb respectively, we can observe that streaming training strategy increases neural models' relative performance between 10% to 20% on dblp and uspt in terms of the classification metrics (aucroc). On imdb and gith, it also improves the performance of neural models in terms of the information retrieval metrics. More specifically, on imdb, comparing bnn_emb with *t*bnn_emb, we can observe a relative gain of near 200% on some metrics (ndcg2, ndcg5, map2, and map5). Moreover, our training strategy increases neural models' relative performance on most of the information retrieval metrics between 100% and 200% on dblp and uspt. On gith, however, we can observe that using the streaming training strategy decreases the models' performance when using pretrained dense vector representation for input skills even though *t*bnn outperforms bnn in most of the information retrieval metrics.

In response to **RQ2**, i.e., whether adding time explicitly to the input of the neural model improves its performance while utilizing the streaming training strategy, from Table 2, comparing *t*bnn_emb with *t*bnn_dt2v_emb, we see that the models that utilize temporal skills in the input gain relative performance of between 30% to 50% in terms of *all* information retrieval metrics on dblp, uspt, and gith and up to 100% on imdb. Comparing neural models with sparse skill input representation (*t*bnn) with the ones that utilize temporal skill embeddings *t*bnn_dt2v_emb, we observe a substantial gain in relative performance in terms of information retrieval metrics between 100% and 500% on dblp and uspt. On gith, we still observe an increase in models' performance, but the gain in performance is not as substantial. Finally, on imdb, *t*bnn outperforms *t*bnn_dt2v_emb on some of the metrics such as pr5 and rec5 and perform on a par on pr10 and rec10. Nonetheless, we observe a relative gain of up to 100% on other information retrieval metrics on imdb. In summary, the temporal dense vector representation of skills always leads to a performance improvement in terms of classification and information retrieval metrics.

Regarding **RQ3**, i.e., whether the impact of our proposed streaming training strategy is consistent across different datasets with distinct distributions of skills and experts, from Table 2, we can see that the degree of the increase in performance of neural models depends on the distributions of experts and skills (Figs. 3 and 4) in teams and the evolution of experts and skills over time (Fig. 2). More concretely, for datasets with a long-tailed distribution of skills in teams (dblp and uspt), utilizing our proposed streaming strategy will help neural models in the prediction of *future* successful teams, which is contrary to datasets with a limited set of skills that are employed almost uniformly by teams (imdb and gith) (Fig. 4). Finally, from Table 2, we can see that the results of our proposed training strategy and incorporation of temporal skills are always superior compared to the temporal recommender system baseline [44] (rrn) on all four datasets for all the metrics.

6 Concluding Remarks

In this paper, we proposed a streaming training strategy for neural models to learn the evolution of experts' skills and collaborative ties to predict future successful teams. We further examined the impact of adding temporal information to the input of neural models. Our experiments on four datasets with distinct distributions of teams over skills and experts in time show that (1) our proposed streaming training strategy improves the predictive power of neural models, (2) neural models that leverage temporal information in the input obtain better performance compared to the lack thereof in most cases, and (3) neural models utilizing our proposed training strategy outperform the temporal recommender system baseline. Possible future directions of our work include spatio-temporal study of team formation where both temporal dynamics of experts and skills in teams as well as their geo-locations are considered to recommend location-based future teams with minimum communication costs.

References

1. Arkin, E., Hassin, R.: Minimum diameter covering problems. Networks **36**, 147–155 (2000)
2. Askari, G., Asghri, N., Gordji, M.E., Asgari, H., Filipe, J.A., Azar, A.: The impact of teamwork on an organization's performance: a cooperative game's approach. Mathematics **8**(10), 1–15 (2020)
3. Baykasoglu, A., Dereli, T., Das, S.: Project team selection using fuzzy optimization approach. Cybern. Syst. Int. J. **38**(2), 155–185 (2007)
4. Bernabé, R.B., Navia, I.A., García-Peñalvo, F.J.: Faat: freelance as a team. In: Proceedings of the 3rd International Conference on Technological Ecosystems for Enhancing Multiculturality, TEEM 2015, pp. 687–694, New York, NY, USA. Association for Computing Machinery (2015)
5. Blundell, C., Cornebise, J., Kavukcuoglu, K., Wierstra, D.: Weight uncertainty in neural network. In: Proceedings of the 32nd International Conference on Machine Learning. Proceedings of Machine Learning Research, vol. 37, pp. 1613–1622, Lille, France, 07–09 Jul 2015. PMLR (2015)
6. Bursic, K.M.: Strategies and benefits of the successful use of teams in manufacturing organizations. IEEE Trans. Eng. Manage. **39**(3), 277–289 (1992)
7. Chen, S.-J., Li, L.: Modeling team member characteristics for the formation of a multifunctional team in concurrent engineering. IEEE Trans. Eng. Manag. **51**(2), 111–124 (2004)
8. Craig, M., McKeown, D.: How to build effective teams in healthcare. Nurs. Times **111**(14), 16–18 (2015)
9. Dashti, A., Samet, S., Fani, H.: Effective neural team formation via negative samples. In: Proceedings of the 31st ACM International Conference on Information & Knowledge Management, CIKM 2022, pp. 3908–3912, New York, NY, USA. Association for Computing Machinery (2022)
10. Dashti, A., Saxena, K., Patel, D., Fani, H.: OpeNTF: a benchmark library for neural team formation. In: Proceedings of the 31st ACM International Conference on Information & Knowledge Management, Atlanta, GA, USA, 17–21 October 2022, pp. 3913–3917. ACM (2022)
11. Durfee, E.H., Boerkoel, J.C., Sleight, J.: Using hybrid scheduling for the semi-autonomous formation of expert teams. Future Gener. Comput. Syst. **31**, 200–212 (2014). Special Section: Advances in Computer Supported Collaboration: Systems and Technologies
12. Fani, H., Bagheri, E., Du, W.: Temporal latent space modeling for community prediction. In: Jose, J.M., et al. (eds.) ECIR 2020. LNCS, vol. 12035, pp. 745–759. Springer, Cham (2020). https://doi.org/10.1007/978-3-030-45439-5_49
13. Fitzpatrick, E., Askin, R.G.: Forming effective worker teams with multi-functional skill requirements. Comput. Ind. Eng. **48**(3), 593–608 (2005)
14. Gu, Y., et al.: Domain-specific language model pretraining for biomedical natural language processing. ACM Trans. Comput. Healthcare **3**(1) 2021
15. Hall, K., et al.: The science of team science: a review of the empirical evidence and research gaps on collaboration in science. Am. Psychol. **73**, 532–548 (2018)

16. Hamilton, W.L., Leskovec, J., Jurafsky, D.: Diachronic word embeddings reveal statistical laws of semantic change. In: ACL 2016 (2016)
17. Jia, H., Liden, R.C.: Making a difference in the teamwork: linking team prosocial motivation to team processes and effectiveness. Acad. Manag. J. **58**, 1102–1127 (2014)
18. A., Kairgalievna, Zayed, N.M.: The effect of teamwork on employee productivity (2021)
19. Kargar, M., An, A.: Discovering top-k teams of experts with/without a leader in social networks. In: Proceedings of the 20th ACM International Conference on Information and Knowledge Management, pp. 985–994 (2011)
20. Kargar, M., An, A.: Efficient top-k keyword search in graphs with polynomial delay. In: 2012 IEEE 28th International Conference on Data Engineering, pp. 1269–1272 (2012)
21. Kargar, M., Golab, L., Srivastava, D., Szlichta, J., Zihayat, M.: Effective keyword search over weighted graphs. IEEE Trans. Knowl. Data Eng. **34**(2), 601–616 (2022)
22. Kargar, M., Zihayat, M., An, A.: Finding affordable and collaborative teams from a network of experts. In: Proceedings of the 2013 SIAM International Conference on Data Mining, pp. 587–595. SIAM (2013)
23. Karp, R.M.: Reducibility among Combinatorial Problems. In: Miller, R.E., Thatcher, J.W., Bohlinger, J.D. (eds) Complexity of Computer Computations. The IBM Research Symposia Series, pp. 85–103. Springer, Boston (1972). https://doi.org/10.1007/978-1-4684-2001-2_9
24. Keane, P., Ghaffar, F., Malone, D.: Using machine learning to predict links and improve steiner tree solutions to team formation problems - a cross company study. Appl. Netw. Sci. **5**(1), 57 (2020)
25. Khan, A., Golab, L., Kargar, M., Szlichta, J., Zihayat, M.: Compact group discovery in attributed graphs and social networks. Inf. Process. Manag. **57**(2), 102054 (2020)
26. Kou, Y., et al.: Efficient team formation in social networks based on constrained pattern graph. In: 36th IEEE International Conference on Data Engineering, ICDE 2020, Dallas, TX, USA, 20–24 April 2020, pp. 889–900. IEEE (2020)
27. Lappas, T., Liu, K., Terzi, E.: Finding a team of experts in social networks. In: SIGKDD 2009, pp. 467–476. ACM (2009)
28. Leahey, E.: From sole investigator to team scientist: trends in the practice and study of research collaboration. Ann. Rev. Sociol. **42**(1), 81–100 (2016)
29. Li, C.-T., Shan, M.-K., Lin, S.-D.: On team formation with expertise query in collaborative social networks. Knowl. Inf. Syst. **42**(2), 441–463 (2015)
30. Liao, S., Liang, S., Meng, Z., Zhang, Q.: Learning dynamic embeddings for temporal knowledge graphs. In: Proceedings of the 14th ACM International Conference on Web Search and Data Mining, WSDM 2021, pp. 535–543, New York, NY, USA. Association for Computing Machinery (2021)
31. McCormick, T.H., Lee, H., Cesare, N., Shojaie, A., Spiro, E.S.: Using twitter for demographic and social science research: tools for data collection and processing. Sociol. Methods Res. **46**(3), 390–421 (2017)
32. Neshati, M., Beigy, H., Hiemstra, D.: Expert group formation using facility location analysis. Inf. Process. Manag. **50**(2), 361–383 (2014)
33. Rad, R.H., Bagheri, E., Kargar, M., Srivastava, D., Szlichta, J.: Retrieving skill-based teams from collaboration networks. In: SIGIR 2021: The 44th International ACM SIGIR Conference on Research and Development in Information Retrieval, Virtual Event, Canada, 11–15 July 2021, pp. 2015–2019. ACM (2021)

34. Rad, R.H., Bagheri, E., Kargar, M., Srivastava, D., Szlichta, J.: Subgraph representation learning for team mining. In: WebSci 2022: 14th ACM Web Science Conference 2022, Barcelona, Spain, 26–29 June 2022, pp. 148–153. ACM (2022)
35. Rad, R.H., Fani, H., Bagheri, E., Kargar, M., Srivastava, D., Szlichta, J.: A variational neural architecture for skill-based team formation. ACM Trans. Inf. Syst. (2023). Just Accepted
36. Rad, R.H., Fani, H., Kargar, M., Szlichta, J., Bagheri, E.: Learning to form skill-based teams of experts. In: CIKM 2020, pp. 2049–2052. ACM (2020)
37. Rad, R.H., Mitha, A., Fani, H., Kargar, M., Szlichta, J., Bagheri, E.: PyTFL: a python-based neural team formation toolkit. In: Demartini, G., Zuccon, G., Culpepper, J.S., Huang, Z., Tong, H. (eds.) CIKM 2021: The 30th ACM International Conference on Information and Knowledge Management, Virtual Event, Queensland, Australia, 1–5 November 2021, pp. 4716–4720. ACM (2021)
38. Rad, R.H., Seyedsalehi, S., Kargar, M., Zihayat, M., Bagheri, E.: A neural approach to forming coherent teams in collaboration networks. In: Proceedings of the 25th International Conference on Extending Database Technology, EDBT 2022, Edinburgh, UK, March 29 - April 1, 2022, pp. 2:440–2:444. OpenProceedings.org (2022)
39. Rahmanniyay, F., Yu, A.J., Seif, J.: A multi-objective multi-stage stochastic model for project team formation under uncertainty in time requirements. Comput. Ind. Eng. **132**, 153–165 (2019)
40. Rosen, M.A., et al.: Teamwork in healthcare: key discoveries enabling safer, high-quality care. Am. Psychol. **73**(4), 433–450 (2018). Cited by: 297. All Open Access, Green Open Access
41. Sapienza, A., Goyal, P., Ferrara, E.: Deep neural networks for optimal team composition. Front. Big Data **2**, 14 (2019)
42. Sherer, P.D.: Leveraging human assets in law firms: human capital structures and organizational capabilities. ILR Rev. **48**(4), 671–691 (1995)
43. Wi, H., Seungjin, O., Mun, J., Jung, M.: A team formation model based on knowledge and collaboration. Expert Syst. Appl. **36**(5), 9121–9134 (2009)
44. Wu, C.-Y., Ahmed, A., Beutel, A., Smola, A.J., Jing, H.: Recurrent recommender networks. WSDM 2017, pp. 495–503, New York, NY, USA. Association for Computing Machinery (2017)
45. Yang, D.-N., Chen, Y.-L., Lee, W.-C., Chen, M.-S.: On social-temporal group query with acquaintance constraint. Proc. VLDB Endow. **4**(6), 397–408 (2011)
46. Younglove-Webb, J., Gray, B., Abdalla, C.W., Thurow, A.P.: The dynamics of multidisciplinary research teams in academia. Rev. High. Educ. **22**(4), 425–440 (1999)
47. ARMEN Zzkarian and Andrew Kusiak: Forming teams: an analytical approach. IIE Trans. **31**(1), 85–97 (1999)

Role-Guided Contrastive Learning for Event Argument Extraction

Chunyu Yao[1], Yi Guo[1,2,3(✉)], Xue Chen[1], Zhenzhen Duan[1], and Jiaojiao Fu[1(✉)]

[1] East China University of Science and Technology, Shanghai, China
y30221051@mail.ecust.edu.cn, {guoyi,fujj}@ecust.edu.cn

[2] Business Intelligence and Visualization Research Center, National Engineering Laboratory for Big Data Distribution and Exchange Technologies, Shanghai, China

[3] Shanghai Engineering Research Center of Big Data and Internet Audience, Shanghai, China

Abstract. Event argument extraction is a subtask of information extraction. Recent efforts have predominantly focused on mitigating the issue of error propagation associated with pipeline methods for extracting event arguments, such as machine reading comprehension and generative approaches. However, these aforementioned methods necessitate the careful design of various templates, and the choice of templates can significantly impact the model's performance. Therefore, we propose a novel approach to extract event arguments using contrastive learning. Our approach aims to maximize the semantic similarity between role name semantics and actual argument semantics while minimizing the similarity between role name semantics and the semantics of other non-argument words, thereby enabling more precise extraction of argument boundaries. We investigate the impact of different templates on event argument extraction, and experimental results demonstrate that template adjustments have limited effects on our model. To attain more precise argument boundaries, we also introduce entity type boundary embeddings, which substantially enhance the effectiveness of event argument extraction.

Keywords: Event argument extraction · Contrastive learning · Information extraction

1 Introduction

Event Argument Extraction (EAE) is a traditional and challenging task in Information Extraction (IE) [9,22]. Its objective is to extract structured event elements from unstructured textual sentences or paragraphs. Figure 1 depicts a Life-die event triggered by the word *killed* and specifies all its arguments along

This work was supported by the Science and Technology Program project of Shanghai Municipal Committee of Science and Technology (Grants: 22511104800 and 22DZ1204903).

N. Goharian et al. (Eds.): ECIR 2024, LNCS 14608, pp. 341–355, 2024.
https://doi.org/10.1007/978-3-031-56027-9_21

with their corresponding roles, which are entities. For instance, the *killed* event encompasses Killer (*The Taliban*); Victim (*members*); Place (*Maidan Wardak Province*); and MedicalIssue arguments, though there is no corresponding role for MedicalIssue in this event.

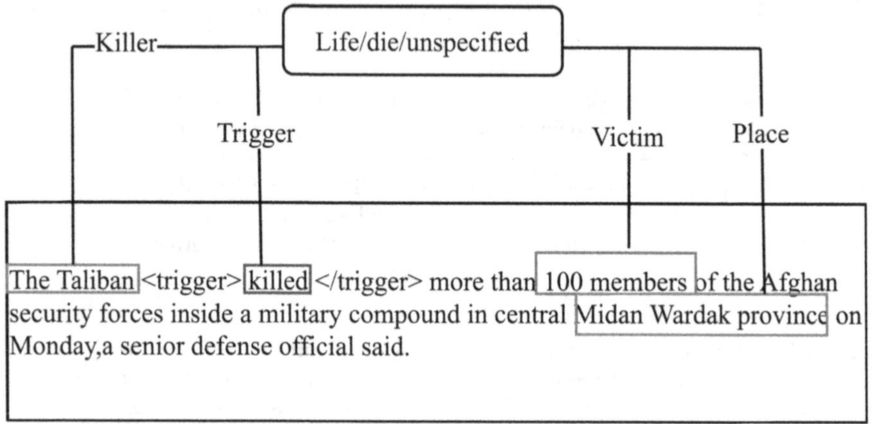

Fig. 1. Example of event argument extraction, where the portions highlighted in colored boxes represent the roles and trigger.

Recent works in event argument extraction can be categorized into the following three classes. The first approach leverages the advantages of pre-trained language model [4,12] to obtain more accurate contextual representations. The first approach decomposes event argument extraction into two separate subtasks: event argument identification and role classification for the identified arguments. This approach can be described as pipeline-based [3], however, heavily rely on the outcomes of argument identification, consequently leading to the issue of error propagation [1,14,21].

The second method, aiming to alleviate error propagation issue, introduces a novel paradigm by modeling event argument extraction as a Question-Answering task [6,13,17,18]. This is achieved by designing templates for posing questions about the arguments and utilizing the machine reading comprehension capabilities of pre-trained language models to answer these questions, thereby identifying the corresponding roles for the arguments. This QA-based approach allows for learning the associative relationships between event categories and argument labels, exhibiting strong generalization capabilities. Meanwhile, the QA-based approach suffers from low inference computational efficiency, as it requires the construction of questions for each argument and separate sentence concatenation for individual assessments, leading to increased computational complexity.

The final approach adopts a novel paradigm combining generative models [15,19] and prompt learning [11,20]. Generative models have the ability to extract event arguments in an end-to-end manner within a sequence-to-structure framework, leveraging weak supervision signals provided by prompt templates.

Both argument-question templates and prompt learning templates can provide some level of weak supervision signals to the model. However, different templates designs can have varying impacts on the performance. The main focus of this paper is to diminish the performance variations introduced by different templates while maintaining strong generalization capabilities.

In this paper, we propose the Contrastive Learning model for Event Argument Extraction. Our approach also requires the design of argument templates to provide weak supervision signals to the contrastive model. Specifically, we encode both the event context and role name templates into the same representation space or hypersphere. Subsequently, we maximize the semantic similarity between role names and the true arguments corresponding to those roles while simultaneously minimizing the semantic similarity between role names and other words. To enable the templates to learn correct sample features, we introduce entity boundary type embeddings. This module significantly enhances the performance of argument extraction for the document level argument extraction.

Our contributions are summarized as follows:

- We introduce contrastive learning into the event argument extraction task, which is less influenced by different template designs and exhibits strong generalization capabilities. It has also demonstrated nice performance in low resource scenarios.
- We incorporate entity boundary type embeddings, which provide precise entity type and boundary information to the contrastive model. The module enables our model to learn relationships between role labels and entity types effectively.
- We conduct extensive experiments on the ACE2005, WikiEvent and RAMS datasets. The results indicate that our model exhibits robustness to template variations and can easily transfer to new events with the provision of role labels.

2 Related Work

2.1 Event Argument Extraction

Event argument extraction is one of the subtasks of information extraction. Initially, early work relied on pipeline-based methods, utilizing convolutional neural network or recurrent neural network [21] to obtain event context representations for extraction. DMCNN [3] introduced a method based on word representation models and dynamic multi-pooling convolutional neural networks. This approach has the ability to automatically extract both word-level and sentence-level features and dynamically capture more valuable information based on different event triggers and arguments. However, this method, as it required separate identification and classification of arguments, introduced a significant issue of error propagation. To mitigate the performance drawbacks of extraction due to error propagation, machine reading comprehension based methods were consequently proposed.

EEQA [6] addressed the aforementioned task by transforming them into a question answering task through the design of context-specific queries for event trigger words and event arguments. This approach fully leveraged the reading comprehensions abilities of pre-trained models. R-GQA [7] further improved the extraction performance by incorporating retrieval-based enhancement on top of QA-based arguments extraction. Liu [16] proposed an unsupervised question generation method to generate contextually relevant questions. However, it is worth noting that machine reading comprehension based methods tend to be slower in terms of inference efficiency since they require posing questions for each argument.

The combination of prompt learning and generative models effectively address the efficiency issues introduced by machine reading comprehension methods. PAIE [20] constructs various prompt templates containing all arguments and decodes the arguments contained in these templates within the decoder. DEGREE [11] adopts an end to end template, providing extensive prior knowledge to the generative model and demonstrating better performance in low-resource scenarios.

2.2 Contrastive Learning

Contrastive Learning [2], as a specialized form of unsupervised learning, aims to learn data representations by maximizing the similarity between related samples and minimizing the similarity between unrelated samples. It has found extensive application in model pretraining. MoCo [10] introduced an efficient contrastive learning framework and demonstrated the immense potential of unsupervised learning by utilizing features learned from MoCo-based unsupervised learning for ImageNet classification.

Zhang and Ye [24] introduced a triplet contrastive training objective in the context of relation extraction and event argument extraction tasks. They employed a generative N-tuple encoder-decoder and a contrastive learning multi-task learning paradigm. CLEVE [23] leveraged abstract semantic parsing (AMR) to obtain pseudo-label pairs of trigger words to arguments from source text. Contrastive learning was applied to pre-train both text encoders and graph encoders, aiding in event extraction and the discovery of new event patterns.

Furthermore, Binder [25] employed a dual encoder approach for establishing contrastive objectives between entity types and text spans in the context of named entity recognition tasks. They achieved optimal results in both distant supervision and supervised named entity recognition scenarios.

3 Methodology

In this section, we provide a comprehensive overview of CLEAE, as shown in Fig. 2. We will sequentially discuss task definition, strategies for selecting event argument label templates, construction strategies for positive and negative sample pairs, training objectives for contrastive learning, and inference strategies.

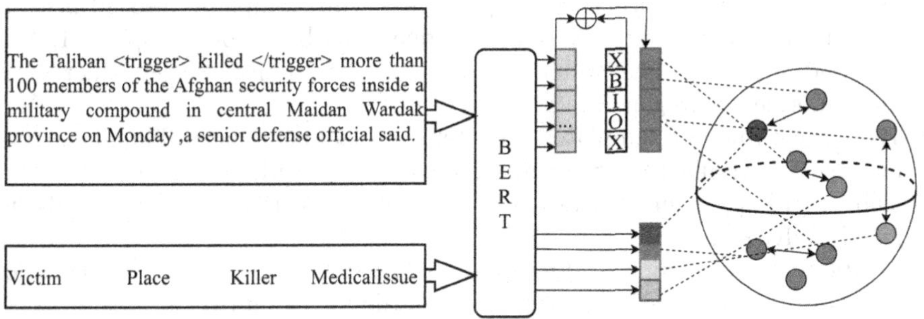

Fig. 2. The architecture of CLEAE. We employ special tokens to specify trigger words, forming the event context. CLEAE then selects specific argument labels into the same representation space. specifically, we append an entity boundary type embedding to the representation of event context, accurately specifying the entity type and boundary corresponding to the argument. Finally, we perform cosine similarity calculations to identify entity positions that are most similar to the argument labels.

3.1 Task Definition

We utilize contrastive learning to formulate the event argument extraction as a role-guided position classification problem on dataset D. For a given sample $(C, t, e, L^e) \in D$, where C refers to the event context determined by the trigger t, with event type e, and L^e represents the argument labels specific to event type e. Our objective is to obtain starting position *start* and ending position *end* in C for each $l \in L^e$.

3.2 Construction Strategies of Label Templates

Both the PAIE [20] and EEQA [6] approaches employed argument templates to provide weak supervision signals to the contrastive model. Consequently, we also adopt role templates and role questions templates, along with role label annotations, as anchor texts in our contrastive sample pairs. We select three declarative templates and one question template, each of which carries relevant semantics, and introduce a new annotation template, resulting in a total of five template categories used in our experiments. In summary, the template constructions are as follows:

- Template 1 (multi-format joint role name:) The template is derived from PAIE [20] and consits solely of the role names corresponding to the event type, such as Victim; Place; Killer; MedicalIssue. However the number of each role in the template is set to the maximum count of that role within the respective event type.

- Template 2 (soft role name:) The template builds upon Template 1 by introducing a learnable marker for each role, represented as <ARG> name </ARG>, serving as a specific pseudo-representation for that role.
- Template 3 (multi-format role question:) This template also utilizes a multi-format approach by posing role questions based on Template 1. For example, questions like "What are the roles of Victim, Place, Killer, and MedicalIssue? " are generated.
- Template 4 (manual template:) This template adopts a natural language format by connecting the arguments, making the template more contextually aligned with the event. For example, "Victim died at Place from MedicalIssue killed by Killer".
- Template 5 (role name + description:) This template annotates each role, following the ACE2005 annotation standards for certain roles. Therefore, it is similar to the argument question template and is formatted as single queries for each role. For roles not present in ACE2005 annotations, we annotate them based on similar standards.

Table 1. Template examples for argument labels.

Template Type	Example
multi-format joint role name	Victim (Victim) Killer (Killer) Place MedicalIssue
soft role name	<ARG> Victim </ARG> <ARG> Killer </ARG> <ARG>Place</ARG> <ARG> MedicalIsssue </ARG>
multi-format role question	what are the roles of Victim; Killer; Place; MedicalIssue?
manual template	Victim died at Place from MedicalIssue killed by Killer
role name and description	Victim: The person(s) who died. Killer: The attacking agent. Place: Where the death takes place. MedicalIssue: A general reference to a health condition.

Table 1 provides illustrative examples of the five templates. We will employ these five templates as anchor texts for contrastive learning experiments.

3.3 Construction Strategy of Positive and Negative Sample Pairs

Once the anchor texts are determined, we need to select corresponding positive and negative samples to create positive pair P and a set of negative pairs N. For the event context C, after encoding with the BERT encoder, it will be tokenized into several tokens by the WordPiece tokenizer, resulting in $C = \{c_1, c_2, ...c_n\}$ where n represents the length of the encoded sequence after tokenization.

For each token c_i in C, there are corresponding start and end positions, which allow us to obtain a list $span = \{(s_1, e_1), (s_2, e_2)...(s_n, e_n)\}$ corresponding to C.

We consider the true argument start and end positions as positive samples, so they should be included in $P = \{(s_{pos}, e_{pos})...\}$. On the other hand, all other start and end positions that do not correspond to arguments are considered negative samples, forming the set $N = \{(s_{neg}, e_{neg})...\}$.

We utilize the same encoder for both the event context C and templates, obtaining token level representations. For the event context C, we add an entity boundary type embedding layer after the event context representation. For each token, it is assigned a corresponding boundary type et_i. We convert the entity boundary type to a vector ET utilizing embedding:

$$\boldsymbol{et_1}, \boldsymbol{et_2}, ..., \boldsymbol{et_n} = Embedding(et_1, et_2, ..., et_n) \tag{1}$$

we have context representations C:

$$\boldsymbol{c_1}, \boldsymbol{c_2}, ..., \boldsymbol{c_n} = BERT(c_1, c_2, ..., c_n) \tag{2}$$

So, we can obtain key representations K:

$$\boldsymbol{k_1}, \boldsymbol{k_2}, ..., \boldsymbol{k_n} = Concat(C, ET) \tag{3}$$

For the templates, we have query representations Q:

$$\boldsymbol{q_1}, \boldsymbol{q_2}, ..., \boldsymbol{q_m} = BERT(l_1, l_2, ..., l_m) \tag{4}$$

In the case of the multi-format templates labeled as Template 1 to Template 4, m represents the maximum number of event roles for that event type. In the case of the single-format Template 5, m represents the number of event role types for that event type.

3.4 Position-Based Training Objective

The objective in event argument extraction is to extract the start and end positions of arguments, which naturally leads to optimizing these positions separately. For the contrastive loss function, we have chosen InfoNCE [10] loss because its fundamental idea is to provide a query and several samples, where only one sample matches the query as positive sample, while the others serve as negative samples. We have observed that this aligns well with our positive and negative sample pairs construction strategy. InfoNCE loss value becomes smaller only when the role-guided query is similar to the ground truth of argument span and dissimilar to other spans.

We consider the contrastive tasks for argument start and end positions as dictionary query problems. Therefore, we need to obtain start query vector s_q^i and end query vector e_q^i for each q_i after average pooling in Q separately for start and end positions. Each q_i also has a different span after being encoded, so the average pooling is used to obtain the representation of q_i:

$$\boldsymbol{s_q^i} = StartLinear(AveragePooling(\boldsymbol{q_i})) \tag{5}$$

$$\boldsymbol{e}_q^i = EndLinear(AveragePooling(\boldsymbol{q_i})) \tag{6}$$

Then, the start query $\boldsymbol{s}_q^i$ and end query $\boldsymbol{e}_q^i$ are separately used to calculate the similarity with the key representation $\boldsymbol{K}$. We employ cosine similarity function as the vector distance measure methods, represented as $cos(a, b)$. Following the formula for InfoNCE loss function, we can obtain the similarity $\boldsymbol{start}_{sim}^i$ between the start query vector $\boldsymbol{s}_q^i$ and the key representation $\boldsymbol{K}$:

$$\boldsymbol{start}_{sim}^i = cos(\boldsymbol{s}_q^i, ContextStartLinear(\boldsymbol{K})) \in R^n \tag{7}$$

and the similarity $\boldsymbol{end}_{sim}^i$ between the end query vector $\boldsymbol{e}_q^i$ and the key representation:

$$\boldsymbol{end}_{sim}^i = cos(\boldsymbol{e}_q^i, ContextEndLinear(\boldsymbol{K})) \in R^n \tag{8}$$

Next, we use the above-mentioned $\boldsymbol{start}_{sim}^i$ and $\boldsymbol{end}_{sim}^i$ to calculate the InfoNCE loss :

$$StartLoss_k(C) = -\log\left(\frac{\exp(start_{sim}^k/\tau)}{\sum_{i=0}^n \exp(start_{sim}^i/\tau)}\right) \tag{9}$$

$$EndLoss_k(C) = -\log\left(\frac{\exp(end_{sim}^k/\tau)}{\sum_{i=0}^n \exp(end_{sim}^i/\tau)}\right) \tag{10}$$

where τ represents the temperature coefficient and define the total loss as :

$$Loss = \sum_{C \in D} \sum_{k}^{m} (StartLoss_k(C) + EndLoss_k(C)) \tag{11}$$

where k represents the positive sample in P, and m represents the length of positive set P.

3.5 Inference

During the inference stage, we adopt heuristic rules to score the spans formed by the start position and subsequent k end positions. We then select the span with the highest score as the final prediction. Specifically, for every q_i we can construct the following score matrix:

$$spans^i(s, e) = start_{sim}^i + end_{sim}^i \tag{12}$$

and then we find the index corresponding to the maximum value in this matrix.

$$prediction^i(s, e) = argmax(spans^i) \tag{13}$$

If the $prediction^i(s, e)$ is (0,0), it indicates that the current query does not correspond to any argument.

4 Experiments

4.1 Datasets and Evaluation Metrics

We conduct experiments on three distinct datasets: ACE2005[1] [5]; WikiEvent [15] and RAMS[2] [8]. ACE2005 is a dataset at the sentence level, while WikiEvent and RAMS are document-level datasets. Furthermore, it's worth noting that only ACE2005 and WikiEvent contain annotations for entities. Consequently, experiments on the RAMS do not include the entity type boundary embedding component.

In the model evaluation section, we employ argument recognition F1 score and argument classification F1 score as the standards. The use of the F1 score allows for a comprehensive consideration of both precision and recall results. A predicted event argument is considered correctly recognized if its span and event type match the span of any ground truth argument for that event type. A predicted event argument is considered correctly classified if its span and event type match the ground truth argument role for that event type. However, these two evaluation metrics have strong constraints on the span, so if the labels are not consistent enough or there are some fuzzy boundary cases, the model may be greatly affected due to small differences.

4.2 Baselines

We compare our model with following methods based on machine reading comprehension or templates based training: **DocMRC** [17] is a reading comprehension based document level event argument extraction method, adept at addressing data scarcity issues; **EEQA**[3] [6] is an end-to-end event extraction model that uses question-answering templates; **BART-Gen**[4] [15] is a model that transforms the argument extraction task into a generation task, using argument names as template; **PAIE**[5] [20] is a model based on prompt learning, which generalizes well for both sentence-level and document-level event argument extraction.

In Table 2, we represent a performance comparison of CLEAE in event argument extraction against baseline models. The asterisk (*) indicates that the F1 score is sourced from the original paper. We replicate the results of PAIE and obtain the latest results on three datasets, using it as one of the baseline models for comparison. For ACE2005 at the sentence-level argument extraction, our model achieve competitive performance compared to PAIE, surpassing other baseline models. In the case of document-level argument extraction on WikiEvent, our model achieve 3% and 3.6% F1 score improvement for ARG-I and ARG-C, respectively. Since the RAMS dataset lacks entity annotations, we only conducted experiments with contrastive learning. On this dataset, our model achieve 1.2% improvement in ARG-1 and 1.3% improvement in ARG-C.

---

[1] https://www.ldc.upenn.edu/.
[2] http://nlp.jhu.edu/rams.
[3] https://github.com/xinyadu/eeqa.
[4] https://github.com/raspberryice/gen-arg.
[5] https://github.com/mayubo2333/PAIE.

Table 2. Performance comparison. We compare the best F1 scores obtained from the training on the five templates with baseline performance. The best scores for each dataset are highlighted in bold. ARG-I indicates argument identification and ARG-C indicates argument classification.

model	PLM	ACE		Wiki		RAMS	
		ARG-I	ARG-C	ARG-I	ARG-C	ARG-I	ARG-C
DocMRC	BERT-b	–	–	–	43.3*	–	45.7*
EEQA	BERT-b	68.2*	65.4*	–	–	–	–
BART-Gen	BART-l	69.9*	66.7*	–	65.1*	–	48.64*
PAIE	BART-b	73.6*	69.8*	68.9*	63.4*	54.7*	49.5 *
PAIE(repetition)	BART-b	**74.2**	**70.7**	69.0	63.6	54.0	49.0
CLEAE (ours)	BERT-b	**74.2**	70.0	**72.0**	**67.2**	**55.2**	**50.3**

Table 3. Ablation experiments with base encoder. - Represents that the RAMS dataset does not have this component, and therefore, there are no experimental records for it.

model	EBT	SPT	ACE		Wiki		RAMS	
			ARG-I	ARG-C	ARG-I	ARG-C	ARG-I	ARG-C
CLEAE-base	T/F	T	74.2	70.0	72.0	67.2	55.2	50.3
w/o EBT	F	T	69.8	65.6	67.5	62.8	–	–
w/o SPT	T/F	F	60.3	56.8	58.3	55.0	49.2	44.1

4.3 Ablation Study

This section investigates the impact of entity boundary type embedding and special token for event trigger on the model's performance. Table 3 presents the performance changes when entity boundary type embedding is removed on ACE and WikiEvent, as well as the performance changes when the special event trigger token is removed on all datasets. The results indicate that entity boundary type embedding and special event trigger token have significant impact on F1 score.

We also use the BART large version with a dimension of 1024 as the encoder and conduct ablation experiments. Table 4 provides the experiments results for the large encoder version. The experimental results indicate that a higher dimensional encoder leads to better template representations for contrastive learning. The experimental results demonstrate that using a larger encoder has limited impact on the model's performance on the ACE dataset. This could be due to the constraints of the event context in this dataset. However, on the RAMS and WIKIEVENT datasets, a larger encoder leads to significant improvements. Documents in these datasets provide rich event context, which can benefit representation learning.

Table 4. Ablation experiments with large encoder. - Represents that the RAMS dataset does not have this component, and therefore, there are no experimental records for it.

model	EBT	SPT	ACE		Wiki		RAMS	
			ARG-I	ARG-C	ARG-I	ARG-C	ARG-I	ARG-C
CLEAE-large	T/F	T	75.0	70.8	74.3	69.5	57.8	53.2
w/o EBT	F	T	72.9	69.0	69.2	63.8	–	–
w/o SPT	T/F	F	66.6	64.5	56.2	52.5	52.0	47.7

5 Analysis

5.1 Template Variants

We conduct experiments with five different templates, including four declarative templates and one interrogative template. We notice that in sentence level argument extraction, Template 1 exhibits higher performance compared to other four templates. Additionally, the performance difference between Template 2 and Template 3 and 5 is relatively small, but it is more sensitive to Template 4. For document level argument extraction, we do not design Template 5 for RAMS because we can not find suitable ACE annotation standards. Nevertheless, the experimental results on the five templates for WikiEvent and Template 1–4 for RAMS suggest that template design has little to no impact on the model's performance. This implies that for new datasets, providing event types and their corresponding role name in the respective domain should be sufficient for the model to perform well.

As shown in Fig. 3, in all template experiments, Template 1 consistently achieves the best results. We speculate that this is because contrastive learning maps the event context representations and the role representations on the same

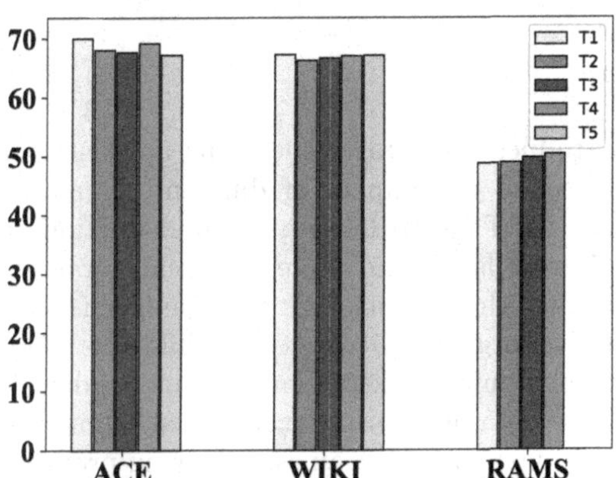

Fig. 3. The impact of different templates on ARG-C. Here, T1 represents Template 1, T2 represents Template 2, and so on.

hypersphere, paying more attention to the semantic distance between the two than to the enhancement of natural semantics to the role, thereby reducing the impact of other natural language variations on the model's performance.

5.2 Low Resource Study

Event argument extraction in low-resource scenarios can be more meaningful, as it reduces annotation costs. We investigate the performance of the model in event argument extraction with different proportions of training samples. Figure 4 illustrates the ARG-C F1 scores at different ratio of training samples:1%; 2%; 5%; 10%; 20%; 50%; 80%; 100%. Through our observations, we notice that as the training samples increase, the model's performance also improves, but the rate of improvement gradually decreases. In document level argument extraction, providing only 10% of the training data can achieve performance close to that of using 80% of the training data, indicating a reduction in annotation cost. The experimental results also demonstrate that our model exhibits good generalization even in low resource scenarios.

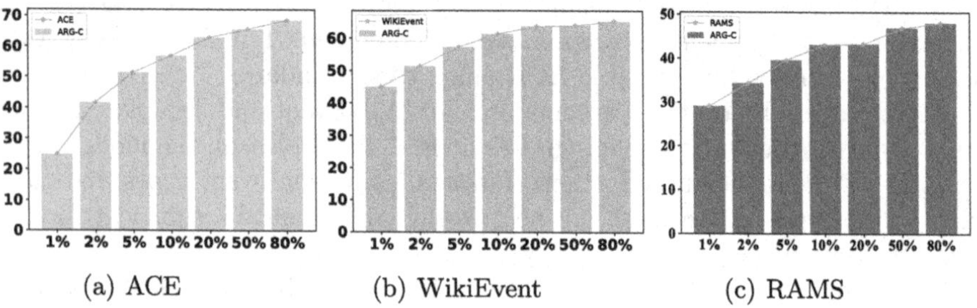

(a) ACE (b) WikiEvent (c) RAMS

Fig. 4. The different training data ratio and their ARG-C scores

6 Conclusion

In this paper, we introduce a contrastive learning approach for event argument extraction. We investigate the impact of different templates providing weak supervision signals on the F1 scores of argument extraction and find that templates with semantic role labels are more beneficial for argument extraction. Furthermore, we discover that the model's performance is less sensitive to templates in document level argument extraction, while it is more sensitive to role names in sentence level argument extraction. Our model demonstrates the capability to generalize to event argument extraction tasks in new domains with only the provision of event type and role names, achieving promising results even in low resource scenarios. During the incorporation of entity information, we have explored the use of named entity recognition as an auxiliary task, but it did not outperform the direct addition of entity boundary type embedding. Therefore, in future work, we plan to explore multi-task contrastive learning to achieve end-to-end event extraction.

Acknowledgements. We thank anonymous reviewers.

References

1. Cao, H., et al.: OneEE: a one-stage framework for fast overlapping and nested event extraction. In: Proceedings of the 29th International Conference on Computational Linguistics, pp. 1953–1964. International Committee on Computational Linguistics, Gyeongju, Republic of Korea (2022). https://aclanthology.org/2022.coling-1.170
2. Chen, T., Kornblith, S., Norouzi, M., Hinton, G.: A simple framework for contrastive learning of visual representations. In: Proceedings of the 37th International Conference on Machine Learning. ICML 2020, JMLR.org (2020)
3. Chen, Y., Xu, L., Liu, K., Zeng, D., Zhao, J.: Event extraction via dynamic multi-pooling convolutional neural networks. In: Proceedings of the 53rd Annual Meeting of the Association for Computational Linguistics and the 7th International Joint Conference on Natural Language Processing (Volume 1: Long Papers), pp. 167–176. Association for Computational Linguistics, Beijing (2015). https://doi.org/10.3115/v1/P15-1017, https://aclanthology.org/P15-1017
4. Devlin, J., Chang, M.W., Lee, K., Toutanova, K.: BERT: pre-training of deep bidirectional transformers for language understanding. In: Proceedings of the 2019 Conference of the North American Chapter of the Association for Computational Linguistics: Human Language Technologies, Volume 1 (Long and Short Papers), pp. 4171–4186. Association for Computational Linguistics, Minneapolis, Minnesota (2019). https://doi.org/10.18653/v1/N19-1423, https://aclanthology.org/N19-1423
5. Doddington, G., Mitchell, A., Przybocki, M., Ramshaw, L., Strassel, S., Weischedel, R.: The automatic content extraction (ACE) program - tasks, data, and evaluation. In: Proceedings of the Fourth International Conference on Language Resources and Evaluation (LREC 2004). European Language Resources Association (ELRA), Lisbon, Portugal (2004). http://www.lrec-conf.org/proceedings/lrec2004/pdf/5.pdf
6. Du, X., Cardie, C.: Event extraction by answering (almost) natural questions. In: Proceedings of the 2020 Conference on Empirical Methods in Natural Language Processing (EMNLP), pp. 671–683. Association for Computational Linguistics, Online (2020). https://doi.org/10.18653/v1/2020.emnlp-main.49, https://aclanthology.org/2020.emnlp-main.49
7. Du, X., Ji, H.: Retrieval-augmented generative question answering for event argument extraction. In: Proceedings of the 2022 Conference on Empirical Methods in Natural Language Processing, pp. 4649–4666. Association for Computational Linguistics, Abu Dhabi, United Arab Emirates (2022). https://doi.org/10.18653/v1/2022.emnlp-main.307, https://aclanthology.org/2022.emnlp-main.307
8. Ebner, S., Xia, P., Culkin, R., Rawlins, K., Van Durme, B.: Multi-sentence argument linking. In: Proceedings of the 58th Annual Meeting of the Association for Computational Linguistics, pp. 8057–8077. Association for Computational Linguistics, Online (2020). https://doi.org/10.18653/v1/2020.acl-main.718, https://aclanthology.org/2020.acl-main.718

9. Grishman, R., Sundheim, B.: Message understanding conference- 6: a brief history. In: COLING 1996 Volume 1: The 16th International Conference on Computational Linguistics (1996). https://aclanthology.org/C96-1079
10. He, K., Fan, H., Wu, Y., Xie, S., Girshick, R.: Momentum contrast for unsupervised visual representation learning. In: 2020 IEEE/CVF Conference on Computer Vision and Pattern Recognition (CVPR), pp. 9726–9735 (2020). https://doi.org/10.1109/CVPR42600.2020.00975
11. Hsu, I.H., et al.: DEGREE: a data-efficient generation-based event extraction model. In: Proceedings of the 2022 Conference of the North American Chapter of the Association for Computational Linguistics: Human Language Technologies, pp. 1890–1908. Association for Computational Linguistics, Seattle, United States (2022). https://doi.org/10.18653/v1/2022.naacl-main.138, https://aclanthology.org/2022.naacl-main.138
12. Lewis, M., et al.: BART: denoising sequence-to-sequence pre-training for natural language generation, translation, and comprehension. In: Proceedings of the 58th Annual Meeting of the Association for Computational Linguistics, pp. 7871–7880. Association for Computational Linguistics, Online (2020). https://doi.org/10.18653/v1/2020.acl-main.703, https://aclanthology.org/2020.acl-main.703
13. Li, F., et al.: Event extraction as multi-turn question answering. In: Findings of the Association for Computational Linguistics: EMNLP 2020, pp. 829–838. Association for Computational Linguistics, Online (2020). https://doi.org/10.18653/v1/2020.findings-emnlp.73, https://aclanthology.org/2020.findings-emnlp.73
14. Li, Q., Ji, H., Huang, L.: Joint event extraction via structured prediction with global features. In: Proceedings of the 51st Annual Meeting of the Association for Computational Linguistics (Volume 1: Long Papers), pp. 73–82. Association for Computational Linguistics, Sofia, Bulgaria (2013). https://aclanthology.org/P13-1008
15. Li, S., Ji, H., Han, J.: Document-level event argument extraction by conditional generation. In: Proceedings of the 2021 Conference of the North American Chapter of the Association for Computational Linguistics: Human Language Technologies, pp. 894–908. Association for Computational Linguistics, Online (2021). https://doi.org/10.18653/v1/2021.naacl-main.69, https://aclanthology.org/2021.naacl-main.69
16. Liu, J., Chen, Y., Liu, K., Bi, W., Liu, X.: Event extraction as machine reading comprehension. In: Proceedings of the 2020 Conference on Empirical Methods in Natural Language Processing (EMNLP), pp. 1641–1651. Association for Computational Linguistics, Online (2020). https://doi.org/10.18653/v1/2020.emnlp-main.128, https://aclanthology.org/2020.emnlp-main.128
17. Liu, J., Chen, Y., Xu, J.: Machine reading comprehension as data augmentation: a case study on implicit event argument extraction. In: Proceedings of the 2021 Conference on Empirical Methods in Natural Language Processing, pp. 2716–2725. Association for Computational Linguistics, Online and Punta Cana, Dominican Republic (2021). https://doi.org/10.18653/v1/2021.emnlp-main.214, https://aclanthology.org/2021.emnlp-main.214
18. Lu, D., Ran, S., Tetreault, J., Jaimes, A.: Event extraction as question generation and answering. In: Proceedings of the 61st Annual Meeting of the Association for Computational Linguistics (Volume 2: Short Papers), pp. 1666–1688. Association for Computational Linguistics, Toronto, Canada (2023). https://doi.org/10.18653/v1/2023.acl-short.143, https://aclanthology.org/2023.acl-short.143

19. Lu, Y., et al.: Text2Event: controllable sequence-to-structure generation for end-to-end event extraction. In: Proceedings of the 59th Annual Meeting of the Association for Computational Linguistics and the 11th International Joint Conference on Natural Language Processing (Volume 1: Long Papers), pp. 2795–2806. Association for Computational Linguistics, Online (2021). https://doi.org/10.18653/v1/2021.acl-long.217, https://aclanthology.org/2021.acl-long.217
20. Ma, Y., et al.: Prompt for extraction? PAIE: prompting argument interaction for event argument extraction. In: Proceedings of the 60th Annual Meeting of the Association for Computational Linguistics (Volume 1: Long Papers), pp. 6759–6774. Association for Computational Linguistics, Dublin, Ireland (2022). https://doi.org/10.18653/v1/2022.acl-long.466, https://aclanthology.org/2022.acl-long.466
21. Nguyen, T.H., Cho, K., Grishman, R.: Joint event extraction via recurrent neural networks. In: Proceedings of the 2016 Conference of the North American Chapter of the Association for Computational Linguistics: Human Language Technologies, pp. 300–309. Association for Computational Linguistics, San Diego, California (2016). https://doi.org/10.18653/v1/N16-1034, https://aclanthology.org/N16-1034
22. Sundheim, B.M.: Overview of the fourth message understanding evaluation and conference. In: Fourth Message Understanding Conference (MUC-4): Proceedings of a Conference Held in McLean, Virginia, June 16–18, 1992 (1992). https://aclanthology.org/M92-1001
23. Wang, Z., et al.: CLEVE: contrastive pre-training for event extraction. In: Proceedings of the 59th Annual Meeting of the Association for Computational Linguistics and the 11th International Joint Conference on Natural Language Processing (Volume 1: Long Papers), pp. 6283–6297. Association for Computational Linguistics, Online (2021). https://doi.org/10.18653/v1/2021.acl-long.491, https://aclanthology.org/2021.acl-long.491
24. Zhang, N., et al.: Contrastive information extraction with generative transformer. IEEE/ACM Trans. Audio Speech Lang. Process. **29**, 3077–3088 (2021). https://doi.org/10.1109/TASLP.2021.3110126
25. Zhang, S., Cheng, H., Gao, J., Poon, H.: Optimizing bi-encoder for named entity recognition via contrastive learning. arXiv preprint arXiv:2208.14565 (2022)

Fine-Tuning CLIP via Explainability Map Propagation for Boosting Image and Video Retrieval

Yoav Shalev(✉) and Lior Wolf

Tel Aviv University, Tel Aviv-Yafo, Israel
yoavshalev@mail.tau.ac.il, wolf@cs.tau.ac.il

Abstract. Recent studies have highlighted the remarkable performance of CLIP for diverse downstream tasks. To understand how CLIP performs these tasks, various explainability methods have been formulated. In this paper, we reveal that the explainability maps associated with CLIP are often focused on a limited portion of the image and overlook objects that are explicitly mentioned in the text. This phenomenon may result in a high similarity score for incongruent image-text pairs, thereby potentially introducing a bias. To address this issue, we introduce a novel fine-tuning technique for CLIP that leverages a transformer explainability method. Unlike traditional approaches that generate a single heatmap using an image-text pair, our method produces multiple heatmaps directly from the image itself. We use these heatmaps both during the fine-tuning process and at inference time to highlight key visual elements, applying them to the features during the image encoding process, steering the visual encoder's attention toward these key elements. This process guides the image encoder across different spatial regions and generates a set of visual embeddings, thereby allowing the model to consider various aspects of the image, ensuring a detailed and comprehensive understanding that surpasses the limited scope of the original CLIP model. Our method leads to a notable improvement in text, image, and video retrieval across multiple benchmarks. It also results in reduced gender bias, making our model more equitable.

Keywords: Image and Video Retrieval · Explainability · CLIP

1 Introduction

The CLIP similarity score [14] is a powerful tool for image and video retrieval [13, 18]. However, being trained with a global contrastive loss, the model can suffer from biases [18], and, as we show, it often neglects regions in the image or parts of the sentence. In this work, we manipulate the image encoder to achieve a more comprehensive representation of each image. We generate multiple salience maps that highlight different regions within an image. This is achieved through an explainability method inspired by gradient-based transformer explainability [1]. By perturbing the output of the CLIP image encoder and back-projecting the resulting gradients, we obtain a set of importance maps using only the image. During the fine-tuning process, the heatmaps

N. Goharian et al. (Eds.): ECIR 2024, LNCS 14608, pp. 356–370, 2024.
https://doi.org/10.1007/978-3-031-56027-9_22

are used to mask visual feature maps. The masked features are reintroduced as residual feature maps during the visual encoding process, resulting in a set of visual embeddings that capture various aspects of the image. Moreover, at inference time, we use multiple representations of each image for retrieval. Our results reveal that our approach outperforms both the original CLIP and a version of CLIP that was fine-tuned on the training split, for retrieval benchmarks across four tasks: image or video retrieval given text, and text retrieval given image or video. Intriguingly, our method not only enhances the performance of CLIP but also helps mitigate gender bias. This bias reduction is achieved as our method promotes a comprehensive understanding of the entire image, moving away from potential biases linked to specific regions or objects within the image. This is due to our use of multiple heatmaps, each highlighting a different area of the image.

2 Related Work

CLIP [14] is a large-scale, transformer-based vision and language model, which uses two separate encoders, one per modality. Visual and language inputs are encoded into a shared latent space, and the model is optimized using a contrastive loss that minimizes or maximizes the distance of the embedding vectors according to their correspondence.

Previous works have demonstrated remarkable performance for a variety of downstream tasks using CLIP. Using predefined prompts depicting target classes, Radford et al. [14] have demonstrated impressive performance for the zero-shot image classification task. Tewel et al. [17] and Su et al. [15] utilized CLIP to score the correspondence between an image and candidate text tokens to generate a caption in an auto-regressive manner. Similarly, Tewel et al. [16] optimized a prefix of pseudo-tokens to generate a caption for images and videos. Wang et al. [18] have demonstrated SOTA results for the zero-shot image retrieval task, by projecting a text query and a set of images into CLIP's latent space and maximizing the cosine similarity to find the target image. Similarly, Quintero et al. [13] represented a video by the average of its encoded frames using CLIP. In addition, they encoded a text query using CLIP and measured the cosine similarity with the videos representations to find the best matching video. They got SOTA results on the zero-shot video retrieval task for the MSVD and MSR-VTT datasets.

Wang et al. [18] introduce a framework for quantifying gender bias in image retrieval. The bias is estimated by calculating the ratio of retrieved images that are labeled as 'male' and 'female' In addition, they suggest a novel post-processing method for reducing the bias, by pruning the text and the visual features that are highly correlated with the gender attribute in CLIP's embedding space. They show an improvement in reducing the bias at the expense of sacrificing some of the model's accuracy.

Explainability methods were developed to shed some light on the way CLIP and other transformer-based models act on their downstream tasks. Goh et al. [5] apply classical gradient-based techniques for feature visualization and estimate the distribution of maximal activating images for a neuron from a dataset. Chefer et al. [1] run a forward pass and calculate relevancy matrices using pre-defined update rules to generate a final heatmap. Zabari et al. [20] utilize CLIP and image augmentations to generate pixel-level pseudo-labels for a given image using the explainability method of Chefer et al. [1]. The pseudo-labels are then post-processed using interactive segmentation and unsupervised clustering algorithms to generate the final segmentation map. They show a major improvement over previous unsupervised segmentation methods. Unlike the explainability method of Chefer et al. [1], which relies on image-text pairs, our method generates explainability maps solely using the image itself. We employ augmented versions of the visual features to generate multiple heatmaps. Within specific layers of CLIP's visual encoder, we use the heatmaps both during the fine-tuning process and at inference time, to mask visual features, subsequently reintroducing the masked features as residuals. This process refines the model's attention, resulting in a set of visual features that are better aligned with the corresponding text representation. Consequently, our model surpasses previous baselines in text, image, and video retrieval tasks while simultaneously reducing gender bias.

3 Method

The proposed method employs heatmaps, generated exclusively from images, both during the fine-tuning process and at inference time to optimize the visual-embedding representation of CLIP's [14] visual encoder E_v. By utilizing these heatmaps, we aim to refine the model's attention, directing it effectively towards the salient objects within a given image I. Unlike the explainability method of Chefer et al. [1], which relies on image-text pairs, our technique generates explainability maps solely using the image itself. This distinction is particularly vital for retrieval tasks where a corresponding text is not available during inference. The generated heatmaps are then utilized as feature masks within specific layers of E_v, guiding and refining the model's attention to a variety of foreground objects, while minimizing attention to other parts of the image.

The heatmap generation process encodes the image I using E_v into a feature vector, which is then normalized using L_2 normalization to obtain $I_e \in \mathbb{R}^{1\times 512}$:

$$I_e = \frac{E_v(I)}{\|E_v(I)\|_2} \tag{1}$$

Then we apply a dropout operation to generate $I_{\text{drop}} \in \mathbb{R}^{1\times 512}$, which is an augmented version of the image features:

$$I_{\text{drop}} = \text{Dropout}(I_e, p)\,, \tag{2}$$

where p is the dropout probability.

Using this augmented version of the image features allows us to explore different heatmap representations at the fine-tuning and inference steps, based on the same image features I_e.

Next, we compute the dot product between the augmented and the original image features as a measure of similarity $S \in \mathbb{R}$ as:

$$S = I_e \cdot I_{\text{drop}}^T, \tag{3}$$

In the forward pass of CLIP's visual transformer, spatial attention probabilities are calculated for each of the $h = 12$ attention heads within every residual block. The input image is divided into a 7×7 grid, leading to a 49×49 matrix of attention probabilities, which allows CLIP's visual encoder to capture pairwise interactions between spatial regions. We extract the attention probabilities $P_i \in \mathbb{R}^{49\times 49}$ for each attention head from the last residual block. For an attention head i, the gradients $\text{Grad}_i \in \mathbb{R}^{49\times 49}$ of S w.r.t. P_i are given by:

$$\text{Grad}_i = \frac{\partial S}{\partial P_i}. \tag{4}$$

The activations that contribute to the heatmap for each attention head i, denoted $A_i \in \mathbb{R}^{49\times 49}$, are then calculated by:

$$A_i = \text{Grad}_i \cdot P_i. \tag{5}$$

After obtaining the activations A_i for all h attention heads, they are clamped to non-negative values and then averaged:

$$A = \frac{1}{h}\sum_{i=1}^{h} \text{clamp}(A_i, \text{min} = 0). \tag{6}$$

From the 49×49 interaction matrix A, we compute a vector by averaging its rows. This vector of length 49 is then reshaped into a 7×7 matrix M, where each entry corresponds to a region of the image. The matrix M is subsequently normalized using min-max normalization to produce the final heatmap H:

$$H = \frac{M - \min(M)}{\max(M) - \min(M)} \tag{7}$$

.

To generate a diverse set of heatmaps that focus on different regions of the image, we use n random dropout values to generate a set of heatmaps $\{H_1, H_2, \ldots, H_n\}$. The heatmaps differ from each other due to the impact of the dropout operation on feature selection. Specifically, the gradients of the similarity scores between the original image features and their augmented counterparts take a different form, which results in distinct subsets of features. Consequently, this introduces variability in the regions displaying high activations. In both the fine-tuning and inference steps, in the forward pass of E_v, we use a generated heatmap $H_i \in \mathbb{R}^{49\times 49}$ at two distinct stages to adjust the intermediate visual features. The first stage follows the initial convolutional layer, ensuring that the early-stage features capture the information prioritized by the heatmap. The second application takes place within the last residual attention block of the transformer, specifically influencing the input of the second linear layer. This dual-stage integration ensures both preliminary and refined features are influenced by the heatmap. To combine H_i with an intermediate visual feature map $F \in \mathbb{R}^{49\times 49}$, we employ the following operation at both stages:

$$F_{\text{modified}} = F + F \odot H_i, \tag{8}$$

where $\odot$ represents element-wise multiplication. The resultant F_{modified} is the modified visual feature map, now influenced by the heatmap, guiding the model to emphasize specific regions during the encoding process. The final image features $I_{e_i} \in \mathbb{R}^{1\times 512}$, influenced by the heatmap H_i, are denoted by:

$$I_{e_i} = \frac{E_v(I, H_i)}{\|E_v(I, H_i)\|_2}. \tag{9}$$

The set of feature vectors $\{I_{e_i}\}_{i=1}^{n}$ generated using the n heatmaps serve as multiple representations for the image both during model optimization in the fine-tuning phase and retrieval tasks during inference. The specific applications of these feature vectors will be introduced in the subsequent sections.

3.1 Optimization

The primary objective of our optimization process consists of two main goals. First, we aim to minimize the cosine distance between I_e, which are the original, unguided embeddings of the image I, as given in Eq. 1, and CLIP's embeddings of the text T. Second, we extend the optimization to minimize the cosine distance between the set of feature vectors $\{I_{e_i}\}_{i=1}^{n}$ and the embeddings of the text T, where I_{e_i} is computed as outlined in Eq. 9. By employing multiple versions of image embeddings, each guided by a different heatmap that focuses on a separate area within the image I, we achieve a more comprehensive alignment with the embeddings of the text T, thereby enhancing the robustness and accuracy of the optimization process.

The text T is encoded into a vector $T_e \in \mathbb{R}^{1\times 512}$ using CLIP's text encoder E_t as:

$$T_e = \frac{E_t(T)}{\|E_t(T)\|_2}. \tag{10}$$

We then calculate the cosine similarities between T_e and the image embeddings I_e (Eq. 1), as well as between T_e and each features vector I_{e_i} (Eq. 9):

$$d_I = e^t \cdot T_e \cdot I_e^T, \quad d_i = e^t \cdot T_e \cdot I_{e_i}^T, \tag{11}$$

where t is a temperature hyper-parameter.

Similarly to Radford et al. [14] we use batch processing to compute pairwise distances among elements in each batch. This allows us to minimize distances between matching image-text pairs while maximizing distances between mismatched pairs. Given a batch of b images and texts, we extend Eq. 11 to calculate $D_I \in \mathbb{R}^{b\times b}$ and $D_i \in \mathbb{R}^{b\times b}$ for the entire batch. Specifically, D_I computes the cosine similarities using the original, unguided image and text embeddings, whereas D_i leverages the heatmap-guided visual embeddings vectors. The optimization process for each batch aims to minimize the cross-entropy loss between the ground-truth target labels and the cosine similarities derived from both, D_I, and each heatmap-guided embeddings vector D_i:

$$L_I = CE(D_I, gt) + CE(D_I^T, gt), \quad L_i = CE(D_i, gt) + CE(D_i^T, gt). \tag{12}$$

where CE represents the cross-entropy loss and gt are the ground-truth target labels. Our final optimization target consists of a weighted sum of the above losses:

$$L = \lambda_1 \cdot L_I + \lambda_2 \cdot \frac{1}{n} \cdot \sum_{i=1}^{n} L_i, \tag{13}$$

where λ_1 and λ_2 are hyper-parameters, balancing the contributions of each component.

3.2 Training Details

We use the CLIP implementation of Radford et al. [14], specifically, the ViT-B/32 model. For the training set, we use the MS-COCO dataset [7] and the Karpathy split.[1]

[1] https://www.kaggle.com/datasets/shtvkumar/karpathy-splits.

We optimize only the attention layers of the last six residual blocks of E_v, excluding the projection layers. The model was initialized with the parameters of the pre-trained model of Radford et al. [14]. For simplicity and lack of resources, we set the values of λ_1 and λ_2 to one. In our experiments, we employ $n = 3$ heatmaps. For each heatmap, the dropout probability p is selected from a set of random values between 0.1 and 0.7. We fine-tuned the model for five epochs with a batch size of 256. We use the Adam optimizer [6] with a learning rate of $5e - 6$, weight decay of 0.2, and beta coefficients of 0.9 and 0.98. We use a multi-step learning rate decay starting from iteration 1,000 to iteration 12,000 with a gamma of 0.9 and a step size of 1,000. We set the temperature parameter t to 100. Fine-tuning takes half an hour on NVIDIA RTX A6000.

4 Experiments

We evaluate our approach with the task of retrieving images and videos using textual queries, and conversely, retrieving texts using visual queries. We also evaluate the image retrieval task with respect to gender bias. We evaluate our approach using multiple configurations of the CLIP model, which differ in their fine-tuning methodologies. The first is $\text{CLIP}_{\text{Base}}$, which is the unaltered CLIP model of Radford et al. [14], that currently sets the standard in zero-shot image-text and video-text retrieval. The second is our proposed model, CLIP_{FTH}, which integrates heatmaps during the fine-tuning stage to enhance and guide the visual encoding process. Lastly, CLIP_{FT} undergoes a fine-tuning process similar to CLIP_{FTH}, but excludes the use of heatmap guidance. At inference time, all configurations can use heatmap guidance in their visual encoding process, which will be examined and ablated in the following sections. Throughout our experiments, we use $n = 3$ heatmaps.

Datasets. Training and evaluation were performed using four different datasets. **MS-COCO** [7] is a large-scale object detection, segmentation, and captioning dataset with 80 object categories. It contains over 330,000 images with over 1,500,000 human-generated captions. We adopt the widely used Karpathy split (see Footnote 1) for training and evaluation. This split consists of 113,287 images for training, 5,000 images for validation, and 5,000 images for testing, where each image is associated with five human-generated captions. **Flickr30K** [12] is a dataset consisting of 31,000 images and over 150,000 human-generated captions collected from Flickr.[2] We use the Karpathy split for evaluation (see Footnote 1). This split consists of 29,000 images for training, 1,000 for validation, and 1,000 for testing. Each image is associated with five human-generated captions. **MSR-VTT** [19] is a large-scale video dataset consisting of 10,000 videos from 20 categories. Each video clip is associated with approximately 20 human-generated captions. The standard split consists of 6,513 clips for training, 497 for validation, and 2,990 for testing. **MSVD** [2] is a video dataset consisting of 1,970 short videos collected from YouTube, with an average of 41 human-generated captions per video. We use the standard split, which consists of 1,200 clips for training, 100 for validation, and 670 for testing.

[2] https://www.flickr.com.

Gender Annotations. Gender annotations are often not explicitly provided in large-scale datasets, such as MS-COCO and Flickr30K. To evaluate retrieval performance in the context of gender bias, we generated the gender annotations for the MS-COCO and the Flickr30K datasets as suggested by Wang et al. [18]. Specifically, the annotations were extracted from the context of the associated caption for each image. Using pre-defined sets of masculine and feminine words, and the set of captions, we classified each image as "male", "female" or "neutral". An image is labeled as "male" if at least one of its captions contains a word from the set of masculine words, e.g. "man", "boy", "father", etc., and does not contain words from the set of feminine words, e.g. "woman", "girl", "mother", etc. Similarly, an image is labeled as "female" if at least one of its captions contains a feminine word and none of the captions contains masculine words. In any other case, the image is labeled as "neutral". We use the sets of masculine and feminine words that were suggested by Wang et al. [18].

Gender Neutral Captions. When measuring bias in image retrieval, gender-neutral text queries are required. We follow the pre-processing procedure of Wang et al. [18] for converting gender-specific text queries into gender-neutral ones. The procedure removes and replaces gender-specific words with the corresponding gender-neutral words from a pre-defined list. For example, the gender-specific caption "A little girl is getting ready to blow out a candle on a small dessert" was converted into the gender-neutral caption "A little child is getting ready to blow out a candle on a small dessert". We exclusively employ these gender-neutral text queries for calculating the metrics associated with our gender bias measurement experiments.

Image and Text Retrieval. Given a dataset of N image-text pairs $D = \{(c_i, v_i)\}_{i=1}^{N}$, where $c_i \in C$ is a caption and $v_i \in V$ is a corresponding image, there are two primary tasks for a vision-language retrieval model: retrieving the best matching image v_i for a given text query c_i, or retrieving the most suitable text c_i for a given image v_i.

In the standard retrieval approach, following Eq. 1 and Eq. 10, an image $v_j \in V$ and a caption $c_i \in C$ are encoded using CLIP's visual and language encoders E_v and E_l, resulting in embeddings vectors v_j^e and c_i^e, respectively. The matching score $S_{ij} \in \mathbb{R}$ is computed as the cosine similarity of the embedding vectors:

$$S_{ij} = \frac{c_i^e \cdot v_j^e}{\|c_i^e\|_2 \|v_j^e\|_2}. \tag{14}$$

For enhanced retrieval using heatmap-guided visual encoding, the image is encoded using each of the generated heatmaps $\{H_k\}_{k=1}^{n}$ to yield n embedding vectors $\{v_{j_k}^e\}_{k=1}^{n}$ as depicted in Eq. 9. For each vector $v_{j_k}^e$, a matching score $S_{ij}^{(k)}$ is computed similarly with the text embeddings vector c_i^e. The final heatmap-guided matching score is the average of these n scores:

$$S_{ij}^h = \frac{1}{n} \sum_{k=1}^{n} S_{ij}^{(k)}. \tag{15}$$

For image retrieval given a text query c_i, and for text retrieval given an image query v_j, the results with the highest scores are retrieved by:

$$v^* = \mathrm{argmax}_{v_j \in V} S_{ij}, \quad c^* = \mathrm{argmax}_{c_i \in C} S_{ij}, \tag{16}$$

Where S_{ij} can represent either the standard matching score or S_{ij}^h.

Table 1. Image and text retrieval results for MS-COCO. 'H' and 'Base' denote with or without heatmap guidance at inference time, respectively.

Model	Inf. Mode	Image Retrieval			Text Retrieval		
		Recall@1	Recall@5	Recall@10	Recall@1	Recall@5	Recall@10
$\text{CLIP}_{\text{Base}}$ [14]	Base	29.44	54.28	65.02	33.54	58.54	68.72
CLIP_{FT}	Base	38.46	65.36	76.12	38.74	65.62	75.94
CLIP_{FTH}	H	**40.26**	**66.94**	**77.60**	**40.6**	**67.7**	**77.98**

Gender Bias in Image Retrieval. Recent vision and language models are trained using a large corpus of image-text pairs crafted from the internet. This type of data contains biases with respect to sensitive attributes. In this work, we evaluate the bias of the CLIP model [14] for the gender attribute, using the image retrieval task. In image retrieval, the text query may be gender-neutral, e.g. "A person is riding a bike". A fair model will treat images equally, i.e. we expect to get approximately the same number of images for each gender. For a given text query $c \in \mathcal{C}$, the image retrieval task provides a similarity matrix, from which we can extract the top K images, denoted as $\mathcal{R}_K(c)$.

The fractions of images labeled as $\text{G} \in \{\text{“}male\text{”}, \text{“}female\text{”}\}$ within the top K results for a text query c are defined as:

$$P_{\text{G}}(c, K) = \tfrac{1}{K} \textstyle\sum_{v \in \mathcal{R}_K(c)} \mathbb{1}[g(v) = \text{G}] \tag{17}$$

Where $\mathbb{1}$ is the indicator function and $g(v)$ is the gender label ground-truth for the image v. Across all text queries c, the average fractions for males and females are:

$$\overline{P}_{\text{G}}(K) = \tfrac{1}{|\mathcal{C}|} \textstyle\sum_{c \in \mathcal{C}} P_{\text{G}}(c, K) \tag{18}$$

The average gender bias for the entire dataset for a specific retrieval size K is:

$$\text{Bias}\ @K = \left|\overline{P}_{\text{male}}(K) - \overline{P}_{\text{female}}(K)\right| \tag{19}$$

Feature Clipping Based on Mutual Information. The CLIP-clip algorithm [18] is a post-processing method for reducing the bias by pruning the text and the visual features that are highly correlated with the gender attribute in CLIP's embedding space. Applying such a post-processing method saves the training time and computational costs required for retraining CLIP. Assuming the visual embedding space of CLIP is d-dimensional, the mutual information of the i-th feature of all images $V^{e_i} \in \mathbb{R}^n$, where n is the number of images, and the gender labels $g(V)$ of all images in V is defined as:

$$I\left(V^{e_i}; g(V)\right) = D_{\text{KL}}\left(\mathbb{P}_{(V^{e_i}, g(V))} \| \mathbb{P}_{V^{e_i}} \otimes \mathbb{P}_{g(V)}\right) \tag{20}$$

where D_{KL} is the KL divergence operator, $\mathbb{P}_{(V^{e_i}, g(V))}$ is the joint distribution, $\mathbb{P}_{V^{e_i}}$ and $\mathbb{P}_{g(V)}$ are the marginals and $\otimes$ is the Hadamarad product. The CLIP-clip algorithm removes the top $m = 100$ visual features that have the highest mutual information with the gender labels $g(V)$. The corresponding language features are also removed, therefore both feature sets are less correlated with the gender attribute. When using heatmaps, we apply the clipping process to all the embeddings $\{v^e_{j_k}\}_{k=1}^n$ defined in Eq. 9.

Table 2. Image and text retrieval results for Flickr30k. 'H' and 'Base' denote with or without heatmap guidance at inference time, respectively.

Model	Inf. Mode	Image Retrieval			Text Retrieval		
		Recall@1	Recall@5	Recall@10	Recall@1	Recall@5	Recall@10
$\text{CLIP}_{\text{Base}}$ [14]	Base	67.0	89.1	93.6	69.6	90.6	95.0
$\text{CLIP}_{\text{Base}}$ [14]	H	69.3	90.2	94.2	73.4	91.2	94.6
CLIP_{FT}	Base	73.5	93.3	96.3	73.4	92.2	96.1
CLIP_{FTH}	Base	75.6	**94.0**	96.9	75.0	93.0	**97.4**
CLIP_{FTH}	H	**78.9**	93.3	**97.0**	**75.5**	**93.5**	97.1

Table 3. Image retrieval and bias results for MS-COCO and Flickr30k using gender-neutral text queries. 'H' and 'C' denote whether heatmap guidance or feature pruning is used at inference time. 'Base' indicates that none of them were used.

Dataset	Model	Inf. Mode	Recall@1	Recall@5	Recall@10	Bias@1	Bias@5	Bias@10
MS-COCO	$\text{CLIP}_{\text{Base}}$ [14]	Base	22.0	41.7	50.3	0.164	0.155	0.153
	$\text{CLIP}_{\text{Base}}$ [14]	C [18]	20.1	38.2	47.3	0.152	0.141	0.139
	CLIP_{FT}	C	26.4	46.7	55.1	0.153	0.147	0.142
	CLIP_{FTH}	CH	**28.0**	**47.8**	**56.3**	**0.106**	**0.130**	**0.132**
Flickr30k	$\text{CLIP}_{\text{Base}}$ [14]	Base	73.6	90.6	94.0	0.056	0.077	0.081
	$\text{CLIP}_{\text{Base}}$ [14]	C [18]	66.0	86.8	91.6	0.014	0.056	0.071
	CLIP_{FT}	C	73.8	91.6	**94.8**	0.026	0.038	0.025
	CLIP_{FTH}	CH	**75.2**	**92.6**	94.6	**0.002**	**0.024**	**0.021**

Results. Our evaluations employ multiple metrics, including Recall@1, Recall@5, and Recall@10, for both image retrieval and text retrieval tasks. We evaluate the MS-COCO and Flickr30K datasets using the Karpathy splits (see Footnote 1). The image and text retrieval results for the MS-COCO and Flickr30K datasets are presented in Table 1 and Table 2, respectively. The inference mode for each configuration is specified: 'Base' represents the standard retrieval approach, where heatmaps are not used in the visual encoding process, and 'H' indicates the enhanced retrieval using heatmap-guided image encoding. For the MS-COCO dataset, the configuration CLIP_{FTH} + 'H' outperforms the baselines $\text{CLIP}_{\text{Base}}$ + 'Base' and CLIP_{FT} + 'Base' for both tasks. To assess the influence of heatmap guidance during training and inference separately, we added two configurations for evaluations on Flickr30K, a dataset on which neither model was trained. The first, $\text{CLIP}_{\text{Base}}$ + 'H', measures the sole impact of heatmap guidance at inference time. Conversely, CLIP_{FTH} + 'Base' was introduced to exclusively examine the effectiveness of heatmap guidance during the fine-tuning process. Notably, using heatmap guidance at inference time with $\text{CLIP}_{\text{Base}}$ improves all metrics, except Recall@10 for the text-retrieval task, demonstrating that even without fine-tuning, the set of visual embeddings produced by our method is highly congruent with the corresponding text embeddings. Each heatmap allows the model to focus on various image aspects and regions, yielding a rich set of embeddings that encapsulate a nuanced comprehension of the image content. Similarly, CLIP_{FTH} + 'Base' outperforms CLIP_{FT} + 'Base', highlighting the benefits of incorporating heatmap guidance during the fine-tuning phase.

Image retrieval and bias results, using gender-neutral text queries for the MS-COCO and Flickr30K datasets are presented in Table 3. For this experiment, we added another inference mode 'C', for leveraging the feature pruning method proposed by Wang et al. [18]. As observed, our method surpasses the baselines in recall and bias metrics for both datasets, except for R@10 on the Flickr30k dataset. By using heatmaps our method effectively directs the attention of the visual encoder toward the main elements in the image, rather than other visual cues that might introduce gender bias.

In Fig. 1 we present a visual analysis. I denotes the input image. Using only the image I and the CLIP_{FTH} model, and based on Eq. (7), we generate the heatmaps m_1, m_2, and m_3. The explainability map, e_c, is derived from the image I and its

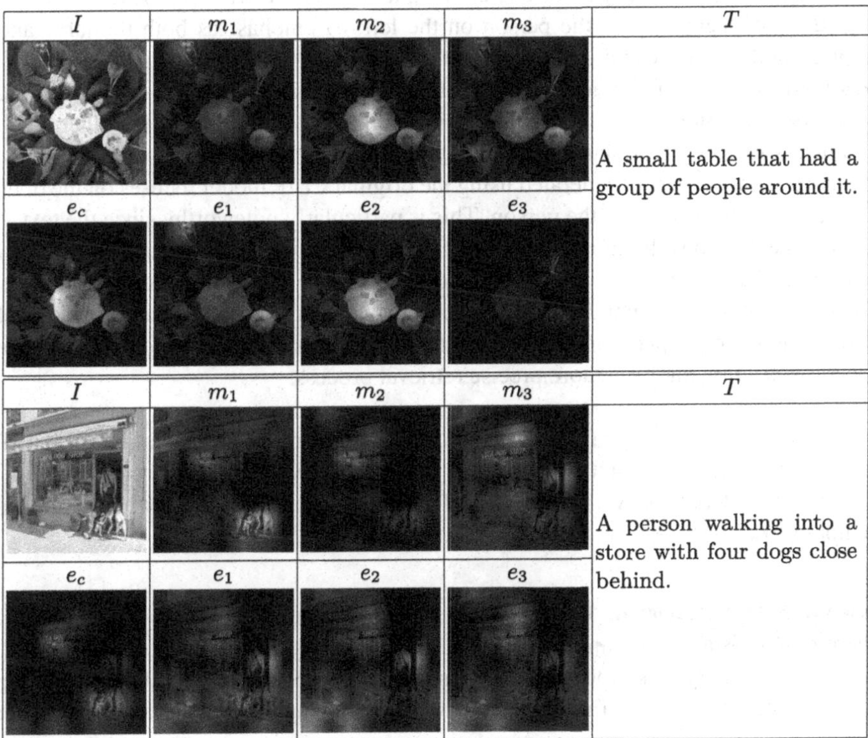

Fig. 1. Visual analysis and comparison of heatmaps and explainability maps. I is the input image. m_1, m_2, and m_3 are the heatmaps generated using our method, demonstrating diverse focus areas in I. e_c is the explainability map of $\text{CLIP}_{\text{Base}}$ for the similarity between I and the text T. e_1, e_2, and e_3 are the explainability maps generated using our method for the similarity between I and the text T. The explainability maps reveal how our method enables nuanced understanding and retrieval by emphasizing varied image elements, thus offering richer context-aware insights compared to the original CLIP model.

corresponding text query T, employing the explainability method of Chefer et al. [1] and the $\text{CLIP}_{\text{Base}}$ model. Lastly, e_1, e_2, and e_3 represent the explainability maps produced using the text query T and the image embeddings I_{e_1}, I_{e_2}, and I_{e_3} respectively. These embeddings are created with the guidance of the heatmaps m_1, m_2, and m_3 using the CLIP_{FTH} model, as described in Eq. (9). The explainability map e_c highlights the regions in the input image I that significantly contribute to the similarity score between the image I and the text T when using the $\text{CLIP}_{\text{Base}}$ model. In contrast, the maps e_1, e_2, and e_3 denote the significant regions when employing our method at the fine-tuning and inference steps. The text T is displayed at the bottom of each row. From the first image, we observe that m_1 emphasizes the person on the top-left, neglecting the table in the center. In contrast, m_2 highlights the table but masks the people. m_3 assigns equal importance to the table, plate, and people. The explainability map e_c generated using the original CLIP model focuses predominantly on the table and masks the people, even though the text T explicitly mentions the people around the table. However, with our method, e_1 concentrates on the person on the left, e_2 emphasizes both the table and the plate, and e_3 focuses on the person on the right. For the second image, m_1 emphasizes both the front of the dog shop and the dogs themselves, but masks the person. m_2 steers attention from the dogs while still highlighting the shop. On the other hand, m_3 attends to the person and the shop, while allocating some emphasis to the dogs. The explainability map e_c generated using the original CLIP model focuses on the dogs and the shop while masking the person. This is particularly noteworthy since the text T refers to the person walking into the store. Conversely, with our method, e_1 gives equal attention to the person, dog, and store, e_2 diminishes the focus on the dogs, and e_3 is primarily concentrated on the store. Using our method, each heatmap enables the model to focus on diverse aspects and regions of the image. This produces a rich set of image embeddings, allowing for a more precise retrieval process.

Runtime. The generation of heatmaps and their application in calculating the CLIP score for a single image-text pair using an NVIDIA RTX A6000 GPU takes approximately 0.061 s. Conversely, this process is faster, taking around 0.019 s when heatmaps are not utilized.

Video and Text Retrieval. Given a dataset of N video-text pairs $D = \{(c_i, V_i)\}_{i=1}^{N}$, where $c_i \in C$ is a caption and $V_i \in V$ is a corresponding video comprised of $\{v_l\}_{l=1}^{M}$ frames, the primary tasks for a video-language retrieval model are twofold: retrieving the best matching video V_i for a text query c_i, or retrieving the most suitable text c_i for a given video V_i.

In the standard retrieval approach, following Eq. 10, a caption $c_i \in C$ is encoded using CLIP's language encoder E_l, yielding an embedding vector c_i^e. For a video $V_j \in V$, each frame v_l is encoded using E_v as depicted in Eq. 1, and following Quintero et al. [13] the embeddings of all frames are averaged to get the video's representation V_j^e:

$$V_j^e = \frac{1}{M} \sum_{l=1}^{M} v_l^e \tag{21}$$

The matching score $S_{ij} \in \mathbb{R}$ is computed as the cosine similarity of the vectors:

Table 4. Text and video retrieval results for MSVD, showing recall and median rank (MdR) scores. 'H' and 'Base' denote with or without heatmap guidance at inference time, respectively.

Method	Inf. Mode	Video Retrieval				Text Retrieval			
		Recall@1	Recall@5	Recall@10	MdR (↓)	Recall@1	Recall@5	Recall@10	MdR (↓)
VSE [10]	N/A	12.3	30.1	42.3	14	34.7	59.9	70.0	3
VSE++ [10]	N/A	15.4	39.6	53.0	9	–	–	–	–
Multi Cues [10]	N/A	20.3	47.8	61.1	6	–	–	–	–
CE [8]	N/A	19.8	49.0	63.8	6	–	–	–	–
Support-set Bottleneck [11]	N/A	28.4	60.0	72.9	4	–	–	–	–
$\text{CLIP}_{\text{Base}}$ [14]	Base	37.0	64.1	73.8	3	59.9	85.2	90.7	**1**
CLIP_{FT}	Base	39.4	67.2	77.0	**2**	60.7	85.4	91.0	**1**
CLIP_{FTH}	H	**41.1**	**68.8**	**78.2**	**2**	**62.7**	**86.7**	**91.9**	**1**

Table 5. Text and video retrieval results for MSR-VTT, showing recall and median ranks (MdR) metrics. 'H' and 'Base' denote with or without heatmap guidance at inference time, respectively.

Method	Inf. Mode	Video Retrieval				Text Retrieval			
		Recall@1	Recall@5	Recall@10	MdR (↓)	Recall@1	Recall@5	Recall@10	MdR (↓)
VSE [10]	N/A	5.0	16.4	24.6	47	7.7	20.3	31.2	28
VSE++ [10]	N/A	5.7	17.1	24.8	65	10.2	25.4	35.1	25
Multi Cues [10]	N/A	7.0	20.9	29.7	38	12.50	32.10	42.4	16
W2VV [3]	N/A	6.1	18.7	27.5	45	11.8	28.9	39.1	21
Dual Enc. [4]	N/A	7.7	22.0	31.8	32	13.0	30.8	43.3	15
E2E [9]	N/A	9.9	24.0	32.4	29.5	–	–	–	–
CE [8]	N/A	10.0	29.0	42.2	16	15.6	40.9	55.2	8.3
$\text{CLIP}_{\text{Base}}$ [14]	Base	21.3	41.1	50.4	10	40.3	69.5	79.5	2
CLIP_{FT}	Base	21.7	42.4	52.2	9	38.5	65.8	75.8	2
CLIP_{FTH}	H	**23.2**	**44.2**	**54.0**	**8**	**42.2**	**70.2**	**80.1**	**2**

$$S_{ij} = \frac{c_i^e \cdot V_j^e}{\|c_i^e\|_2 \|V_j^e\|_2} \tag{22}$$

For enhanced retrieval using heatmap-guided image encoding, each frame v_l is utilized to generate the heatmaps $\{H_k^l\}_{k=1}^n$, which are then used to guide the encoding process to produce n embedding vectors $\{v_{l_k}^e\}_{k=1}^n$. The visual embedding vectors of video V_j, generated using the frames heatmaps $\{H_k^l\}_{l=1}^M$, are averaged to create the video representation $V_j^{(k)}$, and a matching score $S_{ij}^{(k)}$ is computed in a similar way as in Eq. 22. The final heatmap-guided matching score is the average of these n scores:

$$S_{ij}^h = \frac{1}{n} \sum_{k=1}^n S_{ij}^{(k)} \tag{23}$$

For video retrieval by a text query c_i, and for text retrieval by a video query V_j, the results with the highest scores are retrieved by:

$$V^* = \underset{V_j \in V}{\text{argmax}}\, S_{ij} \qquad c^* = \underset{c_i \in C}{\text{argmax}}\, S_{ij} \tag{24}$$

Where S_{ij} can represent either the standard matching score or S_{ij}^h.

Results. We use the test set splits of the MSR-VTT and the MSVD datasets to evaluate our method for the video retrieval task. We calculate the standard video retrieval metrics Recall@1, Recall@5, Recall@10, and the median rank. We compare our method with previous works, including $\text{CLIP}_{\text{Base}}$ and CLIP_{FT}, while the baselines are following the 'Base' inference mode, which means heatmaps are not utilized at inference time. The results for the MSVD and the MSR-VTT datasets are presented in Table 4 and Table 5, respectively. As can be observed, our method consistently outperforms the baselines across all metrics for both datasets. The difference in performance is especially notable for the MSR-VTT dataset. This discrepancy can be attributed to two primary factors: First, videos in the MSVD dataset have an average length of $\sim$10 s, while those in the MSR-VTT dataset average around $\sim$20 s. The extended duration of the latter means they often contain a greater variety of content and more intricate scenes, requiring refined processing for precise retrieval. Second, the MSR-VTT test set is considerably larger, comprising 2290 videos, compared to 670 in MSVD, making the retrieval task inherently more challenging. Our method effectively guides the visual encoder's attention to salient elements within the video stream, providing a distinct advantage in the retrieval task, especially for diverse and large datasets, e.g. MSR-VTT.

5 Conclusions

In our study, we analyzed CLIP's visual-textual alignment and pinpointed its oversight of visual elements w.r.t. a given text. We introduced a method to derive explainability maps solely from images and utilized these maps at the feature level during visual encoding at fine-tuning and inference steps. This approach not only boosts the retrieval accuracy but also reduces biases. By refining CLIP's attention through these maps, we achieved a more nuanced understanding of images and videos. Our work emphasizes the importance of explainability in enhancing model accuracy, suggesting a forward-looking trajectory for research in multimodal systems. Additionally, we note some overlap among the heatmaps in Fig. 1, which may suggest redundancy. This observation is related to the random dropout process applied to the image features to create multiple augmented versions. For future studies, enhancing heatmap diversity could be a key area of focus. A potential strategy is to replace the random dropout with a more targeted technique aimed at reducing similarities between heatmaps. Specifically, this would involve removing features from each heatmap that correspond to a different object, thereby encouraging the production of heatmaps that are more distinct and complementary.

Acknowledgments. This work was supported by a grant from the Tel Aviv University Center for AI and Data Science (TAD). It is part of a Ph.D. research conducted by the first author.

Competing Interests. The authors have no competing interests to declare that are relevant to the content of this article.

References

1. Chefer, H., Gur, S., Wolf, L.: Generic attention-model explainability for interpreting bi-modal and encoder-decoder transformers. In: 2021 IEEE/CVF International Conference on Computer Vision (ICCV), Los Alamitos, CA, USA, October 2021, pp. 387–396. IEEE Computer Society (2021). https://doi.org/10.1109/ICCV48922.2021.00045. https://doi.ieeecomputersociety.org/10.1109/ICCV48922.2021.00045
2. Chen, D., Dolan, W.: Collecting highly parallel data for paraphrase evaluation. In: Proceedings of the 49th Annual Meeting of the Association for Computational Linguistics: Human Language Technologies, Portland, Oregon, USA, June 2011, pp. 190–200. Association for Computational Linguistics (2011). https://aclanthology.org/P11-1020
3. Dong, J., Li, X., Snoek, C.G.M.: Predicting visual features from text for image and video caption retrieval. IEEE Trans. Multimedia **20**(12), 3377–3388 (2018). https://doi.org/10.1109/TMM.2018.2832602
4. Dong, J., et al.: Dual encoding for zero-example video retrieval. In: 2019 IEEE/CVF Conference on Computer Vision and Pattern Recognition (CVPR), pp. 9338–9347 (2019). https://doi.org/10.1109/CVPR.2019.00957
5. Goh, G., et al.: Multimodal neurons in artificial neural networks. Distill (2021). https://doi.org/10.23915/distill.00030. https://distill.pub/2021/multimodal-neurons
6. Kingma, D.P., Ba, J.: Adam: a method for stochastic optimization. arXiv preprint arXiv:1412.6980 (2014)
7. Lin, T., et al.: Microsoft COCO: common objects in context. CoRR abs/1405.0312 (2014). http://arxiv.org/abs/1405.0312
8. Liu, Y., Albanie, S., Nagrani, A., Zisserman, A.: Use what you have: video retrieval using representations from collaborative experts. In: Proceedings of the British Machine Vision Conference (BMVC), September 2019 (2019)
9. Miech, A., Alayrac, J.B., Smaira, L., Laptev, I., Sivic, J., Zisserman, A.: End-to-end learning of visual representations from uncurated instructional videos. In: Proceedings of the IEEE/CVF Conference on Computer Vision and Pattern Recognition, pp. 9879–9889 (2020)
10. Mithun, N.C., Li, J., Metze, F., Roy-Chowdhury, A.K.: Learning joint embedding with multimodal cues for cross-modal video-text retrieval. In: Proceedings of the 2018 ACM on International Conference on Multimedia Retrieval, ICMR 2018, pp. 19–27. Association for Computing Machinery, New York (2018). https://doi.org/10.1145/3206025.3206064
11. Patrick, M., et al.: Support-set bottlenecks for video-text representation learning. arXiv preprint arXiv:2010.02824 (2020)
12. Plummer, B.A., Wang, L., Cervantes, C.M., Caicedo, J.C., Hockenmaier, J., Lazebnik, S.: Flickr30k entities: collecting region-to-phrase correspondences for richer image-to-sentence models. In: Proceedings of the IEEE International Conference on Computer Vision, pp. 2641–2649 (2015)
13. Portillo-Quintero, J.A., Carlos Ortíz-Bayliss, J., Terashima-Marín, H.: A straightforward framework for video retrieval using CLIP. In: Mexican Conference on Pattern Recognition (2021). https://api.semanticscholar.org/CorpusID:232035662
14. Radford, A., et al.: Learning transferable visual models from natural language supervision. In: International Conference on Machine Learning, pp. 8748–8763. PMLR (2021)
15. Su, Y., et al.: Language models can see: plugging visual controls in text generation. arXiv preprint arXiv:2205.02655 (2022)
16. Tewel, Y., Shalev, Y., Nadler, R., Schwartz, I., Wolf, L.: Zero-shot video captioning with evolving pseudo-tokens. arXiv arXiv:2207.11100 (2022). https://api.semanticscholar.org/CorpusID:251018313

17. Tewel, Y., Shalev, Y., Schwartz, I., Wolf, L.: ZeroCap: zero-shot image-to-text generation for visual-semantic arithmetic. In: 2022 IEEE/CVF Conference on Computer Vision and Pattern Recognition (CVPR), pp. 17897–17907 (2021). https://api.semanticscholar.org/CorpusID:244714558
18. Wang, J., Liu, Y., Wang, X.: Are gender-neutral queries really gender-neutral? Mitigating gender bias in image search. In: Proceedings of the 2021 Conference on Empirical Methods in Natural Language Processing, Online and Punta Cana, Dominican Republic, November 2021, pp. 1995–2008. Association for Computational Linguistics (2021). https://doi.org/10.18653/v1/2021.emnlp-main.151. https://aclanthology.org/2021.emnlp-main.151
19. Xu, J., Mei, T., Yao, T., Rui, Y.: MSR-VTT: a large video description dataset for bridging video and language. In: Proceedings of the IEEE Conference on Computer Vision and Pattern Recognition, pp. 5288–5296 (2016)
20. Zabari, N., Hoshen, Y.: Semantic segmentation in-the-wild without seeing any segmentation examples. arXiv preprint arXiv:2112.03185 (2021)

SumBlogger: Abstractive Summarization of Large Collections of Scientific Articles

Pavlos Zakkas[1,2(✉)], Suzan Verberne[1], and Jakub Zavrel[2]

[1] Leiden University, Leiden, The Netherlands
s.verberne@liacs.leidenuniv.nl
[2] Zeta Alpha, Amsterdam, The Netherlands
{zakkas,zavrel}@zeta-alpha.com

Abstract. We propose a prompt-based pipeline for extreme summarization of large collections of scientific articles, which facilitates the consumption of scientific knowledge in high-volume fast-paced fields like AI. Although prompting of generative large language models (LLMs) has been applied to news summarization, its effectiveness in the scientific domain and in multi-document summarization is underexplored. We propose a three-step approach for summarizing a large collection of documents (e.g. hundreds or thousands of papers published in a conference). First, selecting representative papers per document cluster, second, performing single-document summarization (SDS) of the selected papers, and third, aggregating these in a multi-document summarization (MDS) step. Both the single-document summaries and the multi-document summaries are generated with an instruction-tuned LLM. The cluster summaries are used to generate a blog post summarizing a conference. We show that our SDS model achieves better results than strong fine-tuned models on the SciTLDR benchmark. Our two-step approach reaches the performance of state-of-the-art fine-tuned MDS models on the Multi-XScience benchmark. Through a small-scale user study, we find that , although a human-written blog post is clearly preferred over an automatically generated one, the users appreciate the good informativeness and factuality of our pipeline. Our findings demonstrate the potential use of generative LLMs as a way to digest large amounts of scientific papers and help researchers to stay up-to-date with rapidly evolving fields.

1 Introduction

The increasing need to digest large amounts of scientific papers, particularly in fast-paced fields like Artificial Intelligence (AI), has led to the emergence of extreme summarization [28]. This research area, which aims to create concise one-sentence summaries of articles, can be extended to multi-document summarization to help researchers quickly access relevant information. For example, generating an automated summary of a conference with thousands of scientific articles can save significant time and provide a general overview, allowing researchers to focus on their preferred topics.

N. Goharian et al. (Eds.): ECIR 2024, LNCS 14608, pp. 371–386, 2024.
https://doi.org/10.1007/978-3-031-56027-9_23

Advancements in Natural Language Processing (NLP) since the introduction of transformers [40], have improved abstractive summarization methods. Various approaches have been explored, including fine-tuning encoder-decoder transformer models on target datasets [17,32,49], designing transformer variants for longer input contexts [14,31,46] and using generative instruction-tuned Large Language Models (LLMs) for summarizing documents based on prompts [12]. The latter method, however, has been applied mostly to the news domain, with no exploration in the scientific domain, especially when the input document collection is large.

To address this gap, we propose a pipeline for the summarization of a large collection of scientific documents. Our pipeline, called SumBlogger, uses a generative instruction-tuned LLM as the backbone for single-document and multi-document summarization. We first cluster the documents and then investigate the optimal way to perform document selection, single-document summarization, and multi-document summarization. We address the following research questions:

1. How does an instruction-tuned generative LLM compare to fine-tuned encoder-decoder LLMs in Single-Document (SDS) and Multi-Document Summarization (MDS)?
2. How does a two-step approach, summarizing source documents first and then generating the multi-document summary, compare to a one-step approach using the full content of source documents as input?
3. What is the effect of selecting representative documents per cluster on the generation of a summary for a large collection, such as a conference?

We use gpt-3.5-turbo as the backbone LLM, the Zeta Alpha platform [10] to collect AI conference papers, and the Leiden algorithm [38] for clustering. We leverage open-source summarization datasets for evaluation and also conduct a small user study in order to draw our conclusions. Our main contributions are:

- We show that a generative instruction-tuned LLM can be a competitive out-of-the-box solution for both SDS and MDS tasks in the scientific domain, outperforming state-of-the-art finetuned models in the SciTLDR and Multi-XScience datasets.
- We find that using two-step summarization, where the first prompt summarizes each source document separately, and the second prompt aggregates these summaries into a multi-document summary, leads to an improvement in performance, compared to a one-step approach where the full content of the source documents is given in a single prompt.
- We demonstrate the effectiveness of our pipeline by generating the first fully-automated blog post summarizing a whole conference.

We make our code available for reproducibility and follow-up work in GitHub.[1]

[1] https://github.com/zetaalphavector/sumblogger.

2 Related Work

When input documents fit into a vanilla transformer architecture, which is usually the case in the news domain, pre-trained encoder-decoder models can be optimized with fine-tuning [17,24,32,49]. With the rise of general-purpose instruction-tuned LLMs [1,5,36,41], prompt-based approaches have been explored for fluent and informative summaries, leading to zero-shot outputs that may be preferred over fine-tuned models [12,51]. To make the most out of these models, more specific prompting techniques like In-Context Learning (ICL) [7] or Chain-of-Thought (CoT) [42] are commonly applied. Recent work in summarization also investigates iterative refinement of the output using ChatGPT [48] or distilling a gpt-3.5-turbo model into a compact, high-performing model [47].

As the input length is getting larger though, vanilla transformers require truncation, risking information loss. In scientific article summarization, using specific sections as input (e.g., abstract, introduction) can be a sufficient solution for a general-purpose summary, thus this approach has been followed to construct datasets for SDS tasks [6,25,26]. To handle larger contexts, two-step extractive-abstractive approaches can be applied [23,52]. But state-of-the-art results in long document summarization are achieved by modifying the transformer's attention mechanism for larger contexts, up to 16K tokens [14,18,31,34,46]. Closed-source LLMs explore even longer input lengths, like gpt-4 [29] and gpt-3.5-turbo-16k by OpenAI, or Cohere's Summarize API[2], which support respectively up to 8K, 16K tokens, and 50K characters.

The aforementioned approaches can scale the input length, but as we move to even longer inputs as in the cases of book summarization or MDS of large collections of documents, the length issues re-appear. Recent approaches aim to scale the input context of transformers to substantially larger or even unlimited lengths [4,27]. This direction is also explored by generative LLMs such as Claude by Anthopic [1] which claims to support 100K tokens. Alternatively, divide-and-conquer systems can be applied, where parts of a long input are summarized, and then combined in a bottom-up fashion until the full content is covered [45].

Even with the aforementioned methods, providing insights into a large collection of scientific documents remains a challenging task due to the diverse information presented by different papers. Previous approaches focus more on producing topic words and keywords that describe a large corpus [37,53], or generating a set of questions that can be used to summarize a large amount of textual data [35]. These solutions are helpful, but an automated summary is expected to be preferable due to its fluency and high coverage of the collection along with its user-friendliness.

In any case though, and especially in collections of conference papers, it is crucial to divide the content into digestible sections in order to enhance user comprehension. For that reason, clustering the documents into meaningful topics and then summarizing each topic seems to be a worth-exploring idea. In the

[2] https://cohere.com/summarize.

context of MDS, this approach was examined in prior work by using an encoder transformer to cluster the documents and then a decoder to generate the summaries [39]. Our work differs not only in the clustering method, but mostly in enabling the usage of generative LLMs to produce summaries according to the requirements of each specific use case.

3 Methods

The overview of SumBlogger is presented in Fig. 1. Our research focus is on the summarization pipeline which uses a two-step summarization approach to handle extremely long inputs. The components of SumBlogger are explained in detail in the following sections.

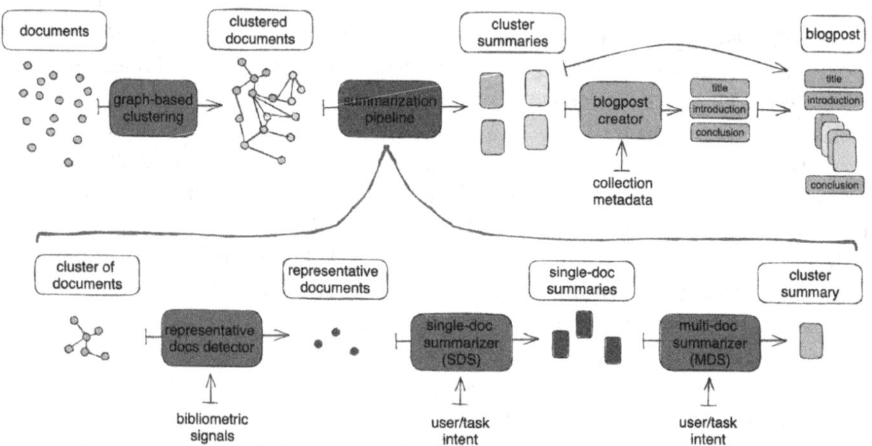

Fig. 1. SumBlogger pipeline: Overview of automatic blog post generation for a large collection of documents, such as a conference. The clustering component is kept fixed, while for the blog post creator component, a simple prompt-based approach was used to generate the blog post title, introduction and conclusion. The focus of our work is given to the summarization module, which is responsible for taking as input a set of clusters of documents and producing a summary for each cluster based on the intent of the user or the task at hand.

3.1 Clustering of Documents

The first component of SumBlogger clusters documents using the VosViewer [8] implementation of the graph-based Leiden algorithm [38]. In order to optimize the modularity of the clustering solution, the algorithm follows an iterative process where the nodes are moved from one cluster to another , until no further improvement can be made [38].

Different types of graphs can be used as input for clustering, in particular citation-based networks where documents are connected to their cited documents [38], and similarity-based networks where documents are connected based on their textual similarity [43]. In our case, since clustering is not the focus of this paper, we use the provided functionality of Zeta Alpha, where the 'More Like This'[3] query functionality of Elastic Search is utilized. This query functionality is term-based, using the terms with the highest TF-IDF score in a document to retrieve a list of similar papers for each paper in the collection. The score returned by Elastic Search is then used as the similarity score between a pair of documents. Edges are created between paper pairs for which the 'More Like This' similarity score exceeds a predefined threshold.[4] The resulting graph with normalized similarity scores is then given to VosViewer which returns the same graph, enhanced with the cluster attributed to each document.

3.2 Selection of Representative Documents

To summarize each cluster, we could input all documents into the LLM at once, up to the token limit of the model. However, our intuition is that it may be difficult for the LLM to process diverse information and produce a robust summary. Instead, we select a set of representative documents per cluster, with the number depending on the desired summary length, since we assume that the length of the summary limits the amount of information and references that can be included. For SumBlogger, we use a reasonable setting of 10 paper summaries per cluster and we generate 100-word cluster summaries based on these 10 paper summaries, with further adjustments left for future work, since the clustering step is not our research focus.

In order to select representative papers, we initially consider the Dominating Set concept in graph theory [30], which is a set of vertices such that any vertex of the graph is either in the set or has a neighbor in the set. However, this approach does not account for weak connections, leading to papers that are not represented sufficiently. For instance, in the graph presented in Fig. 2a, node A belongs to the dominating set, even though its connections are weak. Instead, we would prefer node B, since it has much stronger similarity with its neighbors, although it is connected to one less node compared to A.

An alternative is to assign a similarity strength to each node, also known as node strength [2], which is based on the sum of the similarity scores of its edges, indicating how well a node represents its neighbors. In this case though, subnetworks that are strongly connected would be over-represented (see Fig. 2b), leaving no space for other cluster nodes.

By combining the above two approaches, we create a sequence of iterative steps which overcomes the aforementioned downsides. Firstly, we compute the *Similarity Strength* to each node which is equal to the sum of the similarity scores on its edges. Secondly, we select the node with the highest *Similarity Strength* as a representative of the cluster. Thirdly, we remove this node from

[3] https://www.elastic.co/guide/en/elasticsearch/reference/current/query-dsl-mlt-query.html.

[4] In our experimentation, we used the implementation of 'More Like This' as provided by the Zeta Alpha platform [10] with default parameters through the web interface.

the graph along with its edges. We repeat this process 10 times, which is the target number of papers that are selected. The removal of the edges is the key step of our algorithm since it will lead to a decrease in the similarity strengths of the neighbors of the selected node. This is a desired effect, as these nodes are already likely to be represented by the selected node. In this way, the weaknesses of the original approaches are mitigated, as demonstrated in Figs. 2c, 2d.

Fig. 2. Examples of selecting $k = 3$ representative documents in a cluster, where: (a) uses a minimal dominating set, (b) selects the nodes with the highest similarity strength, and (c), (d) use our suggested algorithm. The size of the nodes indicates their *SimilarityStrength* before the first iteration, where no edge has been removed. The thickness of the edges indicates the strength of the similarity score between two nodes.

3.3 Single-Document Summarization (SDS)

To summarize each paper, the most common solution is to use a pre-trained LLM designed for summarization and finetune it on the target dataset. Such models include BART [17], PEGASUS [49], and T5 [32] which can perform extreme summarization tasks in the scientific domain (e.g. SciTLDR [6]). We replicate some of these models and use them as baselines for our proposed method. Our experimental setting is the use of an instruction-tuned LLM that enables zero-shot summarization without the need for fine-tuning. This allows for adapting the model for the downstream task by modifying the instructions in the given prompt. Thus, besides the document, we can give the user/task intent as input in order to adjust the summary. We use In-Context Learning (ICL) by incorporating document-summary pairs in the prompt as few-shot examples of the target task. Additionally, when using LLMs that that have been optimized for conversations, we experiment with prompting the model with a persona before starting the actual conversation which will contain the task. This is also the suggested approach by Open AI for their chat-based models gpt-3.5-turbo and gpt-4, where with the system message, we can introduce high-level instructions beforehand [44].[5]

Based on the above, we compare the different prompt variants shown in Table 1. First, we use a simple vanilla prompt, previously suggested in the news domain [12] in order to see its effectiveness in the scientific domain. Second, we use our proposed prompt template which contains a setup message for the LLM

[5] https://platform.openai.com/docs/guides/chat.

and task-specific instructions related to the summary style, the target length and the target audience of the summary. Lastly, the latter prompt is enhanced with ICL, by setting up the conversation with a few carefully hand-picked pairs of user and bot messages which correspond to the input documents and their gold summaries respectively.

Table 1. The vanilla and our 0-shot prompts for the SDS task for extreme summarization (e.g. SciTLDR). The '{{ document }}' and '{{ n }}' are the parameters for the input document and the target length in words respectively. For our 2-shot prompt, we initialize the conversation with two user-bot interactions with two document-summary pairs from the training set.

Prompt Method	Template
vanilla	{{document}} Summarize the above article in 1 sentence
0-shot	*System template:* You are the most famous research journalist in writing summaries of scientific articles. Your summaries are not only concise, informative and of high quality but they are also very appealing and pleasant to read. You are also an expert in grammar and vocabulary and you can adapt your writing style following the given instructions. *User message:* Write a short and concise phrase summarizing the provided document in {{ n }} words. The summary should be informative for a reader who is an experienced researcher in this field. Document: {{ document }}

In this work, we use gpt3.5-turbo as the LLM and we adapt the prompts based on the summarization task at hand (SDS, MDS). Also, when the task is not similar to a task defined in a specific dataset, there is no training set to get the gold summaries for the ICL examples. Thus, we pick our ICL examples by iteratively prompting an LLM to adjust the summary of a sample document that is similar to our target collection. We use gpt-4 to generate these ICL examples due to its higher capabilities in following instructions and adapting the output more easily according to our needs.

3.4 Multi-Document Summarization (MDS)

As in the SDS component, a common approach is to use encoder-decoder models, fine-tuned on the MDS task at hand. One challenge lies in the input length of MDS tasks which is usually much larger if we give the full content of all the source documents as input. We select PRIMERA [46] and Pointer-Generator [33] as the baseline models for our MDS experiments since they use architectures designed for long documents and achieve state-of-the-art results in the Multi-XScience dataset [25].

For our instruction-tuned LLM, the suggested two-step approach leads to shorter prompts in the second stage, which allows us to include few-shot examples in order to guide the summary output through ICL. Also, our expectation is that the shorter context length of the MDS stage will make it easier for the LLM to focus on the key points of each paper, since it is already summarized.

In our experimentation, we prompt gpt-3.5-turbo model with manually written MDS task instructions, including the task definition, the output format guidelines, the writing style, the target length and the target audience of the summary. As in SDS, we also use the system message to setup the chatbot persona as an expert in the task at hand [44]. Specifically, when evaluating our summarization components on the Multi-XScience dataset, we define the task as writing a related work paragraph for a paper given a set of other papers as input. On the other hand, when we use SumBlogger to produce a blog post for a large conference, we define the MDS task as generating a paragraph that summarizes the input documents while providing references. Due to their extensive size, these prompt templates can be found on the 'Prompts' section of our github repository[6].

3.5 Conference Summarization

One potential application of SumBlogger is to generate a blog post summarizing a large AI conference. So far, we have the cluster summaries which include references to the representative papers. Those can be used to formulate the sections of our blog post. To expand this into a blog post, we produce a short introductory paragraph for each section, before delving into the details of the referenced papers. To that end, we run our MDS component for a second time in parallel, by slightly adapting the prompt to generate a short introductory paragraph of around 30 words. The introductory paragraph and the cluster summary with references are then concatenated, and a visualization of the VosViewer cluster is also added for a more appealing output.

The next step is to generate the section titles. For that purpose, we prompt the LLM to generate short titles given the introductory paragraphs describing the clusters. The reason for generating all the cluster titles at once was based on our intent to avoid generating the same title for different clusters. Some of the clusters might be semantically close to each other, thus prompting for a title for each of them separately increases the risk to produce the same generic title.

Finally, the blog post surroundings (i.e. title, introduction, conclusion) are generated by prompting again with the introductory paragraphs describing the clusters since those can give the overview of the full content. At this point, we also use the metadata of the collection as input, which in terms of a conference refers to its name, location, website and date. As before, we create the blog post surroundings at once, in order for the model to have access to the already generated parts and produce more consistent content.

[6] https://github.com/zetaalphavector/sumblogger?tab=readme-ov-file#prompts.

4 Experiments and Results

We evaluate the summarization components in isolation, firstly evaluating the SDS stage using the SciTLDR dataset [6], and secondly evaluating our two-step summarization approach on the Multi-XScience benchmark [25].

For fair comparison, we report ROUGE [19], as it is consistently reported in prior work for all summarization models. In addition, we use BLEU [13] and BERTScore [50] (the latter based on SciBERT [3]), which correlate more strongly with human judgment than ROUGE [13,50].

Scaling our experiments for end-to-end evaluation is challenging due to SumBlogger's multiple stages and the lack of large-scale scientific summarization datasets. Even the BigSurvey [21] dataset, with over 50 documents, does not fully meet the needs of large conferences. Therefore, we use SumBlogger to produce an automated blog post for a large AI conference, and we conduct a small user survey based on guidelines suggested by existing evaluation protocols [9,16].

SDS Results. The TLDR generation introduced by the SciTLDR [6] dataset involves high source compression rates and expert background knowledge, which aligns with the purposes of SumBlogger. Thus, we evaluate the SDS stage, where gpt-3.5-turbo is fed with the prompts described in Sect. 3.3, against strong models that were fine-tuned on SciTLDR: the BART-based CATTS method[7] [6], which reported state-of-the-art results, PEGASUS-large[8] [49], and T5-base[9] [32]. All models were configured as suggested by the creators. For CATTS, we also compare our results to the previously reported results on the same dataset, reported by [6].

In Table 2, we observe that our reproduced CATTS version closely matches the authors' reported scores for ROUGE-2 and ROUGE-L [6]. Secondly, we see that switching from the vanilla prompt to our prompt (see Sect. 3.3) substantially improves performance, and using a two-shot setting with static exemplars from the training set leads to the highest scores across all metrics. A paired t-test indicates that the difference between our 2-shot prompt method and the T5 model is statistically significant across all metrics ($p < 0.0001$). The difference between our 2-shot prompt method and the CATTS method is statistically significant across all metrics ($p < 0.05$) except for ROUGE-2.

We also examine how each method approximates the average summary length of 20 words of the training set. CATTS generates shorter summaries, resembling paper headlines, likely due to its initial title generation task. T5 and PEGASUS produce longer summaries, not always adhering to extreme summarization goals, while gpt-3.5-turbo with the vanilla prompt also generates longer summaries despite following the instruction to write one sentence. However, prompting explicitly for 20-word summaries aligns the LLM with the target dataset, showcasing its adaptability.

---

[7] https://huggingface.co/lrakotoson/scitldr-catts-xsum-ao.

[8] https://huggingface.co/alk/pegasus-scitldr.

[9] https://huggingface.co/HenryHXR/t5-base-finetuned-scitldr-only-abstract.

Table 2. SDS results on the SciTLDR dataset. We report max ROUGE scores with stemming and stopword removal, BLEU and BERTScore and average summary length in words. Regarding the gpt-3.5-turbo results, the 'vanilla prompt' was based on a simple prompt for summarization, previously proposed in the news domain [12], '0-shot (ours)' refers to the prompt described in Sect. 3.3, and '2-shot (ours)' refers to the same prompt enhanced with two exemplars from the training set.

Method	ROUGE-1	ROUGE-2	ROUGE-L	BLEU	BERTScore	#words
CATTS (ours)	0.416	0.212	0.359	0.027	0.697	7.48
CATTS [6]	0.443	0.213	0.359	–	–	–
PEGASUS	0.399	0.208	0.346	0.126	0.692	33.62
T5	0.403	0.208	0.351	0.143	0.696	29.21
gpt-3.5-turbo vanilla	0.409	0.199	0.339	0.113	0.699	33.38
0-shot (ours)	0.416	0.200	0.345	0.114	0.700	19.22
2-shot (ours)	**0.445**	**0.223**	**0.375**	**0.144**	**0.708**	18.92

MDS Results. Somebody could argue that for a small collection of source documents such as Multi-XScience, giving the full content in a single prompt could be more effective than our proposed two-step approach, thus we decided to evaluate both variants. Also, it is important to compare against the results achieved by fine-tuned models which report state-of-the-art results in Multi-XSCience, including PRIMERA and Pointer-Generator. For a fair comparison, we use the ROUGE referenced by the authors of the PRIMERA model [46].

The results are shown in Table 3. The table shows that Pointer-Generator obtains the best results in terms of ROUGE_L score, while in terms of ROUGE_1 and ROUGE_2, the gpt-3.5-turbo two-step approach achieves the highest scores. The fine-tuned variant of PRIMERA achieves comparable results to Pointer-Generator, while the zero-shot version is poor compared to the gpt-3.5-turbo variants. Most importantly, our two-step approach leads to a performance gain compared to the one-step approach with regard to all the recorded metrics, which was proven statistically significant with a paired t-test ($p < 0.0001$).

Conference Summarization User Study. In order to assess the effectiveness of selecting representative documents, we evaluate SumBlogger with representative document selection and ICL against a baseline variant of the pipeline, where the component for selecting representative papers is not used, and the prompt includes summaries of all the cluster's papers. As source data, we use the ICLR 2023 conference[10], with over 2300 papers, and for evaluation we conduct a small user study with 22 participants, the majority of whom claim to have significant knowledge of AI. Since no external system generates automated blog posts, we

[10] https://iclr.cc/Conferences/2023.

Table 3. Multi-Document Summarization results on Multi-XScience. Average ROUGE scores with stemming, BLEU and BERTScore are recorded. Regarding the gpt-3.5-turbo results, the 'one-step' approach refers to a single prompt where the full content of all source documents is given at once, whereas the 'two-step' approach refers to an SDS stage followed by an MDS step, which takes as input the summaries produced during the SDS stage and generates the final multi-document summary.

Method	ROUGE-1	ROUGE-2	ROUGE-L	BLEU	BERTScore
Pointer-Generator [46]	0.339	0.068	**0.182**	–	–
PRIMERA (finetuned) [46]	0.319	0.074	0.180	–	–
PRIMERA (0-shot) [46]	0.291	0.046	0.157	–	–
gpt-3.5-turbo one-step	0.347	0.080	0.170	0.070	0.598
gpt-3.5-turbo two-step	**0.354**	**0.084**	0.172	**0.092**	**0.605**

also compare SumBlogger against a human-written gold blog post written by an AI Analyst.[11] We ask for human preference judgments for SumBlogger vs. baseline and for SumBlogger vs. human in terms of the following aspects:

Overall: the general preference of the annotator between the two blog posts
Fluency: the quality of individual sentences
Factuality: Which blog would you trust more that provides a factually consistent summary of the conference?
Coherence: Each blog should build from sentence to sentence to a coherent body of information about a topic. Based on that, which blog is more coherent for you?
Informativeness: Each blog should cover the conference as broadly as possible in order not to miss any potentially relevant and important information. Which blog would you prefer based on that?
Non-redundancy: There should be no unnecessary repetition in the summary. Unnecessary repetition might take the form of whole sentences that are repeated, or repeated facts, or the repeated use of a noun or noun phrase.
Blog Structure: Which blog would you prefer in terms of blog structure (sections, section titles, length, overall structure)?

We also asked respondents to add comments supporting their assessment.

Figure 3a shows that the human-written summary is preferred over SumBlogger's output. Most comments refer to the user-friendliness, the structure and the readability of the human-written blog post, and not to the content of information. Specifically, a positive sign is that for 'Informativeness', 'Non-redundancy', and 'Factuality' around 50% of the respondents believe that SumBlogger was at least as good as the human-written summary. Taking also into account the significant effort and time needed to write such a blog post, SumBlogger is a promising solution, especially if we further adapt the way it presents information.

[11] https://www.zeta-alpha.com/post/a-guide-to-iclr-2023-10-topics-and-50-papers-you-shouldn-t-miss.

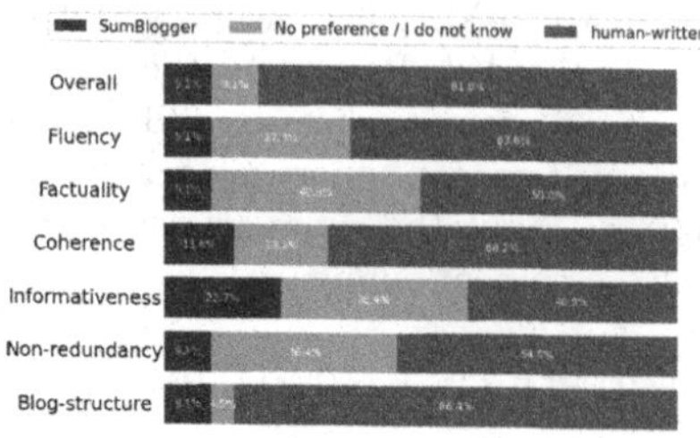

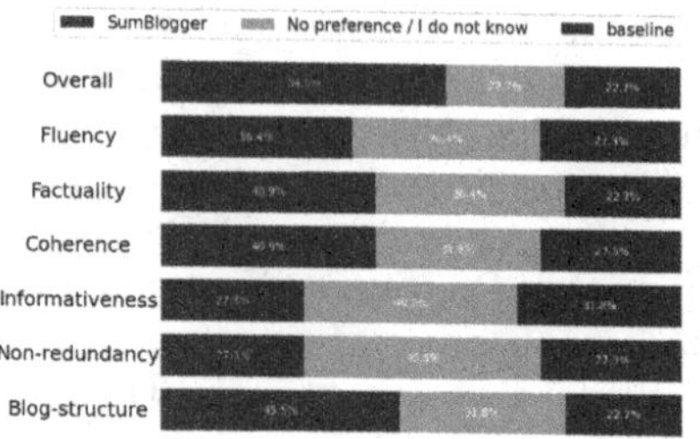

Fig. 3. Preference judgment comparison of SumBlogger against (a) the human-written blog post and (b) the baseline pipeline, where no representative papers were selected.

In Fig. 3b, we see that SumBlogger is preferred over the baseline variant. However, several respondents consider the baseline at least as good as the SumBlogger, mainly because it contained information from a larger number of papers. Those who preferred SumBlogger though point out that the baseline contained too many details which harms readability, usability and informativeness.

5 Discussion

This work shows the capabilities of instruction-tuned LLMs in generating blog posts and summarizing large collections of documents by providing references to the sources. However, there are several limitations and future work directions. Firstly, our research is constrained in utilizing only the closed-source gpt-3.5-turbo LLM, and we encourage the use of open-source LLMs in follow-up research.

Moreover, besides the evaluation of selecting representative papers, the factuality of a generated blog post is another important question [15], and evaluating it is a challenging task, since it is infeasible for human readers to inspect the whole source collection and verify the faithfulness of the summary. Also, using an LLM to evaluate the factuality of a summary [11,22] is not directly applicable in our case since the input context exceeds the tokens limit of the LLMs. In addition, the user study can be further extended in order to examine the preference of more users across a larger set of conference summaries.

A final challenge is the position bias of LLMs to the information in the prompt. To further inspect the effect of this bias on the selection of papers, we produce MDS summaries for 500 random perturbations of 100 papers of one particular cluster, each time changing the order of papers in the prompt. If the selection of the papers by the LLM is robust against prompt order, we would expect the same papers to be picked every time. However, in Fig. 4a, the most cited paper is cited in less than 20% of the generated summaries. On the other hand, if we examine the distribution of the indices – indicating the position of a paper – that are selected by the model independently of which document corresponds to each index (Fig. 4b), we can see that the LLM is highly biased towards the papers presented first. This is further supported by related work, showing that the parts put in the middle of the context are more likely to be

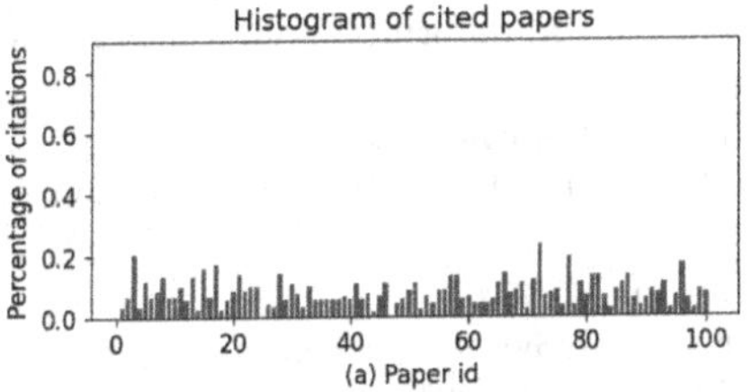

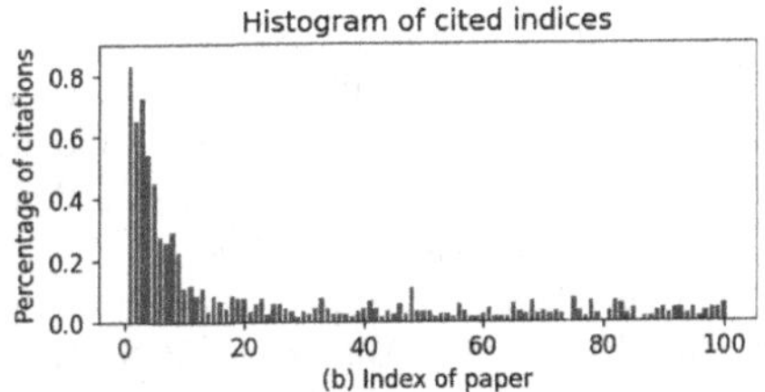

Fig. 4. Histogram of papers (a) and indices (b) that were cited, after 500 executions of the MDS stage in a single cluster, where 100 papers were included in the prompt.

ignored [20]. This reveals the weakness of such LLM to process a large amount of diverse information and select representative papers in a robust way.

6 Conclusion

In this work, we demonstrated that instruction-tuned LLMs like gpt-3.5-turbo show strong summarization abilities and can be adapted out-of-the-box for SDS and MDS tasks without the need for fine-tuning on downstream supervised tasks. The two-step summarization approach of SumBlogger leads to a performance gain over an end-to-end single-step summarization task, which can be attributed to the fact that it will be easier for an LLM to handle a prompt with more concise content. In addition, selecting a handful of representative papers per cluster before generating the summaries leads to blog posts that are preferred by humans, while prompting the model with all the documents of a cluster does not produce robust summaries due to the position bias of the papers in the prompt. Finally, even though the human-written blog post is clearly preferred, the informativeness and the factuality of SumBlogger are assessed equally to human summaries. This observation, combined with the large human effort to write such a blog post, reveals the potential of SumBlogger, especially if the blog structure is further adapted according to the feedback.

References

1. Bai, Y., et al.: Training a helpful and harmless assistant with reinforcement learning from human feedback (2022)
2. Barrat, A., Barthélemy, M., Pastor-Satorras, R., Vespignani, A.: The architecture of complex weighted networks. Proc. Natl. Acad. Sci. U.S.A. **101**(11), 3747–52 (2004). https://doi.org/10.1073/pnas.0400087101
3. Beltagy, I., Lo, K., Cohan, A.: SciBERT: a pretrained language model for scientific text. In: Proceedings of the 2019 Conference on Empirical Methods in Natural Language Processing and the 9th International Joint Conference on Natural Language Processing (EMNLP-IJCNLP), pp. 3615–3620. Association for Computational Linguistics, Hong Kong, China (Nov 2019). https://doi.org/10.18653/v1/D19-1371, https://aclanthology.org/D19-1371

4. Bertsch, A., Alon, U., Neubig, G., Gormley, M.R.: Unlimiformer: Long-range transformers with unlimited length input (2023)
5. Brown, T.B., et al.: Language models are few-shot learners (2020)
6. Cachola, I., Lo, K., Cohan, A., Weld, D.S.: TLDR: Extreme summarization of scientific documents (2020)
7. Dong, Q., et al.: A survey on in-context learning (2023)
8. van Eck, N.J., Waltman, L.: Software survey: VOSviewer, a computer program for bibliometric mapping. Scientometrics **84**, 523–538 (2010). https://doi.org/10.1007/s11192-009-0146-3
9. Fabbri, A.R., Kryściński, W., McCann, B., Xiong, C., Socher, R., Radev, D.: SummEval: re-evaluating Summarization Evaluation. Trans. Assoc. Comput. Linguist. **9**, 391–409 (2021). https://doi.org/10.1162/tacl_a_00373
10. Fadaee, M., Gureenkova, O., Barrera, F., Schnober, C., Weerkamp, W., Zavrel, J.: A new neural search and insights platform for navigating and organizing AI research (2020)
11. Fu, J., Ng, S.K., Jiang, Z., Liu, P.: Gptscore: Evaluate as you desire (2023)
12. Goyal, T., Li, J.J., Durrett, G.: News summarization and evaluation in the era of gpt-3 (2022)
13. Graham, Y.: Re-evaluating automatic summarization with BLEU and 192 shades of ROUGE. In: Proceedings of the 2015 Conference on Empirical Methods in Natural Language Processing, pp. 128–137. Association for Computational Linguistics, Lisbon, Portugal (Sep 2015). https://doi.org/10.18653/v1/D15-1013, https://aclanthology.org/D15-1013
14. Guo, M., et al.: Longt5: Efficient text-to-text transformer for long sequences (2022)
15. Ji, Z., et al.: Survey of hallucination in natural language generation. ACM Comput. Surv. **55**(12), 1–38 (Mar 2023). https://doi.org/10.1145/3571730
16. Kryscinski, W., Keskar, N.S., McCann, B., Xiong, C., Socher, R.: Neural text summarization: A critical evaluation. In: Proceedings of the 2019 Conference on Empirical Methods in Natural Language Processing and the 9th International Joint Conference on Natural Language Processing (EMNLP-IJCNLP), pp. 540–551. Association for Computational Linguistics, Hong Kong, China (Nov 2019). https://doi.org/10.18653/v1/D19-1051, https://aclanthology.org/D19-1051
17. Lewis, M., et al.: Bart: Denoising sequence-to-sequence pre-training for natural language generation, translation, and comprehension (2019)
18. Li, M., Hovy, E., Lau, J.H.: Towards summarizing multiple documents with hierarchical relationships (2023)
19. Lin, C.Y.: ROUGE: a package for automatic evaluation of summaries. In: Text Summarization Branches Out, pp. 74–81. Association for Computational Linguistics, Barcelona, Spain (Jul 2004). https://aclanthology.org/W04-1013
20. Liu, N.F., et al.: Lost in the middle: How language models use long contexts (2023)
21. LIU, S., Cao, J., Yang, R., Wen, Z.: Generating a structured summary of numerous academic papers: Dataset and method. In: Raedt, L.D. (ed.) Proceedings of the Thirty-First International Joint Conference on Artificial Intelligence, IJCAI-22, pp. 4259–4265. International Joint Conferences on Artificial Intelligence Organization (7 2022). https://doi.org/10.24963/ijcai.2022/591
22. Liu, Y., Iter, D., Xu, Y., Wang, S., Xu, R., Zhu, C.: G-eval: Nlg evaluation using gpt-4 with better human alignment (2023)
23. Liu, Y., Lapata, M.: Text summarization with pretrained encoders (2019)
24. Liu, Y., Liu, P., Radev, D., Neubig, G.: Brio: Bringing order to abstractive summarization (2022)

25. Lu, Y., Dong, Y., Charlin, L.: Multi-xscience: A large-scale dataset for extreme multi-document summarization of scientific articles (2020)
26. Mao, Y., Zhong, M., Han, J.: Citesum: citation text-guided scientific extreme summarization and domain adaptation with limited supervision (2022)
27. Martins, P.H., Marinho, Z., Martins, A.F.T.: ∞-former: Infinite memory transformer (2022)
28. Narayan, S., Cohen, S.B., Lapata, M.: Don't give me the details, just the summary! topic-aware convolutional neural networks for extreme summarization. In: Proceedings of the 2018 Conference on Empirical Methods in Natural Language Processing, pp. 1797–1807. Association for Computational Linguistics, Brussels, Belgium (Oct-Nov 2018). https://doi.org/10.18653/v1/D18-1206, https://aclanthology.org/D18-1206
29. OpenAI: Gpt-4 technical report (2023)
30. Ore, Ø.: Theory of Graphs. No. pt. 1 in American Mathematical Society colloquium publications, American Mathematical Society (1962). https://books.google.nl/books?id=g85UAAAAYAAJ
31. Phang, J., Zhao, Y., Liu, P.J.: Investigating efficiently extending transformers for long input summarization (2022)
32. Raffel, C., et al.: Exploring the limits of transfer learning with a unified text-to-text transformer (2020)
33. See, A., Liu, P.J., Manning, C.D.: Get to the point: Summarization with pointer-generator networks (2017)
34. Shen, C., Cheng, L., Nguyen, X.P., You, Y., Bing, L.: A hierarchical encoding-decoding scheme for abstractive multi-document summarization (2023)
35. Surita, G., Nogueira, R., Lotufo, R.: Can questions summarize a corpus? using question generation for characterizing covid-19 research (2020)
36. Taori, R., et al.: Stanford alpaca: an instruction-following llama model. https://github.com/tatsu-lab/stanford_alpaca (2023)
37. Thielmann, A., Seifert, Q., Reuter, A., Bergherr, E., Säfken, B.: Topics in the haystack: Extracting and evaluating topics beyond coherence (2023). https://arxiv.org/abs/2303.17324
38. Traag, V.A., Waltman, L., van Eck, N.J.: From louvain to leiden: guaranteeing well-connected communities. Sci. Reports **9**(1) (mar 2019). https://doi.org/10.1038/s41598-019-41695-z, https://www.nature.com/articles/s41598-019-41695-z
39. Trabelsi, M., Uzunalioglu, H.: Absformer: Transformer-based model for unsupervised multi-document abstractive summarization (2023)
40. Vaswani, A., et al.: Attention is all you need (2017)
41. Wei, J., et al.: Finetuned language models are zero-shot learners (2022)
42. Wei, J., et al.: Chain-of-thought prompting elicits reasoning in large language models (2023)
43. Whissell, J.S., Clarke, C.L.: Improving document clustering using okapi bm25 feature weighting. Inf. Retrieval **14**, 466–487 (2011)
44. White, J., et al.: A prompt pattern catalog to enhance prompt engineering with chatgpt (2023)
45. Wu, J., et al.: Recursively summarizing books with human feedback (2021)
46. Xiao, W., Beltagy, I., Carenini, G., Cohan, A.: Primera: Pyramid-based masked sentence pre-training for multi-document summarization (2022)
47. Xu, Y., et al: Inheritsumm: a general, versatile and compact summarizer by distilling from GPT (2023)
48. Zhang, H., Liu, X., Zhang, J.: Summit: Iterative text summarization via chatgpt (2023)

49. Zhang, J., Zhao, Y., Saleh, M., Liu, P.J.: Pegasus: pre-training with extracted gap-sentences for abstractive summarization (2020)
50. Zhang, T., Kishore, V., Wu, F., Weinberger, K.Q., Artzi, Y.: Bertscore: Evaluating text generation with bert (2020)
51. Zhang, T., Ladhak, F., Durmus, E., Liang, P., McKeown, K., Hashimoto, T.B.: Benchmarking large language models for news summarization (2023)
52. Zhang, X., Wei, F., Zhou, M.: Hibert: Document level pre-training of hierarchical bidirectional transformers for document summarization (2019)
53. Zhang, Z., Fang, M., Chen, L., Rad, M.R.N.: Is neural topic modelling better than clustering? an empirical study on clustering with contextual embeddings for topics (2022). https://aclanthology.org/2022.naacl-main.285/

Attend All Options at Once: Full Context Input for Multi-choice Reading Comprehension

Runda Wang(✉), Suzan Verberne(✉), and Marco Spruit

Leiden University, Leiden, The Netherlands
r.wang.8@outlook.com, {s.verberne,m.r.spruit}@liacs.leidenuniv.nl

Abstract. This paper proposes a method to capture the relations between options in Multiple-choice Machine Reading Comprehension (MMRC) tasks. MMRC is a form of question answering (QA) in which the question is about a given text, and multiple answers are provided as options. Capturing the relations between options is especially important for options with information references between them that cannot stand alone as responses to the questions, such as "None of the above". Our method 1) takes the whole sample including identification of the passage, question, and all options as input for pre-trained language models, and 2) adds a fuser network to emphasize the information interaction between options. Experimental results show that our method improves over the common encoding approaches on COSMOS-QA, an MMRC dataset with between-option references, while having a relatively small impact on other MMRC datasets without references between the options. We conclude that our method actually helps to capture the necessary relationships between options. In addition, our method can reduce the memory usage required for training, and the model can be easily transferred to other domains and models.

Keywords: Machine Reading Comprehension · Question Answering · Pre-trained Language Models

1 Introduction

Machine Reading Comprehension (MRC) is a series of question answering tasks used for measuring the ability of computational methods to read and understand natural language text. MRC typically comes in four different task forms [14]: cloze-style, multi-choice, span-prediction, and free-form. Among these tasks, Multi-choice Machine Reading Comprehension (MMRC) contains three components: a passage, a question, and an option set, and requires the machine to select one correct answer from the option set given the passage and question. An example is shown in Table 1. In this paper, we are specifically interested in the dependence between options in MMRC tasks and devise a method to capture this relationship.

N. Goharian et al. (Eds.): ECIR 2024, LNCS 14608, pp. 387–402, 2024.
https://doi.org/10.1007/978-3-031-56027-9_24

Table 1. Example of an MMRC task from the COSMOS-QA dataset.

Passage:
Until we went to a playdate two weeks ago. Thea's mom is Serbian and crepes are apparently as common in Serbia as they are in France. We discussed the batter, the texture, the cooking process, the topping options and Dee generally brought me up to speed. Being not brave enough to just start throwing ingredients in a bowl as she did, I got a recipe of the internet for general proportions, ended up not using nearly as much water as was called for and successfully made crepes.
Question:
Why did they discuss crepes?
Options:
A. Because Thea's mom is Serbian.
B. Because the writer got a recipe from the internet.
C. Because the writer is interested in learning how to cook crepes.
D. None of the above choices.

The relations between options and references from one option to another can be relevant for selecting the correct answer. In Table 1, we can see an example from COSMOS-QA [5], an MMRC dataset, with the option *'none of the above choices.'* which must be assessed in conjunction with all other options because for the answer to be correct, the reader has to analyze that the statements in the alternatives are incorrect. In other words, the other options constitute the context for assessing the correctness of the fourth option. In many prior approaches to MMRC [3,6,27–29], however, the answer options are modelled by the language models in relation to the question and the passage, but not to each other. These methods take a triple consisting of a passage, a question, and one of the options as input, and classify the triple as being correct or not. As a result, when one option refers to information in other options, this is not modelled.

To address this issue, we propose a method which is feeding the model the passage, question, and all options at once to enable direct access to the full context of the MMRC task. Our contributions are twofold: (1) We propose a new input format to provide the model with complete context for every option to allow the model to process the relations between them; (2) We propose a fuser network to strengthen the information interaction within the option set with transformer encoder blocks. Our experimental results indicate that both complete context and the fuser network are beneficial to the modelling of relations between options. We release our code for follow-up research.[1]

2 Related Work

The publication of the MCTest dataset [16] has greatly promoted the research interest in machine learning methods for MMRC tasks since 2013 [17,23,25]

[1] https://github.com/WongDaDa/AOAO.

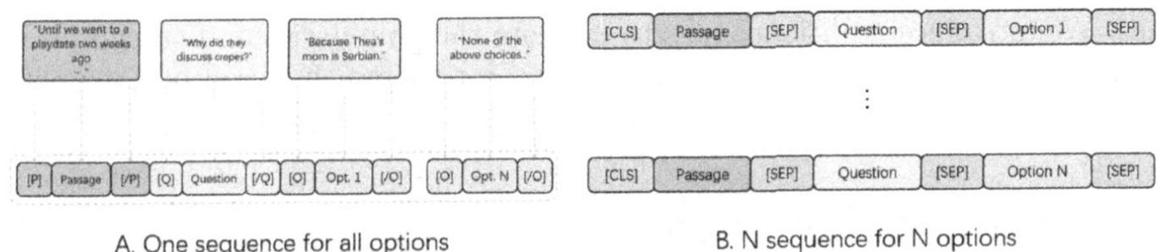

Fig. 1. Differences in model input. We propose A to replace the conventional input format B. Tags are introduced to mark the boundaries and categories of different parts of a full context input sequence.

and has spawned a large number of related datasets [5,9,13,19]. MMRC tasks consist of three types of components: a passage **P**, a question **Q** and answer options $\mathbf{O} = O_1, O_2, ..., O_n$. The purpose of the task is to select the best matching answer O_i corresponding to question **Q** from information provided by **P**.

With the development of deep learning methods, the solution to MMRC tasks has gradually shifted to deep neural networks [11], especially pre-trained language models (PLMs) [2,4,10,12] based on the Transformer architecture [24]. These PLMs have been applied to the MMRC tasks by taking the concatenation of the passage $P = [w_1^P, ...w_{L_P}^P]$ with L_P tokens, the question $Q = [w_1^Q, ..., w_{L_Q}^Q]$ with L_Q tokens and one of the candidate options $O_i = [w_1^{O_i}, ..., w_{L_{O_i}}^{O_i}]$ with L_{O_i} tokens as input $S_i = [P, Q, O_i]$, where $i = 1, 2, ..., N$ is the index of N candidate options, shown as Fig. 1B. The representation of every S_i is fed to the downstream classifier resulting in a probability distribution over N options.

Subsequent work mainly focuses on strengthening the modelling of relationships between different parts of S_i with incremental modules. The Multistep Attention Network [6] makes an effort to model the relationship between the passage and the question-option pair. Similarly, the Dual Multi-head Co-Attention model (DUMA) [29] compares the information between the passage and the question-option pair with a cross-attention mechanism. The Dual Co-Matching Network (DCMN+) [27] goes one step further and models the passage-option, passage-question, and question-option relationships while the Human Reading Comprehension Attention model (HRCA+) [28] models all correspondence within the triple.

The Option Comparison Network (OCN) [15] has been proposed to take into account the pair-wise relationship between options based on their representation extracted from each triple. So, as opposed to our approach, OCN does not process all options in one sequence, and therefore does not take advantage of self-attention between the tokens in the different options.

3 Methods and Model

We propose to constitute a so-called *full context input* for the model to recognize and attend to all components including the passage, question, and all options at once. We introduce additional tags to indicate the boundaries and categories of components, and strengthen the information interaction between options with an extra module.

3.1 Tags for Full Context Input

As discussed in the previous section, a PLM can create representations for all tokens in an input sequence that takes into account the other tokens in the context. Therefore, we propose to put all options into one sequence to make them be the context for each other.

Most of the current methods use the special tokens $[CLS]$ and $[SEP]$ to mark the boundary of different components. Formally, these tokens jointly create an explicit pattern in the input sequence for helping the PLM to recognize three distinct parts. This inspires us to add additional information to the sequence indicating not only the boundary but also the category of every component. To this end, we add multiple tags as identifiers by inserting particular trainable special tokens between the component strings and use them to mark the start and end of each type of component in the sequence, shown in Fig. 1A. Through this, we expect the PLM to be able to identify passages, questions, and options in the input sequence, to encode them accordingly and promote conditioned information interaction (attention) between options to make better predictions for answers that depend on other answers.

Through the form of the full context input, we modify the modelling to be the probability distribution over options conditional to the passage, the question and all options $p(O_i|P, Q, O_1, ..., O_N)$. Here, $i = 1, 2, ..., N$ is the index of N candidate options for corresponding passage P and question Q. It is worth mentioning that the tags for all options are identical to ensure the impact of tags on each option is equivalent and it does not require adjustment on the number of introduced extra tag tokens when the number of options changes.

3.2 Model Architecture

In addition to the input tagging method outlined above, we use transformer encoders to achieve our modelling goal. As shown in Fig. 2, our model consists of three parts: an encoder network, a fuser network, and a discriminator network.

Encoder Network. The encoder network is a PLM for encoding the input, denoted as $E(.)$. By feeding the tagged full context input sequence $S_{fc} = concate(P, Q, O_1, ..., O_n)$ into it, we obtain the corresponding contextual representations of all tokens. We regard it as a tag-oriented representation learner and take the embedding of tags as the representation of each component, defined as $r_s^P, r_e^P = E(t_s^P, t_e^P|S_{fc})$, $r_s^Q, r_e^Q = E(t_s^Q, t_e^Q|S_{fc})$ and $r_{i,s}^O, r_{i,e}^O = E(t_{i,s}^O, t_{i,e}^O|S_{fc})$, where the $t_s^P, t_e^P, t_s^Q, t_e^Q$ are the start and end tags of the passage and question, and $t_{i,s}^O, t_{i,e}^O$ are the start and end tags of the i-th

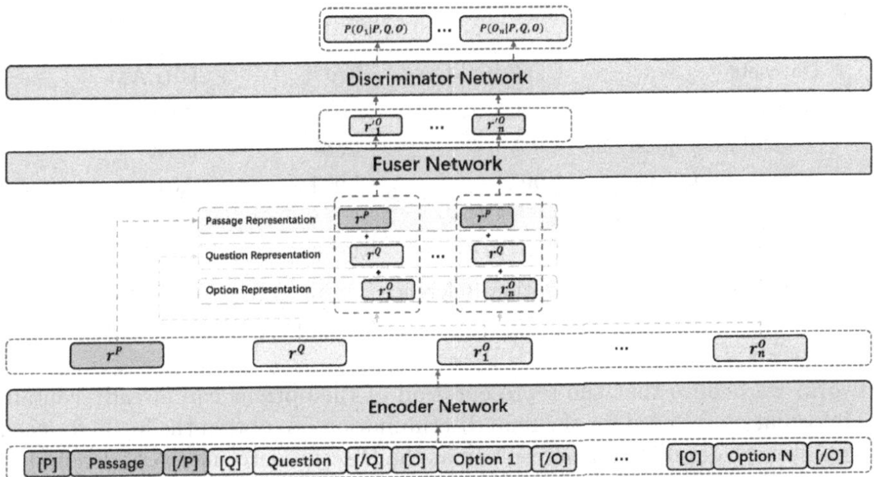

Fig. 2. Model architecture. The Encoder Network is a pre-trained language model, the Fuser Network is a stack of transformer encoder blocks and the Discriminator Network is a multi-layer perceptron.

option. By training on the MMRC tasks, we assume they can carry the task-relevant information in the co-occurrences of full context components thus adopting the embedding of tags as the representation of their corresponding annotated components, with $r^P = r^P_s + r^P_e$, $r^Q = r^Q_s + r^Q_e$, and $r^O_i = r^O_{i,s} + r^O_{i,e}$.

Fuser Network. On top of the Encoder network, we insert another stack of transformer encoders as a Fuser Network. Based on the structure of the transformer encoder, we assume that this network creates representations of each option by attending to the representation of other options in the option set. Henceforth, we refer to this module as a fuser network while it fuses the information of different options. The Fuser network models the relationship between all options at the representation level. We first use the previously obtained r^P, r^Q, r^O_i to reinforce the contextual representation of options as $r_i^{comp} = r^P + r^Q + r^O_i$, where $i = 1, 2, ..., N$ are index of N options. This additive representation is derived from the initialization of token embeddings in PLMs. It is worth mentioning that the set $O^R = [r_1^{comp}, r_2^{comp}, ..., r_n^{comp}]$, consisting of the representation of options, is actually permutation-invariant, so the position-independent modelling can be performed using a transformer encoder without position-wise information. Taking this set as input, the Fuser network refines the representation of every option by referring to each other: $r_i^{'O} = F(r_i^{comp}|O^R)$, where $F(.)$ denotes the Fuser network and $r_i^{'O}$ is the new representation for option i in N options.

Discriminator Network. After encoding them through the Encoder network and strengthening the information interaction between them through the Fuser

Table 2. Statistics of experiment datasets. Lengths are in the number of words.

Datasets	COSMOS-QA	RACE	DREAM
# train questions	25,262	87,866	6,116
# validation questions	2,985	4,887	2,040
# test questions	6,963	4,934	2,041
# options per question	4	4	3
Avg. length (P/Q/O)	70.3/10.6/8.1	321.9/10.0/5.3	85.9/8.6/5.3
Characteristics	Option Reference	Ordinary	Dialogue

network, we believe that the representation of the options can already contain the information we need for the final discrimination. To obtain the logits for each option, we use a multi-layer perceptron as the Discriminator network, denoted as $D(.)$, which takes the representation of each option as input features to generate a logit $L_i^O = D(r_i^{'O})$ of option i. Normalized by softmax, the probability can be formed: $p(O_i|P, Q, O_1, ..., O_n) = \frac{e^{L_i^O}}{\sum_{i=1}^{n} e^{L_i^O}}$.

4 Experiment

4.1 Datasets

In order to validate our method, we use COSMOS-QA [5], a commonsense-based reading comprehension dataset with between-option references, as our main target. Its test set is hosted online with the leaderboard and there is a frequency limit for submissions.[2] Therefore, we can only run our experiments on the COSMOS test set a single time. In addition, to verify the universality of our method on other MMRC datasets, we also apply it to DREAM [19], a multiple-choice dialogue-based reading comprehension examination dataset, and RACE [9], a more general dataset collected from English examinations in China designed for middle school and high school students. Statistics of these datasets are shown in Table 2.

4.2 Experimental Settings

The maximum input length is set to 512 tokens to match the settings of pre-trained encoder models. We truncate the passages that exceed the length limit into segments and each of them is combined with the corresponding question and options as input (see appendix for statistics on the truncation). The representation of passage, question, and options from different segments are averaged to maintain dimension alignment.

We apply the *AdamW* optimizer and adopt the warming up and linear learning rate decay strategy in our training. Only the pre-trained weights are loaded

[2] https://leaderboard.allenai.org/cosmosqa/submissions/public.

Table 3. Accuracy (%) on RACE, COSMOS-QA, and DREAM test sets. Results for RACE and DREAM are averages of 5 models trained with different random seeds. The COSMOS test set is hosted online and can only be evaluated a single time. The baselines are either collected from previous publications ([5] (a), [22] (b), [10] (c), [12] (d), [4] (e), [6] (f)) or implemented by ourselves (∗), with the method mentioned in Sect. 4.2.

Model	COSMOS-QA	RACE	DREAM
Baselines			
BERT-large	67.1[a]	72.0[c]	66.8[f]
RoBERTa-large	80.6[b]	83.2[d]	85.0[f]
ALBERT-xxlarge	82.3[b]	86.5[c]	88.5[f]
DeBERTa-large	84.3*	86.8[e]	87.3*
Our Methods			
BERT-large	63.2 (**−3.9**)	68.8 (**−3.2**)	55.1 (**−11.7**)
RoBERTa-large	83.9 (**+3.3**)	84.1 (**+0.9**)	84.0 (**−1.0**)
ALBERT-xxlarge	86.0 (**+3.7**)	87.2 (**+0.7**)	87.9 (**−0.6**)
DeBERTa-large	86.3 (**+2.0**)	86.9 (**+0.1**)	87.0 (**−0.3**)

for the encoder network using the *Transformers* library[3] and the weights of the fuser network and discriminator network are initialised for every training. For evaluation, we use accuracy as the indicator to measure the quality of our methods and models for all tasks, as is common in MMRC tasks.

The baseline method for comparison in our experiment is the method of direct application of PLMs on MMRC tasks, which (1) takes the concatenation of P, Q, O_i with $[SEP]$ tokens between them as input sequence for each O_i as input, and (2) then takes the embedding of the $[CLS]$ token for discrimination into correct/incorrect with a fully connected network.

4.3 Results

We assess the performance changes brought by our method to the baseline setting with the same PLMs.

Main Results. Table 3 shows that incorporating options into the context and enhancing their relational modelling improves RoBERTa, ALBERT and DeBERTa on COSMOS-QA by 3.3% point, 3.7% point and 2.0% point respectively. At the same time, it has an inconsistent impact on RACE and DREAM, which do not have explicit option relations. Our method improves RoBERTa, ALBERT and DeBERTa on RACE by 0.9% point, 0.7% point, and 0.1% point while decreasing on DREAM by 1.0%, 0.6%, and 0.3% point. From this, we can say that our method can improve the quality of MMRC on datasets that need to model the relationship between options.

[3] https://huggingface.co/.

Table 4. Ablation study of COSMOS-QA dataset with RoBERTa-large. Accuracy (%) in this table is calculated on the validation set and expressed as a percentage. The results are averages of 10 runs trained with different random seeds. Standard deviations over the runs are also indicated. The numbers do not match Table 3 because those were obtained on the official test set and the ablation study was done on the validation set.

Method	Accuracy (stdev)
Baseline	80.82 ($\sigma = 0.56$)
Full Context & Tags	81.76 ($\sigma = 0.54$)
+Extra Encoder Layers	81.56 ($\sigma = 0.36$)
+Fuser network	82.23 ($\sigma = 0.28$)

The table also shows that BERT-large is the lowest performing model compared to RoBERTa, ALBERT, and DeBERTa, both in the baseline setting and our method. Also, in the experiments with BERT as the encoder network, our method causes a performance loss on all three datasets. Since the architectures of BERT-large and RoBERTa-large are identical, we speculate that this is caused by the pre-training tasks and possibly also the scale of pre-training data: previous studies [7,10,12] suggest that the Next Sentence Prediction (NSP) task in pre-training would damage the performance of many downstream tasks, and the ablation study in XLNet [26] shows that the NSP task improves the performance of the baseline method on RACE.

Setting apart the weakly performing BERT baseline, our model improves 2 to 4% points on COSMOS over the much stronger baselines RoBERTa, ALBERT, and DeBERTa.

Ablation Study. To explore the contribution of different parts of our method, we apply it module by module, using the RoBERTa-large encoder as a basis. This way, we can assess their impact on our architecture on COSMOS-QA. As the results in Table 4 show, full context input directly improves the performances on COSMOS-QA compared to the baseline method. A two-sided t-test for independent samples ($n = 10$ runs) indicates that the difference between the baseline and the full-context setting is significantly different ($p = 0.001$). When the fuser network is added on top of the encoder with full context input, it improves the performance compared to simply deepening the encoder network. The difference between these two settings is also significant ($p = 0.0004$). This indicates that the improvement brought by the fuser network is not only due to the increase in model depth.

Analysis of Effect on 'None of the Above' Options. To more intuitively display the changes brought about by our method for determining the *'none of the above choices'* option, we selected the samples containing such an option into a subset and calculated the precision and recall of our method to compare with the baseline. In the training set, 17,509 questions (69.3% of the total) contain an option 'none of the above choices' option and 1465 of them take it as the correct answer (8.4% are positive). In the evaluation set, 1937 questions (65.0% of the total) contain the 'none of the above choices' option and 279 of them take it as the correct answer (14.4% are positive). Table 5 shows that, although the 'none of the above choices' option accounts for a low proportion of correct answers (14.4%), our method can lead to a salient improvement in recall (+8.5% point) with minor precision loss (−0.8% point). Combining two indicators, our method has a higher F1-score, indicating stronger reliability than the baseline method.

Table 5. Precision and recall of the *'None of the above choices.'* option with RoBERTa-large. Results in this table are calculated on the validation set.

Method	Precision	Recall	F1-score
Baseline	66.7%	62.6%	0.646
Ours	65.9%	71.1%	0.684

The results of current experiments suggest that in most cases our method is effective for COSMOS-QA, which has obvious option relevance, and it could be affected by the pre-training tasks and data of PLM encoders. As for other MMRC datasets, the impact of our method is somewhat inconsistent, but small.

5 Discussion

In this section, we discuss the effects that occurred in our experiments and try to give explanations for them with additional analysis. Furthermore, we show the transfer capability and memory occupation of our method to comprehensively illustrate its advantages.

5.1 Visualization of Inferences

Although we see the influence of full context components in our experimental results, we cannot directly conclude that the performance changes are caused by attending option relations. To analyze the modelling of options relations by our network, we apply Integrated Gradients (IG) [21] to our fine-tuned model and input samples for analyzing the contribution of tokens to the likelihood of options. We use the *Captum*[4] library and its default visualization scheme for this purpose.

The IG algorithm attributes the prediction of a network to its input features by approximating the integral of the model output gradient along a path from a given reference to the input. In short, this algorithm analyzes the sensitivity to input changes and assigns a corresponding importance score as the contribution to each input token using the interpolation transformation from the reference to the actual sample.

[4] https://captum.ai/.

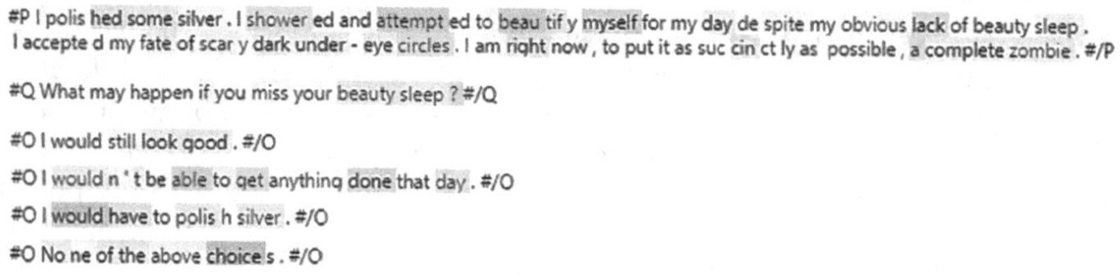

#P I polis hed some silver . I shower ed and attempt ed to beau tif y myself for my day de spite my obvious lack of beauty sleep . I accepte d my fate of scar y dark under - eye circles . I am right now , to put it as suc cin ct ly as possible , a complete zombie . #/P

#Q What may happen if you miss your beauty sleep ? #/Q

#O I would still look good . #/O

#O I would n ' t be able to get anything done that day . #/O

#O I would have to polis h silver . #/O

#O No ne of the above choice s . #/O

Fig. 3. Attribution for *'None of the above choices'* option in a COSMOS-QA sample.

An example is shown in Fig. 3, with *'None of the above choices'* being the correct option. Tokens marked in green indicate positive contribution and in red indicate negative contribution to the option, and the brighter the colour, the larger the absolute value. It can be easily seen from this figure that when determining the relevance of this option, not only tokens in the passage and the question but also tokens in the other options have been attributed to the correct option. This situation is consistent among more samples and independent of the option characteristics of involved training data, which is with or without option relevance. From this, we can conclude that our method indeed processes the information relations between options and can model the option relevance.

5.2 Depth of the Fuser Network

The effect of the depth of the Fuser network on the performance change on the COSMOS-QA dataset is shown in Fig. 4. We can see there is no clear pattern in either accuracy or standard deviation for deepening the fuser network. This result is unexpected and also a bit unwanted since it does not give much information about the optimal depth, meaning that the depth of the network will remain a hyperparameter which needs to be adjusted according to the dataset. The only thing we can be sure of is that adding a fuser network can improve the model performance with full context input.

5.3 Transfer Capability

Our model is designed to adapt to a variety of different datasets without modification. This characteristic allows it to be easily transferred to other datasets after training on MMRC tasks. To verify its transfer capability, we jointly train it on RACE, COSMOS-QA and DREAM, and test it on other MMRC tasks in different domains and languages.

Table 6 shows the three non-English target datasets that we experimented with: (1) C^3 [20], which contains dialogues and more formally written mixed-genre texts in Chinese; (2) SweQUAD-MC [8], a small dataset with Swedish QA items; and (3) MuSeRC [18], which requires multi-sentences reasoning in Russian. MuSeRC was designed initially as a binary classification task; we combined the options corresponding to identical passages and questions to make it multi-choice.

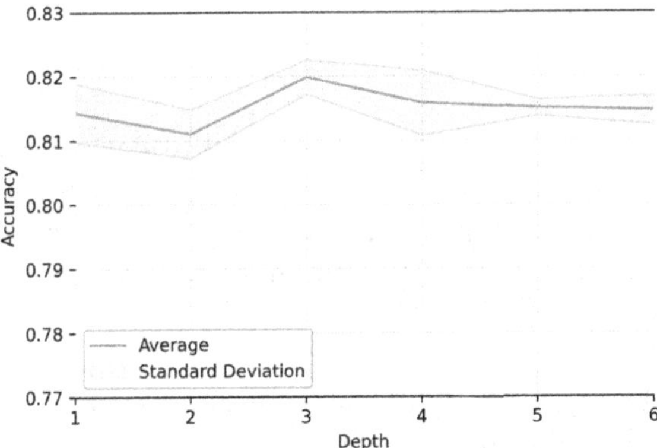

Fig. 4. Accuracy on COSMOS-QA validation set with different depth of Fuser Network.

Table 6. Statistics of the non-English target datasets. Lengths are word counts.

Dataset	C^3	SweQUAD-MC	MuSeRC
# of train	11,869	962	5,380
# of validation	3,816	126	993
# of test	3,892	102	–
Option Num	2/3/4	4	2/3/4
Avg. length (P/Q/O)	116.9/12.2/5.5	379.4/7.8/4.3	203.9/7.6/5.3
Language	Chinese	Swedish	Russian

We conduct our experiments with a multilingual PLM, XLM-RoBERTa [1], to support the processing of multiple languages. In the training step, we sample from these three datasets to construct each mini-batch, and use the following sampling strategy to balance the number of samples from different datasets:

$$c_i = \frac{r_i^\alpha}{\sum_{k=1}^{N} r_k^\alpha} \text{ with } r_i = \frac{n_i}{\sum_{k=1}^{N} n_k}, \tag{1}$$

where n_i, n_k are the number of training samples in datasets i and k respectively; r_i, r_k denote the ratio of sample amounts in dataset i and k respectively to the total sample capacity in all datasets; and c_i denotes the chance of a sample in dataset i being selected. In our experiments, we set $\alpha = 0.5$.

The results are shown in Table 7, which includes the accuracy of single-task fine-tuning on these target datasets with our method, zero-shot transfer, and further fine-tuning on them. In two of the three target datasets, C^3 and SweQUAD-MC, the zero-shot performance slightly surpasses the performance of single-task training. With post-fine-tuning, the performance on all datasets is improved.

This result indicates that our method learns to model the relationships between different components in the sample and can be widely applied to various MMRC tasks. In addition, this shows that our method and model have

Table 7. Transfer learning performances on XLM-RoBERTa. The accuracy of MuSeRC is calculated on the validation dataset. The results (%) are averages of 5 models trained with different random seeds.

Settings	C^3	SweQUAD-MC	MuSeRC
Single-task fine-tuning	77.58	92.08	91.78
Zero-shot transfer	78.17 (**+0.59**)	92.16 (**+0.08**)	86.61 (**−5.17**)
Further fine-tuning	82.73 (**+5.15**)	98.29 (**+6.21**)	94.39 (**+2.61**)

the potential to be applied to larger-scale training on more MMRC datasets for constructing a universal model for such tasks.

5.4 Reduction of Memory Occupancy

When we consider the memory occupancy of the intermediate result of our model, we mainly focus on the activation result of the transformer encoder blocks, which can be calculated as $M_{activation} = D_h \times L_s \times B \times N_l$, where $M_{activattion}$ is the memory occupancy caused by activation result, D_h is the dimensionality of the hidden layer, L_s is the length of the input sequence, B is the batch size, and N_l is the number of layers.

For a sample with n options and the length of the passage, question, and each option p_l, q_l, and o_l, the total memory occupancy by this sample is linear to $n * (p_l + q_l + o_l)$ when the baseline method is applied and it would be reduced to linear to $p_l + q_l + n * o_l$ with our method. We roughly apply the average length of all components to have a cursory estimation of reduction in each layer: 74% for RACE, 68% for COSMOS-QA, 63% for DREAM, 72% for C^3, 74% for SweQUAD-MC, and 73% for MuSeRC.

This means that taking the full context input could reduce memory occupancy, especially when the samples have long passage texts. As a result, our method allows the RoBERTa-Large model to be trained on the COSMOS-QA dataset with batch size 16 on a single NVIDIA RTX3090 GPU without any memory trick.

6 Conclusions

Our work mainly focuses on modelling the relationship between options for MMRC tasks with option relevance. By putting all components in a sample into the input sequence and marking their boundaries and categories with tags, we draw support from the self-attention mechanism in Pretrained Language Models (PLMs) to obtain the representation of each option with the passage, question and other options as context. In addition, we introduce a new module, the Fuser network, to strengthen the information interaction among the options in the option set.

We experiment with our method on four encoder PLMs. For the three best-performing models, our method outperforms the baselines by 2–4% point; the exception is the BERT model, which already gives much lower baseline performance than its cousins RoBERTa, ALBERT, and DeBERTa, and therefore seems much less suitable for MMRC tasks. By comparing the results on different datasets, we conclude that our method can improve the performance of data with option relevance. As evidence, we verify that our method can substantially improve recall on options that require information about the option set and provide visualizations of inference. Additionally, our method allows the transfer of the model to other domains and languages without modification, which could be applied to general models for MMRC and even text classification tasks, especially with its memory occupancy reduction for improving training efficiency.

A Appendix

A.1 Truncation Statistics

As we mentioned in Sect. 4.2, we truncated the passages that exceed the length limit into segments and each of them is combined with the corresponding question and options as input. The statistics of truncation are shown in Table 8.

Table 8. Statistics of truncation in experiments.

Split	Statistics	RACE	COSMOS-QA	DREAM	C^3	SewQUAD-MC	MuSeRC
Train	Full	68648	25262	5855	11219	461	4897
	Truncated	19218	0	261	650	506	483
	Avg. truncation count	1.14	0	1.6	1.12	2.11	1.39
	Max. truncation count	6	0	4	3	14	5
Eval	Full	3839	2985	1962	3574	66	883
	Truncated	1048	0	78	242	61	110
	Avg. truncation count	1.12	0	1.42	1.12	1.79	1.22
	Max. truncation count	4	0	4	2	13	3
Test	Full	3983	6963	1943	3709	42	–
	Truncated	951	0	98	183	60	–
	Avg. truncation count	1.16	0	1.52	1.07	1.6	–
	Max. truncation count	4	0	3	2	9	–

A.2 Hyper-parameters Settings

The hyper-parameters for the major experiments in Table 3 and Table 4 are summarized in Table 9.

For the transfer learning experiment on XLM-RoBERTa-large, the hyper-parameters for intermediate training and post-fine-tuning are shown in Table 10.

Table 9. Hyper-parameters setting for COSMOS-QA.

Params	BERT-large	RoBERTa-large	ALBERT-xxlarge	DeBERTa-large
Encoder LR	1e-5	1e-5	1e-5	6e-6
Fuser LR	1e-3	1e-3	1e-3	1e-4
Discriminator LR	1e-3	1e-3	1e-3	1e-4
Weight Decay	0.01	0.01	0.01	0.01
LR Decay	Linear	Linear	Linear	Linear
Epochs	8	8	8	10
Warming Up	0.125	0.125	0.125	0.1
Fuser Depth	3	3	3	3
Discriminator Depth	2	2	2	2
Batch Size	16	16	16	16

Table 10. Hyper-parameters setting for XLM-RoBERTa-Large in transfer experiment.

	Intermediate Task	Target Tasks		
	COSMOS-QA, RACE, DREAM	C^3	SweQUAD-MC	MuSeRC
Batch Size	16	16	8	8
Encoder LR	1e-5	1e-5	1e-5	1e-5
Fuser LR	1e-3	1e-5	1e-5	1e-5
Discriminator LR	1e-3	1e-5	1e-5	1e-5
Weight Decay	0.01	0.01	0.01	0.01
LR Decay	Linear	Linear	Linear	Linear
Total Iters	35,000	5 Epochs	5 Epochs	5 Epochs
Warming Up Iters	5,000	1 Epoch	1 Epoch	1 Epoch
Fuser Depth	3	3	3	3
Discriminator Depth	2	2	2	2

References

1. Conneau, A., et al.: Unsupervised cross-lingual representation learning at scale. CoRR abs/1911.02116 (2019). http://arxiv.org/abs/1911.02116
2. Devlin, J., Chang, M.W., Lee, K., Toutanova, K.: BERT: pre-training of deep bidirectional transformers for language understanding. In: Proceedings of the 2019 Conference of the North American Chapter of the Association for Computational Linguistics: Human Language Technologies (Long and Short Papers), vol. 1, pp. 4171–4186. Association for Computational Linguistics, Minneapolis, Minnesota, June 2019. https://doi.org/10.18653/v1/N19-1423. https://aclanthology.org/N19-1423
3. Ghosal, D., Majumder, N., Mihalcea, R., Poria, S.: Two is better than many? Binary classification as an effective approach to multi-choice question answering (2022)
4. He, P., Liu, X., Gao, J., Chen, W.: DeBERTa: decoding-enhanced BERT with disentangled attention. In: International Conference on Learning Representations (2021). https://openreview.net/forum?id=XPZIaotutsD

5. Huang, L., Bras, R.L., Bhagavatula, C., Choi, Y.: COSMOS QA: machine reading comprehension with contextual commonsense reasoning. CoRR abs/1909.00277 (2019). http://arxiv.org/abs/1909.00277
6. Jin, D., Gao, S., Kao, J., Chung, T., Hakkani-Tür, D.: MMM: multi-stage multi-task learning for multi-choice reading comprehension. CoRR abs/1910.00458 (2019). http://arxiv.org/abs/1910.00458
7. Joshi, M., Chen, D., Liu, Y., Weld, D.S., Zettlemoyer, L., Levy, O.: SpanBERT: improving pre-training by representing and predicting spans. CoRR abs/1907.10529 (2019). http://arxiv.org/abs/1907.10529
8. Kalpakchi, D., Boye, J.: BERT-based distractor generation for Swedish reading comprehension questions using a small-scale dataset. CoRR abs/2108.03973 (2021). https://arxiv.org/abs/2108.03973
9. Lai, G., Xie, Q., Liu, H., Yang, Y., Hovy, E.: RACE: large-scale reading comprehension dataset from examinations. arXiv preprint arXiv:1704.04683 (2017)
10. Lan, Z., Chen, M., Goodman, S., Gimpel, K., Sharma, P., Soricut, R.: ALBERT: a lite BERT for self-supervised learning of language representations. CoRR abs/1909.11942 (2019). http://arxiv.org/abs/1909.11942
11. Liu, S., Zhang, X., Zhang, S., Wang, H., Zhang, W.: Neural machine reading comprehension: methods and trends. CoRR abs/1907.01118 (2019). http://arxiv.org/abs/1907.01118
12. Liu, Y., et al.: RoBERTa: a robustly optimized BERT pretraining approach. CoRR abs/1907.11692 (2019). http://arxiv.org/abs/1907.11692
13. Ostermann, S., Modi, A., Roth, M., Thater, S., Pinkal, M.: MCScript: a novel dataset for assessing machine comprehension using script knowledge. In: Proceedings of the Eleventh International Conference on Language Resources and Evaluation (LREC 2018). European Language Resources Association (ELRA), Miyazaki, Japan, May 2018. https://aclanthology.org/L18-1564
14. Qiu, B., Chen, X., Xu, J., Sun, Y.: A survey on neural machine reading comprehension. CoRR abs/1906.03824 (2019). http://arxiv.org/abs/1906.03824
15. Ran, Q., Li, P., Hu, W., Zhou, J.: Option comparison network for multiple-choice reading comprehension. CoRR abs/1903.03033 (2019). http://arxiv.org/abs/1903.03033
16. Richardson, M.: MCTest: a challenge dataset for the open-domain machine comprehension of text. In: Proceedings of the 2013 Conference on Empirical Methods in Natural Language Processing (EMNLP 2013), October 2013. https://www.microsoft.com/en-us/research/publication/mctest-challenge-dataset-open-domain-machine-comprehension-text/
17. Sachan, M., Dubey, K., Xing, E., Richardson, M.: Learning answer-entailing structures for machine comprehension. In: Proceedings of the 53rd Annual Meeting of the Association for Computational Linguistics and the 7th International Joint Conference on Natural Language Processing (Volume 1: Long Papers), pp. 239–249 (2015)
18. Shavrina, T., et al.: RussianSuperGLUE: a Russian language understanding evaluation benchmark. arXiv preprint arXiv:2010.15925 (2020)
19. Sun, K., Yu, D., Chen, J., Yu, D., Choi, Y., Cardie, C.: DREAM: a challenge dataset and models for dialogue-based reading comprehension. CoRR abs/1902.00164 (2019). http://arxiv.org/abs/1902.00164
20. Sun, K., Yu, D., Yu, D., Cardie, C.: Probing prior knowledge needed in challenging Chinese machine reading comprehension. CoRR abs/1904.09679 (2019). http://arxiv.org/abs/1904.09679

21. Sundararajan, M., Taly, A., Yan, Q.: Axiomatic attribution for deep networks. CoRR abs/1703.01365 (2017). http://arxiv.org/abs/1703.01365
22. Tian, Z., Zhang, Y., Liu, K., Zhao, J., Jia, Y., Sheng, Z.: Scene restoring for narrative machine reading comprehension. In: Proceedings of the 2020 Conference on Empirical Methods in Natural Language Processing (EMNLP), pp. 3063–3073. Association for Computational Linguistics, November 2020. https://doi.org/10.18653/v1/2020.emnlp-main.247. https://aclanthology.org/2020.emnlp-main.247
23. Trischler, A., Ye, Z., Yuan, X., He, J., Bachman, P., Suleman, K.: A parallel-hierarchical model for machine comprehension on sparse data. arXiv preprint arXiv:1603.08884 (2016)
24. Vaswani, A., et al.: Attention is all you need. In: Advances in Neural Information Processing Systems 30 (2017)
25. Wang, H., Bansal, M., Gimpel, K., McAllester, D.: Machine comprehension with syntax, frames, and semantics. In: Proceedings of the 53rd Annual Meeting of the Association for Computational Linguistics and the 7th International Joint Conference on Natural Language Processing (Volume 2: Short Papers), pp. 700–706 (2015)
26. Yang, Z., Dai, Z., Yang, Y., Carbonell, J.G., Salakhutdinov, R., Le, Q.V.: XLNet: generalized autoregressive pretraining for language understanding. CoRR abs/1906.08237 (2019). http://arxiv.org/abs/1906.08237
27. Zhang, S., Zhao, H., Wu, Y., Zhang, Z., Zhou, X., Zhou, X.: DCMN+: dual co-matching network for multi-choice reading comprehension. In: Proceedings of the AAAI Conference on Artificial Intelligence, vol. 34, pp. 9563–9570 (2020)
28. Zhang, Y., Yamana, H.: HRCA+: advanced multiple-choice machine reading comprehension method. In: LREC (2022)
29. Zhu, P., Zhao, H., Li, X.: Dual multi-head co-attention for multi-choice reading comprehension. CoRR abs/2001.09415 (2020). https://arxiv.org/abs/2001.09415

Zero-Shot Generative Large Language Models for Systematic Review Screening Automation

Shuai Wang[1(✉)], Harrisen Scells[2], Shengyao Zhuang[3], Martin Potthast[2,4], Bevan Koopman[3], and Guido Zuccon[1]

[1] The University of Queensland, Brisbane, Australia
{shuai.wang2,g.zuccon}@uq.edu.au
[2] Leipzig University, Leipzig, Germany
{harry.scells,martin.potthast}@uni-leipzig.de
[3] CSIRO, Canberra, Australia
{shengyao.zhuang,b.koopman}@csiro.com
[4] ScaDS.AI, Leipzig, Germany

Abstract. Systematic reviews are crucial for evidence-based medicine as they comprehensively analyse published research findings on specific questions. Conducting such reviews is often resource- and time-intensive, especially in the screening phase, where abstracts of publications are assessed for inclusion in a review . This study investigates the effectiveness of using zero-shot large language models (LLMs) for automatic screening. We evaluate the effectiveness of eight different LLMs and investigate a calibration technique that uses a predefined recall threshold to determine whether a publication should be included in a systematic review. Our comprehensive evaluation using five standard test collections shows that instruction fine-tuning plays an important role in screening, that calibration renders LLMs practical for achieving a targeted recall, and that combining both with an ensemble of zero-shot models saves significant screening time compared to state-of-the-art approaches.

Keywords: Systematic Reviews · Document Classification · Large Language Models

1 Introduction

Systematic reviews are used extensively in medicine to comprehensively summarise all research finding s on a specific question. Systematic reviews ensure a high level of rigour by including all and only those publications that meet predefined criteria, called the set of 'included documents'.[1] The selection of included documents starts with searching relevant databases such as PubMed [58] and the Cochrane Library [17]. This search returns a list of 'candidate documents', which are then screened for relevance and quality using the researchers' explicit inclusion and exclusion criteria.

[1] Other commonly used terms are 'studies', 'research publications', and 'references'.

N. Goharian et al. (Eds.): ECIR 2024, LNCS 14608, pp. 403–420, 2024.
https://doi.org/10.1007/978-3-031-56027-9_25

Systematic reviews are labour-intensive and time-consuming, with most resources being invested in screening candidate documents, a process that can take months. While there are various methods to assist in optimizing the creation of systematic reviews (Sect. 2), one particular line of work focuses on minimising the number of documents that need to be manually screened. This has previously been pursued with classifiers to filter out documents that are not relevant, which may include manually labelling a significant number of the candidate documents to tune the classifier to the screening task at hand. Meanwhile, instruction-based generative large language models (LLMs), such as OpenAI's ChatGPT,[2] Llama [46], and Alpaca [44], have demonstrated a remarkable ability to generate high-quality results in response to user instructions that often do not require task-specific tuning [44,63]. In automating systematic reviews, these models have been fine-tuned for query formulation [55,56], as well as document classification and ranking [4,35,43,56].

In this paper, we focus specifically on the use of *zero-shot* large language models for the automatic screening of documents in systematic reviews (Sect. 3). By 'zero-shot', we mean using generative LLMs without explicitly optimising them for the screening task, which has the potential to relieve medical experts of any additional labelling burden. We examine two settings of our approach, an *uncalibrated* and a *calibrated* one. Both approaches prompt the model and use the probability of the next predicted (target) tokens to categorise documents as either 'included' or 'excluded'; the former directly uses the token with higher probability between 'yes' and 'no', the latter introduces the hyperparameter θ as a new decision boundary of the classifier, calculated from the difference of the two tokens instead; θ is adjusted based on starting documents or previous systematic reviews.

In our evaluation, we address four research questions to investigate the factors that influence the effectiveness of the proposed zero-shot generative LLM-based automated screening method for systematic reviews (Sect. 4):

RQ1 How does the architecture and size of the LLMs influence effectiveness?
RQ2 How does instruction-based fine-tuning influence effectiveness?
RQ3 How does the calibration of the classifier's decisions with respect to the target tokens' likelihoods influence effectiveness?
RQ4 How does ensembling LLM-based classifiers and current strong neural baselines influence effectiveness?

Our evaluation results (Sect. 5) show that LlaMa2-7b-ins is currently the best model for this task, much better than the 13b parameter variant. In general, instruction-based fine-tuning always outperforms the base models that have not been fine-tuned, and models based on LlaMa2 consistently outperform the baseline BERT-based method. Our approach also slightly outperforms (i.e. is competitive with) the fine-tuned Bio-SIEVE baseline. The calibrated setting of our method with ensembling achieves the best result overall and approaches the predefined recall target for the test topics, which indicates practical use.

[2] https://chat.openai.com/.

2 Related Work

It is a requirement for high-quality systematic reviews to retrieve literature using a Boolean query [11,14]; the set of all retrieved documents must then be fully screened (assessed) for inclusion in the systematic review [11]. Research has explored the automatic creation of effective Boolean queries [38–41,55] (also with respect to the use of controlled vocabularies such as MeSH [50,51,53]), and the ranking of the set of documents retrieved by the Boolean query (a task called "screening prioritisation") [1–3,12,27,28,28,32,36,37,59,65], in order to begin downstream processes of the systematic review earlier [33], e.g., acquiring the full-text of studies or results extraction. The datasets that we consider in our experiments, including the CLEF TAR datasets [22–24], specifically considered the task of screening prioritisation. In our paper, we consider a different task, the one of automating the screening phase of the systematic review; we discuss previous work related to this direction next.

Popular methods for automating the document screening phase are based on text classification [45]: a classifier is learned for an individual systematic review, typically in a supervised manner using labels obtained on a subset of the documents to be screened. Methods include traditional machine learning models like SVM [15,49], as well as classifiers based on encoder-based LLMs like BERT/BioBERT [6,10,35]. Text classification methods are typically trained incrementally (acquiring labels through cycles of automatic classification) and often using active learning [5,9,20,21,31,42,48,62]. It is important to note that all of the methods above requires fine-tuning using labelled data specific to systematic review text classification in order to be effective.

In our work, we take a step further by considering the latest developments in generative LLMs to enhance the screening process. At the same time of developing this work, others have also explored similar directions. [43] employed ChatGPT for document screening, finding that ChatGPT's effectiveness is poor if the set of documents to be screened is imbalanced – which is often the case in systematic reviews (i.e., typically, there are many more excluded documents than included among those that have been screened). Higher classification accuracy than ChatGPT was displayed by Bio-SIEVE [35], a model fine-tuned from the Guanaco checkpoint [19]– which in turn is based on the Llama architecture. However, Bio-SIEVE also displayed severe consistency issues across review topics. Importantly, both these works have notable limitations. The first study [43] focused solely on the closed-sourced ChatGPT model. In addition, the evaluation was limited to only five systematic review topics and did not consider publicly available datasets with a broader range of review topics used in previous work. The second study [35] required to fine-tune the LLM, and relied on a self-constructed dataset for evaluation[3], limiting comparison with previous work. Furthermore, it only reported evaluation with respect to precision, recall, and accuracy; thus: (i) there is no account for the effect of class imbalance, (ii) there

[3] Although the dataset is described to be public, it currently only contains the DOIs of the systematic review topics but not the labels, making reproduction difficult.

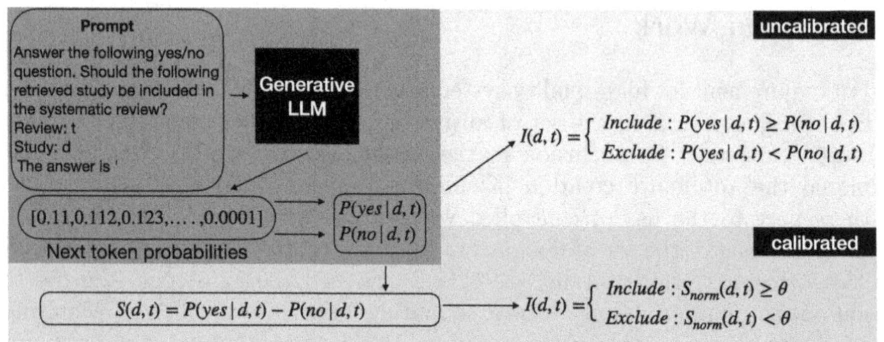

Fig. 1. Our framework for automatic document screening using generative LLMs. $P(\texttt{yes}|d,t)$ ($P(\texttt{no}|d,t)$) is the likelihood of the yes (no) token in the next token probability list, and θ is the decision boundary(threshold) used by the calibrated setting.

is no account that high-recall is considered essential in practice when conducting a systematic review. Conversely, in our work, we (1) consider open-sourced LLMs in a zero-shot setup, where further fine-tuning is not required, (2) take into account class imbalance and the high-recall nature of the task when evaluating methods, (3) rely on publicly available datasets that have been extensively used in previous work, thus facilitating comparison and reproduction.

3 Generative LLMs for Automatic Document Screening

Our framework for using a generative LLM for automatic document screening is shown in Fig. 1. The LLM considers a candidate document $d \in D$ for the systematic review topic $t \in T$; document screening is modelled as a classification task, using the function $I(d,t) : D, T \rightarrow \{0,1\}$. Document d is included for systematic review t when $I(d,t)$ is 1, and otherwise excluded. The function $I(d,t)$ is computed with respect to the output of the LLM for the prompt containing d and t. We investigate two instantiations of $I(d,t)$, uncalibrated and calibrated, which we explain below.

Uncalibrated Screening. To determine whether a document should be included or not, uncalibrated screening directly compares the absolute values of the token likelihoods $P(\texttt{yes}|d,t)$ and $P(\texttt{no}|d,t)$ as generated by the LLM:

$$I(d,t) = \begin{cases} 1, & \text{if } P(\texttt{yes}|d,t) \geq P(\texttt{no}|d,t) \\ 0, & \text{otherwise.} \end{cases}$$

To ensure deterministic output, we forgo actual text generation with LLM. Instead, we represent the model decision using solely the probability of the next predicted token either to be 'yes' or 'no'. In this setting, the LLM returns an answer to the provided prompt of the decision with respect to the highest likelihood from the two tokens.

Calibrated Screening. Building upon our uncalibrated instantiation, we calculate the difference between the likelihood of the next token to be `yes`, or `no`; then, we use a threshold to determine the inclusion of the document. We begin by computing the score $S(d,t)$ as the difference between the `yes` and `no` token likelihoods:

$$S(d,t) = \begin{cases} P(\texttt{yes}|d,t) - P(\texttt{no}|d,t), & \text{if } P(\texttt{yes}|d,t) \geq P(\texttt{no}|d,t) \\ 0, & \text{otherwise.} \end{cases}$$

However, the probability distribution of the tokens depends on the individual documents, and thus is different across the documents. We then use min-max normalisation to normalise scores across all documents for a review topic t:

$$S_{\text{norm}}(d,t) = \frac{S(d,t) - \text{Min}(\{\forall d_i \in D : S(d_i,t)\}))}{\text{Max}(\{\forall d_i \in D : S(d_i,t)\}) - \text{Min}(\{\forall d_i \in D : S(d_i,t)\})}$$

Next, we identify a threshold θ using training data; θ is determined such that when used as the lower bound on scores for inclusion decisions, it ensures a minimum recall rate k. Finally, we use θ to decide if a candidate document should be included:

$$I(d,t) = \begin{cases} 1, & \text{if } S_{\text{norm}}(d,t) \geq \theta \\ 0, & \text{otherwise.} \end{cases}$$

The intuition behind exploring a calibrating screening approach is twofold. First, in the context of systematic review document screening, recall is of paramount importance. For automation techniques to be used in practice, they must ensure the identification of all (or most) of the documents that should be included in the review. This is crucial because failing to capture all relevant documents may compromise the integrity of the review's conclusions and miss the main objective of a systematic review, that is its comprehensiveness. However, this focus on recall may not be naturally accounted for by LLMs, especially when accuracy is used to train/fine-tune classification models in the presence of highly imbalanced classes. Second, the inherent biases in different LLMs can lead to varying outcomes; some models may be naturally more inclusive, capturing a broader array of documents, while others may be more exclusive, being overly selective in their output. To account for these biases and to allow for customisation based on specific review needs, the calibrated instantiation of $I(d,t)$ offers a more adaptable and nuanced approach.

Ensembling of Screening Methods. We also consider an ensemble of screening methods. In particular, in our experiments we will ensemble the two most zero-shot effective LLMs and the BERT-based method we use as a comparative baseline. We use CombSUM to fuse the individual methods' decisions [26].

For Uncalibrated Screening, we directly combine the likelihoods of the model outputs. The decision rule $I(d,t)$ is formulated as follows:

$$I(d,t) = \begin{cases} 1, & \text{if } \sum_{m \in \text{Methods}} P_m(\text{yes}|d,t) \geq \sum_{m \in \text{Methods}} P_m(\text{no}|d,t) \\ 0, & \text{otherwise.} \end{cases}$$

For Calibrated Screening, we normalize S_{norm} to make the decisions:

$$I(d,t) = \begin{cases} 1, & \text{if } \sum_{m \in \text{Methods}} S_{\text{norm}}(d,t) \geq \theta \\ 0, & \text{otherwise.} \end{cases}$$

4 Experimental Setup

4.1 Considered LLMs

We employ an array of zero-shot generative LLMs that differ in architecture, training steps, and size (model parameters) to extensively evaluate their effectiveness for automatic systematic review document screening.

LlaMa: The LlaMa series offers an open-sourced suite of decoder models with parameter sizes ranging from 7B to 65B. Exceptional in its zero-shot capabilities, LlaMa outperforms GPT-3 across multiple NLP benchmarks. These models leverage a rich and diverse training dataset of approximately 1.4 trillion tokens, harvested from various sources including web pages, code repositories, and Wikipedia [46].

Alpaca: Alpaca has been fine-tuned on the 7B-parameter LlaMa model according to the self-instruct methodology [57]. Alpaca's training corpus originates from the text-davinci-003 model[4], initialized with 175 unique tasks. Preliminary assessments suggest that Alpaca, through instruction-based fine-tuning, achieves similar effectiveness to the OpenAI's text-davinci-003 model [44].

Guanaco: The Guanaco models stem from the LlaMa base models and are obtained through the memory-efficient 4-bit QLoRA fine-tuning on the OASST1 dataset [25,60]. This represents a different fine-tuning strategy than that used in the other considered LLMs. Guanaco models have demonstrated competitive performance against commercial systems on the Vicuna and OpenAssistant benchmarks [13,25].

[4] https://platform.openai.com/docs/models/gpt-3-5.

Falcon: Falcon is available in two variants: Falcon-7B and Falcon-40B. These models were trained on large-scale corpora of 1 and 1.5 trillion tokens, respectively, primarily sourced from the RefinedWeb dataset [34]. Notably, the Falcon family includes specialized "instruct" versions — Falcon-7B-Instruct and Falcon-40B-Instruct — that excel in assistant-style tasks through fine-tuning on instructional and conversational datasets.

LlaMa2: LlaMa2 extends the original LlaMa family, and comes in three parameter sizes: 7B, 13B, and 70B. Despite maintaining architectural similarity with its predecessor, LlaMa2 is trained on an expanded dataset of 2 trillion tokens, a 40% increase from LlaMa [46,47]. LlaMa2 also includes a specialized "Chat" variant, LlaMa2 Chat, which incorporates advanced fine-tuning techniques such as "Ghost Attention" for multi-turn dialogue consistency and an array of reinforcement learning methods [47].

Overall, we select eight models in our study: LlaMa-7b, Alpaca-7b-ins, Guanaco-7b-ins, LlaMa2-7b, LlaMa2-13b, Falcon-7b-ins, LlaMa2-7b-ins, LlaMa2-13b-ins.[5] Table 1 demonstrates the prompts used for LLM-based automatic screening. Note that we do not include special tokens in the prompt due to page limit, specific prompt for each model are adapted based on their special token setup. While we could have considered other models like the popular ChatGPT, their use can be financially prohibitive for our task. The predicted cost will be USD$4,000 and USD$80,000 if we use GPT-3.5-turbo and GPT-4, respectively. In our experiments, all employed models were configured to have a maximum token limit of 2048. This adjustment was particularly applied to the Alpaca model (original model has a limit of 512) and LlaMa2 models(original model has a limit of 4096) to ensure uniformity across all models. Consequently, we observed no instances of truncation in the experimental data.

4.2 Datasets

We experiment on the CLEF TAR datasets and the Seed Collection dataset. Four datasets were released as part of CLEF TAR [22–24], covering different types of systematic reviews. The 2017 dataset contains 50 Diagnostic Test Accuracy (DTA) topics; 2018 adds 30 more; while in 2019, a dataset consisted of 8 DTA topics, while another included 40 intervention review (Int) topics. These datasets contain relevance assessments for about 600,000 documents in total, and for each topic, the review title and the protocol file are also provided. These datasets are distributed with standard train-test splits; however, because we consider the zero-shot capabilities of the investigated models, we do not use these splits and instead test on all available topics.

[5] Note that for consistency of the paper, we name all instruction-tuned models with *-ins*; The original names are: Alpaca-7b-ins: alpaca; Guanaco-7b-ins: guanaco-7b; Falcon-7b-ins: falcon-7b-instruct; LlaMa2-7b-ins: LlaMa2-7b-chat; LlaMa2-13b-ins: LlaMa2-13b-chat;.

Table 1. Input types and prompts designed for each model. Italicised text indicates values that are replaced with respective content.

Model	Prompt

Alpaca	`Below is an instruction that describes a task, paired with an input that provides further context. Write a response that appropriately completes the request.`
	`### Instruction:`
	`Answer 'yes' or 'no' to Judge if the following retrieved study should be included by the systematic review?`
	`### Input:`
	`Review:` *review_title*
	`Study:` *candidate_document*
	`### Response:`

All Other Models	`Answer 'yes' or 'no' to Judge if the following retrieved study should be included by the systematic review?`
	`Review:` *review_title*
	`Study:` *candidate_document*
	`The answer is '`

The Seed Collection dataset consists of 39 review topics and over 50,000 candidate documents [52].[6] For each topic, the review title and inclusion/inclusion labels are provided along with a set of "seed documents": documents that were provided to the researcher designing the search strategy (query) for the review and that provide examples of documents related to the review (most are likely to meet the inclusion criteria, but it is possible some do not). In our experiment, we only evaluate based on the retrieved documents; included documents that are not in retrieved document set are removed.

4.3 Baseline

We compare the effectiveness of zero-shot LLMs against a baseline that relies on the BERT architecture but uses a domain-specific variant as backbone: BioBERT [29,54]. BioBERT employs the same architecture as BERT, but the corpus used for self-supervised training contains biomedical text (instead of general domain text like for BERT). BioBERT has been shown effective across a range of applications related to health tasks, including for screening prioritisation on medical systematic reviews on the datasets we consider [54], and thus is a strong baseline. To use BioBERT in our text classification task, we concatenate the topic title with the candidate document to form the input to the backbone. A classification head based on a sigmoid activation function is then used to determine the inclusion of a candidate document for the specified topic.[7]

[6] We removed topic 18 as no relevant document exited in the candidate document list (the topic only contains one relevant document).

[7] In the uncalibrated setting for BioBERT, we established a decision threshold of 0.5 to determine the inclusion of a document in a review topic. Specifically, a document is included if the BioBERT output satisfies the condition $output \geq 0.5$; otherwise, it is excluded.

4.4 Evaluation Measures

We use set-based metrics for evaluation: precision, recall, and F-3, which emphasize the importance of recall over precision. Additionally, we adopt balanced accuracy (B-AC) as a pivotal metric, as it particularly suits the nature of the systematic review document screening task, where excluded documents substantially outnumber included ones; B-AC $= \frac{1}{2}\left(\frac{\text{TP}}{\text{TP+FN}} + \frac{\text{TN}}{\text{TN+FP}}\right)$. We also report the success rate, which quantifies the fraction of topics achieving a pre-specified target recall. We adopt a representative target recall of 0.95, a standard threshold for systematic review document screening [7,8,18]: often systems that do not achieve at least 0.95 recall are deemed of no practical use for systematic review automation. Lastly, we gauge the efficiency of automatic document screening using the Work Saved by Sampling at a specific recall level (WSS) [16]. This is expressed as: $WSS = \frac{TN+FN}{N} - (1-r)$ where N denotes the total sample count and r signifies the recall level; we set r to 1, representing total recall.

4.5 Threshold Setting

For the calibration setting, the threshold θ value needs to be set. We devise two approaches to determine θ:

1. **Extrapolation from Collection**: we perform a leave-one-out experiment across all systematic review topics in a dataset. We identify threshold values that have consistently yielded robust results in the sample topics (all other topics except the target topic)—optimizing for a high recall rate—using the median score of candidate documents that achieved the target recall. The obtained threshold is then applied to the target topic under consideration. Note that cross-validation is used to determine the θ value only: the LLMs are still zero-shot. This is, however, a somewhat artificial setting, in that if training material was available for determining θ, then it could also be used to tune the LLMs (though computational costs may prevent this but our method does not require training of the LLM itself). We will consider a more appropriate option next.
2. **Calibration with Seed Studies**: we employ the uncalibrated LLM to generate inclusion scores for a set of seed studies (exemplar documents that are often identified prior to searching and screening). If the lowest score for a seed study is below the classifier's threshold for inclusion (decision boundary), then the threshold is lowered to the score obtained by that seed study: we use this as the new threshold for the calibrated LLM. This adjustment aims to improve recall. Typical targets for recall for systematic review are 0.95 or 1; we then experiment with these values to determine θ.

Table 2. Comparison of uncalibrated results between baseline method and generative large language models. Statistical significance, determined by a Student's two-tailed paired t-test with Bonferroni correction ($p < 0.05$), between the top-performing method *LlaMa2-7b-ins* and others is marked by *.

Dataset	Model	P	R	B-AC	F3	Suc	WSS
	BioBERT	0.06	0.95*	0.61*	0.30	0.74*	0.26*
CLEF-2017	LlaMa-7b	0.04*	0.92*	0.48*	0.24*	0.46*	0.03*
CLEF-2017	LlaMa2-7b	0.07	0.50*	0.60*	0.23*	0.02*	0.70*
CLEF-2017	LlaMa2-13b	0.04*	1.00*	0.50*	0.25*	0.98*	0.00*
CLEF-2017	Falcon-7b-ins	0.05*	0.92*	0.52*	0.25*	0.44	0.12*
CLEF-2017	Alpaca-7b-ins	0.04*	0.92*	0.51*	0.25*	0.38	0.11*
CLEF-2017	LlaMa2-7b-ins	0.08	0.87	**0.72**	**0.35**	0.26	0.56
CLEF-2017	LlaMa2-13b-ins	**0.19***	0.41*	0.66*	0.31	0.04*	**0.91***
CLEF-2017	Guanaco-7b-ins	0.04*	**1.00***	0.50*	0.25*	**1.00***	0.00*
	BioBERT	0.06	0.97*	0.59*	0.9	0.87*	0.19*
CLEF-2018	LlaMa-7b	0.05*	0.92*	0.48*	0.25*	0.33	0.04*
CLEF-2018	LlaMa2-7b	0.07	0.49*	0.59*	0.22*	0.03*	0.69*
CLEF-2018	LlaMa2-13b	0.05*	1.00*	0.50*	0.26	**1.00***	0.00*
CLEF-2018	Falcon-7b-ins	0.05*	0.92	0.51*	0.25*	0.40	0.11*
CLEF-2018	Alpaca-7b-ins	0.05*	0.91	0.51*	0.25*	0.30	0.11*
CLEF-2018	LlaMa2-7b-ins	0.09	0.88	**0.75**	**0.37**	0.27	0.59
CLEF-2018	LlaMa2-13b-ins	**0.26***	0.36*	0.66*	0.30	0.00*	**0.94***
CLEF-2018	Guanaco-7b-ins	0.05*	**1.00***	0.50*	0.26	**1.00***	0.00*
	BioBERT	0.07	0.99	0.58	0.30	0.88	0.18*
CLEF-2019-dta	LlaMa-7b	0.07	0.93	0.48*	0.27	0.25	0.03*
CLEF-2019-dta	LlaMa2-7b	0.08	0.48*	0.58*	0.23	0.00*	0.68
CLEF-2019-dta	LlaMa2-13b	0.07	1.00	0.50*	0.28	**1.00**	0.00*
CLEF-2019-dta	Falcon-7b-ins	0.07	0.95	0.54*	0.29	0.50	0.12*
CLEF-2019-dta	Alpaca-7b-ins	0.07	0.91	0.52*	0.28	0.25	0.12*
CLEF-2019-dta	LlaMa2-7b-ins	0.09	0.92	**0.71**	**0.35**	0.62	0.49
CLEF-2019-dta	LlaMa2-13b-ins	**0.19**	0.49*	0.69	0.32	0.00*	**0.87***
CLEF-2019-dta	Guanaco-7b-ins	0.07	**1.00**	0.50*	0.28	**1.00**	0.00*
	BioBERT	0.10	0.98*	0.58*	0.32	0.90*	0.16*
CLEF-2019-Int	LlaMa-7b	0.05*	0.86	0.47*	0.26	0.30	0.08*
CLEF-2019-Int	LlaMa2-7b	0.08	0.30*	0.55*	0.18*	0.05*	0.80*
CLEF-2019-Int	LlaMa2-13b	0.05	1.00*	0.50*	0.29	0.97*	0.00*
CLEF-2019-Int	Falcon-7b-ins	0.05	0.91	0.50*	0.27	0.57	0.09*
CLEF-2019-Int	Alpaca-7b-ins	0.05	0.87	0.49*	0.27	0.30	0.12*
CLEF-2019-Int	LlaMa2-7b-ins	0.08	0.90	**0.70**	**0.35**	0.42	0.48
CLEF-2019-Int	LlaMa2-13b-ins	**0.17***	0.45*	0.67	0.33	0.05*	**0.87***
CLEF-2019-Int	Guanaco-7b-ins	0.05	**1.00***	0.50*	0.29	**1.00***	0.00*
	BioBERT	0.04	0.93	0.54*	0.24	0.77*	0.16*
Seed Collection	LlaMa-7b	0.04	0.89	0.48*	0.21	0.56	0.07*
Seed Collection	LlaMa2-7b	0.04	0.29*	0.53*	0.15*	0.03*	0.78*
Seed Collection	LlaMa2-13b	0.04	**1.00***	0.50*	0.23	**1.00***	0.00*
Seed Collection	Falcon-7b-ins	0.04	0.93	0.50*	0.22	0.69	0.07*
Seed Collection	Alpaca-7b-ins	0.04	0.90	0.50*	0.22	0.49	0.10*
Seed Collection	LlaMa2-7b-ins	0.05	0.90	0.66	0.27	0.54	0.40
Seed Collection	LlaMa2-13b-ins	**0.13***	0.48*	**0.67**	**0.28**	0.05*	**0.85***
Seed Collection	Guanaco-7b-ins	0.04	**1.00***	0.50*	0.23	**1.00***	0.00*

5 Results

RQ1: Architecture and Size of Model. Consider the results reported in Table 2. For *model architecture*, we compare four models: *Falcon-7b-ins*, *Alpaca-7b-ins*, *LlaMa2-7b-ins* and *Guanaco-2-7b-ins* — all of which have the same number of parameters. The results indicate that *LlaMa2-7b-ins* is the most effective for the task, outperforming the others across all evaluation metrics except recall and success rate. Specifically, this model obtained a high WSS while incurring only a marginal drop in recall: a significant loss was observed only on CLEF-2017. Concerning success rate, *LlaMa2-7b-ins* exhibited comparable performance to its counterparts, showing no statistically significant differences.

For *model size*, we consider two variants of the *LlaMa2-ins* architecture: one with 7 billion parameters (*LlaMa2-7b-ins*) and another with 13 billion parameters (*LlaMa2-13b-ins*). Our findings suggest a trade-off between recall and WSS. Specifically, the 7-billion parameter variant obtains significantly higher recall, but this comes at the expense of reduced savings, evidenced by significantly lower WSS. Regarding B-AC, *LlaMa2-7b-ins* generally outperforms its larger counterpart across multiple datasets, except for the Seed Collection. Statistically significant differences in B-AC were only noted for CLEF-2017 and CLEF-2018.

RQ2: Impact of Instruction Fine-Tuning. Consider again Table 2. We contrast instruction-fine-tuned models against their base counterparts: *LlaMa2-7b-*

ins VS. *LlaMa2-7b*, *LlaMa2-13b-ins* VS. *LlaMa2-13b*, *Alpaca-7b-ins* VS. *LlaMa-7b*. Across all differences, a significant improvement in B-AC is observed. Nevertheless, the models exhibit divergent behaviours in other metrics. For *LlaMa-7b* and *LlaMa2-13b*, fine-tuning leads to higher WSS at the expense of reduced recall. Conversely, *LlaMa2-7b-ins* exhibits a significant decline in WSS but obtains higher recall, success rate, and F3 except in the CLEF-2019-dta, where the F3 improvement is not statistically significant. We also conducted a comparative evaluation with *Guanaco-7b-ins*, a QLoRA fine-tuned model. While it does outperform *LlaMa-7b* in B-AC, the model classifies all candidate documents as relevant, nullifying any practical applicability for systematic review screening.

In summary, our analyses suggest that instruction-based fine-tuning is generally beneficial for improving document screening accuracy. However, the specific gains — whether in savings or recall — depend on the base model's architecture. Our experiments also suggest that QLoRa fine-tuning does not yield an effective model for this particular task.

RQ3: Impact of Calibration. Consider Table 3. We find that calibrated models reliably meet their pre-set recall targets and provide an attractive solution for practical implementation for automatic document screening. Specifically, in our tests that considered the extrapolation from collection calibration, approximately 50% of the topics met the pre-set recall target of 0.95 by comparing success rates obtained in each dataset (note that success rate in our experiments is set to measure a 0.95 recall level). This further improves (success rate between 0.56 and 1.00) when the target recall for determining the threshold is set to 1. We further compare the performance of three calibrated models, *BioBERT*, *LlaMa2-7b-ins* and *LlaMa2-13b-ins*. Generally, the 7-billion parameter LlaMa2 model significantly outperforms the two other models in both B-AC and WSS. As for success rate and recall, the models exhibit similar effectiveness; *LlaMa2-7b-ins* performs the same or better in 60% of the cases for success rate and in 40% of cases for average recall.

The calibration with seed documents method could only be tested on the Seed Collection, as CLEF datasets have no seed studies. In this case, *LlaMa2-7b-ins* consistently obtains higher recall: 70% of topics achieved perfect recall, compared to only 50% using the other calibration method. Although calibration with seed studies generally improves recall, our analysis indicates that *LlaMa2-13b-ins* displays more volatile effectiveness in this setting, possibly due to the varying quality and quantity of seed documents across different topics.

Ensemble of Automatic Screening Methods. Consider Table 3 with respect to the Ensemble results, obtained by ensambling *LlaMa2-7b-ins*, *LlaMa2-13b-ins* and the *BioBERT* baseline. The Ensemble strategy yields consistently higher B-AC and WSS, when calibrated. Moreover, when pitted against individual generative LLMs calibrated with the same threshold recall, the Ensemble method obtains higher WSS, precision, and F3. Exceptions are observed in CLEF-2018 and Seed Collection, where the Ensemble strategy registers lower success rates.

Table 3. Comparison between the Calibrated (Cal) and Uncalibrated (Unc) approaches using the BioBERT model, LlaMa2-7b-ins model (7b-ins), the LlaMa2-13b-ins model (13b-ins) and the Ensemble of the three models (Ensemb). The calibrated method's number or character in the bracket () denotes the pre-set target recall (0.95 & 1) or using seed documents (S). Statistical significance for each generative model across different datasets is assessed using a Student's two-tailed paired t-test with a Bonferroni correction ($p < 0.05$) with respect to the uncalibrated approach, denoted by *. The highest evaluated scores for *each dataset* are bolded.

Dataset	Model	Setting	P	R	B-AC	F3	Suc	WSS
CLEF-2017	BioBERT	Unc	0.06	0.95	0.61	0.30	0.74	0.26
		Cal(0.95)	0.06	0.92	0.64	0.31	0.50*	0.34*
		Cal(1)	0.06	0.97	0.60	0.29	0.82	0.23
	7b-ins	Unc	0.08	0.87	0.72	0.35	0.26	0.56
		Cal(0.95)	0.06*	0.92*	0.69*	0.32	0.52	0.44
		Cal(1)	0.05*	**0.99***	0.60*	0.28	0.96	0.20
	13b-ins	Unc	0.19	0.41	0.66	0.31	0.04	0.91
		Cal(0.95)	0.06*	0.93	0.59*	0.28	0.50*	0.25*
		Cal(1)	0.05*	0.98	0.53*	0.26	0.88*	0.08*
	Ensemb	Unc	**0.31**	0.13	0.56	0.13	0.00	**0.98**
		Cal(0.95)	0.08	0.93*	**0.72**	**0.35***	0.52*	0.50*
		Cal(1)	0.06	0.97*	0.63	0.30	**0.90***	0.29*
CLEF-2018	BioBERT	Unc	0.06	0.97	0.59	0.29	0.87	0.19
		Cal(0.95)	0.07	0.91*	0.63	0.30	0.57*	0.33*
		Cal(1)	0.06	0.97	0.59	0.29	0.87	0.21
	7b-ins	Unc	0.09	0.88	0.75	0.37	0.27	0.59
		Cal(0.95)	0.08*	0.94*	0.71*	0.35*	0.50	0.46
		Cal(1)	0.06*	**0.99***	0.62*	0.30	**1.00**	0.24
	13b-ins	Unc	0.26	0.36	0.66	0.30	0.00	0.94
		Cal(0.95)	0.06	0.94*	0.59*	0.29	0.47*	0.22*
		Cal(1)	0.05	0.97	0.53*	0.27	0.80*	0.08*
	Ensemb	Unc	**0.35**	0.12	0.54	0.12	0.00	**0.95**
		Cal(0.95)	0.09*	0.94*	**0.75**	**0.38***	0.50*	0.54*
		Cal(1)	0.06	0.99*	0.64	0.32*	0.93*	0.28*
CLEF-2019-dta	BioBERT	Unc	0.07	0.99	0.58	0.30	0.88	0.18
		Cal(0.95)	0.08	0.89	0.59	0.26	0.50	0.27
		Cal(1)	0.08	0.91	0.59	0.27	0.62	0.25
	7b-ins	Unc	0.09	0.92	0.71	**0.35**	0.62	0.49
		Cal(0.95)	0.10*	0.91*	0.71*	0.34	0.50	0.50
		Cal(1)	0.08*	0.97*	0.66	0.32	0.75	0.34
	13b-ins	Unc	0.19	0.49	0.69	0.32	0.00	0.87
		Cal(0.95)	0.08	0.95	0.56	0.29	0.50*	0.16*
		Cal(1)	0.07	0.99	0.51	0.28	0.88*	0.03*
	Ensemb	Unc	**0.31**	0.21	0.59	0.19	0.00	**0.96**
		Cal(0.95)	0.10	0.91	**0.73**	0.34*	0.50*	0.52*
		Cal(1)	0.09*	**0.99**	0.64	0.32*	**1.00***	0.28*
CLEF-2019-Int	BioBERT	Unc	0.10	0.98	0.58	0.32	**0.90**	0.16
		Cal(0.95)	0.10	0.87*	0.59	0.29	0.50*	0.31*
		Cal(1)	0.10	0.90*	0.59	0.30	0.62*	0.27*
	7b-ins	Unc	0.08	0.90	0.70	0.35	0.42	0.48
		Cal(0.95)	0.08*	0.91*	0.67*	0.34	0.50	0.42
		Cal(1)	0.07*	0.93*	0.64*	0.33	0.65	0.34
	13b-ins	Unc	0.17	0.45	0.67	0.33	0.05	0.87
		Cal(0.95)	0.07*	0.90	0.58	0.30	0.50*	0.25*
		Cal(1)	0.06*	0.94	0.55	0.29	0.62*	0.16*
	Ensemb	Unc	**0.35**	0.23	0.58	0.22	0.05	**0.92**
		Cal(0.95)	0.09*	0.93*	**0.70**	**0.37***	0.50*	0.45*
		Cal(1)	0.08*	**0.96***	0.67	0.35*	0.68*	0.35*
Seed Collection	BioBERT	Unc	0.04	0.93	0.54	0.24	0.77	0.16
		Cal(0.95)	0.05	0.80*	0.55	0.22	0.50*	0.29*
		Cal(1)	0.05	0.83	0.55	0.23	0.53*	0.26
		Cal (S)	0.04	0.93	0.54	0.23	0.76	0.15
	7b-ins	Unc	0.05	0.90	0.66	0.27	0.54	0.40
		Cal(0.95)	0.05*	0.90*	0.66	0.28*	0.51	0.41
		Cal(1)	0.05*	0.92*	0.65	0.27*	0.56	0.38
		Cal (S)	0.05	0.97*	0.6*	0.26	0.77*	0.22*
	13b-ins	Unc	0.13	0.48	0.67	0.28	0.05	0.85
		Cal(0.95)	0.06*	0.87	0.64*	0.27*	0.51*	0.39*
		Cal(1)	0.05*	0.93	0.59*	0.26	0.59*	0.25*
		Cal (S)	0.06*	0.87*	0.63	0.29	0.54*	0.38*
	Ensemb	Unc	**0.16**	0.18	0.52	0.14	0.00	**0.86**
		Cal(0.95)	0.07*	0.86*	**0.71**	0.31*	0.49*	0.53*
		Cal(1)	0.07*	0.88*	0.70	**0.30***	0.56*	0.49*
		Cal (S)	0.04*	**1.00***	0.55	0.25*	**0.97***	0.10*

Interestingly, the Ensemble's performance dips in recall when not calibrated. This decline may be attributed to the model's aggressive document exclusion strategy, as evidenced by its consistently high WSS across datasets. Overall, our findings indicate that a calibrated Ensemble approach generally outperforms single generative LLMs.

6 Discussion and Outlook

Comparison with Fine-Tuned LLMs. Although this study aimed to investigate the effectiveness of zero-shot generative LLMs in systematic review document screening, we are also interested in comparing our method to the state-of-the-art fine-tuned model. For this comparison, we consider the Bio-SIEVE

Table 4. Comparison of Fine-tuned baseline to our method; Statistical significance, determined by a Student's two-tailed paired t-test with Bonferroni correction ($p < 0.05$), between Uncalibrated *Bio-SIEVE* method and others is marked by *.

Model	Setting	P	R	B-AC	F3	Suc	WSS
Bio-SIEVE	Original/Calibrated	**0.232**	0.576	0.727	**0.429**	0.111	0.858
	Calibrated(Recall=0.95)	0.102*	0.877*	0.683	0.348	0.481*	0.471*
	Calibrated (Recall=1)	0.088*	0.945*	0.666	0.339	0.704*	0.369*
LlaMa2-7b-ins	Uncalibrated	0.078*	0.920*	0.725	0.359	0.333	0.513*
	Calibrated (Recall=0.95)	0.068*	0.935*	0.685	0.333	0.481*	0.421*
	Calibrated(Recall=1)	0.059*	**0.990***	0.621*	0.311	**1.000***	0.241*
Ensemble	Uncalibrated	0.400*	0.204*	0.594*	0.199*	0.037	**0.972***
	Calibrated (Recall=0.95)	0.095*	0.937*	**0.729**	0.373	0.519*	0.500*
	Calibrated (Recall=1)	0.068*	0.981*	0.630*	0.322	0.889*	0.266*

approach, a fine-tuned model for systematic review document screening, and compare it with our best methods in Table 4.[8] We also apply our calibration approach to Bio-SIEVE. Surprisingly the most effective model, LlaMa2-7b-ins, obtains a B-AC comparable to Bio-SIEVE, and our Ensemble method is even more effective than Bio-SIEVE, although differences are not significant.

Another noteworthy observation is Bio-SIEVE's low recall and success rate, especially when not calibrated (original). These results raise concerns regarding Bio-SIEVE's practical utility for the screening task, as a low recall is often not accepted by the researchers conducting the review as it translates into missing important studies. While calibration improves Bio-SIEVE's recall, this is still inferior to our zero-shot model under the same calibration setting. This finding suggests that although fine-tuning can improve effectiveness, it requires careful calibration for systematic review document screening. Looking forward, fine-tuning remains an interesting avenue for research but may necessitate alternative calibration strategies for practical utility for this task.

Variation in Model Input Prompt. While we only considered one type of prompt for each model, it is important to highlight that generative LLMs are sensitive to prompt formulation [30,61,64]. Due to page constraints, we could not deeply discuss the effects of alternative prompt formulations, such as those based on inclusion/exclusion criteria or seed studies. However, preliminary investigations into these aspects show a similar trend to what is observed when solely using review topic titles as prompts. These additional results are provided in a supplementary digital appendix for completeness.[9]

7 Conclusion

We comprehensively evaluated zero-shot LLMs for systematic review document screening and introduced a calibration method for tuning the model output.

[8] Comparison is however not straightforward as Bio-SIEVE used most of the datasets we consider here for fine-tuning; we then evaluate effectiveness using the only 27 topics from CLEF-TAR that were not used to fine-tune Bio-SIEVE.

[9] https://github.com/ielab/ECIR-2024-llm-screening.

We further explored the utility of an ensemble method that combines the top zero-shot LLMs with the BioBERT baseline.

Our results highlight the importance of output calibration when applying generative LLMs to systematic review document screening. This calibration maintains review quality and reliably by meeting pre-set recall targets, thus offering the flexibility to adjust the model to the specific requirements of a systematic review. Furthermore, when calibrated, our ensemble method outperforms the current state-of-the-art fine-tuned model, Bio-SIEVE [35]. We also emphasized the role of instruction-based fine-tuning in effectively leveraging generative LLMs for this application, while we showed that QLoRa-tuning does not yield effective results for this task.

The findings reported in the paper suggest that LLM-based methods can be created for automatically screening documents for systematic reviews, leading to considerable savings in manual effort. Furthermore, this can be done without requiring expensive fine-tuning (both in terms of labelling and computation). The fact that a high recall level can be obtained across a large number of different types of reviews suggests that these methods might be mature enough for actual adoption in systematic review workflows.

References

1. Abualsaud, M., Ghelani, N., Zhang, H., Smucker, M.D., Cormack, G.V., Grossman, M.R.: A system for efficient high-recall retrieval. In: Proceedings of the 41st Annual International ACM SIGIR Conference on Research and Development in Information Retrieval, pp. 1317–1320 (2018)
2. Alharbi, A., Briggs, W., Stevenson, M.: retrieving and ranking studies for systematic reviews: University of Sheffield's Approach to CLEF eHealth 2018 Task 2. In: CEUR Workshop Proceedings: Working Notes of CLEF 2018: Conference and Labs of the Evaluation Forum. vol. 2125. CEUR Workshop Proceedings (2018)
3. Alharbi, A., Stevenson, M.: Ranking abstracts to identify relevant evidence for systematic reviews: the university of sheffield's approach to clef ehealth 2017 Task 2. In: CEUR Workshop Proceedings: Working Notes of CLEF 2017: Conference and Labs of the Evaluation Forum (2017)
4. Alshami, A., Elsayed, M., Ali, E., Eltoukhy, A.E., Zayed, T.: Harnessing the power of chatgpt for automating systematic review process: methodology, case study, limitations, and future directions. Systems **11**(7), 351 (2023)
5. Anagnostou, A., Lagopoulos, A., Tsoumakas, G., Vlahavas, I.P.: Combining inter-review learning-to-rank and intra-review incremental training for title and abstract screening in systematic reviews. In: CEUR Workshop Proceedings: Working Notes of CLEF 2017: Conference and Labs of the Evaluation Forum (2017)
6. Aum, S., Choe, S.: srbert: automatic article classification model for systematic review using BERT. Syst. Contr. Found. Appl. **10**(1), 1–8 (2021)
7. Bramer, W.M., Rethlefsen, M.L., Kleijnen, J., Franco, O.H.: Optimal database combinations for literature searches in systematic reviews: a prospective exploratory study. Syst. Contr. Found. Appl. **6**, 1–12 (2017)
8. Callaghan, M.W., Müller-Hansen, F.: Statistical stopping criteria for automated screening in systematic reviews. Syst. Contr. Found. Appl. **9**(1), 1–14 (2020)

9. Carvallo, A., Parra, D., Lobel, H., Soto, A.: Automatic document screening of medical literature using word and text embeddings in an active learning setting. Scientometrics **125**, 3047–3084 (2020)
10. Carvallo, A., Parra, D., Rada, G., Perez, D., Vasquez, J.I., Vergara, C.: Neural language models for text classification in evidence-based medicine. arXiv preprint arXiv:2012.00584 (2020)
11. Chandler, J., Cumpston, M., Li, T., Page, M.J., Welch, V.A.: Cochrane Handbook for Systematic Reviews of Interventions. John Wiley & Sons (2019)
12. Chen, J., et al.: ECNU at 2017 eHealth task 2: technologically assisted reviews in empirical medicine. In: CEUR Workshop Proceedings: Working Notes of CLEF 2017: Conference and Labs of the Evaluation Forum (2017)
13. Chiang, W.L., et al.: Vicuna: An open-source chatbot impressing gpt-4 with 90%* chatgpt quality. See https://vicuna. lmsys. org (Accessed 14 April 2023) (2023)
14. Clark, J.: Systematic reviewing: introduction, locating studies and data abstraction. In: Doi, S.A.R., Williams, G.M. (eds.) Methods of Clinical Epidemiology, pp. 187–211. Springer Berlin Heidelberg, Berlin, Heidelberg (2013). https://doi.org/10.1007/978-3-642-37131-8_12
15. Cohen, A.M., Ambert, K., McDonagh, M.: A prospective evaluation of an automated classification system to support evidence-based medicine and systematic review. In: AMIA annual symposium proceedings. vol. 2010, p. 121. American Medical Informatics Association (2010)
16. Cohen, A., Hersh, W., Peterson, K., Yen, P.: Reducing workload in systematic review preparation using automated citation classification. J. Am. Med. Inform. Assoc. **13**(2), 206–219 (2006)
17. Collaboration, C.: The cochrane library. Database available on disk and CDROM. Oxford, UK, Update Software (2002)
18. Crumley, E.T., Wiebe, N., Cramer, K., Klassen, T.P., Hartling, L.: Which resources should be used to identify rct/ccts for systematic reviews: a systematic review. BMC Med. Res. Methodol. **5**, 1–13 (2005)
19. Dettmers, T., Pagnoni, A., Holtzman, A., Zettlemoyer, L.: Qlora: efficient finetuning of quantized llms. arXiv preprint arXiv:2305.14314 (2023)
20. Di Nunzio, G.M., Beghini, F., Vezzani, F., Henrot, G.: An interactive two-dimensional approach to query aspects rewriting in systematic reviews. IMS unipd at CLEF eHealth task 2. In: CEUR Workshop Proceedings: Working Notes of CLEF 2017: Conference and Labs of the Evaluation Forum (2017)
21. Di Nunzio, G.M., Ciuffreda, G., Vezzani, F.: Interactive sampling for systematic reviews. IMS unipd at CLEF 2018 eHealth task 2. In: CEUR Workshop Proceedings: Working Notes of CLEF 2018: Conference and Labs of the Evaluation Forum (2018)
22. Kanoulas, E., Li, D., Azzopardi, L., Spijker, R.: CLEF 2017 technologically assisted reviews in empirical medicine overview. In: CEUR Workshop Proceedings: Working Notes of CLEF 2017: Conference and Labs of the Evaluation Forum (2017)
23. Kanoulas, E., Li, D., Azzopardi, L., Spijker, R.: CLEF 2019 technology assisted reviews in empirical medicine overview. In: CEUR Workshop Proceedings: Working Notes of CLEF 2018: Conference and Labs of the Evaluation Forum. vol. 2380 (2019)
24. Kanoulas, E., Spijker, R., Li, D., Azzopardi, L.: CLEF 2018 technology assisted reviews in empirical medicine overview. In: CEUR Workshop Proceedings: Working Notes of CLEF 2018: Conference and Labs of the Evaluation Forum (2018)
25. Köpf, A., Kilcher, Y., et al.: Openassistant conversations-democratizing large language model alignment. arXiv preprint arXiv:2304.07327 (2023)

26. Kozorovitsky, A.K., Kurland, O.: From"identical"to"similar": fusing retrieved lists based on inter-document similarities. J. Artif. Intell. Res. **41**, 267–296 (2011)
27. Lagopoulos, A., Anagnostou, A., Minas, A., Tsoumakas, G.: Learning-to-rank and relevance feedback for literature appraisal in empirical medicine. In: Bellot, P., et al. (eds.) Experimental IR Meets Multilinguality, Multimodality, and Interaction: 9th International Conference of the CLEF Association, CLEF 2018, Avignon, France, September 10-14, 2018, Proceedings, pp. 52–63. Springer International Publishing, Cham (2018). https://doi.org/10.1007/978-3-319-98932-7_5
28. Lee, G.E., Sun, A.: Seed-driven document ranking for systematic reviews in evidence-based medicine. In: Proceedings of the 41st Annual International ACM SIGIR Conference on Research and Development in Information Retrieval, pp. 455–464 (2018)
29. Lee, J., et al.: Biobert: a pre-trained biomedical language representation model for biomedical text mining. Bioinformatics **36**(4), 1234–1240 (2020)
30. Lu, Y., Bartolo, M., Moore, A., Riedel, S., Stenetorp, P.: Fantastically ordered prompts and where to find them: overcoming few-shot prompt order sensitivity. arXiv preprint arXiv:2104.08786 (2021)
31. Minas, A., Lagopoulos, A., Tsoumakas, G.: Aristotle university's approach to the technologically assisted reviews in empirical medicine task of the 2018 CLEF eHealth lab. In: CEUR Workshop Proceedings: Working Notes of CLEF 2018: Conference and Labs of the Evaluation Forum (2018)
32. Miwa, M., Thomas, J., O'Mara-Eves, A., Ananiadou, S.: Reducing systematic review workload through certainty-based screening. J. Biomed. Inform. **51**, 242–253 (2014)
33. Norman, C.R., Leeflang, M.M., Porcher, R., Névéol, A.: Measuring the impact of screening automation on meta-analyses of diagnostic test accuracy. Syst. Contr. Found. Appl. **8**(1), 243 (2019)
34. Penedo, G., et al.: The refinedweb dataset for falcon llm: outperforming curated corpora with web data, and web data only. arXiv preprint arXiv:2306.01116 (2023)
35. Robinson, A., et al.: Bio-sieve: exploring instruction tuning large language models for systematic review automation. arXiv preprint arXiv:2308.06610 (2023)
36. Scells, H., Zuccon, G.: You can teach an old dog new tricks: rank fusion applied to coordination level matching for ranking in systematic reviews. In: Proceedings of the 42nd European Conference on Information Retrieval, pp. 399–414 (2020)
37. Scells, H., Zuccon, G., Deacon, A., Koopman, B.: QUT ielab at CLEF eHealth 2017 technology assisted reviews track: initial experiments with learning to rank. In: CEUR Workshop Proceedings: Notes of CLEF 2017: Conference and Labs of the Evaluation Forum (2017)
38. Scells, H., Zuccon, G., Koopman, B.: Automatic boolean query refinement for systematic review literature search. In: Proceedings of the 28th World Wide Web Conference, pp. 1646–1656 (2019)
39. Scells, H., Zuccon, G., Koopman, B.: A comparison of automatic boolean query formulation for systematic reviews. Information Retrieval Journal, pp. 1–26 (2020)
40. Scells, H., Zuccon, G., Koopman, B.: A computational approach for objectively derived systematic review search strategies. In: Proceedings of the 42nd European Conference on Information Retrieval, pp. 385–398 (2020)
41. Scells, H., Zuccon, G., Koopman, B., Clark, J.: Automatic boolean query formulation for systematic review literature search. In: Proceedings of the 29th World Wide Web Conference, pp. 1071–1081 (2020)

42. Singh, J., Thomas, L.: IIIT-H at CLEF eHealth 2017 task 2: Technologically assisted reviews in empirical medicine. In: CEUR Workshop Proceedings: Working Notes of CLEF 2017: Conference and Labs of the Evaluation Forum (2017)
43. Syriani, E., David, I., Kumar, G.: Assessing the ability of chatgpt to screen articles for systematic reviews. arXiv preprint arXiv:2307.06464 (07 2023)
44. Taori, R., et al.: Stanford alpaca: an instruction-following llama model. https://github.com/tatsu-lab/stanford_alpaca (2023)
45. Thomas, J., Harden, A.: Methods for the thematic synthesis of qualitative research in systematic reviews. BMC Med. Res. Methodol. **8**(1), 45 (2008)
46. Touvron, H., et al.: Llama: Open and efficient foundation language models. arXiv preprint arXiv:2302.13971 (2023)
47. Touvron, H., et al.: Llama 2: Open foundation and fine-tuned chat models. arXiv preprint arXiv:2307.09288 (2023)
48. Wallace, B.C., Small, K., Brodley, C.E., Lau, J., Trikalinos, T.A.: Deploying an interactive machine learning system in an evidence-based practice center: Abstrackr. In: Proceedings of the 2nd ACM International Health Informatics Symposium, pp. 819–824 (2012)
49. Wallace, B.C., Trikalinos, T.A., Lau, J., Brodley, C., Schmid, C.H.: Semi-automated screening of biomedical citations for systematic reviews. BMC Bioinform. **11**(1), 55 (2010)
50. Wang, S., Li, H., Scells, H., Locke, D., Zuccon, G.: Mesh term suggestion for systematic review literature search. In: Proceedings of the 25th Australasian Document Computing Symposium, pp. 1–8 (2021)
51. Wang, S., Li, H., Zuccon, G.: Mesh suggester: a library and system for mesh term suggestion for systematic review boolean query construction. In: Proceedings of the Sixteenth ACM International Conference on Web Search and Data Mining, pp. 1176–1179 (2023)
52. Wang, S., Scells, H., Clark, J., Koopman, B., Zuccon, G.: From little things big things grow: A collection with seed studies for medical systematic review literature search. In: Proceedings of the 45th International ACM SIGIR Conference on Research and Development in Information Retrieval, pp. 3176–3186 (2022)
53. Wang, S., Scells, H., Koopman, B., Zuccon, G.: Automated mesh term suggestion for effective query formulation in systematic reviews literature search. Intell. Syst. Appl. 200141 (2022)
54. Wang, S., Scells, H., Koopman, B., Zuccon, G.: Neural rankers for effective screening prioritisation in medical systematic review literature search. In: Proceedings of the 26th Australasian Document Computing Symposium, pp. 1–10 (2022)
55. Wang, S., Scells, H., Koopman, B., Zuccon, G.: Can chatgpt write a good boolean query for systematic review literature search? In: Proceedings of the 46th International ACM SIGIR Conference on Research and Development in Information Retrieval, pp. 1426–1436. SIGIR '23, Association for Computing Machinery, New York, NY, USA (2023). https://doi.org/10.1145/3539618.3591703
56. Wang, S., Scells, H., Potthast, M., Koopman, B., Zuccon, G.: Generating natural language queries for more effective systematic review screening prioritisation. arXiv preprint arXiv:2309.05238 (2023)
57. Wang, Y., et al.: Self-instruct: Aligning language model with self generated instructions. arXiv preprint arXiv:2212.10560 (2022)
58. White, J.: Pubmed 2.0. Medical reference services quarterly **39**(4), 382–387 (2020)
59. Wu, H., Wang, T., Chen, J., Chen, S., Hu, Q., He, L.: Ecnu at 2018 ehealth task 2: technologically assisted reviews in empirical medicine. Methods-a Companion Methods Enzymol. **4**(5), 7 (2018)

60. Xu, Y., et al.: Qa-lora: Quantization-aware low-rank adaptation of large language models. arXiv preprint arXiv:2309.14717 (2023)
61. Yang, C., et al.: Large language models as optimizers. arXiv preprint arXiv:2309.03409 (2023)
62. Yang, E., MacAvaney, S., Lewis, D.D., Frieder, O.: Goldilocks: just-right tuning of BERT for technology-assisted review. In: Hagen, M., et al. (eds.) ECIR 2022. LNCS, vol. 13185, pp. 502–517. Springer, Cham (2022). https://doi.org/10.1007/978-3-030-99736-6_34
63. Zhang, R., Wang, Y.S., Yang, Y.: Generation-driven contrastive self-training for zero-shot text classification with instruction-tuned gpt. arXiv preprint arXiv:2304.11872 (2023)
64. Zhao, Z., Wallace, E., Feng, S., Klein, D., Singh, S.: Calibrate before use: Improving few-shot performance of language models. In: International Conference on Machine Learning, pp. 12697–12706. PMLR (2021)
65. Zou, J., Li, D., Kanoulas, E.: Technology assisted reviews: finding the last few relevant documents by asking Yes/No questions to reviewers. In: Proceedings of the 41st Annual International ACM SIGIR Conference on Research and Development in Information Retrieval, pp. 949–952 (2018)

Cross-Modal Retrieval for Knowledge-Based Visual Question Answering

Paul Lerner[1(✉)], Olivier Ferret[2], and Camille Guinaudeau[3]

[1] Sorbonne Université, CNRS, ISIR, 75005 Paris, France
lerner@isir.upmc.fr
[2] Université Paris-Saclay, CEA, List, 91120 Palaiseau, France
olivier.ferret@cea.fr
[3] Université Paris-Saclay, CNRS, LISN, 91400 Orsay, France
camille.guinaudeau@lisn.upsaclay.fr

Abstract. Knowledge-based Visual Question Answering about Named Entities is a challenging task that requires retrieving information from a multimodal Knowledge Base. Named entities have diverse visual representations and are therefore difficult to recognize. We argue that cross-modal retrieval may help bridge the semantic gap between an entity and its depictions, and is foremost complementary with mono-modal retrieval. We provide empirical evidence through experiments with a multimodal dual encoder, namely CLIP, on the recent ViQuAE, InfoSeek, and Encyclopedic-VQA datasets. Additionally, we study three different strategies to fine-tune such a model: mono-modal, cross-modal, or joint training. Our method, which combines mono- and cross-modal retrieval, is competitive with billion-parameter models on the three datasets, while being conceptually simpler and computationally cheaper.

Keywords: Visual Question Answering · Multimodal · Cross-modal Retrieval · Named Entities

1 Introduction

The work we present in this article takes place in the context of Multimodal Information Retrieval, a field at the intersection between Information Retrieval (IR), Computer Vision, and Machine Learning. More precisely, we focus on Knowledge-based Visual Question Answering about named Entities (KVQAE), which has two specificities in regards to multimodal modeling [4,6,20,57]: (i) images represent named entities; (ii) multimodal interactions are complex and may be combined as both questions and retrieved passages are (text, image) pairs. Indeed,

We thank the anonymous reviewers for their helpful comments, as well as Antoine Chaffin for fruitful discussions about CLIP and cross-modal retrieval. Paul Lerner did this work during his PhD at LISN. This work was supported by the ANR-19-CE23-0028 MEERQAT project. This work was granted access to the HPC resources of IDRIS under the allocation 2021-AD011012846 made by GENCI.

N. Goharian et al. (Eds.): ECIR 2024, LNCS 14608, pp. 421–438, 2024.
https://doi.org/10.1007/978-3-031-56027-9_26

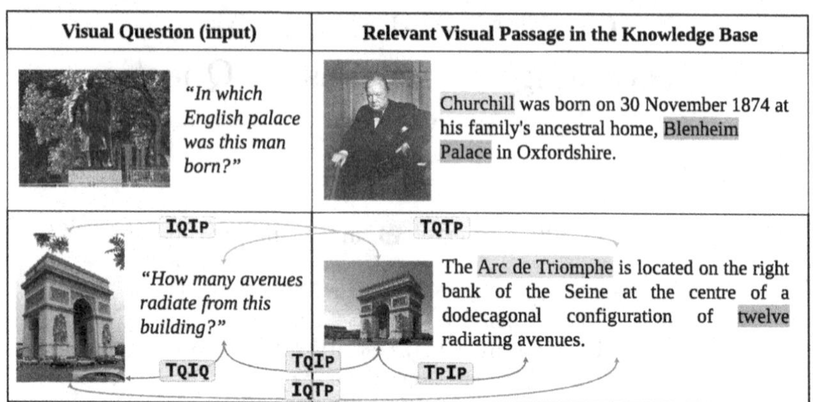

Fig. 1. Two visual questions from the ViQuAE dataset along with relevant visual passages from its Knowledge Base. The different types of mono- and cross-modal interactions studied are also shown for the second question. The acronyms of the interactions are composed of the letters T (Text), I (Image), Q (question) and P (passage).

KVQAE consists in answering questions about named entities grounded in a visual context [31,43]. We focus here on Entity Retrieval based on this visual context, similarly to Visual Named Entity Linking [46]. Figure 1 shows two examples of visual questions along with corresponding relevant visual passages from the ViQuAE dataset [31] and its multimodal Knowledge Base (KB), i.e. the set of multimedia documents in which the answers to the questions are searched.

The first example shows how heterogeneous depictions of named entities can be: Winston Churchill is depicted through a *statue* in the visual question and by a *standard photograph* in the KB. This heterogeneity makes mono-modal image retrieval difficult. On the other hand, cross-modal retrieval may bridge the semantic gap between the two representations by using a more abstract representation of the entity, e.g. its name, here *Winston Churchill.*

We formalize these different multimodal interactions in the framework exemplified in Fig. 1. The cross-modal interaction between the image of the question and the text of the passage is noted IQTP, while the mono-modal interaction between the two images is noted IQIP. This work is inspired by [30] who studied early multimodal fusion methods, also modeling the TQIQ (resp. TPIP) interaction within the visual question (resp. passage), but found that IQTP was the most important multimodal interaction.

KVQAE differs from standard Visual Question Answering (VQA [3]), which targets the content of the image (e.g., "*What color is the car?*"), and therefore does not require IR. Commonsense VQA [37,42] falls in between standard VQA and KVQAE but (i) focuses on Commonsense Knowledge; (ii) is limited to coarse-grained object categories, e.g., *person* and *building*, instead of *Winston Churchill* and *Arc de Triomphe*, which makes image retrieval straightforward using an object detector [18].

KVQAE was introduced in [43] and received an increased interest recently, with a shift towards unstructured KBs [30,31] and later Large Language Models (LLMs), which do not use an explicit KB but rather generate an answer from the knowledge implicitly stored in their parameters [8,24,33,38]. Given the results of [8,38] and the various caveats of LLMs for factual information generation (hallucinations, lack of generalization and updatability [26,56]), our work adopts a more classical Question Answering architecture, also exploited by [30,31], in which a first IR step is followed by an answer extraction step.

More precisely, we focus on Entity Retrieval and propose to use a multimodal dual encoder [16], namely CLIP [41], for both mono- and cross-modal retrieval, i.e. modeling IqIp and IqTp, respectively. Multimodal dual encoders like CLIP are used as foundation models for a set of diverse tasks such as multimodal analogy [10], Visual Named Entity Linking [46], Cross-modal Question Answering [35], and Commonsense VQA [19]. We show that both mono- and cross-modal retrieval are complementary and can be simply yet effectively combined. We provide empirical evidence through experiments on the ViQuAE, InfoSeek, and Encyclopedic-VQA datasets, being as such the first comparative study of these recently introduced datasets. Furthermore, we study three different strategies to fine-tune such a model, which has been pre-trained in a cross-modal fashion, in this context: mono-modal, cross-modal, or joint training.

2 Related Work

In this section, we present a review of datasets and methods for KVQAE.

Datasets. KVQA was the first KVQAE dataset proposed in [43]. Despite its apparent large size, it has several limitations as pointed out by [31]: (i) only one entity type is considered, namely *person*; (ii) it is generated automatically, and thus, has a limited diversity of topics, lexicon, and syntax. Another key difference with the other datasets is that KVQA was designed for structured KBs, in particular Wikidata, from which it was generated, and not an unstructured KB like the following works. To address the limitations of KVQA, ViQuAE was introduced in [31]. It has fewer visual questions but they are manually annotated and it covers a broad range of topics, lexicon, and syntax, as showed in Table 1. Above all, ViQuAE comprises a large number of different entity types, including for example landmarks and organizations in addition to persons. Recently, two other datasets were proposed, aiming at larger size than ViQuAE and with fewer textual bias: InfoSeek [8] and Encyclopedic-VQA (EVQA [38]). InfoSeek is split into two subsets according to the annotation method: manual (ISM) or automatic (ISA). Unfortunately, since neither ISM nor the test set of ISA is available at the time of writing, we can evaluate our model only on the validation set of ISA. As its annotation is automatic, it shares part of the caveats of KVQA but covers more diverse entity types. EVQA alleviates these by using more sophisticated question generation techniques than templates. However, it is sometimes biased towards text, with questions such as "*Which republic celebrated the vendémiaire*

Table 1. Key features of different KVQAE datasets: ViQuAE [31], InfoSeek [8], Encyclopedic-VQA (EVQA [38]), and KVQA [43]. InfoSeek is split into two subsets according to the annotation method: manual (ISM) or automatic (ISA). *Computed on a subset of 500 questions by [8].

	ViQuAE	ISM	ISA	EVQA	KVQA
# Visual questions	3,700	8,900	1,356,000	1,036,000	183,000
# Unique questions (text-only)	3,562	2,022	1,498	175,000	8,310
# Unique POS sequences	2,759	1,056	267	91,945	376
# Questions per image	1.1	1.0	1.4	2.0	7.4
Vocabulary (# words)	4,700	1,307	725	40,787	8,400
Average question length (# words)	12.4	7.8	8.9	11.6	10.1
Answer prior	0.3%	–	0.6%	0.4%	15.9%
Answer overlap	25.3%	–	48.1%	59.6%	89.4%
Entity overlap	18.1%	–	20.1%	82.0%	40.6%
# Questions per entity	1.5	11.0	117.6	62.5	9.7
# Entity types	980	527	2,739	–	1
Requires knowledge*	95.2%	95.6%	–	–	–

in the month that the growing season ends for this tree?", a type of overspecified questions that were typically filtered by the manual annotation in ViQuAE [31]. Some key features of these datasets are summarized in Table 1. Question length is expressed in number of words provided by spaCy's English tokenizer. Answer prior is computed as the most likely answer in the training set, independently of the question. All datasets are limited to the English language.

Methods. Because the KVQA dataset is limited to *person*-named entities, it was addressed through face recognition in [43]: a Wikidata subgraph is constructed from the recognized entities and processed by a *memory network* to output an answer [50]. A few other studies were carried out on KVQA but the comparison with the rest of the state of the art is made difficult as their systems take the image *caption* as input, making the image *itself* redundant [17,22,48].

Our work is closer to [31], which uses an unstructured KB, a collection of visual passages (as in Fig. 1). The authors tackle the task in two steps, where Reading Comprehension follows IR. Their retrieval is a combination of two mono-modal retrievals: textual with DPR [28] and visual with a combination of CLIP, ArcFace [12], and a ResNet model trained on ImageNet [11,21]. We aim at simplifying this system by (i) removing the dependency on ArcFace and ImageNet, two supervised models that provide *a priori* less generic representations than CLIP; (ii) taking full advantage of CLIP by combining mono-modal and cross-modal retrieval. After the IR step, answers are extracted using Multi-passage BERT [49]. This work was then extended in [1], by combining the text retrieval of DPR with Wikidata embeddings, but in doing so, it sets aside multimodal interactions and the image of the visual question. For their part, [30] have, like us, focused on IR. In order to model cross-modal interactions, they jointly represent text and image using a multimodal Transformer [16,29]. However, this model requires an expensive pre-training and the authors ultimately suggest that it mostly leverages the IQTP interaction. Our conclusions converge because our model outperforms theirs — without additional pre-training — by explicitly modeling IQTP via CLIP, as described in the next section.

Very recently, following the overall trend in our domains, there has been a handful of works aiming to tackle KVQAE with (multimodal) LLMs, directly generating an answer from the visual question, without IR [8,24,33,38]. The same conclusions are reached in [8,38]: multimodal LLMs suffer from the same caveats as text-only LLMs and underperform compared to retrieval-augmented models. As a consequence, a sophisticated planning method using a tool-augmented LLM as agent was proposed in [24]. However, [24,38] share the same experimental protocol problem: they query the whole Web for image or text retrieval through Google APIs, although the images of the visual questions are public and indexed by Google, which leads to overoptimistic and non-reproducible results. On the contrary, we follow the methodology of [8,30,31], using a controlled, publicly available KB. As for [33], they tackle KVQAE with a multimodal LLM, which is only able to generate long explanatory answers. Therefore, [33] evaluate it on ViQuAE using ROUGE-L [34], after paraphrasing the ground-truth answers with ChatGPT. For that reason, their results are unfortunately not comparable with the rest of the state of the art.

3 Entity Retrieval from Visual Context

3.1 Method

Before being able to extract the answer to the question from a visual passage, or even retrieve such a passage, we focus here on Entity Retrieval, given the image of the question $\mathbf{i_q}$ and a collection of entities $(\mathbf{t_p}, \mathbf{i_p})$, where $\mathbf{t_p}$ denotes the name of the entity and $\mathbf{i_p}$ its reference image. To do so, we define the following similarity function, which combines mono- and cross-modal similarities:

$$s(\mathbf{i_q}, \mathbf{t_p}, \mathbf{i_p}) = \alpha_I s_I(\mathbf{i_q}, \mathbf{i_p}) + \alpha_C s_C(\mathbf{i_q}, \mathbf{t_p}) \quad (1)$$

where the parameters $\alpha_{\{I,C\}}$ weigh each similarity. We focus on CLIP, a multimodal dual encoder, to implement $s_I(\mathbf{i_q}, \mathbf{i_p})$ and $s_C(\mathbf{i_q}, \mathbf{t_p})$, which models the IQIP and IQTP interactions, respectively (see Fig. 1). The objective is thus to bring the image of the question closer to the image of this entity in the KB (*mono-modal training*), or to its name (*cross-modal training*), or both jointly.

More formally, the objective underlying our IR model is to maximize $s(\mathbf{i_q}, \mathbf{t_p}, \mathbf{i_p})$ if the two images $\mathbf{i_q}$ and $\mathbf{i_p}^{(+)}$ depict the same entity, named with the textual form $\mathbf{t_p}^{(+)}$, and to minimize it otherwise. In such a contrastive approach, the other entities of a batch, for which the textual and visual representations are respectively noted $\mathbf{t_p}^{(j)}$ and $\mathbf{i_p}^{(j)}$, are used as negatives. To implement this approach, we jointly train $s_I(\mathbf{i_q}, \mathbf{i_p})$ and $s_C(\mathbf{i_q}, \mathbf{t_p})$ for each $\mathbf{i_q}$ image of the batch by minimizing the following objective, given the temperature τ:

$$-\log \frac{\exp\left(s(\mathbf{i_q}, \mathbf{t_p}^{(+)}, \mathbf{i_p}^{(+)})e^{\tau}\right)}{\exp\left(s(\mathbf{i_q}, \mathbf{t_p}^{(+)}, \mathbf{i_p}^{(+)})e^{\tau}\right) + \sum_j \exp\left(s(\mathbf{i_q}, \mathbf{t_p}^{(j)}, \mathbf{i_p}^{(j)})e^{\tau}\right)} \tag{2}$$

Since we implement $s_C(\mathbf{i_q}, \mathbf{t_p})$ with CLIP, we have:

$$s_C(\mathbf{i_q}, \mathbf{t_p}) = \cos\left(\text{CLIP}_V(\mathbf{i_q}), \text{CLIP}_T(\mathbf{t_p})\right) \tag{3}$$

If $\alpha_I = 0$ and $\alpha_C = 1$ (cross-modal training only), the objective is equivalent to the one used during the pre-training of CLIP, except that it is asymmetric (the softmax function expresses the probabilities according to $\mathbf{i_q}$ and not according to $\mathbf{t_p}$). Since $\mathbf{i_q}$, $\mathbf{t_p}$, and $\mathbf{i_p}$ are encoded independently, this objective leverages all the other images and texts of the batch in a highly efficient way (we only need a matrix product to compute the denominator of Eq. 2). We implement $s_I(\mathbf{i_q}, \mathbf{i_p})$ in a similar way: $s_I(\mathbf{i_q}\mathbf{i_p}) = \cos\left(\text{CLIP}_V(\mathbf{i_q}), \text{CLIP}_V(\mathbf{i_p})\right)$. The same method could be applied to any multimodal dual encoder [16].

3.2 Data

As mentioned in the introduction, our evaluations are performed on the ViQuAE, ISA, and EVQA datasets. For ViQuAE and ISA, we use the KB proposed in [31], which consists of 1.5 million Wikipedia articles and images of corresponding Wikidata entities. Unfortunately, the KB proposed by [8] has yet to be made available; so our results on ISA will not be directly comparable to theirs. Indeed, 11.5% of ISA entities are missing from our KB, which filters down the training set by 28%. On the contrary, only a few entities from ViQuAE are missing from the KB. For EVQA, we use the corresponding KB of [38], which consists of 2 million Wikipedia articles and corresponding images in WIT [45].

ViQuAE contains 3,700 visual questions about 2,400 different entities, randomly divided into equal-sized sets for training, validation, and testing, with no overlap between images. As a result, the overlap between entities in the training and test sets is quite small, only 18%. Likewise, the entity overlap in ISA is of 20%. Our models must therefore learn to generalize not only to new images but also to new entities. On the contrary, the entity overlap of EVQA is of 82%.

3.3 Hyperparameters

We use the `ViT-B/32` version of CLIP unless otherwise mentioned. To take full advantage of the entities associated with the other images in the batch $\mathbf{t_p}^{(j)}$ and $\mathbf{i_p}^{(j)}$, we use a batch of the largest possible size, here 3,072 $(\mathbf{i_q}, \mathbf{t_p}^{(+)}, \mathbf{i_p}^{(+)})$ triples, i.e., more than the whole training set of ViQuAE. We use a single NVIDIA V100 GPU with 32 GB of RAM. The large batch size is partly enabled by gradient checkpointing.

Because the training set of ViQuAE is so small, training is very cheap: our best model converges, i.e., starts to overfit, after 11 steps/epochs, in less than

15 min, which is negligible compared to the pre-training of 8,000 steps in three days of [30] with the same hardware ([30] reports a carbon footprint of 1.7 kgCO2e for 3 d of GPU power consumption). On the larger ISA and EVQA datasets, our models converge roughly after 500 steps in 5 h.

We use a very small learning rate of 2×10^{-6}, increasing linearly for 4 steps and then decreasing for 50 steps on ViQuAE (or 1,000 on ISA and EVQA) if training is not interrupted before. We use the AdamW optimizer [36], with a weight decay of 0.1. For joint training, we initialize $\alpha_I = \alpha_C = 0.5$ and assign them a learning rate of 0.02, much larger than the rest of the model. Like [41], the temperature τ remains trainable but, given the small learning rate, it remains close to its initial value, 4.6.[1] These hyperparameters were set manually through experiments on the validation set of ViQuAE.

Early stopping is done according to the *in-batch* mean reciprocal rank on the validation set, i.e., by reranking the images or texts of the batch according to the similarity score s, to avoid computing the representations of the whole KB at each epoch.

Our implementation is based on Lightning,[2] PyTorch [39], and Transformers [53] for training, and Datasets [32], Faiss [27], and Ranx [5] for IR, based on the codebase of [31]. Our code is freely available at https://github.com/PaulLerner/ViQuAE to ensure the reproducibility of our results.

3.4 Results

We evaluate Entity Retrieval according to the relevance of the Wikipedia article associated with the target entity, which is determined automatically according to the presence of the answer after standard preprocessing (lowercasing, stripping articles, and punctuation). Additionally, because ISA contains a large portion of numerical answers, we follow the same soft matching method as [8] for ISA (years can be off by one and there is a 10% tolerance for measures and various numerical answers). We focus on the single-hop questions of EVQA, following [38]. The metrics used are Precision at 1 (P@1) and Mean Reciprocal Rank (MRR).[3]

We first explore in Table 2 three training strategies and three ways of using a multimodal dual encoder through experiments conducted on the validation set. These three strategies can be defined from Eq. 1:

- Mono-modal (image-image) retrieval/training, i.e., $\alpha_I = 1$, $\alpha_C = 0$;
- Cross-modal (image-text) retrieval/training, i.e., $\alpha_I = 0, \alpha_C = 1$;
- Hybrid retrieval or joint training, i.e., $\alpha_I > 0, \alpha_C > 0$.

For hybrid retrieval, the weights $\alpha_{\{I,C\}}$ are set through a grid search over the validation set to maximize the mean reciprocal rank while constraining their

[1] We kept the formulation of [41] but the temperature is usually expressed as $\frac{1}{\tau'}$ and not e^{τ}, which would be equivalent to $\tau' = \frac{1}{100}$ here.

[2] https://www.pytorchlightning.ai/.

[3] The results are consistent with precision and recall at higher cutoffs, which we omit for the sake of space.

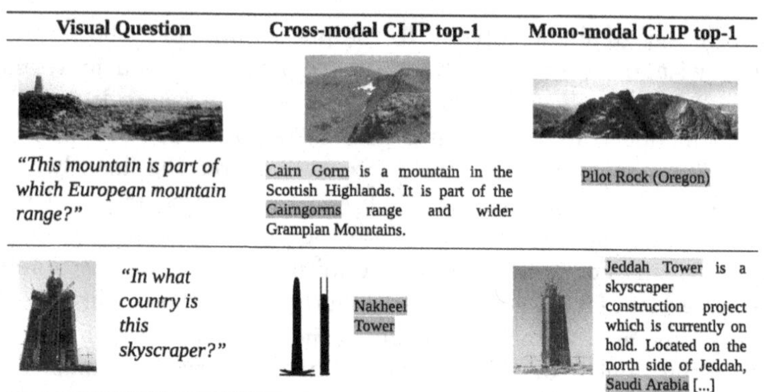

Visual Question
Cross-modal CLIP top-1
Mono-modal CLIP top-1
"This mountain is part of which European mountain range?"
Cairn Gorm is a mountain in the Scottish Highlands. It is part of the Cairngorms range and wider Grampian Mountains.
Pilot Rock (Oregon)
"In what country is this skyscraper?"
Nakheel Tower
Jeddah Tower is a skyscraper construction project which is currently on hold. Located on the north side of Jeddah, Saudi Arabia [...]

Fig. 2. Strengths and weaknesses of mono- and cross-modal retrieval exemplified through CLIP results (not fine-tuned) on ViQuAE's validation set.

sum to 1, so as to fairly compare joint training with mono- and cross-modal training. Note that the retrieval is independent from the training strategy, as shown in Table 2. Recall that CLIP's pre-training is only cross-modal [41], as most multimodal dual encoders [16].

Mono- or Cross-Modal Retrieval? Before comparing the different training methods, we can first notice that cross-modal IR outperforms[4] mono-modal IR on both ViQuAE and ISA,[5] especially without fine-tuning (first lines of each block in Table 2), which may seem curious since proper nouns are not *a priori* very meaningful. Therefore, it is surprising that CLIP generalizes[6] to new entity names. Nevertheless, some names carry meaning. For example, a name can indicate the gender of a person or suggest their nationality.[7] Moreover, we are working here with titles of Wikipedia articles, which are also likely to contain the nature of the entity (e.g., the profession of a person or the type of a monument). These features can thus be mapped to visual attributes.

Foremost, we mainly attribute the success of cross-modal IR to its adequacy with the pre-training of CLIP: the representation space of CLIP is organized to bring together similar texts and images, which the mono-modal proximity of images is only an indirect consequence of. We show examples of successes and failures in Fig. 2. In line with the results of [30], we observe that mono-modal retrieval may be more sensitive to superficial image details (color vs. black-and-

[4] Significantly according to Fisher's randomization test [15,44] with $p \leq 0.01$.

[5] An exception is EVQA, for which mono-modal retrieval outperforms cross-modal retrieval. This is surprising as both EVQA and ISA stem from the iNaturalist [47] and Google Landmarks [51] datasets. Further investigations are required.

[6] Unless its pre-training dataset contains enough entities from ViQuAE and ISA so that it circumvents generalization. We develop this discussion in Sect. 5.

[7] An interactive visualization is provided at https://paullerner.github.io/ViQuAE/#text-embedding-cross-modal.

Table 2. Entity Retrieval with a multimodal dual encoder, CLIP, on the validation subsets of ViQuAE, InfoSeek-Automatic (ISA), and EVQA (single-hop). Mono- and cross-modal retrieval model the IqIp and IqTp interactions, respectively. The best results are marked in bold for each type of retrieval. Hybrid retrieval of disjoint training combines *mono-modal trained* mono-modal retrieval and *cross-modal trained* cross-modal retrieval.

Retrieval	Training	ViQuAE		ISA		EVQA	
		MRR	P@1	MRR	P@1	MRR	P@1
Mono-modal	–	29.4	21.8	28.3	18.1	26.1	15.2
	Mono-modal	30.0	21.8	**31.4**	**20.5**	**32.6**	**21.7**
	Cross-modal	29.8	21.4	29.0	18.3	29.7	18.9
	Joint	**30.4**	**22.0**	30.5	19.9	30.7	19.6
Cross-modal	–	32.7	23.1	32.8	22.4	20.9	12.2
	Mono-modal	31.6	21.9	33.0	22.0	20.5	12.0
	Cross-modal	**37.1**	**26.9**	**34.7**	**23.8**	**23.2**	**13.8**
	Joint	30.8	21.3	31.1	20.3	22.4	12.9
Hybrid	–	39.6	30.6	36.2	25.8	28.7	18.7
	Mono-modal	40.1	31.8	38.2	27.4	33.8	22.9
	Cross-modal	**44.1**	**34.9**	38.5	27.8	33.8	23.3
	Joint	41.0	32.6	37.6	26.9	34.0	23.3
	Disjoint	43.7	34.5	**40.0**	**29.6**	**37.4**	**27.8**

white photography, subject pose...). Here, the two photographs at the top of two mountains, showing the horizon, are judged to be similar even though they are different mountains. In contrast, the mono-modal retrieval is more effective in the second example, where the two photographs of the Jeddah Tower are taken from similar vantage points. These qualitative results support our hypothesis that cross-modal retrieval might help addressing the heterogeneity of visual representations of named entities.

Why Choose? We show that mono- and cross-modal retrievals are complementary: their results can be simply combined at the score level (as in Eq. 1). Thus, without fine-tuning (first lines of each block in Table 2), fusing the two retrievals brings a relative improvement of 32% in P@1 for ViQuAE (and 15% for ISA, 23% for EVQA) compared to the best single retrieval (significant with Fisher's $p \leq 0.01$). It would be interesting to study whether these results generalize to other tasks. For example, this method could benefit Content-based Image Retrieval in a Web browsing context. Overall, hybrid retrieval gives the best performance, on all three datasets.

How to Fine-Tune Multimodal Dual Encoders? We see that fine-tuning with a given strategy (e.g. mono-modal) always enhances the performance of retrieval with the same strategy. However, it also sometimes decreases retrieval with another strategy (e.g. cross-modal). Therefore, we find it best to combine models trained disjointly: a *mono-modal trained* mono-modal retrieval and a *cross-modal trained* cross-modal retrieval.

4 Retrieving Passages and Extracting Answers

4.1 Methods

While we have focused on Entity Retrieval through cross-modal retrieval, we are ultimately interested in answering questions about these entities. To do so, we follow the same framework as [31], where Entity Retrieval results are mapped to the corresponding passages to enable fusion with a text passage retrieval method, such as DPR.

This implies redefining s as follows:

$$s(\mathbf{t_q}, \mathbf{i_q}, \mathbf{t_p}, \mathbf{i_p}) = \alpha_T s_T(\mathbf{t_q}, \mathbf{t_p}) + \alpha_I s_I(\mathbf{i_q}, \mathbf{i_p}) + \alpha_C s_C(\mathbf{i_q}, \mathbf{t_p}) \tag{4}$$

where $s_T(\mathbf{t_q}, \mathbf{t_p})$ models the TQTP interaction between the text of the question and of the passage and is implemented with DPR. We note this model DPR_{V+T} as it combines DPR, CLIP_V, and CLIP_T, or $\text{DPR}_{\mathbf{V+T}}$ (in bold font) when CLIP is fine-tuned.[8] The weights $\alpha_{\{T,I,C\}}$ are set through a grid search on the validation set like in the previous section (see Fig. 3 for an illustration of the impact of these hyperparameters on MRR). DPR is a dual encoder model that combines two BERT encoders, one for the question and one for the passage [28].

Answers are then extracted from these passages using Multi-passage BERT [49], which also models the TQTP interaction.

4.2 Data and Implementation

The 1.5 (resp. 2) million articles of the KB of ViQuAE [31] (resp. EVQA [38]) are divided into 12 (resp. 27) million 100-word passages, while preserving sentence boundaries, as in [31].

Both DPR and Multi-passage BERT are pre-trained on TriviaQA, filtered out of all questions used in [31] to generate ViQuAE,[9] before being fine-tuned on the downstream KVQAE dataset, following [31]. Both models are built upon the uncased version of BERT-base [13]. We refer the reader to [31] for further implementation details.

4.3 Baselines

We compare our approach to the DPR_{V+R+A} model of [31], which combines DPR, CLIP_V, ArcFace, and an ImageNet-trained ResNet model. The results of

[8] DPR is always fine-tuned as described in the next section.

[9] https://huggingface.co/datasets/PaulLerner/triviaqa_for_viquae.

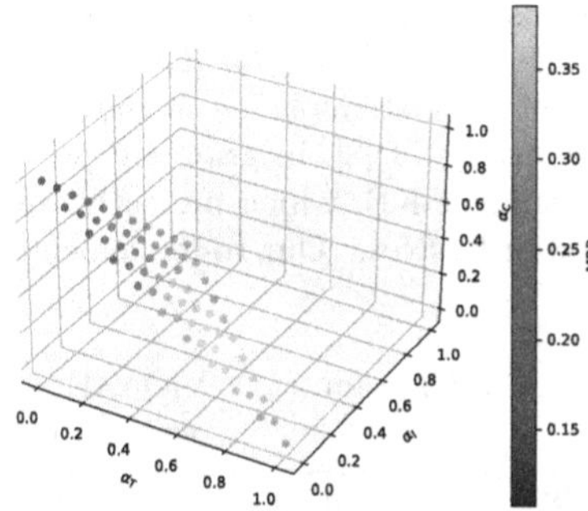

Fig. 3. Passage-level MRR on the validation set of ViQuAE depending on the $\alpha_{\{T,I,C\}}$ hyperparameters.

the four models are combined in the same way as in Eq. 4, where DPR implements $s_T(\mathbf{t_q}, \mathbf{t_p})$, CLIP$_V$, ArcFace, and ImageNet compose $s_I(\mathbf{i_q}, \mathbf{i_p})$, and there is no cross-modal similarity, i.e., $s_C(\mathbf{i_q}, \mathbf{t_p}) = 0$.

We also compare our methods to the ECA$_V$ and ILF$_V$ models of [30]. ECA (Early Cross-Attention) early-fuses modalities through an attention mechanism. The similarity is computed as $s(\mathbf{t_q}, \mathbf{i_q}, \mathbf{t_p}, \mathbf{i_p}) = \text{ECA}(\mathbf{t_q}, \mathbf{i_q}) \cdot \text{ECA}(\mathbf{t_p}, \mathbf{i_p})$ and thus combines all the multimodal interactions shown in Fig. 1. ILF (Intermediate Linear Fusion) fuses modalities with a simple linear projection and thus has, like our method, neither TQIQ nor TPIP interactions since the similarity can be reduced to:

$$s(\mathbf{t_q}, \mathbf{i_q}, \mathbf{t_p}, \mathbf{i_p}) = s_T(\mathbf{t_q}, \mathbf{t_p}) + s_{C'}(\mathbf{t_q}, \mathbf{i_p}) + s_I(\mathbf{i_q}, \mathbf{i_p}) + s_C(\mathbf{i_q}, \mathbf{t_p}) \quad (5)$$

Note that [30,31] use CLIP$_V$ with the ResNet architecture while we use ViT [14] in most of our experiments (but compare the two in the next section and find no significant difference).

Moving away from the Retrieval+Extraction framework of [31], we compare our results to [8,38], who both use the PaLM LLM [9], either as is or augmented with the image caption and in-context learning examples (denoted PromptCap [23]). [8] also experiment with FiD [25], augmented with CLIP retrieval results.

4.4 Results

Metrics. Extracted answers are evaluated using Exact Match (EM) and token-level F1 score on ViQuAE following [31], using the soft matching score defined by [8] on ISA (see Sect. 3.4), and using both F1 and BEM [7] on EVQA. The results for these three benchmarks are reported in Table 3.

Hybrid Retrieval Effectiveness. When comparing the DPR$_V$ and DPR$_{\mathbf{V+T}}$ models, we see that the effectiveness of combining mono- and cross-modal retrieval observed earlier indeed translates to more accurate answers, on all three

Table 3. Reading Comprehension results on the test set of ViQuAE, the validation set of ISA, and the test single-hop questions of EVQA. As in [28], the reader takes as input the top-24 of different IR systems listed in the "Method" column (except for the methods of [8,38]). The results of [8], in gray, are provided as reference but use a different, yet unavailable, smaller KB, which perfectly covers ISA. *CLIP is based on ViT's architecture instead of ResNet. †Our re-implementation of the reader, which fixes the loss function.

Method	# Param. (M)	ViQuAE		ISA	EVQA	
		EM	F1	Soft Match	BEM	F1
PaLM few-shot (text-only) [8]	540,000	31.5	–	4.8	–	–
CLIP + FiD [8]	1,170	–	–	20.9	–	–
PaLM zero-shot (text-only) [38]	540,000	–	–	–	19.7	–
PromptCap + PaLM [38]	540,870	–	–	–	29.7	–
DPR (text-only) [31]	330	16.9	20.1	–	–	–
DPR_V [30]	432	19.0	22.3	–	–	–
DPR_V* (baseline)	481	19.7	23.3	–	–	–
DPR_V*† (baseline)	481	26.4	29.1	7.7	27.4	25.4
DPR_{V+R+A} [31]	500	22.1	25.4	–	–	–
ECA_V [30]	432	20.6	24.4	–	–	–
ILF_V [30]	432	21.3	25.4	–	–	–
$\text{DPR}_{\mathbf{V+T}}$* (this work)	481	24.7	28.7	–	–	–
$\text{DPR}_{\mathbf{V+T}}$*† (this work)	481	**30.9**	**34.3**	**12.4**	**29.1**	**26.6**
Oracle retrieval + FiD [8]	Oracle + 770	–	–	52.5	–	–
Oracle retrieval† (this work)	Oracle + 110	68.3	72.7	46.8	65.3	59.7

datasets. Therefore, our model also outperforms the previously proposed models of [30,31] on ViQuAE, while being conceptually simpler and computationally cheaper (emitting hundred times less CO2 than [30]). Furthermore, we found a bug in the implementation of the reader's loss provided by [31]. Fixing it consistently improved results, for both DPR_V and $\text{DPR}_{\mathbf{V+T}}$. Our model is also competitive with the method of [38] on EVQA, while using 1,000 times less parameters.[10]

Knowledge Base Incompleteness. The results of [8] are provided as reference but are hardly comparable to the others. Apart from PaLM being three order of magnitude greater than the other models and partly trained on ViQuAE's test set,[11] they use a different KB. This KB, yet unavailable, is fifteen times

[10] We focus on the single-hop subset of EVQA following [38]. On the two-hop questions, the model using $\text{DPR}_{\mathbf{V+T}}$ achieves 31.1 BEM and 25.6 F1, and 9.8 BEM/3.8 F1 on the multi-answer questions.

[11] According to [9], around 20% of TriviaQA is contained in PaLM's pre-training dataset. ViQuAE was derived from TriviaQA [31].

smaller than ours, so contains less distractors, and covers 100% of the entities and questions of ISA. In contrast, our KB lacks 11.5% of ISA entities and is not guaranteed to contain the answers for the 88.5% remaining, because of differences between the Wikipedia versions.

Oracle Retrieval. We conduct additional experiments in an "oracle retrieval" setting, where the reader only takes relevant passages as input, similarly to the oracle experiments of [8,31]. In agreement with their results, we find a large gap between our best retrieval model and the oracle, showing that IR is still the main bottleneck of KVQAE. Compared to the FiD model of [8], with 770M parameters, we approach its performance, although Multi-passage BERT is seven times smaller and our KB does not fully cover ISA.

5 Conclusion

This paper studies cross-modal retrieval and its combination with mono-modal retrieval for Knowledge-based Visual Question Answering about named Entities (KVQAE). Retrieval is carried out with a multimodal dual encoder, namely CLIP. Our results demonstrate the superiority of cross-modal retrieval over mono-modal retrieval, but also the complementarity of the two, which can be easily combined.

We argue that cross-modal retrieval may help addressing the heterogeneity of visual representations of named entities, consistently with prior work. It would be interesting to study whether these results generalize to other tasks. For example, this method could benefit Content-based Image Retrieval, in a Web browsing context.

Although it was the abundance of cross-modal data that enabled CLIP's training in the first place, which would have been difficult with a mono-modal annotation, this limits our results because it is difficult to control such a large amount of data and thus to estimate CLIP's generalization capabilities. We hypothesize that mono-modal retrieval is better suited to generalize to new entities.

We show that the effectiveness of cross-modal retrieval leads to more accurate answers, on all three studied datasets. Therefore, our method outperforms our baseline (mono-modal retrieval) but also the methods of [30,31], while being conceptually simpler and computationally cheaper. Furthermore, it is competitive with billion-scale parameters models on ISA and EVQA. As such, this is the first comparative study of the recently introduced ViQuAE, ISA, and EVQA datasets. We find that ISA is more challenging as it is less biased towards text, but advocate for further studies on all three datasets — which all have their pros and cons — with diverse methods.

Consistently with [8,31], we find a large gap between our best retrieval model and oracle retrieval, showing that entity retrieval is the main bottleneck of KVQAE. For future work, we plan to combine our unstructured KB with a structured one, such as Wikidata, to enable the modeling of links between the entities

[2,40,52,54], which would further address the heterogeneity of their visual representations. A more IR perspective on the matter could cast KVQAE as a query expansion problem, with an initial ambiguous textual query which would benefit from pseudo-relevant feedback [55].

References

1. Adjali, O., Grimal, P., Ferret, O., Ghannay, S., Le Borgne, H.: Explicit knowledge integration for knowledge-aware visual question answering about named entities. In: Proceedings of the 2023 ACM International Conference on Multimedia Retrieval, pp. 29–38. ICMR '23, Association for Computing Machinery, New York, NY, USA (2023). https://doi.org/10.1145/3591106.3592227
2. Alberts, H., et al.: VisualSem: a high-quality knowledge graph for vision and language. In: Proceedings of the 1st Workshop on Multilingual Representation Learning, pp. 138–152. Association for Computational Linguistics, Punta Cana, Dominican Republic (Nov 2021). https://doi.org/10.18653/v1/2021.mrl-1.13, https://aclanthology.org/2021.mrl-1.13
3. Antol, S., Agrawal, A., Lu, J., Mitchell, M., Batra, D., Zitnick, C.L., Parikh, D.: VQA: Visual Question Answering. In: 2015 IEEE International Conference on Computer Vision (ICCV), pp. 2425–2433. IEEE, Santiago, Chile (Dec 2015). https://doi.org/10.1109/ICCV.2015.279, http://ieeexplore.ieee.org/document/7410636/
4. Baltrušaitis, T., Ahuja, C., Morency, L.P.: Multimodal machine learning: a survey and taxonomy. IEEE Trans. Pattern Anal. Mach. Intell. 41(2), 423–443 (2019). https://doi.org/10.1109/TPAMI.2018.2798607, conference Name: IEEE Transactions on Pattern Analysis and Machine Intelligence
5. Bassani, E.: ranx: a blazing-fast python library for ranking evaluation and comparison. In: Hagen, M., Verberne, S., Macdonald, C., Seifert, C., Balog, K., Nørvåg, K., Setty, V. (eds.) Advances in Information Retrieval, pp. 259–264. Lecture Notes in Computer Science, Springer International Publishing, Cham (2022). https://doi.org/10.1007/978-3-030-99739-7_30
6. Bokhari, M.U., Hasan, F.: Multimodal information retrieval: challenges and future trends. Int. J. Comput. Appl. **74**(14) (2013), publisher: Foundation of Computer Science
7. Bulian, J., Buck, C., Gajewski, W., Börschinger, B., Schuster, T.: Tomayto, tomahto. beyond token-level answer equivalence for question answering evaluation. In: Proceedings of the 2022 Conference on Empirical Methods in Natural Language Processing, pp. 291–305. Association for Computational Linguistics, Abu Dhabi, United Arab Emirates (Dec 2022). https://doi.org/10.18653/v1/2022.emnlp-main.20, https://aclanthology.org/2022.emnlp-main.20
8. Chen, Y., et al.: Can pre-trained vision and language models answer visual information-seeking questions? (Feb 2023). https://doi.org/10.48550/arXiv.2302.11713, http://arxiv.org/abs/2302.11713, arXiv:2302.11713 [cs]
9. Chowdhery, A., et al.: Palm: scaling language modeling with pathways. J. Mach. Learn. Res. **24**(240), 1–113 (2023)
10. Couairon, G., Douze, M., Cord, M., Schwenk, H.: Embedding arithmetic of multimodal queries for image retrieval. In: Proceedings of the IEEE/CVF Conference on Computer Vision and Pattern Recognition (CVPR) Workshops, pp. 4950–4958 (June 2022)

11. Deng, J., Dong, W., Socher, R., Li, L.J., Li, K., Fei-Fei, L.: ImageNet: a large-scale hierarchical image database. In: 2009 IEEE Conference on Computer Vision and Pattern Recognition, pp. 248–255 (Jun 2009). https://doi.org/10.1109/CVPR.2009.5206848, iSSN: 1063-6919
12. Deng, J., Guo, J., Xue, N., Zafeiriou, S.: Arcface: additive angular margin loss for deep face recognition. In: Proceedings of the IEEE/CVF Conference on Computer Vision and Pattern Recognition (CVPR) (June 2019). https://openaccess.thecvf.com/content_CVPR_2019/html/Deng_ArcFace_Additive_Angular_Margin_Loss_for_Deep_Face_Recognition_CVPR_2019_paper.html
13. Devlin, J., Chang, M.W., Lee, K., Toutanova, K.: BERT: Pre-training of deep bidirectional transformers for language understanding. In: Proceedings of the 2019 Conference of the North American Chapter of the Association for Computational Linguistics: Human Language Technologies, Volume 1 (Long and Short Papers), pp. 4171–4186. Association for Computational Linguistics, Minneapolis, Minnesota (Jun 2019). https://doi.org/10.18653/v1/N19-1423, https://aclanthology.org/N19-1423
14. Dosovitskiy, A., et al.: An image is worth 16x16 words: transformers for image recognition at scale. In: International Conference on Learning Representations (2021). https://openreview.net/forum?id=YicbFdNTTy
15. Fisher, R.A.: The design of experiments. The design of experiments. (2nd Ed) (1937). https://www.cabdirect.org/cabdirect/abstract/19371601600, publisher: Oliver & Boyd, Edinburgh & London
16. Gan, Z., Li, L., Li, C., Wang, L., Liu, Z., Gao, J.: Vision-language pre-training: basics, recent advances, and future trends. Found. Trends. Comput. Graph. Vis. **14**(3–4), 163–352 (dec 2022). https://doi.org/10.1561/0600000105
17. Garcia-Olano, D., Onoe, Y., Ghosh, J.: Improving and diagnosing knowledge-based visual question answering via entity enhanced knowledge injection. In: Companion Proceedings of the Web Conference 2022, pp. 705–715. WWW '22, Association for Computing Machinery, New York, NY, USA (2022). https://doi.org/10.1145/3487553.3524648
18. Gardères, F., Ziaeefard, M.: ConceptBert: Concept-Aware Representation for Visual Question Answering. Findings of the Association for Computational Linguistics: EMNLP 2020, pp. 10 (2020). https://aclanthology.org/2020.findings-emnlp.44/
19. Gui, L., Wang, B., Huang, Q., Hauptmann, A., Bisk, Y., Gao, J.: KAT: a knowledge augmented transformer for vision-and-language. In: Proceedings of the 2022 Conference of the North American Chapter of the Association for Computational Linguistics: Human Language Technologies, pp. 956–968. Association for Computational Linguistics, Seattle, United States (Jul 2022), https://aclanthology.org/2022.naacl-main.70
20. Guo, W., Wang, J., Wang, S.: Deep multimodal representation learning: a survey. IEEE Access **7**, 63373–63394 (2019). https://doi.org/10.1109/ACCESS.2019.2916887, conference Name: IEEE Access
21. He, K., Zhang, X., Ren, S., Sun, J.: Deep residual learning for image recognition. In: Proceedings of the IEEE Conference On Computer Vision and Pattern Recognition, pp. 770–778 (2016), https://openaccess.thecvf.com/content_cvpr_2016/papers/He_Deep_Residual_Learning_CVPR_2016_paper.pdf
22. Heo, Y.J., Kim, E.S., Choi, W.S., Zhang, B.T.: Hypergraph Transformer: Weakly-supervised multi-hop reasoning for knowledge-based visual question answering. In: Proceedings of the 60th Annual Meeting of the Association for Computational Lin-

guistics (Volume 1: Long Papers), pp. 373–390. Association for Computational Linguistics, Dublin, Ireland (May 2022). https://doi.org/10.18653/v1/2022.acl-long.29, https://aclanthology.org/2022.acl-long.29
23. Hu, Y., Hua, H., Yang, Z., Shi, W., Smith, N.A., Luo, J.: Promptcap: prompt-guided task-aware image captioning (2023)
24. Hu, Z., et al.: AVIS: Autonomous Visual Information Seeking with Large Language Models (Jun 2023). http://arxiv.org/abs/2306.08129, arXiv:2306.08129 [cs]
25. Izacard, G., Grave, E.: leveraging passage retrieval with generative models for open domain question answering. In: Proceedings of the 16th Conference of the European Chapter of the Association for Computational Linguistics: Main Volume, pp. 874–880. Association for Computational Linguistics, Online (Apr 2021). https://doi.org/10.18653/v1/2021.eacl-main.74, https://aclanthology.org/2021.eacl-main.74
26. Ji, Z., et al.: Survey of hallucination in natural language generation. ACM Comput. Surv. **55**(12), 248:1–248:38 (Mar 2023). https://doi.org/10.1145/3571730, https://dl.acm.org/doi/10.1145/3571730
27. Johnson, J., Douze, M., Jégou, H.: Billion-scale similarity search with GPUs. IEEE Trans. Big Data **7**(3), 535–547 (2019). https://doi.org/10.1109/TBDATA.2019.2921572
28. Karpukhin, V., et al.: Dense passage retrieval for open-domain question answering. In: Proceedings of the 2020 Conference on Empirical Methods in Natural Language Processing (EMNLP). pp. 6769–6781. Association for Computational Linguistics, Online (Nov 2020), https://www.aclweb.org/anthology/2020.emnlp-main.550
29. Khan, S., Naseer, M., Hayat, M., Zamir, S.W., Khan, F.S., Shah, M.: Transformers in vision: a survey. ACM Comput. Surv. **54**(10s) (sep 2022). https://doi.org/10.1145/3505244
30. Lerner, P., Ferret, O., Guinaudeau, C.: Multimodal inverse cloze task for knowledge-based visual question answering. In: Advances in Information Retrieval (ECIR 2023), pp. 569–587. Springer Nature Switzerland, Cham (2023). https://doi.org/10.1007/978-3-031-28244-7_36
31. Lerner, P., et al.: ViQuAE, a dataset for knowledge-based visual question answering about named entities. In: Proceedings of The 45th International ACM SIGIR Conference on Research and Development in Information Retrieval. SIGIR '22, Association for Computing Machinery, New York, NY, USA (2022). https://doi.org/10.1145/3477495.3531753, https://hal.archives-ouvertes.fr/hal-03650618
32. Lhoest, Q.,et al.: Datasets: a community library for natural language processing. In: Proceedings of the 2021 Conference on Empirical Methods in Natural Language Processing: System Demonstrations. pp. 175–184. Association for Computational Linguistics, Online and Punta Cana, Dominican Republic (Nov 2021), https://aclanthology.org/2021.emnlp-demo.21
33. Li, L., et al.: M^{3}IT: A Large-Scale Dataset towards Multi-Modal Multilingual Instruction Tuning (Jun 2023). https://doi.org/10.48550/arXiv.2306.04387, http://arxiv.org/abs/2306.04387, arXiv:2306.04387 [cs]
34. Lin, C.Y.: Rouge: a package for automatic evaluation of summaries. In: Text Summarization Branches Out, pp. 74–81 (2004)
35. Liu, Z., Xiong, C., Lv, Y., Liu, Z., Yu, G.: Universal vision-language dense retrieval: learning a unified representation space for multi-modal retrieval. In: The Eleventh International Conference on Learning Representations (2023). https://openreview.net/forum?id=PQOlkgsBsik
36. Loshchilov, I., Hutter, F.: Decoupled weight decay regularization. In: International Conference on Learning Representations (2019), https://openreview.net/forum?id=Bkg6RiCqY7

37. Marino, K., Rastegari, M., Farhadi, A., Mottaghi, R.: OK-VQA: a visual question answering benchmark requiring external knowledge. In: Proceedings of the IEEE Conference on Computer Vision and Pattern Recognition, pp. 3195–3204 (2019), https://ieeexplore.ieee.org/document/8953725/
38. Mensink, T., et al.: Encyclopedic vqa: Visual questions about detailed properties of fine-grained categories. In: Proceedings of the IEEE/CVF International Conference on Computer Vision (ICCV), pp. 3113–3124 (October 2023)
39. Paszke, A., et al.: PyTorch: an imperative style, high-performance deep learning library. In: Advances in Neural Information Processing Systems 32 (2019). https://papers.nips.cc/paper/2019/hash/bdbca288fee7f92f2bfa9f7012727740-Abstract.html
40. Pezeshkpour, P., Chen, L., Singh, S.: Embedding multimodal relational data for knowledge base completion. In: Proceedings of the 2018 Conference on Empirical Methods in Natural Language Processing, pp. 3208–3218 (2018)
41. Radford, A., , et al.: Learning transferable visual models from natural language supervision. In: International Conference on Machine Learning, pp. 8748–8763. PMLR (2021)
42. Schwenk, D., Khandelwal, A., Clark, C., Marino, K., Mottaghi, R.: A-OKVQA: a benchmark for visual question answering using world knowledge. In: Avidan, S., Brostow, G., Cissé, M., Farinella, G.M., Hassner, T. (eds.) Computer Vision – ECCV 2022: 17th European Conference, Tel Aviv, Israel, October 23–27, 2022, Proceedings, Part VIII, pp. 146–162. Springer Nature Switzerland, Cham (2022). https://doi.org/10.1007/978-3-031-20074-8_9
43. Shah, S., Mishra, A., Yadati, N., Talukdar, P.P.: KVQA: knowledge-aware visual question answering. In: Proceedings of the AAAI Conference on Artificial Intelligence. 33, pp. 8876–8884, 2019. https://144.208.67.177/ojs/index.php/AAAI/article/view/4915
44. Smucker, M.D., Allan, J., Carterette, B.: A comparison of statistical significance tests for information retrieval evaluation. In: Proceedings of the sixteenth ACM conference on Conference on information and knowledge management, pp. 623–632. CIKM '07, Association for Computing Machinery, New York, NY, USA (Nov 2007). https://doi.org/10.1145/1321440.1321528
45. Srinivasan, K., Raman, K., Chen, J., Bendersky, M., Najork, M.: Wit: Wikipedia-based image text dataset for multimodal multilingual machine learning. In: Proceedings of the 44th International ACM SIGIR Conference on Research and Development in Information Retrieval, pp. 2443–2449. SIGIR '21, Association for Computing Machinery, New York, NY, USA (2021). https://doi.org/10.1145/3404835.3463257
46. Sun, W., Fan, Y., Guo, J., Zhang, R., Cheng, X.: Visual named entity linking: a new dataset and a baseline. In: Findings of the Association for Computational Linguistics: EMNLP 2022. pp. 2403–2415. Association for Computational Linguistics, Abu Dhabi, United Arab Emirates (Dec 2022). https://doi.org/10.18653/v1/2022.findings-emnlp.178, https://aclanthology.org/2022.findings-emnlp.178
47. Van Horn, G., et al.: The iNaturalist species classification and detection dataset. In: Proceedings of the IEEE Conference on Computer Vision and Pattern Recognition (CVPR), pp. 8769–8778 (2018). https://openaccess.thecvf.com/content_cvpr_2018/html/Van_Horn_The_INaturalist_Species_CVPR_2018_paper.html
48. Vickers, P., Aletras, N., Monti, E., Barrault, L.. In: Factuality: efficient integration of relevant facts for visual question answering. In: Proceedings of the 59th Annual Meeting of the Association for Computational Linguistics and the

11th International Joint Conference on Natural Language Processing (Volume 2: Short Papers), pp. 468–475. Association for Computational Linguistics, Online (Aug 2021). https://doi.org/10.18653/v1/2021.acl-short.60, https://aclanthology.org/2021.acl-short.60

49. Wang, Z., Ng, P., Ma, X., Nallapati, R., Xiang, B.: Multi-passage BERT: a globally normalized bert model for open-domain question answering. In: Proceedings of the 2019 Conference on Empirical Methods in Natural Language Processing and the 9th International Joint Conference on Natural Language Processing (EMNLP-IJCNLP), pp. 5878–5882. Association for Computational Linguistics, Hong Kong, China (Nov 2019). https://doi.org/10.18653/v1/D19-1599, https://www.aclweb.org/anthology/D19-1599
50. Weston, J., Chopra, S., Bordes, A.: Memory networks (2014). https://doi.org/10.48550/ARXIV.1410.3916, https://arxiv.org/abs/1410.3916
51. Weyand, T., Araujo, A., Cao, B., Sim, J.: Google landmarks dataset v2 - A large-scale benchmark for instance-level recognition and retrieval. In: Proceedings of the IEEE/CVF Conference on Computer Vision and Pattern Recognition (CVPR), pp. 2575–2584 (2020), https://openaccess.thecvf.com/content_CVPR_2020/html/Weyand_Google_Landmarks_Dataset_v2_-_A_Large-Scale_Benchmark_for_Instance-Level_CVPR_2020_paper.html
52. Wilcke, W.X., Bloem, P., de Boer, V., Veer, R.H.v.t., van Harmelen, F.A.H.: End-to-End Entity Classification on Multimodal Knowledge Graphs. arXiv:2003.12383 [cs] (Mar 2020). http://arxiv.org/abs/2003.12383, arXiv: 2003.12383
53. Wolf, T., et al.: HuggingFace's Transformers: State-of-the-art Natural Language Processing. arXiv:1910.03771 [cs] (Jul 2020), http://arxiv.org/abs/1910.03771
54. Xie, R., Liu, Z., Luan, H., Sun, M.: Image-embodied knowledge representation learning. In: Proceedings of the 26th International Joint Conference on Artificial Intelligence, pp. 3140–3146. IJCAI'17, AAAI Press, Melbourne, Australia (Aug 2017)
55. Xu, J., Croft, W.B.: Query expansion using local and global document analysis. In: Proceedings of the 19th Annual International ACM SIGIR Conference on Research and Development in Information Retrieval, pp. 4–11. SIGIR '96, Association for Computing Machinery, New York, NY, USA (1996). https://doi.org/10.1145/243199.243202
56. Zamani, H., Diaz, F., Dehghani, M., Metzler, D., Bendersky, M.: Retrieval-enhanced machine learning. In: Proceedings of the 45th International ACM SIGIR Conference on Research and Development in Information Retrieval, pp. 2875–2886. SIGIR '22, Association for Computing Machinery, New York, NY, USA (Jul 2022). https://doi.org/10.1145/3477495.3531722
57. Zhang, D., Cao, R., Wu, S.: Information fusion in visual question answering: a survey. Information Fusion **52**, 268–280 (2019). https://www.sciencedirect.com/science/article/pii/S1566253518308893

WebSAM-Adapter: Adapting Segment Anything Model for Web Page Segmentation

Bowen Ren[1], Zefeng Qian[1], Yuchen Sun[1], Chao Gao[3], and Chongyang Zhang[1,2](✉)

[1] School of Electronic Information and Electrical Engineering, Shanghai Jiao Tong University, Shanghai 200240, China
{renbowen,zefeng_qian,sunyc22,sunny_zhang}@sjtu.edu.cn
[2] MoE Key Lab of Artificial Intelligence, AI Institute, Shanghai Jiao Tong University, Shanghai 200240, China
[3] China Pacific Insurance (Group) Co., Ltd., Shanghai 200010, China

Abstract. With the advancement of internet technology, web page segmentation, which aims to divide web pages into semantically coherent units, has become increasingly crucial for web-related applications. Conventional purely visual web page segmentation approaches, which depend on traditional edge detection, face challenges in generalizing across complex web pages. Recently, the Segment Anything Model (SAM) represents remarkable visual understanding and segmentation abilities. This inspires us that SAM can also demonstrate great potential in Web Page Segmentation. However, due to the lack of web-specific training data, its direct adaptation to web page segmentation domain has been hindered. To address this challenge, we propose WebSAM-Adapter, an effective adaptation of SAM, featuring a three-module architecture specifically tailored for web page segmentation with minimal additional trainable parameters. First, we propose a patch embedding tune module for adjusting the frozen patch embedding features, which is crucial for modifying the distribution of the original model. Second, an edge components tune module is designed to learn significant structural features within each web page. Finally, the outputs of these specialized modules are sent into our key Adapter module, which employs a lightweight multi-layer perceptron (MLP) to amalgamate these enriched features and generate webpage-specific knowledge. To the best of our knowledge, our method is the first successful adaptation of a large visual model like SAM to web page segmentation. Empirical evaluations on the comprehensive Webis-WebSeg-20 dataset demonstrate our model's state-of-the-art performance.

Keywords: Web Page Segmentation · Segment Anything Model · Adapter

1 Introduction

Web page segmentation is the process of identifying semantically coherent units, essentially serving as the inverse operation to web page layout [1]. This critical

N. Goharian et al. (Eds.): ECIR 2024, LNCS 14608, pp. 439–454, 2024.
https://doi.org/10.1007/978-3-031-56027-9_27

analytical procedure is foundational in web-related applications and is widely applied in information retrieval [2–4], software engineering [5–7] and assistive technology for visually impaired users [8,9].

Within the domain of web page segmentation, previous methods are predominantly categorized into three distinct approaches: DOM-based [10–13], Integrated Visual-DOM [14–18], and purely visual-based methods [19,20]. The DOM-based methods [10–13] exclusively utilize the Document Object Model (DOM) tree of a web page's source code for segmenting and categorizing web elements. While these methods offer structural precision, the tree structure and the relationships between nodes don't always align with the user's visual perception of the rendered page.

Differently, Integrated Visual-DOM methods [14–18] aim for richer segmentation by integrating DOM attributes with visual cues of web pages. A prime example is the Visual Page Segmentation (VIPS) [14] algorithm. However, the dependency on both visual and structural elements can limit their applicability, especially when access to source code is restricted. Conversely, purely visual based methods, such as those by Cao et al. [19] and Cormier et al. [20], focusing only on the rendered web pages and eliminating the need for source code. Although these approaches represent advancements through traditional computer vision edge detection, they face difficulties in generalizing across complex web pages, as illustrated in Fig. 1. This underscores the compelling need for further exploration of purely visual segmentation methods that leverage only the rendered web page, ensuring enhanced robustness and comprehensive applicability.

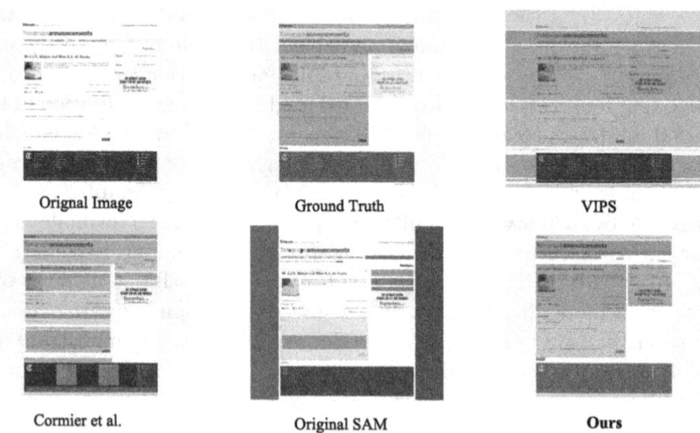

Fig. 1. A comparison of several segmentation methods. In dealing with a complex web page, VIPS [14], Cormier et al. [20], and Original SAM [21] face difficulties in performing accurate and effective segmentation. In contrast, our WebSAM-Adapter is capable of dividing a web page into more precise regions.

To address the aforementioned challenges, particularly the intricate structures and diverse styles unique to web pages compared to natural images, we think that models endowed with enhanced visual understanding and segmentation capabilities are better equipped to tackle these issues. The recent Segment Anything Model (SAM) [21] has demonstrated remarkable visual understanding and segmentation abilities, achieving outstanding results when adapted to various vision segmentation tasks [22–26]. However, despite SAM's claim to "Segment Anything" and its impressive achievements in natural images, our empirical evaluations indicate that its direct application to web page segmentation is ineffective, as illustrated in Fig. 1. The primary reason for SAM's ineffectiveness in segmenting web pages is attributed to the lack of training data specific to web usage. To this end, this paper delves further into SAM to mitigate its limitations in web page segmentation and accordingly propose WebSAM-Adapter.

Building on this, our key insight is to leverage individual image features for webpage-specific knowledge, utilizing the comprehensive semantic understanding inherent in the foundational SAM model. The WebSAM-Adapter is crafted with a three-module architecture, specifically tailored for web page segmentation. The functions of our model can be concisely summarized as:

1. **Aligning Data Distribution through Patch Embedding Tune**: This module refines the alignment between pre-trained and target datasets by tuning the patch embedding features, thereby enhancing model generalization.
2. **Learning Structural Features through Edge Components Tune**: This module identifies and leverages significant structural features in web pages, enriching the model's representation and capturing layout information traditionally sourced from the DOM tree.
3. **Fusing Enriched Features through Adapter**: This module amalgamates the enriched outputs from the Patch Embedding Tune and Edge Components Tune modules using a lightweight MLP. It serves as an adaptable auxiliary network, infusing webpage-specific knowledge from web page samples into the foundational SAM with minimal additional trainable parameters.

To validate our approach, we conduct experiments on the Webis-WebSeg-20 dataset [1], the most comprehensive dataset currently available for this domain. Our results demonstrate that the WebSAM-Adapter not only mitigates SAM's limitations but also achieves state-of-the-art performance.

In summary, our main contributions are as follows:

1. We propose WebSAM-Adapter, an effective web page segmentation method that leverages a purely visual approach to overcome the limitations of existing techniques relying on the DOM tree or traditional computer vision edge detection methods.
2. To the best of our knowledge, we are the first to propose an adaptation strategy for web page segmentation that acquires webpage-specific knowledge through image embeddings and edge components, with minimal additional trainable parameters.
3. Our approach achieves state-of-the-art performance on the most comprehensive dataset currently available for web page segmentation tasks.

2 Related Work

2.1 Web Page Segmentation

Web page segmentation techniques can be broadly categorized into three approaches: DOM-based, Integrated Visual-DOM, and purely visual-based methods. DOM-based methods exploit the Document Object Model (DOM) to segment and categorize web page elements [10–13]. These methods necessitate the source code of the web page and specifically employ DOM attributes, tags, and subtree structures for segmentation and analysis. While offering a structured paradigm, these methods focus predominantly on the DOM's tree structure, limiting their broader applicability by failing to account for the visual content that users actually perceive on-screen.

Another category, Integrated Visual-DOM segmentation techniques [14–18], incorporates visual attributes for a more contextually rich segmentation. Notably, these methods adopt a hybrid approach that combines both DOM attributes and visual cues. For instance, Cai et al. [14] introduce the Visual Page Segmentation (VIPS) algorithm, which merges DOM structure with visual cues to form a hierarchical content model. Similarly, Bajammal et al. [16] present Cortex, which also integrates DOM attributes with visual elements to achieve superior segmentation precision.

Conversely, a few algorithms [19,20] exclusively depend on the rendered web page, eliminating the need for source code and providing a purely visual strategy for web page segmentation. Cao et al. [19] employ edge detection algorithms such as Canny in the preprocessing stage, subsequently shrinking and splitting the web page image into visually congruent sub-images. In contrast, Cormier et al. [20] utilize a Bayesian framework with Sobel operators for edge detection to achieve the most probable segmentation. However, while these approaches showcase advancements in web page segmentation through traditional computer vision edge detection methods, they face limitations in generalizing across complex web pages. Therefore, the creation of a universally applicable, visually-driven web page segmentation algorithm is essential. The emergence of large models like SAM offers a promising direction for overcoming this challenge, harnessing the power of deep learning for improved adaptability and performance.

2.2 Segment Anything Model

Recently, Meta AI Research introduced the Segment Anything Model (SAM) [21], which has received considerable academic attention. Designed as a foundational model for image segmentation, SAM frequently rivals or even outperforms fully supervised methods in zero-shot scenarios. Its robust generalization capabilities are highlighted by its training on a vast dataset SA-1B [21], encompassing over 11 million natural images and their corresponding 1 billion masks.

The model has shown an impressive range of applicability, extending beyond generic image segmentation to specialized tasks. These include medical image segmentation [22–24], remote sensing instance segmentation [25], and camouflage image segmentation [26,27].

2.3 Adapter-Based Fine-Tuning

The Adapter methodology is inspired by Parameter-Efficient Fine-Tuning (PEFT) techniques prevalent in Natural Language Processing [28]. Unlike traditional full fine-tuning methods, the Adapter approach [29–32] incorporates parameter-efficient modules into the original model while keeping the majority of the parameters frozen. This facilitates efficient learning and faster updates, thereby addressing computational and memory constraints in large-scale model fine-tuning. Liu et al. [29] introduce the Explicit Visual Prompting (EVP) methodology for low-level structure segmentation, which integrates explicit visual cues into the adapter module to get task-specific information and outperforms existing tuning protocols.

Recent studies have extended Adapter methodologies to SAM, targeting specialized segmentation challenges. Wu et al. [23] introduced the Medical SAM Adapter (MSA), which augments SAM's capabilities in medical image segmentation by incorporating domain-specific knowledge through Adapter modules. Similarly, Chen et al. [26] developed the SAM-Adapter, enhancing SAM's performance in specialized tasks like shadow and camouflaged object detection through domain-specific visual prompts. In this paper, we introduce WebSAM-Adapter, an effective approach that capitalizes on the strengths of existing Adapter methodologies to address the unique challenges of web page segmentation.

3 Method

3.1 The Preliminary of Segment Anything Model

The objective of the WebSAM-Adapter is to capitalize on the knowledge acquired from the SAM architecture. To begin with, we provide an overview of the SAM architecture before discussing our modifications. SAM is composed of three principal components: a large-scale image encoder, a prompt encoder, and a lightweight mask decoder. This framework allows for different masks to be generated for the same image based on different prompts. The image encoder employs a pre-trained Visual Transformer (ViT) [33] to process high-resolution inputs, subsequently outputting feature maps at a scale reduced to 1/16 of the original image. The prompt encoder supports a variety of input types, including both sparse (points, boxes, text) and dense (masks). The mask decoder is a specialized Transformer decoder block, equipped with a dynamic mask prediction head.

3.2 The Architecture of WebSAM-Adapter

In this section, we introduce the WebSAM-Adapter, a specialized framework designed to address the challenge of web page segmentation. Inspired by preceding Adapter-based methodologies and tailored for the unique demands of web page segmentation, the WebSAM-Adapter represents a strategic confluence of existing methods and specific web page segmentation requirements. Built upon the foundational elements of the Segment Anything Model (SAM), the architecture incorporates three modules that are purposefully designed for web page segmentation tasks. Our key insight is to leverage individual image features from both patch embeddings and edge components for webpage-specific knowledge, a process facilitated by an Adapter module. The model is lightweight, adaptable to limited data, and serves as an auxiliary network, infusing webpage-specific knowledge from web page segmentation samples into the foundational SAM with minimal additional trainable parameters.

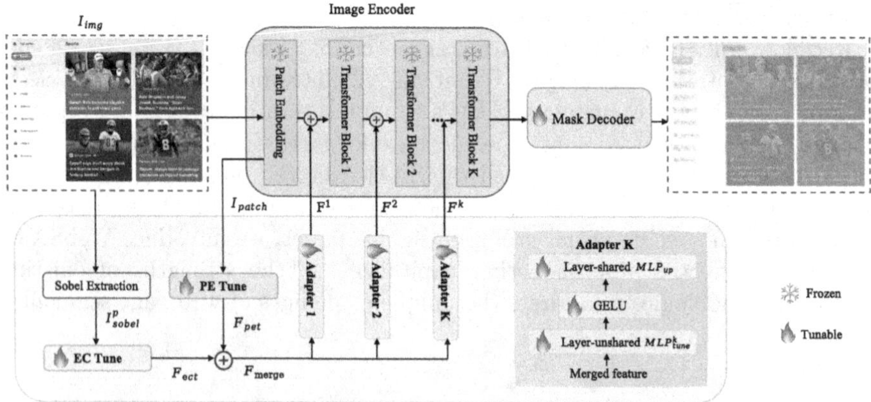

Fig. 2. The architecture of the proposed WebSAM-adapter.The Patch Embedding Tune (PE Tune) and the Edge Component Tuning (EC Tune) are utilized to refine the extracted features. The Adapter is meticulously designed to amalgamate these features, facilitating the acquisition of webpage-specific knowledge.

Framework Overview. As illustrated in Fig. 2, our image encoder, with its weights frozen from the SAM pre-trained model, incorporates three specialized modules designed for web page segmentation: Patch Embedding Tuning (PC Tune), Edge Component Tuning (EC Tune), and an Adapter module. The Patch Embedding Tune module adjusts the frozen patch embedding features, crucial for modifying the distribution of the original model. Next, the Edge Components Tune module is designed to learn significant structural features within each web page. These specialized modules then feed their outputs into our key Adapter module, employing a lightweight multi-layer perceptron (MLP) to amalgamate these enriched features and generate webpage-specific knowledge. Furthermore, the mask decoder of SAM is employed and initialized with weights from a pre-trained SAM model, which we fine-tune during the training phase. Considering that real-world applications of web page segmentation do not necessitate interactive prompts, we have intentionally excluded the use of a prompt encoder and refrained from providing prompts to SAM's original mask decoder.

Patch Embedding Tune. In the pre-trained SAM image encoder [21], the patch embedding layer partitions each 1024×1024 input image I_{img} into a 64×64 grid of non-overlapping patches. These patches are then transformed into embeddings I_{patch} with dimensions $[B, P_{\mathrm{H}}, P_{\mathrm{W}}, E]$, where B represents the batch size, P_H and P_W are the counts of patches along the height and width, respectively, and E is the embedding dimension of 768 for *vit-b*. While this approach is effective for natural images, it falls short when applied to web page images, which often exhibit unique structural and semantic complexities.

To address this limitation, we introduce the Patch Embedding Tune module, which is positioned subsequent to SAM's patch embedding layer. This module serves two critical functions. First, it amplifies SAM's semantic understanding of web page images, aligning it more closely with the task-specific requirements of web page segmentation. Second, it refines the input patch encodings to be fed into subsequent layers, including the Adapter module, which further leverages these enriched embeddings for fine-tuning.

The Patch Embedding Tune module utilizes a tunable linear layer L_{pet} to project the original image embeddings I_{patch} from an E-dimensional space into a feature space $F_{pet} \in \mathbb{R}^e$ with reduced dimensionality e. The extent of this dimensionality reduction can be adjusted by a scale factor μ, which is employed to regulate the number of learnable parameters.

$$F_{\mathrm{pet}} = L_{\mathrm{pet}}(I_{\mathrm{patch}}), \text{with } e = \frac{E}{\mu} \tag{1}$$

Edge Components Tune. The unique layout of web pages, predominantly composed of rectangular blocks and complex images, necessitates a tailored approach in segmentation algorithms. Recognizing this, our Edge Components Tune module is specifically designed to leverage these distinctive structural characteristics inherent to web pages [20]. Traditionally, such features are extracted from the DOM tree, which plays a pivotal role in enhancing the adaptability and performance of our model in web page segmentation tasks.

Initially, the input image I_{img} is preprocessed through inverse normalization and converted to grayscale I_{gray}. We then apply the Sobel operator to compute the gradient magnitude, denoted as I_{sobel}, using the following unified equation:

$$I_{\mathrm{sobel}} = \sqrt{(\mathrm{Sobel}_x * I_{\mathrm{gray}})^2 + (\mathrm{Sobel}_y * I_{\mathrm{gray}})^2} \tag{2}$$

Here, Sobel_x and Sobel_y are the Sobel kernels for the x and y directions, respectively, and $*$ denotes the convolution operation.

The computed Sobel features I_{sobel} are then partitioned into non-overlapping patches I^p_{sobel}, in alignment with the SAM model [21]. These patches are projected into an e-dimensional feature space $F_{ect} \in \mathbb{R}^e$ using a tunable linear layer L_{ect}.

$$F_{\mathrm{ect}} = L_{\mathrm{ect}}(I^p_{sobel}) \tag{3}$$

Importantly, the edge features captured by this module can be changed to emphasize key features in other domains, thereby extending its applicability beyond web page segmentation tasks.

Adapter. The Adapter module serves as an integral component within the WebSAM-Adapter framework, engineered to synthesize webpage-specific knowledge by fusing features from both the Patch Embedding Tune and Edge Components Tune modules. It fulfills two primary objectives: enhancing the model's generalization capabilities by aligning image embeddings with web page structures, and augmenting the feature space tailored for web page segmentation tasks.

To achieve these goals, the Adapter employs a lightweight multi-layer perceptron (MLP) to generate webpage-specific knowledge. For the k_{th} Adapter, the inputs F_{pet} and F_{ect} are directly added together to form F_{merge}, which is then utilized to produce webpage-specific knowledge, denoted as F^k.

$$F^k = \mathrm{MLP}_{up}(\mathrm{GELU}(\mathrm{MLP}^k_{tune}(F_{merge})), with\ F_{merge} = F_{pet} + F_{ect} \tag{4}$$

In this configuration, GELU [34] serves as the the activation function. MLP^k_{tune} is a linear layer responsible for generating different webpage-specific knowledge, while MLP_{up} is an up-projection layer designed to harmonize the dimensions of the transformer features across all Adapters. Finally, F^k represents the webpage-specific knowledge output integrated into each transformer block. Notably, k corresponds to the number of transformer blocks in the SAM image encoder, with $k = 12$ for SAM's base model.

4 Experiments

4.1 Datasets

Although the task of web page segmentation has been proposed for nearly two decades, its foundational resources, particularly datasets and evaluation methods, have been fragmented and often incompatible with each other. Prior datasets were frequently tailored to specific downstream tasks, leading to a proliferation of datasets with varying standards and metrics, most of which are not publicly available or standardized.

The Webis-WebSeg-20 dataset, introduced by Kiesel et al. [1], emerges as a pivotal solution and provides a more comprehensive basis for evaluation. This dataset comprises 42,450 crowdsourced segmentations across 8,490 web pages.

This includes consensus annotations from five human annotators, web archival files for historical rendering, HTML DOM for partial rendering, visual screenshots, and coordinate-mapped DOM nodes. The richness and diversity of this dataset make it an ideal choice for our research, enabling a more thorough and nuanced analysis of web page segmentation.

To more effectively evaluate our model's performance, we employed two distinct types of evaluation metrics. The primary set comprises three metrics: P_{B^3}, R_{B^3} and $F^*_{B^3}$, proposed by Kiesel et al. [1]. These metrics are adaptations of the extended BCubed measures derived from clustering theory [35]. Focusing on these metrics, P_{B^3} is calculated by dividing the elements co-segmented in both the ground-truth and algorithm-generated segmentations by the algorithm-segmented elements. R_{B^3} follows a similar approach but considers the total elements in the ground-truth segmentation as the denominator. F_{B^3} is the harmonic mean of P_{B^3} and R_{B^3} for individual pages, while $F^*_{B^3}$ is the harmonic mean of their averages across all pages. For this evaluation, we conducted experiments on each of the five types of atomic elements defined by Kiesel et al. [1]: all pixels (*pixels*), all pixels at visual edges in both coarse ($edges_C$) and fine settings ($edges_F$), all visible DOM nodes (*nodes*), and all textual characters (*chars*). The second evaluation approach utilizes F_1 score and *Pixel Accuracy* to evaluate the performance.

4.2 Implementation Details

Our method is implemented in PyTorch and runs on a single NVIDIA RTX 3090 GPU with 24 GB of memory. Due to memory constraints, we focus solely on fine-tuning the base model of SAM in this study. We employ the AdamW optimizer [36], starting with an initial learning rate of 2e-4, and train the model for a total of 20 epochs. Cosine decay is applied to adjust the learning rate during the training process. The images are resized to a resolution of 1024×1024 pixels. The loss function that supervises the mask predictions combines Binary Cross-Entropy (BCE) loss and Intersection Over Union (IOU) loss in a linear fashion.

Table 1. Comparison of six segmentation algorithms, highlighting the type of input documents they are designed for, the features they utilize, and the format of their output segmentation.

Method	Document	Features	Output
VIPS [14]	Web page	DOM tree, style, location	Rectangle tree
Cormier et al. [20]	Web page	Screenshot	Rectangle tree
HEPS [18]	Web page	DOM tree, style	Node set
MMDetection [37]	Photo	Screenshot	Pixel masks
Kiesel. [38]	Web page	DOM tree, style, location	Rectangle tree
Ours	Web page	Screenshot	Pixel masks

4.3 Main Results

In our experiments, we aim to validate the effectiveness of our method by comparing it against the benchmarks established by the empirical study conducted by Kiesel et al. [38]. This study, which evaluates five different web page segmentation algorithms across various paradigms, serves as a valuable reference for comparison in our research. To ensure a fair comparison, we strictly adhere to the testing methods outlined in Kiesel et al.'s study. We employ standard 10-fold cross-validation and utilize the 10-fold partitioning method they provided. For the sake of academic rigor, we neither modify nor re-implement the algorithms from Kiesel et al. [38], relying solely on their reported results for comparative analysis.

Table 2. The evaluation results for six algorithms on the Webis-WebSeg-20 dataset [1]: VIPS [14], Cormier et al. [20], HEPS [18], MMDetection [37], Kiesel et al. [38], and our WebSAM-Adapter. Evaluation metrics are the average F_1-score (F_{B^3}), precision (P_{B^3}), recall (R_{B^3}), and the harmonic mean of the averaged precision and recall ($F^*_{B^3}$) for each type of atomic element. The ground truth contains an average of 9.1 segments. The highest score in each row is highlighted in bold.

Measure		VIPS	Corm.	HEPS	MMD.	Kiesel.	ours
Segments		16.1	15.3	36.1	23.0	18.7	**8.9**
pixels	F_{B^3}	0.38	0.36	0.33	0.42	0.39	**0.62**
	$F^*_{B^3}$	0.47	0.53	0.44	0.54	0.50	**0.70**
	P_{B^3}	0.36	0.39	0.36	0.51	0.38	**0.63**
	R_{B^3}	0.67	**0.80**	0.56	0.57	0.72	0.79
edges$_F$	F_{B^3}	0.59	0.51	0.48	0.53	0.56	**0.71**
	$F^*_{B^3}$	0.68	0.65	0.58	0.61	0.66	**0.76**
	P_{B^3}	0.66	0.55	0.61	**0.73**	0.61	0.69
	R_{B^3}	0.69	0.80	0.55	0.53	0.71	**0.85**
edges$_C$	F_{B^3}	0.61	0.53	0.49	0.54	0.57	**0.72**
	$F^*_{B^3}$	0.68	0.66	0.59	0.62	0.67	**0.77**
	P_{B^3}	0.67	0.56	0.62	**0.74**	0.63	0.69
	R_{B^3}	0.70	0.80	0.56	0.53	0.72	**0.86**
nodes	F_{B^3}	0.63	0.52	0.43	0.52	0.54	**0.69**
	$F^*_{B^3}$	0.70	0.65	0.54	0.61	0.65	**0.75**
	P_{B^3}	0.69	0.53	0.63	**0.74**	0.64	**0.74**
	R_{B^3}	0.71	**0.82**	0.46	0.51	0.65	0.77
chars	F_{B^3}	0.67	0.61	0.50	0.61	0.62	**0.77**
	$F^*_{B^3}$	0.75	0.71	0.60	0.69	0.71	**0.82**
	P_{B^3}	0.77	0.61	0.73	**0.79**	0.72	0.78
	R_{B^3}	0.72	0.84	0.51	0.60	0.71	**0.87**

Table 1 provides an overview of the algorithms used in the experiments, and Table 2 displays the performance of each algorithm on the Webis-WebSeg-20 dataset [1]. Our algorithm outperforms all the algorithms in the crucial comprehensive indicators F_{B^3} and F^*_{B3} across the five types of atomic elements, demonstrating an average increase of 18.2%. However, Our P_{B^3} is marginally lower than the MMDetection [37] method in the three atomic elements $edges_F$, $edges_C$, and $chars$. This is attributed to MMDetection's propensity to detect more smaller segments, which results in a higher P_{B^3} but a significantly lower R_{B^3} compared to ours. Similarly, our R_{B^3} is marginally lower than the Cormier [20] method in the two atomic elements $pixels$ and $nodes$. This algorithm tends to detect larger areas, resulting in a higher R_{B^3}, but its P_{B^3} is notably lower. Overall, our method achieves competitive results. A visual comparison of the results is represented in the Fig. 3. This visual evidence not only supports but also strengthens our claim of the method's efficacy.

Simultaneously, our semantic segmentation results achieved an F_1 score of 0.892 and a Pixel Accuracy of 0.881, underscoring the strong performance of our model.

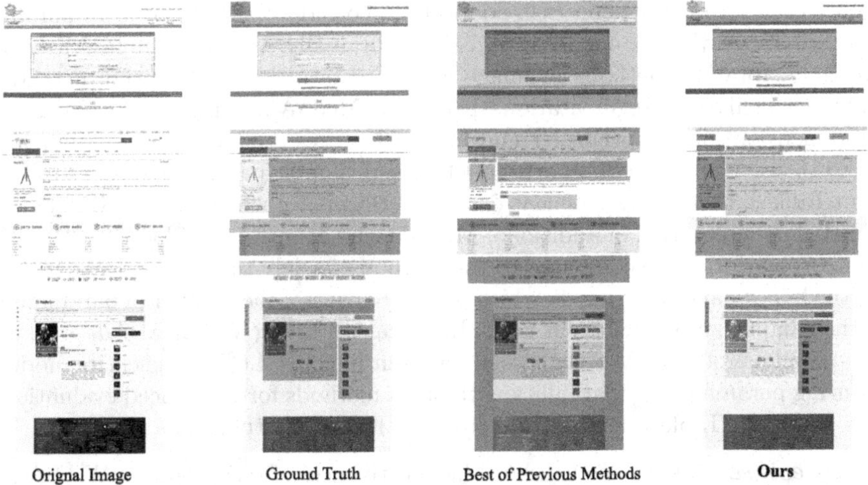

Fig. 3. Comparison of Original Image, Ground Truth, Best of Previous Methods, and Our Result, across three web page images from the dataset. The column "Best of Previous Methods" showcases the most effective result among the five algorithms previously evaluated, as referenced in Table 2. The clear differentiation in segmentation quality among these columns visually underscores the superior performance of our proposed method.

4.4 Ablation Study

We conduct an ablation study on the Webis-WebSeg-20 dataset [1] to demonstrate the effectiveness of each component. Unless specified otherwise, the experiments are performed with the scaling factor $\mu = 8$.

Table 3. Comparison with efficient tuning approaches.

Method	Trainable Param.(M)	F_1 score	Pixel Accuracy
Only Decoder	4.06	0.833	0.798
MedSAM-Adapter [23]	4.67	0.883	0.865
AdaptFormer [31]	4.36	0.879	0.862
Ours	**4.34**	**0.892**	**0.881**

Comparison with Efficient Tuning Approaches. We evaluate our method by comparing it to the approach of exclusively tuning the decoder, a strategy widely used for downstream task adaptation, considering that the total parameter size of SAM's base model is 93.7M. Additionally, We also compare it with similar adaptation methods in image classification, namely AdaptFormer [31], and in medical image segmentation, specifically MedSAM-Adapter [23]. The middle dimension for AdaptFormer and hidden features for MedSAM-Adapter are set to 16 to ensure a fair comparison in terms of tunable parameters. As observed from Table 3, there is a significant drop in performance when only the decoder is tuned. The integration of additional MLPs in the Transformer block [23,31] also improves performance compared to the method of solely tuning the decoder. We set the hyper-parameter $\mu = 8$, which is used to control the number of parameters of the Adapter, as described in Eq. 1. Importantly, in our WebSAM-Adapter, we fine-tuned merely 4.34 M parameters, representing only 4.6% of the entire model, ensuring parameter comparability with other methods for a balanced evaluation. The results in Table 3 demonstrate that our method outperforms others.

Architecture Design. To validate the effectiveness of our proposed WebSAM-Adapter architecture, we modify it into different variants. As indicated in Table 4, the application of shared MLP_{tune} in various Adapters preserves only a limited number of parameters and results in a substantial decline in performance. The incorporation of different MLP_{up} in diverse Adapters does not guarantee consistent improvement and, furthermore, increases the number of parameters by 19%. In contrast, the omission of Patch Embedding Tune (PET) or Edge Components Tune (ECT) leads to a decrease in performance, underscoring their significance in visual information. The proposed WebSAM-Adapter (Decoder + PET + ECT + Adapter) exhibits enhanced efficacy.

Scale Factor μ. In Sect. 3.2 of the main paper, we introduce μ to regulate the number of learnable parameters in Eq. 1. A larger μ yields fewer parameters

Table 4. Ablation on the architecture designs described in Fig. 2.

Method	Trainable Param.(M)	F_1 score	Pixel Accuracy
Only Decoder	4.06	0.833	0.798
Ours w/o ECT	4.31	0.871	0.850
Ours w/o PET	4.27	0.872	0.852
Ours w/Shared MLP_{tune}	4.24	0.866	0.842
Ours w/Unshared MLP_{up}	5.16	0.875	0.857
Ours	**4.34**	**0.892**	**0.881**

available for tuning. As depicted in Table 5, the performance improves as μ decreases from 64 to 8. However, when μ is further reduced to 2 or 1, there is no sustained significant improvement in performance, despite the increase in model size. This observation suggests that $\mu = 8$ represents an optimal choice for balancing performance with model size.

Table 5. Ablation on the parameter scale factor μ.

scale factor μ	Trainable Param.(M)	F_1 score	Pixel Accuracy
1	12.5	**0.893**	**0.884**
2	6.52	0.892	0.884
4	4.85	0.892	0.881
8	4.34	0.892	0.881
16	4.17	0.874	0.871
32	4.11	0.867	0.863
64	4.08	0.869	0.849

5 Conclusion

In conclusion, this study presents the application of the WebSAM-Adapter, an effective implementation of SAM, to address the specific challenges of web page segmentation. Utilizing a tailored three-module architecture, it effectively acquires webpage-specific knowledge, demonstrating state-of-the-art performance on the Webis-WebSeg-20 dataset [1]. This research not only mitigates SAM's limitations in this domain but also lays the groundwork for future innovations in purely visual web page segmentation methods.

Acknowledgements. This work was financially supported by the Joint Research Fund of China Pacific Insurance (Group) Co. and SJTU-Artificial Intelligence Institute.

References

1. Kiesel, J., Kneist, F., Meyer, L., Komlossy, K., Stein, B., Potthast, M.: Web page segmentation revisited: evaluation framework and dataset. In: Proceedings of the 29th ACM International Conference on Information & Knowledge Management, pp. 3047–3054 (2020)
2. Cai, D., He, X., Wen, J.-R., Ma, W.-Y.: Block-level link analysis. In: Proceedings of the 27th Annual International ACM SIGIR Conference on Research and Development in Information Retrieval, pp. 440–447 (2004)
3. Bing, L., Guo, R., Lam, W., Niu, Z.-Y., Wang, H.: Web page segmentation with structured prediction and its application in web page classification. In: Proceedings of the 37th International ACM SIGIR Conference on Research & Development in Information Retrieval, pp. 767–776 (2014)
4. Akpinar, M.E., Yesilada, Y.: Vision based page segmentation algorithm: extended and perceived success. In: Sheng, Q.Z., Kjeldskov, J. (eds.) ICWE 2013. LNCS, vol. 8295, pp. 238–252. Springer, Cham (2013). https://doi.org/10.1007/978-3-319-04244-2_22
5. Saar, T., Dumas, M., Kaljuve, M., Semenenko, N.: Browserbite: cross-browser testing via image processing. Softw. Pract. Exp. **46**(11), 1459–1477 (2016)
6. Mahajan, S., Abolhassani, N., McMinn, P., Halfond, W.G.: Automated repair of mobile friendly problems in web pages. In: Proceedings of the 40th International Conference on Software Engineering, pp. 140–150 (2018)
7. Geng, G.-G., Lee, X.-D., Zhang, Y.-M.: Combating phishing attacks via brand identity and authorization features. Secur. Commun. Netw. **8**(6), 888–898 (2015)
8. Cormier, M., Cohen, R., Mann, R., Rahim, K., Wang, D.: A robust vision-based framework for screen readers. In: Agapito, L., Bronstein, M.M., Rother, C. (eds.) ECCV 2014. LNCS, vol. 8927, pp. 555–569. Springer, Cham (2015). https://doi.org/10.1007/978-3-319-16199-0_39
9. Cormier, M., Moffatt, K., Cohen, R., Mann, R.: Purely vision-based segmentation of web pages for assistive technology. Comput. Vis. Image Underst. **148**, 46–66 (2016)
10. Sanoja, A., Gançarski, S.: Block-o-matic: a web page segmentation framework. In: 2014 International Conference on Multimedia Computing and Systems (ICMCS), pp. 595–600. IEEE (2014)
11. Vineel, G.: Web page dom node characterization and its application to page segmentation. In: 2009 IEEE International Conference on Internet Multimedia Services Architecture and Applications (IMSAA), pp. 1–6. IEEE (2009)
12. Chen, Y., Ma, W.-Y., Zhang, H.-J.: Detecting web page structure for adaptive viewing on small form factor devices. In: Proceedings of the 12th International Conference on World Wide Web, pp. 225–233 (2003)
13. Rajkumar, K., Kalaivani, V.: Dynamic web page segmentation based on detecting reappearance and layout of tag patterns for small screen devices. In: 2012 International Conference on Recent Trends in Information Technology, pp. 508–513. IEEE (2012)
14. Cai, D., Yu, S., Wen, J.-R., Ma, W.-Y.: Vips: a vision-based page segmentation algorithm (2003)
15. Zeleny, J., Burget, R., Zendulka, J.: Box clustering segmentation: a new method for vision-based web page preprocessing. Inf. Process. Manag. **53**(3), 735–750 (2017)
16. Bajammal, M., Mesbah, A.: Page segmentation using visual adjacency analysis. arXiv preprint arXiv:2112.11975 (2021)

17. Andrew, J., Ferrari, S., Maurel, F., Dias, G., Giguet, E.: Web page segmentation for non visual skimming. In: The 33rd Pacific Asia Conference on Language, Information and Computation (PACLIC 33) (2019)
18. Manabe, T., Tajima, K.: Extracting logical hierarchical structure of html documents based on headings. In: Proceedings of the VLDB Endowment, pp. 1606–1617 (2015). http://dx.doi.org/10.14778/2824032.2824058
19. Cao, J., Mao, B., Luo, J.: A segmentation method for web page analysis using shrinking and dividing. Int. J. Parallel Emergent Distrib. Syst. **25**(2), 93–104 (2010)
20. Cormer, M., Mann, R., Moffatt, K., Cohen, R.: Towards an improved vision-based web page segmentation algorithm. In: 2017 14th Conference on Computer and Robot Vision (CRV), pp. 345–352. IEEE (2017)
21. Kirillov, A., et al.: Segment anything. arXiv preprint arXiv:2304.02643 (2023)
22. Ma, J., Wang, B.: Segment anything in medical images. arXiv preprint arXiv:2304.12306 (2023)
23. Wu, J., et al.: Medical sam adapter: adapting segment anything model for medical image segmentation. arXiv preprint arXiv:2304.12620 (2023)
24. Shaharabany, T., Dahan, A., Giryes, R., Wolf, L.: Autosam: adapting sam to medical images by overloading the prompt encoder. arXiv preprint arXiv:2306.06370 (2023)
25. Chen, K., et al.: Rsprompter: learning to prompt for remote sensing instance segmentation based on visual foundation model. arXiv preprint arXiv:2306.16269 (2023)
26. Chen, T., et al.: Sam fails to segment anything?-sam-adapter: adapting sam in underperformed scenes: Camouflage, shadow, and more. arXiv preprint arXiv:2304.09148 (2023)
27. Tang, L., Xiao, H., Li, B.: Can sam segment anything? when sam meets camouflaged object detection. arXiv preprint arXiv:2304.04709 (2023)
28. Zaken, E.B., Ravfogel, S., Goldberg, Y.: Bitfit: simple parameter-efficient fine-tuning for transformer-based masked language-models. arXiv preprint arXiv:2106.10199 (2021)
29. Liu, W., Shen, X., Pun, C.-M., Cun, X.: Explicit visual prompting for low-level structure segmentations. In: Proceedings of the IEEE/CVF Conference on Computer Vision and Pattern Recognition, pp. 19 434–19 445 (2023)
30. He, X., Li, C., Zhang, P., Yang, J., Wang, X.E.: Parameter-efficient model adaptation for vision transformers. arXiv preprint arXiv:2203.16329 (2022)
31. Chen, S., et al.: Adaptformer: adapting vision transformers for scalable visual recognition. Adv. Neural Inf. Process. Syst. **35**, 16 664–16 678 (2022)
32. Chen, Z., et al.: Vision transformer adapter for dense predictions. arXiv preprint arXiv:2205.08534 (2022)
33. Dosovitskiy, A., et al.: An image is worth 16×16 words: transformers for image recognition at scale. arXiv preprint arXiv:2010.11929 (2020)
34. Hendrycks, D., Gimpel, K.: Gaussian error linear units (gelus). Cornell University - arXiv (2016)
35. Amigó, E., Gonzalo, J., Artiles, J., Verdejo, F.: A comparison of extrinsic clustering evaluation metrics based on formal constraints. Inf. Retr. 461–486 (2009). https://doi.org/10.1007/s10791-008-9066-8
36. Kingma, D.P., Ba, J.: Adam: a method for stochastic optimization. arXiv preprint arXiv:1412.6980 (2014)

37. Chen, K., et al.: Mmdetection: open mmlab detection toolbox and benchmark. arXiv Computer Vision and Pattern Recognition (2019)
38. Kiesel, J., Meyer, L., Kneist, F., Stein, B., Potthast, M.: An empirical comparison of web page segmentation algorithms. In: Hiemstra, D., Moens, M.-F., Mothe, J., Perego, R., Potthast, M., Sebastiani, F. (eds.) ECIR 2021. LNCS, vol. 12657, pp. 62–74. Springer, Cham (2021). https://doi.org/10.1007/978-3-030-72240-1_5

A Phrase-Level Attention Enhanced CRF for Keyphrase Extraction

Shinian Li, Tao Jiang, and Yuxiang Zhang(✉)

School of Computer Science and Technology, Civil Aviation University of China, Tianjin, China
{2021052051,yxzhang}@cauc.edu.cn

Abstract. Since sequence labeling-based methods take into account the dependencies between neighbouring labels, they have been widely used for keyphrase prediction. Existing methods mainly focus on the word-level sequence labeling over the word-level features, and fail to capture the phrase-level information (*i.e.*, inner properties of multi-word keyphrases). In this paper, we concentrate on how to effectively capture the phrase-level features and then integrate them with the word-level features to improve the performance of keyphrase extraction in the sequence labeling-based method. Specifically, we propose a *p*hrase-level *a*ttention *e*nhanced *c*onditional *r*andom *f*ield (PAE-CRF) model for keyphrase extraction, which consists of two major modules: a *phrase-level attention module* that captures phrase-level features, and a *phrase-level attention enhanced CRF module* that integrates the phrase-level attention information with the word-level features into CRF to extract keyphrases. Finally, these two modules are jointly trained to help them learn complementary information from each other. Compared with the recent state-of-the-art methods, our model can achieve better results through experiments on four benchmark datasets. The code and keyphrase prediction results of our model are available in public at https://github.com/pae-crf/PAE-CRF.

Keywords: Keyphrase extraction · Deep neural networks · Phrase-level attention mechanism · Conditional random field

1 Introduction

Keyphrase extraction is the task to automatically select a set of representative phrases that are related to the main topics discussed in a document. Since keyphrases can provide a high-level topic description of a document, they are beneficial for a wide range of natural language processing tasks such as information extraction, text summarization and question answering [31,36]. However, the performance of existing methods is still far from being satisfactory.

Existing supervised approaches for keyphrase extraction can be roughly divided into classification-based and *sequence labeling-based methods*. Specifically, classification-based methods usually treat the keyphrase extraction as a binary classification task, in which a classifier is trained on the features of labeled keyphrases to determine whether a candidate phrase is a keyphrase [13,19,27]. Sequence labeling-based methods formulate keyphrase extraction as a sequence labeling task, where linear-chain CRFs are used

N. Goharian et al. (Eds.): ECIR 2024, LNCS 14608, pp. 455–469, 2024.
https://doi.org/10.1007/978-3-031-56027-9_28

to solve this task. Compared with the former which classifies the labels of each candidate phrase independently, the latter conditions each label prediction on the previously predicted label (*i.e.*, they take into account the *dependencies* between neighbouring labels) and improves the performance over baseline models for this task [2,24].

Recently, researchers have proposed many sequence labeling-based approaches for the keyphrase extraction task, including traditional CRF methods such as EK-CRF [11] and DM-SMCRFs [20], and neural CRF methods such as Bi-LSTM-CRF [2], Memory-CRF [43] and Dake [24]. However, almost all these existing works focus on the word-level sequence labeling over the word-level features, and fail to capture the phrase-level information (*i.e.*, inner properties of multi-word keyphrase). Although in order to utilize the phrase-level features, DM-SMCRFs [20] employed semi-Markov conditional random fields (Semi-CRFs) [25] to directly classify the phrase as keyphrase or non-keyphrase, it is lacking of an automatic phrase-level feature encoder.

This paper makes the first attempt to develop a phrase-level attention enhanced CRF (PAE-CRF) for the keyphrase extraction task, which not only effectively captures the phrase-level features but also integrates them with word-level features into CRF to gain better performance in keyphrase extraction task. Figure 1 illustrates the architecture of our framework. Firstly, we design a phrase-level attention mechanism to capture the phrase-level features. Then, a phrase-level attention transformation module is proposed to transform the phrase-level attention scores into the word-level feature scores. Finally, we integrate the transformed word-level features and original word-level features into the neural CRF to extract keyphrases.

To summarize, our main contributions are as follows:

- To the best of our knowledge, this is the first attempt to incorporate the phrase-level features into the neural CRF through automatically encoding the phrase-level features to extract keyphrases.
- We propose a novel phrase-level attention enhanced CRF for keyphrase extraction, which not only effectively leverages the global dependencies to predict the single-word keyphrases, but also utilizes the local dependencies within phrase to predict the multi-word keyphrases.
- We compare our PAE-CRF with the eight state-of-the-art methods on four publicly-available datasets for keyphrase extraction. Experimental results show our model can achieve better results.

2 Related Work

As mentioned in Sect. 1, supervised keyphrase extraction methods can be roughly divided into classification-based and sequence labeling methods. Since the sequence labeling methods can model the dependencies between neighbouring labels, they have been widely used and proven to be effective in the keyphrase extraction task [2,24]. This work is mainly related to neural sequence labeling models for keyphrase extration. Thus, we give a short review of this research area and distinguish our work from the existing approaches.

Zhang *et al.* [39] is the first to utilize CRFs extracting keyphrases, which provides a way to identify each candidate word by modeling the dependencies between neighbouring labels and exploring the traditional features (*e.g.*, part-of-speech and term frequency-inverse document frequency). Following this work, Bhaskar *et al.* [3] used CRFs trained mainly on linguistic features such as part-of-speech, chunking and named-entity tags for keyphrase extraction. Gollapalli *et al.* [11] also utilized CRFs to extract keyphrases from research papers, which was trained on token-based features incorporating linguistic, document-structure information and expert knowledge.

More recently, deep neural networks have been widely used in the keyphrase prediction task. CopyRNN [21] is the first to employ the attentional sequence to sequence (seq2seq) framework [30] with the copying mechanism [12] to generate both present and absent keyphrases (that do not appear in the given document). In order to generate multiple keyphrases and determine the appropriate number of keyphrases at a time for a target document, its modification CatSeq [38] introduced a new One2Seq training paradigm in the seq2seq framework. To further eliminate the bias caused by the predefined order in One2Seq paradigm [38], One2Set training paradigm [37] was proposed to predict the keyphrases as a set. Numerous seq2seq extensions have been proposed to boost its generation ability under these three training paradigms. Some studies incorporated different types of side information into seq2seq networks to improve keyphrase generation, such as correlation among keyphrases [5], title of source document [7], syntactic constraints [42] and topic information [34,40]. Some works focused on improving the decoding process of seq2seq networks, such as designing an exclusive hierarchical decoder [6] and extractor-generator to jointly extract and generate keyphrases [1]. Some researches utilized the hyperbolic deep networks to effectively model the hierarchical semantic relations expressed in a document for keyphrase prediction [26,41].

In addition, deep neural networks are integrated with sequence labeling models for keyphrase prediction. In order to capture the long-term dependencies and semantic relationships hidden in text, Al-Zaidy *et al.* [2] proposed Bi-LSTM-CRF sequence labeling for keyphrase extraction, which combines a bi-directional long short-term memory (Bi-LSTM) layer to model the sequential text data with a CRF layer to model dependencies in the output. In this method, the hand-engineering input features have been replaced with the deep features. Santosh *et al.* [24] augmented Bi-LSTM-CRF with a document-level attention to leverage additional supporting information within a given document. Zhou *et al.* [43] proposed a multi-level memory network with CRFs to capture both the long-range and local contextual information in text. In addition, Lu *et al.* [20] used semi-Markov CRFs to avoid the post-processing to transform keyword into keyphrase in previous sequence labeling methods.

We observe that almost all these existing sequence labeling approaches focus on the word-level sequence labeling over the word-level features, and fail to consider the phrase-level information (*i.e.*, inner properties of multi-word keyphrases) using deep neural networks. In this paper, we first integrate the phrase-level features into the neural CRF through automatically encoding the phrase-level features to extract keyphrases, which can ensure that the local dependencies within phrase are captured to better predict the multi-word keyphrases.

3 Methodology

In this section, we formulate keyphrase extraction as a sequence labeling task and present the details of the proposed model.

Table 1. Word-level and phrase-level labeling on an input sequence

Position	1	2	3	4	5	6	7	8	9
Sequence	**Fixed**	**points**	of	**correspondences**	defined	on	**cone**	**metric**	**space**
Word-level label	B_{kp}	E_{kp}	O	SW	O	O	B_{kp}	I_{kp}	E_{kp}
Phrase-level label	MW		O	SW	O	O	MW		

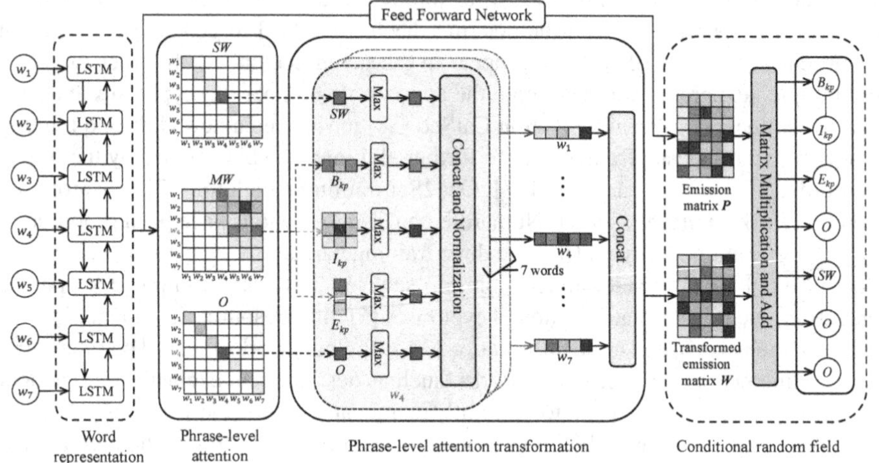

Fig. 1. The overall architecture of the proposed PAE-CRF model

3.1 Problem Formulation and Framework

Keyphrase extraction can be formulated as a task of sequence labeling, predicting the label sequence for the input text. Specifically, we denote an input document as a sequence $X = \{x_1, x_2, ..., x_n\}$ (also denoted as $X[1:n]$), where x_i represents i-th input word and n is the length of the sequence. The *word-level* and *phrase-level* sequence labeling methods are defined as follows: (1) The goal of word-level sequence labeling method is to predict a sequence of labels $Y_w = \{y_1, y_2, ..., y_n\}$ for X, where y_i is an element of the word-level label set $L_w = \{\mathrm{B}_{kp}, \mathrm{I}_{kp}, \mathrm{E}_{kp}, \mathrm{SW}, \mathrm{O}\}$. B_{kp}, I_{kp} and E_{kp} respectively denote the beginning, inside and ending label of a keyphrase. SW and O refer to single-word keyphrase (*i.e.*, keyword) label and non-keyphrase label, respectively. (2) The goal of phrase-level sequence labeling method is to predict a sequence of consecutive spans, $Y_p = \{s_1, s_2, ..., s_m\}$ for the input sequence X, where a label s_i is a tuple consisting of (y_i, b_i, e_i). y_i is an element of the phrase-level label set $L_p = \{\mathrm{MW}, \mathrm{SW}, \mathrm{O}\}$, where MW denotes the multi-word keyphrase label. b_i and e_i represent the beginning position and ending position of s_i, respectively. Table 1 shows the difference between word-level labeling and phrase-level labeling.

The overall architecture of our proposed method is shown in Fig. 1, which consists of four particular modules: (1) a word representation layer where a Bi-LSTM [9] is employed to yield the contextualized word representations from an input text sequence, (2) a phrase-level attention model mechanism that is used to capture the phrase-level features and compute attention feature scores, (3) a phrase-level attention transformation module that transforms the phrase-level attention scores into the word-level feature scores, and (4) a phrase-level attention enhanced CRF module that integrates the transformed and original word-level features into neural CRF to extract keyphrases.

3.2 Technical Background: Neural CRF

Conditional random field (CRF) [18] has been shown to be effective for many sequence labeling tasks and here we leverage it to jointly model the sequence of labels for the keyphrase extraction task. Given an input text sequence $X = \{x_1, x_2, ..., x_n\}$ with n words and an output label sequence $Y = \{y_1, y_2, ..., y_n\}$, the probability of the predicted label sequence Y given X defined by CRF is modeled as:

$$p(Y|X) = \frac{\exp(score(X,Y))}{\sum_{\tilde{Y} \in Y_X} \exp(score(X,\tilde{Y}))} \tag{1}$$

where Y_X represents all possible label sequences given input sequence X, and the scoring function $score(X, Y)$ is defined as:

$$score(X,Y) = \sum_{i=0}^{n} A_{y_i,y_{i+1}} + \sum_{i=1}^{n} P_{i,y_i} \tag{2}$$

where $\boldsymbol{A}$ is a transition matrix in which $A_{y_i,y_{i+1}}$ represents the score of a transition from label y_i to label y_{i+1}, and $\boldsymbol{P}$ is an emission matrix in which P_{i,y_i} represents the score of label y_i at the i-th position.

In neural CRF models for keyphrase ectraction such as Bi-LSTM-CRF [2] and Memory-CRF [43], the emission matrix $\boldsymbol{P}$ is provided by deep neural networks such as bidirectional LSTM (Bi-LSTM) [9] and memory network [15]. In other words, the last hidden layer in neural networks is directly used to predict confidence scores for the word having each of the possible labels. Such structures combine the complementary strengths of neural networks and CRF for keyphrase extraction.

3.3 Word-Level Neural CRF

Given an input sequence $X = \{x_1, x_2, ..., x_n\}$ with n words, the word representation $\boldsymbol{h}$ is encoded as follows:

$$\boldsymbol{e}_1, ..., \boldsymbol{e}_n = \texttt{PTE}(x_1, ..., x_n) \tag{3}$$

$$\boldsymbol{h}_1, ..., \boldsymbol{h}_n = \texttt{BiLSTM}(\boldsymbol{e}_1, ..., \boldsymbol{e}_n) \tag{4}$$

where $\texttt{PTE}(\cdot)$ is the contextualized pre-trained embeddings BERT [10], which has been widely used in various natural language processing tasks and achieved good performance. $\texttt{BiLSTM}(\cdot)$ is a bidirectional long short-term memory (Bi-LSTM) [9], which is employed to learn the long-range semantic dependencies within the target text. Here, we end up with a new matrix $\boldsymbol{H} = \{\boldsymbol{h}_1, ..., \boldsymbol{h}_n\} \in \mathbb{R}^{n \times d_h}$, where d_h denotes the dimension of a word representation.

Then, a neural CRF presented in Subsect. 3.2 is adopted, in which the emission matrix $\boldsymbol{P}$ is obtained by the linear transformation $\boldsymbol{P} = \boldsymbol{H}\boldsymbol{W}$, where $\boldsymbol{H}$ is given above, and $\boldsymbol{W} \in \mathbb{R}^{d_h \times k}$ (k denotes the size of labels, and the word-level label set is defined in Subsect. 3.1 and shown in Table 1) is a learnable weight matrix.

3.4 Phrase-Level Attention

For any text span $X[i:j]$ in an input sequence $X[1:n]$, here we compute the attention score to identify a particular label α from the candidate label set $L_p = \{\text{MW}, \text{SW}, \text{O}\}$ described in Subsect. 3.1 and shown in Table 1. Firstly, an upper triangular matrix is constructed to traverse all valid spans, as shown in Fig. 1 (phrase-level attention), where each grid corresponds to a phrase span. Given the representations $\boldsymbol{h}_i$ and $\boldsymbol{h}_j$ at begin position i and end position j, and a particular label α, the query vector $\boldsymbol{q}_{i,\alpha}$ and key vector $\boldsymbol{k}_{i,\alpha}$ of the two are obtained through two feedforward layers:

$$\boldsymbol{q}_{i,\alpha} = \boldsymbol{W}_{q,\alpha}\boldsymbol{h}_i + \boldsymbol{b}_{q,\alpha}, \tag{5}$$

$$\boldsymbol{k}_{j,\alpha} = \boldsymbol{W}_{k,\alpha}\boldsymbol{h}_j + \boldsymbol{b}_{k,\alpha}. \tag{6}$$

Then, the score of span $X[i:j]$ predicted as a phrase of label α is calculated as:

$$s_\alpha(i,j) = \boldsymbol{q}_{i,\alpha}^\top \boldsymbol{k}_{j,\alpha}. \tag{7}$$

In order to leverage the relative position information, the rotary position embedding (RoPE) mechanism [28] is introduced to compute the phrase-level attention score $s_\alpha(i,j)$. More specifically, RoPE is applied to calculate two orthogonal matrices $\boldsymbol{R}_i$ and $\boldsymbol{R}_j$ with respect to positions i and j, which satisfies $\boldsymbol{R}_i^\top \boldsymbol{R}_j = \boldsymbol{R}_{j-i}$. Thus, the attention score of span $X[i:j]$ for label α, which contains the relative position information, is calculated as:

$$s_\alpha(i,j) = (\boldsymbol{R}_i \boldsymbol{q}_{i,\alpha})^\top (\boldsymbol{R}_j \boldsymbol{k}_{j,\alpha}) = \boldsymbol{q}_{i,\alpha}^\top \boldsymbol{R}_{j-i} \boldsymbol{k}_{j,\alpha}. \tag{8}$$

3.5 Phrase-Level Attention Enhanced CRF

In this subsection, we first transform the phrase-level attention scores into the word-level feature scores, which will be integrated with the original word-level feature scores and inputted into CRF to improve the accuracy of keyphrase extraction. Specifically, for a given word x_k $(1 \le i \le k \le j \le n)$ in a span $X[i:j]$, we transform its attention scores of the three phrase-level labels into feature scores of the five word-level labels. The transformed feature scores $w_{k,\beta}$ are defined as follows.

If the phrase-level label α is the same as the word-level label β and it is one of the single-word keyphrase label SW and non-keyphrase label O (such that $k = i = j$), then $w_{k,\beta}$ is defined by:

$$w_{k,\beta} = s_\alpha(k,k), \qquad \alpha = \beta \in \{\text{SW}, \text{O}\} \tag{9}$$

If the phrase-level label α is the multi-word keyphrase label MW, then there are three types of transformations, which are defined as follows:

$$w_{k,\text{B}_{kp}} = \texttt{max}(\{s_\alpha(k,[k+1:n])\}) \qquad k = i \tag{10}$$

$$w_{k,\text{I}_{kp}} = \texttt{max}(\{s_\alpha([1:k-1],[k+1:n])\}) \qquad i < k < j \tag{11}$$

$$w_{k,\text{E}_{kp}} = \texttt{max}(\{s_\alpha([1:k-1],k)\}) \qquad k = j \tag{12}$$

where n is the length of the input document sequence X.

Then, we concatenate and normalize the transformed feature scores of the word-level labels to obtain the transformed emission matrix ***W***. Additionally, we employ the neural CRF mentioned in Subsect. 3.2 to obtain the emission matrix ***P***. Finally, we integrate the transformed emission matrix ***W*** with the original emission matrix ***P*** into CRF to enhance the accuracy of the keyphrase extraction. Thus, in our PAE-CRF, $score(X, Y)$ is calculated as:

$$score(X,Y) = \sum_{k=0}^{n} A_{y_k,y_{k+1}} + \sum_{k=1}^{n} (W_{k,y_k} P_{k,y_k} + P_{k,y_k}). \tag{13}$$

3.6 Joint Training

Since the phrase-level attention module and the word-level neural CRF module both aim to extraction keyphrases from input documents, we jointly train the two modules to help them learn complementary information from each other. The loss function of our model consists of two parts. The multi-label classification cross entropy loss [29] is used for the phrase-level attention module when the phrase-level attention scores have been calculated, defined as:

$$\mathcal{L}_{PA} = \log\left(1 + \sum_{(i,j)\in P_\alpha} e^{-S_\alpha(i,j)}\right) + \log\left(1 + \sum_{(i,j)\in Q_\alpha} e^{S_\alpha(i,j)}\right), \tag{14}$$

where P_α is a set of spans with the phrase-level label α, and Q_α is a set of spans whose the phrase-level label is not α.

The negative log-likelihood loss is used for CRF, defined as:

$$\mathcal{L}_{CRF} = -\log p(Y|X). \tag{15}$$

This objective function and its gradients can be efficiently computed by dynamic programming algorithm. The final overall loss of the entire framework's training objective is the linear combination of the two parts, defined as

$$\mathcal{L} = \mathcal{L}_{PA} + \mathcal{L}_{CRF}. \tag{16}$$

4 Experiments

In this section, we will describe the datasets used for the experiments, then introduce the comparative methods and evaluation metrics, and finally evaluate the performance of the model in extracting keyphrase.

4.1 Datasets

Following previous researches [22,38], we split the KP20k dataset collected by Meng et al. [22] into training, validation and test sets. We use the training set to train our model, and employ the validation set to find the optimal hyper-parameters during the training process. In order to evaluate our models comprehensively, we test the models on

Table 2. Summary of the training, validation and testing datasets

Dataset		#Abs	#Avg.Kps	#Avg.PKps	#Avg.SWs	#Avg.MWs	%SWs	%MWs
Train	KP20k	509820	5.27	3.31	1.30	2.01	39.4	60.6
Valid	KP20k	19990	5.27	3.31	1.30	2.01	39.4	60.6
Test	KP20k	20000	5.27	3.31	1.30	2.01	39.4	60.6
	Inspec	500	9.93	7.20	1.02	6.18	14.2	85.8
	NUS	211	10.80	5.64	1.87	3.77	33.2	66.8
	SemEval	100	14.43	6.12	1.60	4.52	26.1	73.9

four widely-adopted public datasets, including KP20k [22], Inspec [14], NUS [23] and SemEval [16]. These datasets contain a large amount of high-quality scientific metadata in the computer science domain from various online digital libraries, in which each example contains a title and an abstract of a scientific article as input source text, and multiple author-assigned keywords as target keyphrases.

The detailed statistic information of these benchmark datasets are summarized in Table 2, along with the number of abstracts (#Abs), the average number of keyphrases, present keyphrases, single-word keyphrases and multi-word keyphrases per document (#Avg.Kps, #Avg.PKps, #Avg.SWs and #Avg.MWs), and the percentage of single-word keyphrases and multi-word keyphrases (%SWs and %MWs).

4.2 Comparative Methods

To comprehensively evaluate the performance of our method, we compare PAE-CRF with eight state-of-the-art keyphrase prediction methods, as follows:

- **CatSeq** [38] uses a gate recurrent unit (GRU) [8] based encoder-decoder framework to predict keyphrases, in which the One2Seq training paradigm is introduced to generate keyphrases and decide the suitable number of keyphrases at a time for a target document.
- **CatSeqTG** [7] is a simple extension of both CatSeq and TG-Net [7], which leverages the highly summative information in the titles of papers with the One2Seq training paradigm to guide keyphrase generation.
- **CatSeq-RL** [4] is also a simple extension of Catseq using reinforcement learning with adaptive rewards during the decoding phase to generate sufficient and accurate keyphrases.
- **ExHiRD-h** [6] aslo uses GRU as the backbone network and designs an exclusive hierarchical decoder with soft and hard exclusion mechanism to avoid generating duplicated keyphrases.
- **SEG-Net** [1] employs Transformer [32] as the backbone network, and designs a sentence-selector to select the salient sentences in a document and an extractor-generator to jointly extract and generate keyphrases from the selected sentences.
- **One2Set** [37] is also a Transformer-based model with the One2Set training paradigm which generates keyphrases as a set.

- **DualCopyNet** [33] applies a neural CRF for identifying some important words (seed words) which further guide to generate keyphrases using Bi-GRUs [8] with dual copy mechanisms.
- **Prompt-KG** [35] is a Transformer-based model in which a sequence labeling module is used to extract present keyphrases and a prompt-based learning module is used to generate absent keyphrases.

4.3 Evaluation Metrics

For fairly comparing different approaches, we follow the literature and adopt top-N macro-averaged *precision*, *recall* and F_1-measure as the evaluation metrics. In particular, precision is defined as the number of correctly predicted keyphrases over the number of all predicted keyphrases, recall is defined as the number of correctly predicted keyphrases over the total number of data records, and F_1 is the harmonic mean of precision and recall.

Note that $F_1@k$ is used in almost all existing works on the keyphrase prediction, in which k (usually 5 or 10) is a fixed number of top-k predictions. Since top-5 is very close to the number of gold keyphrases for big dataset KP20k, we adopt $F_1@5$. When the prediction number is less than five in our PAE-CRF based on CRF, we randomly append incorrect keyphrases until it obtains five predictions. Therefore, it is unfair to compare our PAE-CRF with the existing different methods using this metric $F_1@5$. In order to avoid this unfairness, we also adopt $F_1@M$ proposed in CatSeq [38], which compares all the keyphrases predicted by the model with the ground-truth keyphrases and is more suitable for variable-number extraction than $F_1@5$.

4.4 Implementation Details

We follow the previous works [22,38] to pre-process the experimental data, including lowercasing, tokenizing, etc. We use BERT-base [10] and a single-layer Bi-LSTM [9] to obtain word representations, with the dimension $d_h = 768$. The size of the query vector and key vector $\boldsymbol{q}_{i,\alpha}$ and $\boldsymbol{k}_{j,\alpha}$ in the phrase-level attention are set as 64.

In the training process, we adopt the Adam optimizer [17] to optimize the model parameters. The initial learning rate is set as 10^{-4} for all layers with a batch size of 16. We utilize dropout to avoid over-fitting, and the drop rate is set to 0.1. We stop training if the model performance does not improve for three successive iterations. PAE-CRF is trained on a GTX 2080Ti GPU.

Our implementation of the neural network model is based on a modified version of the implementation developed by bert4torch[1].

4.5 Performance Comparison

We compare the performance of PAE-CRF with the baselines on four datasets, and the experimental results are shown in Table 3. From the results, we can see that PAE-CRF outperforms all the baseline methods on all four datasets in term of the metric $F_1@M$. Specifically, PAE-CRF achieves the improvement of 2.1 $F_1@M$ points on KP20k over

[1] https://github.com/Tongjilibo/bert4torch.

Table 3. Results of keyphrase prediction of different approaches on four datasets. The best performing score in each column is highlighted with bold, and the second-best is highlighted with underline.

Model	KP20k		Inspec		NUS		SemEval	
	F_1@M	F_1@5	F_1@M	F_1@5	F_1@M	F_1@5	F_1@M	F_1@5
CatSeq [38]	0.367	0.291	0.262	0.225	0.397	0.323	0.283	0.242
CatSeqTG [7]	0.366	0.292	0.270	0.229	0.393	0.325	0.290	0.246
CatSeq-RL [4]	0.386	0.321	0.301	0.253	0.433	0.375	0.329	0.287
ExHiRD-h [6]	0.374	0.311	0.291	0.253	-	-	0.335	0.284
SEG-Net [1]	0.379	0.311	0.265	0.216	<u>0.461</u>	0.396	0.332	0.283
One2Set [37]	<u>0.392</u>	**0.358**	0.324	0.285	0.450	0.406	<u>0.357</u>	<u>0.331</u>
DualCopyNet [33]	0.337	0.312	<u>0.342</u>	0.284	0.395	0.379	0.339	0.315
Prompt-KG [35]	0.355	<u>0.351</u>	0.260	<u>0.294</u>	0.439	<u>0.412</u>	0.356	0.329
PAE-CRF	**0.413**	0.316	**0.417**	**0.350**	**0.463**	**0.416**	**0.367**	**0.333**

Table 4. Evaluation on predicting the correct number of keyphrases on KP20k dataset. #Avg.EKps indicates the average number of extracted keyphrases per document. MAE is mean absolute error. Oracle is a model that predicts the ground-truth keyphrases.

Model	Oracle	CatSeq	CatSeqTG	CatSeq-RL	SEG-Net	One2Set	PAE-CRF
#Avg.EKps	3.31	3.69	3.77	3.86	3.79	5.10	4.01
MAE	0.00	2.27	2.28	2.20	2.19	2.94	**2.16**

the best baselines, of 7.5 $F_1@M$ points on Inspec, of 0.2 $F_1@M$ points on NUS, and of 1.0 $F_1@M$ points on SemEval, respectively. These results show our PAE-CRF can achieve the average increase of 2.7 points on this metric, which is a significant improvement in the current keyphrase prediction task. It is worth mentioning that the noticeable performance improvement 7.5 $F_1@M$ points on Inspec, which consists of more multi-word keyphrases (see #Avg.MWs and %MWs in Table 2), benefits from capturing the phrase-level features of multi-word keyphrases.

For the metric $F_1@5$, PAE-CRF outperforms all the baseline methods on three out of four datasets (5.6 $F_1@5$ points on Inspec, 0.4 $F_1@5$ points on NUS and 0.2 $F_1@5$ points on SemEval). Similarly to the result on $F_1@M$, we also observe the noticeable performance improvement 5.6 $F_1@5$ points on Inspec. However, on KP20k dataset, PAE-CRF performs worse than some baselines. This slight performance drop may be caused by the fixed number of predicted keyphrases metric $F_1@5$. In particular, when the prediction number is less than five in our PAE-CRF, we randomly append incorrect keyphrases until it obtains five predictions. In order to illustrate the ability of predicting an accurate number of keyphrases on KP20k dataset, we further compare PAE-CRF with other baseline approaches. We measure the mean absolute error (MAE, the lower the better) between the number of extracted keyphrases and the number of ground-truth keyphrases for all documents in KP20k dataset, and also report the average number of generated keyphrases per document (denoted as #Avg.EKps). The results for the

KP20k dataset are presented in Table 4. The lower MAEs of PAE-CRF indicate it better understands documents' semantic than other comparative methods. Therefore, we can infer that PAE-CRF appends more incorrect keyphrases in term of $F_1@5$, resulting in worse $F_1@5$ score.

Table 5. Ablation on PAE-CRF without CRF and phrase-level attention. We preclude one design choice at a time.

	Model	KP20k		Inspec		NUS		SemEval	
		F_1@M	F_1@5	F_1@M	F_1@5	F_1@M	F_1@5	F_1@M	F_1@5
TotalK	PAE-CRF	0.413	0.316	0.417	0.350	0.463	0.416	0.367	0.333
	w/o CRF	0.399	0.290	0.382	0.315	0.450	0.378	0.362	0.321
	w/o PA	0.374	0.245	0.172	0.123	0.389	0.280	0.250	0.188
MW	PAE-CRF	0.376	0.273	0.440	0.343	0.423	0.354	0.362	0.319
	w/o CRF	0.362	0.249	0.379	0.295	0.390	0.305	0.345	0.293
	w/o PA	0.327	0.200	0.157	0.110	0.322	0.216	0.224	0.161
SW	PAE-CRF	0.083	0.051	0.087	0.066	0.131	0.095	0.077	0.058
	w/o CRF	0.076	0.046	0.073	0.054	0.131	0.092	0.075	0.052
	w/o PA	0.079	0.048	0.020	0.014	0.099	0.065	0.038	0.028

4.6 Ablation Study

To analyze the relative contributions of different components to the model performance in predicting total keyphrases (TotalK), multi-word keyphrases (MW) and single-word keyphrases (SW), we compare our full model PAE-CRF with the following ablated variants: (1) *w/o CRF* where we utilize only the phrase-level attention scores to extract keyphrases, (2) *w/o phrase-level attention* (PA) where only neural CRF based on the word-level features is leveraged to extract keyphrases. The results of ablation study are shown in Table 5.

From these results, we have the following observations: (1) PAE-CRF consistently outperforms its ablated variants on predicting three types of keyphrases (*i.e.*, TotalK, MW and SW) across all datasets in terms of all the metrics, indicating that combining these two different modules is useful for keyphrase extraction. (2) Compared with PAE-CRF w/o PA, PAE-CRF w/o CRF gains better performance on predicting both TotalK and MW keyphrases across all datasets in terms of all the metrics. However, on predicting SW keyphrases, PAE-CRF w/o CRF underperforms slightly PAE-CRF w/o PA on KP20k dataset in terms of all the metrics, and outperforms slightly PAE-CRF w/o PA on other three datasets in terms of all the metrics. This observation indicates that the phrase-level attention mechanism is more helpful than CRF module to improve the performance of keyphrase prediction, especially for MW keyphrases. However, the effect of CRF module is even better than of the phrase-level attention where the percentage of SW keyphrases is large (*e.g.*, KP20k with 39.4% SW, see Table 2). (3) PAE-CRF w/o CRF outperforms PAE-CRF w/o PA on Inspec, NUS and SemEval datasets (in which the percentage of MW keyphrases is large, especially for Inspec with 85.8%

MW) in terms of all the metrics, indicating that the phrase-level attention can effectively capture the local dependencies within phrase and help to predict the multi-word keyphrases. (4) It is worth pointing out that an extreme variance in the performance of the proposed method and its ablated variants between extracting single-word versus multi-word keyphrases, indicating that our proposed models are good at extracting multi-word keyphrases rather than single-word keyphrases. We will explore the possible reasons for this result in future research.

Table 6. An example of predicted keyphrases by different approaches. Author-assigned (*i.e.*, Gold) keyphrases are shown in bold, absent keyphrases are labeled by the *, and overlapping words between correctly-predicted keyphrases and incorrectly-predicted keyphrases are labeled by the underline.

Example (#203 in KP20k)
Title: **Fixed**(1) **points**(2) of(3) **correspondences**(4) defined(5) on(6) **cone**(7) **metric**(8) **spaces**(9) .(10) eos(11)
Abstract: In(12) the(13) present(14) note(15) ,(16) we(17) investigate(18) the(19) **fixed**(20) **points**(21) of(22) **correspondences**(23) defined(24) on(25) **cone**(26) **metric**(27) **spaces**(28) satisfying(29) a(30) conditionally(31) contractive(32) condition(33) .(34)
Gold: **fixed point; cone metric space; correspondence; banach lattice***
PAE-CRF: ***fixed point*****;** ***cone metric space*****;** ***correspondence***
CatSeq-RL: **fixed points**; **cone metric spaces**; **correspondence**; fixed point matching; cone metric
ExHiRD-h: **fixed point**; **cone metric space**; point set; cone; <digit>
One2Set: **cone metric space**; **fixed point**; fixed point of correspondences; conditionally contractive mapping

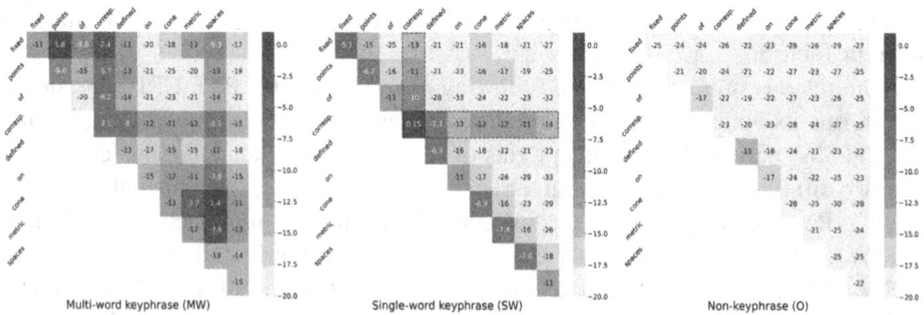

Fig. 2. Phrase-level attention visualization for predicting multi-word keyphrase (MW), single-word keyphrase (SW) and non-keyphrase (O) on the title of paper #203 in KP20k. Deeper shading denotes higher value. "corresp." stands for correspondences.

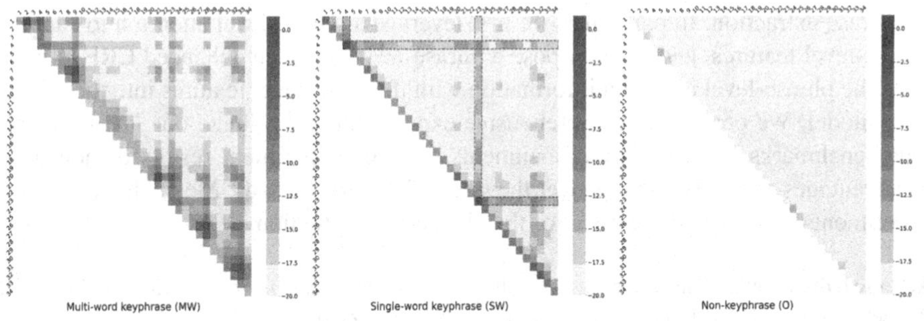

Fig. 3. Phrase-leve attention visualization for predicting MW, SW and O on the title and abstract of paper #203 in KP20k dataset.

4.7 Case Study and Visualization

Here we select one anecdotal example of research papers from KP20k dataset shown in Table 6. The predictions predicted by different methods along with author-assigned keyphrases are listed in this table. As can be seen from this table, PAE-CRF can exactly extract the present keyphrases. In addition, overlapping words between correctly-predicted keyphrases and incorrectly-predicted keyphrases appear in all other comparative methods, but do not appear in our PAE-CRF. This example can indicate that other comparative methods can not accurately determine the boundary of a keyphrase. By contrast, PAE-CRF directly leverages the boundary information by adding the position embedding RoPE into the phrase-level attention, and the word-level sequence labeling (*e.g.*, B_{kp} and E_{kp}) in CRF module of PAE-CRF may help to determine the boundary of a keyphrase. Therefore, our PAE-CRF is capable of determining boundaries of keyphrases and gains better performance in keyphrase prediction.

We then visualize the attention map of this example generated by the phrase-level attention mechanism. The attention visualizations for predicting multi-word keyphrase (MW), single-word keyphrase (SW) and non-keyphrase (O) of its title and title+abstract are shown in Fig. 2 and Fig. 3, respectively. From the attention map of predicting multi-word keyphrases, we can see that PAE-CRF pays more attention to the local dependencies within a keyphrase, as illustrated by the small sections framed by red solid line (which correspond to "fixed points" and "cone metric spaces"). From the attention map of predicting single-word keyphrases, we can see that PAE-CRF pays more attention to the self-dependency, as illustrated by the small sections framed by magenta solid line (which correspond to "correspondences"), and that the single-word keyphrases have a stronger long-distance dependencies, as illustrated by the small sections framed by magenta dashed line. These observations indicate that the phrase-level attention is very helpful for recognizing multi-word and single-word keyphrases.

5 Conclusions

In this study, we propose a phrase-level attention enhanced conditional random field model (PAE-CRF) for keyphrase extraction, which incorporates the phrase-level features with the word-level features into the CRF model to improve the performance of keyphrase extraction. In particular, we first leverage a phrase-level attention to capture phrase-level features, and then propose a phrase-level attention enhanced CRF to integrate the phrase-level attention information with the word-level features into the neural CRF model. We conducted comprehensive experiments to evaluate our PAE-CRF on four benchmarks from scientific documents, and the experiment results demonstrate its advantages and effectiveness over the state-of-the-art methods. Meanwhile, ablation experiments show a positive effect of the phrase-level attention.

Acknowledgements. This work was partially supported by grants from the Scientific Research Project of Tianjin Educational Committee (Grant No. 2021ZD002).

References

1. Ahmad, W., Bai, X., Lee, S., Chang, K.W.: Select, extract and generate: neural keyphrase generation with layer-wise coverage attention. In: Proceedings of ACL, pp. 1389–1404 (2021)
2. Alzaidy, R., Caragea, C., Giles, C.L.: Bi-LSTM-CRF sequence labeling for keyphrase extraction from scholarly documents. In: Proceedings of WWW, pp. 2551–2557 (2019)
3. Bhaskar, P., Nongmeikapam, K., Bandyopadhyay, S.: Keyphrase extraction in scientific articles: a supervised approach. In: Proceedings of COLING, pp. 17–24 (2012)
4. Chan, H.P., Chen, W., Wang, L., King, I.: Neural keyphrase generation via reinforcement learning with adaptive rewards. In: Proceedings of ACL, pp. 2163–2174 (2019)
5. Chen, J., Zhang, X., Wu, Y., Yan, Z., Li, Z.: Keyphrase generation with correlation constraints. In: Proceedings of EMNLP, pp. 4057–4066 (2018)
6. Chen, W., Chan, H.P., Li, P., King, I.: Exclusive hierarchical decoding for deep keyphrase generation. In: Proceedings of ACL, pp. 1095–1105 (2020)
7. Chen, W., Gao, Y., Zhang, J., King, I., Lyu, M.R.: Title-guided encoding for keyphrase generation. In: Proceedings of AAAI, pp. 6268–6275 (2019)
8. Cho, K., van Merriënboer, B., Bahdanau, D., Bengio, Y.: On the properties of neural machine translation: Encoder-decoder approaches. In: Proceedings of SSST, pp. 103–111 (2014)
9. Chung, J., Gulcehre, C., Cho, K., Bengio, Y.: Empirical evaluation of gated recurrent neural networks on sequence modeling. In: NIPS 2014 Workshop on Deep Learning (2014)
10. Devlin, J., Chang, M.W., Lee, K., Toutanova, K.: BERT: Pre-training of deep bidirectional transformers for language understanding. In: Proceedings of NAACL, pp. 4171–4186 (2019)
11. Gollapalli, S.D., Li, X.L., Yang, P.: Incorporating expert knowledge into keyphrase extraction. In: Proceedings of AAAI, pp. 3180–3187 (2017)
12. Gu, J., Lu, Z., Li, H., Li, V.O.: Incorporating copying mechanism in sequence-to-sequence learning. In: Proceedings of ACL. pp. 1631–1640 (2016)
13. Hasan, K.S., Ng, V.: Automatic keyphrase extraction: a survey of the state of the art. In: Proceedings of ACL, pp. 1262–1273 (2014)
14. Hulth, A.: Improved automatic keyword extraction given more linguistic knowledge. In: Proceedings of EMNLP, pp. 216–223 (2003)
15. Weston, J., Sumit Chopra, A.B.: Memory networks. In: Proceedings of ICLR (2015)
16. Kim, S.N., Medelyan, O., Kan, M.Y., Baldwin, T.: Automatic keyphrase extraction from scientific articles. Lang. Resour. Eval. **47**(3), 723–742 (2013)
17. Kingma, D., Ba, J.: Adam: a method for stochastic optimization. In: Proceedings of ICLR (2015)
18. Lafferty, J.D., McCallum, A., Pereira, F.C.N.: Conditional random fields: Probabilistic models for segmenting and labeling sequence data. In: Proceedings of ICML, pp. 282–289 (2001)

19. Liu, T., Iwaihara, M.: Supervised learning of keyphrase extraction utilizing prior summarization. In: Proceedings of ICADL, pp. 157–166 (2021)
20. Lu, X., Chow, T.W.S.: Duration modeling with semi-Markov conditional random fields for keyphrase extraction. IEEE Trans. Knowl. Data Eng. **33**(4), 1453–1466 (2021)
21. Meng, R., Zhao, S., Han, S., He, D., Brusilovsky, P., Chi, Y.: Deep keyphrase generation. In: Proceedings of ACL, pp. 582–592 (2017)
22. Meng, R., Zhao, S., Han, S., He, D., Brusilovsky, P., Chi, Y.: Deep keyphrase generation. In: Proceedings of ACL, pp. 582–592 (2017)
23. Nguyen, T.D., Kan, M.Y.: Keyphrase extraction in scientific publications. In: Proceedings of ICADL, pp. 317–326 (2007)
24. Santosh, T.Y.S.S., Sanyal, D.K., Bhowmick, P.K., Das, P.P.: Dake: document-level attention for keyphrase extraction. In: Proceedings of ECIR, pp. 392–401 (2020)
25. Sarawagi, S., Cohen, W.W.: Semi-Markov conditional random fields for information extraction. In: Proceedings of NeurIPS, pp. 1185–1192 (2004)
26. Song, M., Liu, H., Jing, L.: Hyperrank: hyperbolic ranking model for unsupervised keyphrase extraction. In: Proceedings of EMNLP, pp. 16070–16080 (2023)
27. Sterckx, L., Caragea, C., Demeester, T., Develder, C.: Supervised keyphrase extraction as positive unlabeled learning. In: Proceedings of EMNLP, pp. 1924–1929 (2016)
28. Su, J., Ahmed, M., Lu, Y., Pan, S., Bo, W., Liu, Y.: Roformer: enhanced transformer with rotary position embedding. Neurocomputing **568**, 127063 (2023)
29. Su, J., et al.: Global pointer: novel efficient span-based approach for named entity recognition (2022). https://arxiv.org/abs/2208.03054
30. Sutskever, I., Vinyals, O., Le, Q.V.: Sequence to sequence learning with neural networks. In: Proceedings of NIPS, pp. 3104–3112 (2014)
31. Tang, Y., et al.: Qalink: enriching text documents with relevant q&a site contents. In: Proceedings of CIKM, pp. 1359–1368 (2017)
32. Vaswani, A., et al.: Attention is all you need. In: Proceedings of NeurIPS, pp. 6000–6010 (2017)
33. Wang, S., Jiang, J., Huang, Y., Wang, Y.: Automatic keyphrase generation by incorporating dual copy mechanisms in sequence-to-sequence learning. In: Proceedings of COLING, pp. 2328–2338 (2022)
34. Wang, Y., Li, J., Chan, H.P., King, I., Lyu, M.R., Shi, S.: Topic-aware neural keyphrase generation for social media language. In: Proceedings of ACL, pp. 2516–2526 (2019)
35. Wu, H., Ma, B., Liu, W., Chen, T., Nie, D.: Fast and constrained absent keyphrase generation by prompt-based learning. In: Proceedings of AAAI, pp. 11495–11503 (2022)
36. Yang, T., Hu, L., Shi, C., Ji, H., Li, X., Nie, L.: HGAT: heterogeneous graph attention networks for semi-supervised short text classification. ACM Trans. Inf. Syst. **39**(3), 1–29 (2021)
37. Ye, J., Gui, T., Luo, Y., Xu, Y., Zhang, Q.: One2Set: generating diverse keyphrases as a set. In: Proceedings of ACL, pp. 4598–4608 (2021)
38. Yuan, X., et al.: One size does not fit all: Generating and evaluating variable number of keyphrases. In: Proceedings of ACL, pp. 7961–7975 (2020)
39. Zhang, C., Wang, H., Liu, Y., Wu, D., Liao, Y.P., Wang, B.: Automatic keyword extraction from documents using conditional random fields. J. Comput. Inf. Syst. **4**(3), 1169–1180 (2008)
40. Zhang, Y., Jiang, T., Yang, T., Li, X., Wang, S.: Htkg: deep keyphrase generation with neural hierarchical topic guidance. In: Proceedings of SIGIR, pp. 1044–1054 (2022)
41. Zhang, Y., Yang, T., Jiang, T., Li, X., Wang, S.: Hyperbolic deep keyphrase generation. In: Proceedings of ECML-PKDD, pp. 521–536 (2022)
42. Zhao, J., Zhang, Y.: Incorporating linguistic constraints into keyphrase generation. In: Proceedings of ACL, pp. 5224–5233 (2019)
43. Zhou, T., Zhang, Y., Zhu, H.: Multi-level memory network with CRFs for keyphrase extraction. In: Proceedings of PAKDD, pp. 726–738 (2020)

TWOLAR: A TWO-Step LLM-Augmented Distillation Method for Passage Reranking

Davide Baldelli[1,3(✉)], Junfeng Jiang[2], Akiko Aizawa[3], and Paolo Torroni[1]

[1] University of Bologna, Bologna, Italy
davide.baldelli4@studio.unibo.it
[2] University of Tokyo, Tokyo, Japan
[3] National Institute of Informatics, Tokyo, Japan

Abstract. In this paper, we present TWOLAR: a two-stage pipeline for passage reranking based on the distillation of knowledge from Large Language Models (LLM). TWOLAR introduces a new scoring strategy and a distillation process consisting in the creation of a novel and diverse training dataset. The dataset consists of 20K queries, each associated with a set of documents retrieved via four distinct retrieval methods to ensure diversity, and then reranked by exploiting the zero-shot reranking capabilities of an LLM. Our ablation studies demonstrate the contribution of each new component we introduced. Our experimental results show that TWOLAR significantly enhances the document reranking ability of the underlying model, matching and in some cases even outperforming state-of-the-art models with three orders of magnitude more parameters on the TREC-DL test sets and the zero-shot evaluation benchmark BEIR. To facilitate future work we release our data set, finetuned models, and code (Code: https://github.com/Dundalia/TWOLAR; Models and Dataset: https://huggingface.co/Dundalia).

Keywords: Information Retrieval · Reranking · Knowledge distillation · Large Language Model

1 Introduction

Text ranking, a foundational task in search engines and question-answering systems, involves ordering textual documents based on their relevance to a given query or context.

The state-of-the-art text rerankers are traditional cross-encoders like monoBERT [27], monoT5 [28,35], and RankT5 [44], and more recently rerankers based on Large Language Models (LLMs) like RankGPT [36] and PRP [31].

Cross-encoders are computationally efficient but rely on costly human-annotated labels for training, while LLM-based methods bypass the need of in-domain fine-tuning. However, a significant downside is their substantial size and computational footprint, which could render them unsuitable for real-time

N. Goharian et al. (Eds.): ECIR 2024, LNCS 14608, pp. 470–485, 2024.
https://doi.org/10.1007/978-3-031-56027-9_29

inference. However, the knowledge of an LLM could be distilled to produce a student model with performance comparable to the teacher model, but a size several orders of magnitude smaller.

In this paper, we present **TWOLAR**, a **TWO**-step **L**LM-**A**ugmented distillation method for passage **R**eranking. The distillation consists in exploiting the capabilities of an LLM as a reranker to produce high-quality annotations. The annotations are applied to a dataset of queries generated artificially, either as cropped sentences or again by a specialized language model. In this way, we obtain a compact model that ranks among top performing supervised, zero-shot, and LLM-based distillation methods in various popular benchmarks.

The paper is structured as follows: Sect. 2 provides background on text ranking methods and benchmarks. Section 3 details our approach, subdivided into scoring and distillation strategies. Section 4 covers the experimental setup, including datasets, training, baselines, and results. Section 5 discusses the results and ablation studies. Section 6 concludes the paper.

2 Background

Formally, given a query and a passage from a large text collection, text ranking requires returning a ranked list of the n most relevant texts according to the relevance scores of a model.

The early text ranking methods relied mainly on statistical lexical features, like BM25 [34] and TF-IDF, and used heuristic methods for retrieval [42]. Based on this scheme, each query-document relevance score is computed according to the lexical similarity. Subsequently, statistical language modeling has been widely explored for text ranking [41]. With the development of machine learning, supervised approaches, which still rely on hand-crafted features as well as lexical features, have been proposed [19,22]. Further progress was made with the adoption of neural networks mapping pieces of text into low-dimensional vectors to obtain better representations [13,14,25]. Recently, a new paradigm emerged [9,20,30,42,43], consisting of multiple stages: using a first-stage retriever that aims to reduce the candidate space by retrieving a subset of relevant candidates, often numbering in the hundreds or thousands, and then refining these initial results with a second-stage reranker.

The advent of pretrained language models (PLMs) [8,32,33] and large-scale human annotated datasets [18,26,40] marked a significant advancement in the field. Models such as DPR [17], SPLADE [10–12], and DRAGON [21] emerged as powerful first-stage retrievers. Complementing these retrievers, models like monoBERT [27], monoT5 [28,35], and RankT5 [44] have been conceived as second-stage rerankers, designed explicitly to refine and optimize the results provided by the initial retrieval stage.

Recently, large language models (LLMs) have begun to play an influential role in text reranking [24,43]. The latest approaches in text reranking utilize LLMs in various ways. For instance, InPars [2] and InParsV2 [16] leverage GPT-3 Curie [3] and GPT-J [39] respectively, for data augmentation, generating synthetic queries to adapt a reranking model to different reranking tasks and domains. Other approaches instead consist in directly prompting the LLM to permute a set of documents given a query. In this vein, RankGPT [36], LRL [24], and PRP [31] have demonstrated the potential of LLMs as zero-shot rerankers. Moreover, RankGPT [36] demonstrates how the ranking abilities of ChatGPT could be distilled into a more efficient DeBERTa [15]. For a comprehensive survey on LLMs for Information Retrieval, the interested readers can refer to [43].

As for the benchmarks, The largest annotated dataset for information retrieval is the MS MARCO passage reranking dataset [26]. It contains around 530K train queries and 6.8K 'dev' queries. The corpus is composed of more than 8.8M passages. For each query, relevant passages are annotated as 1 and others are annotated as 0. TREC-DL2019 [7] and TREC-DL2020 [6] are standard benchmarks derived from MS MARCO that provide dense human relevance annotations for each of their 43 and 54 queries. BEIR [37] is a heterogeneous benchmark containing 18 retrieval datasets, covering different retrieval tasks and domains, designed for zero-shot evaluation.

We believe that the potential of LLM distillation for text ranking has not been fully exploited yet. Our contribution aims at bridging this gap by the methodological construction of a training dataset.

3 Approach

Our reranking model is based on Flan-T5 [5], which is a text-to-text model developed as an instructed version of T5 [33]. For our task, we use the following input template:

Query: [Q] Document: [D] Relevant:

where [Q] and [D] are the query and document texts, respectively, similar to the one adopted in monoT5 [28,35].

3.1 Scoring Strategy

Flan-T5 can be straightforwardly applied to various tasks due to its text-to-text nature, such as summarization, translation, and classification. However, adapting to the ranking task is not trivial, because for each query-document pair, we usually ask models to answer with a score representing the degree of relevance. The state-of-the-art rerankers, monoT5 [28,35] and RankT5 [44], which are specialized text-to-text models, suffer from this limitation.

MonoT5 finetunes T5 on a binary classification task: given a query-document pair, the model is optimized to produce the words '`true`' if the document is relevant to the query and '`false`' otherwise. At inference time, the ranking score is obtained from the logits of the '`true`' and '`false`' tokens as follows:

$$s = \frac{e^{z_{\texttt{true}}}}{e^{z_{\texttt{true}}} + e^{z_{\texttt{false}}}} \tag{1}$$

where $z_{\texttt{true}}, z_{\texttt{false}}$ are the logits of '`true`' and '`false`', respectively.

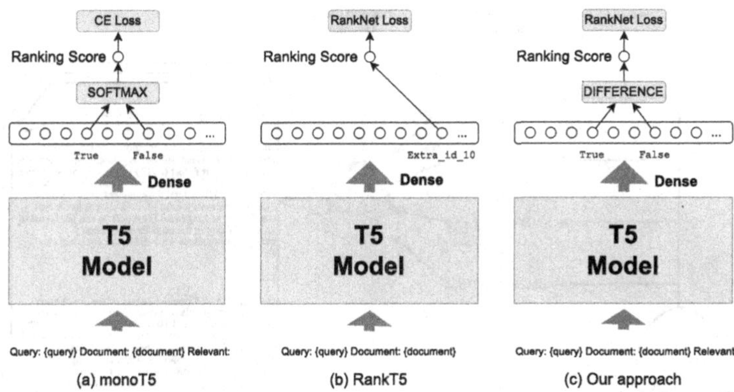

Fig. 1. Illustration of the score strategies from monoT5, RankT5 and our proposed approach.

RankT5 directly learns to rank by optimizing a ranking-based loss function. This family of loss functions requires the model to directly output the ranking score for each query-document pair at training time, so that the unnormalized logit of a special unused token ('`extra_id_10`') in the vocabulary is used as ranking score.

On one hand, monoT5 is not directly finetuned as a ranking model, which may not optimize its ranking performance. On the other hand, RankT5 does not exploit the learned representation in the language modeling head. To overcome both limitations, we propose a new approach. Our idea consists of using the difference between the unnormalized logits corresponding to the '`true`' and '`false`' tokens. In this way, the model is able to output a score directly at training time, and since it is optimized on top of the learned representations of the two tokens, we can make full use of the knowledge from the PLMs. An illustration of these scoring strategies is shown in Fig. 1.

Accordingly, we adopt the RankNet loss [4], a pairwise loss function that models the probability of one document being more relevant than another given a query. RankNet loss has shown compelling results in information retrieval [36, 44] and provided a solid foundation for our optimization process.

Given a query q and M passages $(p_1, \ldots, p_M)$ ordered by relevance $R = (r_1, \ldots, r_M)$, where $r_i \in \{1, 2, \ldots, M\}$ is the rank of the passage p_i (if $r_i = 1$, it means that p_i is the most relevant passage), our model takes as input each query-document pair (q, p_i) and outputs a relevance score s_i.

Therefore, we optimize our model with the following loss function measuring the correctness of relative passage orders:

$$\mathcal{L}_{RankNet} = \sum_{i=1}^{M} \sum_{j=1}^{M} \mathbb{I}_{r_i < r_j} \log(1 + e^{s_i - s_j}) \tag{2}$$

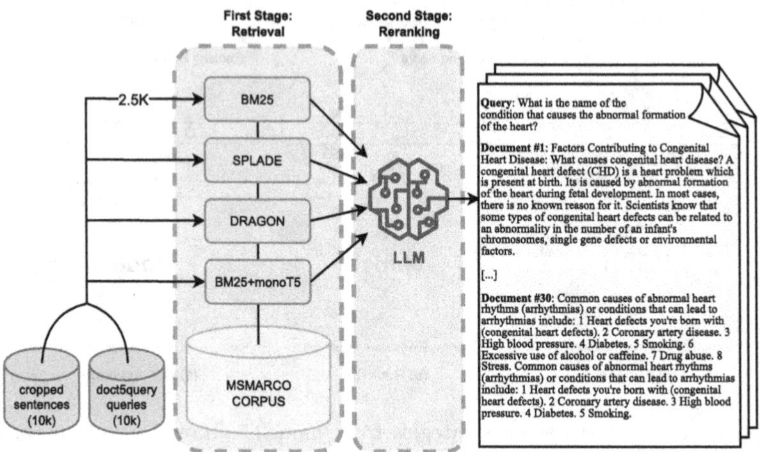

Fig. 2. Illustration of the method used to build the dataset.

It should be noted that the adoption of different ranking loss functions could potentially lead to alternative outcomes, but exploring their potential differences is not the main purpose of this paper. Thus, we utilize the RankNet Loss here and leave the comparison of different ranking loss functions as future work.

3.2 Distillation Strategy

Our distillation strategy aims to capture the reranking capability of LLMs, in our case ChatGPT, through constructing a query-document dataset. The core design principle is the synthesis of suitable artificial queries by query augmentation, and the subsequent use of multiple retrieval models and stages of distillation.

Query Augmentation. Our query augmentation method is inspired by DRAGON [21], whereby two approaches are combined to amplify the size of training queries from a given corpus: sentence cropping and pseudo query generation. The former can readily scale up the size of query sets without any computationally expensive operation. The latter generates high-quality human-like queries using LLMs. The combination of the two strategies increases the diversity of the dataset and accordingly the challenge and complexity of the task.

Following DRAGON, we randomly sampled 10K queries from DRAGON's collection of cropped sentences, drawn from the MS MARCO corpus [26]. Simultaneously, we sampled an additional 10K queries from the query pool created by docT5query [29], a specialized T5 model that generates queries based on a given passage.

First-Stage Distillation: Retrieval. The initial phase of our distillation process (see Fig. 2) involves splitting each of the two sets of 10K queries, one set composed of cropped sentences and another set composed of docT5query-generated queries, into four subsets of 2.5K queries each. To retrieve documents for these queries, we chose four distinct retrieval models:

- **BM25** [34]: A state-of-the-art bag-of-words approach that relies primarily on word overlap to match documents to queries. Consequently, its hard negatives are expected to challenge the language model on lexical-level matches.
- **DRAGON** [21]: A dense retrieval model designed to detect semantic similarity between queries and passages. It pushes the language model towards understanding deeper semantic relations and contexts. We have chosen the DRAGON+ version.
- **SPLADE** [10–12]: It serves as a kind of midpoint between BM25's focus on word overlap and DRAGON's emphasis on semantic similarity. It introduces a different level of complexity by considering interactions between the tokens of the document or query and all the tokens from the vocabulary. We have chosen the SPLADE++ version.
- **monoT5** [28,35]: A combination of BM25 and monoT5 where the top-100 documents retrieved by BM25 are re-ranked using monoT5. It introduces negatives that are influenced by the ranking capabilities of a cross-encoder.

In all cases, we retrieve the top 30 documents for each query.

This methodology is designed to provide high-quality results and to diversify the types of challenges and contexts presented to ChatGPT in the subsequent distillation stage.

Table 1 gives a quantitative account of the diversity of the documents retrieved by the four distinct models by computing the intersection rate between the sets of documents obtained from any two sources of supervision.[1] The low mean intersection rates between different sources provide a clear evidence of the diversity among the retrieved document sets for both types of queries.

Table 1. Average intersection rate between each pair of sources. The upper triangular part of the table represents the intersection rate for cropped sentences and the lower triangular part represents the intersection rate for docT5query-generated queries.

doct5query\cropped sentence (%)	BM25	SPLADE	DRAGON	monoT5
BM25	\	20.0	29.0	49.8
SPLADE	17.8	\	35.8	26.0
DRAGON	25.0	41.0	\	38.4
monoT5	46.4	27.2	38.5	\

[1] The average intersection rates were then calculated to provide a comprehensive view of the overall overlap among the retrieved documents from all sources:

$$\frac{1}{|\mathcal{Q}|}\sum_{q\in\mathcal{Q}}\frac{|S_q^1 \cap S_q^2|}{N} \tag{3}$$

where $\mathcal{Q}$ is the whole query set, S_q^1 and S_q^2 represent the retrieved document set from two sources given the query q. This process was carried out separately for both types of queries: the cropped sentence queries and the docT5query-generated queries.

Second-Stage Distillation: Reranking. For reranking we used ChatGPT, in particular the checkpoint '`gpt-3.5-turbo-16k-0613`'. We prompted the model with the same prompt design proposed by RankGPT [36], including all the 30 documents per query to rerank in a single prompt.

We used the prompt made available by the RankGPT public repository.[2] We fed each of the 20K queries and their corresponding top 30 retrieved documents to ChatGPT, asking it to provide permutations of the indices of these documents, ordered according to their relevance to the associated query.

This approach requires significant computational resources due to the complexity of the task and the vast number of queries and documents involved. However, the total cost of this reranking operation using the ChatGPT API amounted to $212, demonstrating the feasible financial aspect of employing a large-scale language model in creating such a diverse and complex dataset.

Train-Validation Split. We divided the dataset into training and validation splits. We included 1,000 queries in the validation set: 500 queries generated by docT5query and the rest extracted as cropped sentences. The remaining 19,000 samples were allocated to the training set.

4 Experimental Setup

4.1 Datasets

We evaluate our approach using TREC-DL2019, TREC-DL2020 and BEIR for the zero-shot evaluation. All comparisons on TREC-DL2019 and TREC-DL2020 are based on the reranking of top-100 passages retrieved by BM25 [34] for each query. The evaluation on the BEIR benchmark is based on the reranking of the top-100 passages retrieved by three different retrievers: BM25, SPLADE++ [11], and DRAGON+ [21]. The objective is to evaluate the adaptability of the rerankers to different retrievers. We also present the evaluation by reranking the top-1000 documents retrieved by BM25, to give a broad view of the performances in different settings.

4.2 Training

We initialized our model with pretrained Flan-T5-xl checkpoint [5]. We set the maximum input sequence length to 500 as monoT5. The batch size is set to 32, meaning that the parameters are updated every 32×30 query-document pairs. We utilize the AdamW [23] optimizer with a constant learning rate of $5e-5$.

4.3 Baselines

We evaluate our model on the TREC-DL2019 and TREC-DL2020 competitions against several baselines including three supervised methods: monoBERT [27],

[2] https://github.com/sunnweiwei/RankGPT.

monoT5-3B [28,35], RankT5 [44]; two zero-shot LLM-based methods: the *list-wise* prompting based approach of RankGPT [36], using both `gpt-3.5-turbo` and `gpt-4`, and the sliding window approach performed only for 10 passes of PRP [31], using Flan-T5-xl (3B), Flan-T5-xxl (11B) and Flan-UL2 (20B); and a representative distilled model based on DeBERTav3 proposed in [36] as the only other LLM distillation method other than ours. Regarding the zero-shot evaluation on the BEIR benchmark, we evaluate our models comparing with three different rerankers, including InParsV2 [2,16], monoT5-3B [28,35], and the distilled DeBERTav3 model proposed in [36].

4.4 Results

Our results on the TREC-DL2019 and TREC-DL2020 benchmarks are summarized in Table 2. Tables 3 and 4 instead summarize respectively the results on the BEIR benchmark by reranking the top-100 and the top-1000 documents.

Table 2. Results on TREC-DL2019 and TREC-DL2020 datasets by reranking top 100 documents retrieved by BM25. The column titled '#Calls' indicates the exact number of inference times of LLM when reranking the top 100 documents. The 'Input Size' column uses the notation $|q| + n|d|$: $|q|$ represents one query and $n|d|$ indicates the number of documents included. For instance, $|q| + 20|d|$ signifies an input of one query with 20 documents. Best model is highlighted in boldface and the second best is underlined for each metric. All the reported results apart from the LLM distillation Methods are taken from the original papers.

Method	LLM	Size	#Calls	Input Size	TREC-DL2019			TREC-DL2020		
					nDCG@1	nDCG@5	nDCG@10	nDCG@1	nDCG@5	nDCG@10
BM25	–	–	–	–	54.26	52.78	50.58	57.72	50.67	47.96
Supervised Methods										
monoBERT	BERT	340M	100	$\|q\| + \|d\|$	79.07	73.25	70.50	78.70	70.74	67.28
monoT5	T5-xl	3B	100	$\|q\| + \|d\|$	79.07	73.74	71.83	80.25	72.32	68.89
RankT5	T5-xl	3B	100	$\|q\| + \|d\|$	77.38	73.94	71.22	<u>80.86</u>	72.99	69.49
LLM distillation Methods										
RankGPT	DeBertaV2	184M	100	$\|q\| + \|d\|$	78.68	69.77	66.56	59.26	59.83	59.43
TWOLAR-large	Flan-T5-large	783M	100	$\|q\| + \|d\|$	79.84	75.94	72.82	79.94	71.35	67.61
TWOLAR-xl	Flan-T5-xl	3B	100	$\|q\| + \|d\|$	78.29	<u>76.71</u>	<u>73.51</u>	80.25	73.73	**70.84**
Zero-shot LLM Methods										
RankGPT	gpt-3.5-turbo	154B[a]	10	$\|q\| + 20\|d\|$	<u>82.17</u>	71.15	65.80	79.32	66.76	62.91
RankGPT	gpt-4	1T[a]	2[b]	$\|q\| + 20\|d\|$	**82.56**	**79.16**	**75.59**	78.40	<u>74.11</u>	<u>70.56</u>
PRP-Sliding-10	Flan-T5-xl	3B	990	$\|q\| + 2\|d\|$	75.58	71.23	68.66	75.62	69.00	66.59
PRP-Sliding-10	Flan-T5-xxl	11B	990	$\|q\| + 2\|d\|$	64.73	69.49	67.00	75.00	70.76	67.35
PRP-Sliding-10	Flan-UL2	20B	990	$\|q\| + 2\|d\|$	78.29	75.49	72.65	**85.80**	**75.35**	70.46

[a] OpenAI has not publicly released the amount of parameters and the numbers are based on public estimates [1,38].

[b] In [36], gpt-4 reranks the top-30 passages reranked by gpt-3.5-turbo.

5 Discussion

In this section, We will delve into a comprehensive discussion of our experimental results, and in the following section, we will explore the ablation study in detail.

Table 3. Results on the BEIR Benchmark by reranking the top 100 documents with different retrievers. Best model is in boldface and second best is underlined for each dataset. Evaluation for InPars on CQADupStack is missing due to its unavailability on the Hugging Face hub. We computed statistical tests comparing our biggest model against the baselines. The results revealed no significant difference compared to InPars ($p = 0.477$), but indicated significant improvements over MonoT5-3B ($p = 4.20 \times 10^{-4}$) and RankGPT-Deberta ($p = 2.82 \times 10^{-13}$). It's noteworthy that while our model operates in a zero-shot manner, the InPars models have been fine-tuned for each BEIR dataset.

Retriever	BM25						SPLADE						DRAGON					
Reranker	-	MonoT5-3B	InPars	RankGPT-Deberta	TWOLAR-xl	TWOLAR-large	-	MonoT5-3B	InPars	RankGPT-Deberta	TWOLAR-xl	TWOLAR-large	-	MonoT5-3B	InPars	RankGPT-Deberta	TWOLAR-xl	TWOLAR-large
	nDCG@10																	
TREC-COVID	59.5	79.8	82.5	79.4	82.7	84.3	72.8	82.9	85.7	80.1	85.2	**86.9**	75.8	82.8	84.8	82.6	84.6	86.8
NFCorpus	32.2	37.4	35.0	33.3	36.6	35.7	34.8	39.2	38.8	33.2	37.3	35.5	33.9	**39.7**	39.3	33.2	37.9	35.7
FiQA-2018	23.6	46.1	46.2	32.7	41.9	41.1	34.8	50.0	50.0	33.7	44.8	43.8	35.7	**51.2**	50.9	43.1	45.3	44.8
ArguAna	30.0	33.4	32.8	21.1	32.9	34.7	38.8	31.7	31.2	18.6	32.9	34.6	**46.9**	41.5	40.9	25.7	42.8	45.5
Tóuche-2020	**44.2**	31.6	29.6	37.7	37.1	33.4	24.6	29.8	28.7	36.4	35.2	30.4	26.3	30.6	29.4	38.2	36.0	31.5
Quora	78.9	84.1	84.8	78.8	87.2	86.0	83.5	84.3	85.1	80.3	87.4	86.0	**87.5**	83.5	84.4	78.7	87.2	85.7
SCIDOCS	14.9	19.0	19.2	16.1	19.5	18.3	15.9	19.9	**20.9**	16.4	20.2	18.8	15.9	19.8	20.7	16.4	20.2	18.8
SciFact	67.9	76.4	73.5	70.5	**76.5**	75.6	70.2	76.4	76.0	69.1	75.6	74.7	67.8	76.0	75.7	69.4	75.6	74.7
NQ	30.6	56.8	57.8	46.1	58.0	57.7	53.7	65.9	66.4	50.6	66.8	65.8	53.8	65.1	66.6	50.6	**66.9**	66.2
HotpotQA	63.3	74.2	76.5	69.9	76.7	75.9	68.7	74.1	77.1	70.5	**77.7**	76.4	66.2	72.9	75.7	69.8	76.4	75.3
DBPedia	31.8	44.8	44.0	41.9	48.0	47.8	43.6	48.2	51.1	45.9	**52.9**	51.6	41.9	47.2	50.3	44.9	52.1	51.3
FEVER	65.1	83.2	85.5	80.2	84.9	83.4	79.3	85.0	88.0	81.8	87.5	85.4	78.0	84.7	**87.7**	81.7	87.2	85.2
Climate-FEVER	16.5	27.4	30.1	24.2	26.9	26.1	22.9	28.7	**32.8**	25.9	28.9	27.9	22.7	28.6	32.5	25.9	28.6	27.4
CQADupStack	30.2	41.5	-	34.7	41.2	40.6	33.4	43.7	-	35.9	43.6	42.7	35.4	**44.4**	-	36.0	44.2	43.4
Robust04	40.8	56.6	58.7	52.8	57.9	58.3	46.7	62.1	**64.3**	57.3	64.9	65.2	48.1	61.3	63.2	56.6	63.4	63.7
Signal-1M	33.1	32.2	32.9	33.4	**33.8**	33.9	30.0	29.4	30.3	30.0	30.1	30.5	30.0	29.7	30.4	29.4	30.2	30.1
BioASQ	52.3	57.2	**59.8**	53.0	56.2	56.0	49.7	54.1	57.2	49.5	54.6	53.8	43.4	51.9	54.4	48.0	51.9	50.8
TREC-NEWS	39.5	48.5	49.8	51.8	52.7	50.8	41.5	50.0	50.9	53.4	53.3	50.7	44.4	49.5	50.8	52.1	**53.8**	50.0
	avg nDCG@10																	
BEIR 18	41.9	51.7	-	47.6	52.8	52.2	46.9	53.0	-	48.3	54.4	53.4	47.4	53.4	-	48.5	**54.7**	53.7
BEIR 17	42.6	52.3	52.9	48.4	53.5	52.9	47.8	53.5	55.0	49.0	55.0	54.0	48.1	53.9	55.2	49.2	**55.3**	54.3

On TREC-DL. In our evaluation on TREC-DL2019 and TREC-DL2020, our model demonstrated outstanding performance, consistently outperforming established supervised methods and LLM-distilled baselines. When set against zero-shot LLM baselines, our model either matches or exceeds their performance. Although the results are not directly comparable because we are using a specific checkpoint, we find it remarkable that our model outperforms even the teacher LLM used for the distillation process, i.e. `gpt-3.5-turbo`. We take it as an indication that our distillation strategy is well conceived. The sole model that distinctly outperformed ours was `gpt-4`. This performance difference suggests that leveraging a more advanced LLM for distillation within our methodology might lead to even superior outcomes. Importantly, this is achieved with significantly reduced computational overhead during inference since we distilled LLMs to obtain a much smaller task-specific model. It is worthwhile noticing the difference in size between the models used for comparison with TWOLAR. Remarkably the largest of the TWOLAR models is several orders of magnitude smaller than the largest RankGPT model.

On BEIR Benchmark. In our evaluations using the BEIR benchmark, TWOLAR consistently outperformed most existing baselines when reranking the top-100 documents, as shown in Table 3. This is particularly significant when compared with the approach of models such as InPars. InPars employs a strategy of fine-tuning a monot5-3B on generated, topic-specific data tailored for each of the 18 datasets within the BEIR benchmark. This strategy means that, for each dataset, their model has been exposed to data related to the topic in question. In contrast, TWOLAR has never been exposed to any topic-specific data, making it genuinely zero-shot when facing new topics and tasks. Furthermore, our

Table 4. Results on the BEIR Benchmark by reranking the top 1000 BM25 retrieved documents. The best model is highlighted in boldface, and the second best is underlined for each dataset. All results, apart from TWOLAR-xl, are from [16].

	BM25	monoT5-3B	InPars-v2	RankT5	TWOLAR-xl
nDCG@10					
TREC-COVID	59.5	80.1	**84.6**	82.3	<u>84.3</u>
NFCorpus	32.2	38.3	<u>38.5</u>	**39.9**	37.3
FiQA-2018	23.6	**50.9**	<u>50.9</u>	49.3	45.2
ArguAna	30.0	<u>37.9</u>	36.9	**40.6**	32.7
Tóuche-2020	<u>44.2</u>	30.9	29.1	**48.6**	35.9
Quora	78.9	83.5	<u>84.5</u>	81.9	**87.3**
SCIDOCS	14.9	19.7	**20.8**	19.1	<u>20.3</u>
SciFact	67.9	**77.4**	<u>77.4</u>	76.0	76.8
NQ	30.6	62.5	63.8	**64.7**	<u>64.2</u>
HotpotQA	63.3	76.0	<u>79.1</u>	75.3	**79.5**
DBPedia	31.8	47.2	<u>49.8</u>	45.9	**52.0**
FEVER	65.1	84.8	**87.2**	84.8	<u>86.7</u>
Climate-FEVER	16.5	<u>28.8</u>	**32.3**	27.5	27.8
CQADupStack	30.2	**44.9**	<u>44.8</u>	–	43.8
Robust04	40.8	61.5	<u>63.2</u>	–	**64.2**
Signal-1M	**33.1**	30.2	30.8	<u>31.9</u>	31.5
BioASQ	52.3	56.6	**59.5**	<u>57.9</u>	56.0
TREC-NEWS	39.5	47.7	<u>49.0</u>	–	**53.2**
avg nDCG@10					
BEIR 18	41.9	53.3	**54.5**	–	<u>54.4</u>
BEIR 15	42.9	53.7	<u>54.9</u>	**55.0**	54.5

method does not require fine-tuning for different applications and is thus more economical.

It is worthwhile noticing the performance variations across different datasets within BEIR. In datasets with a specific focus, such as BioASQ, InPars tends to perform better due to its targeted fine-tuning on artificial topic-specific data. However, in datasets where queries are centered around general knowledge, like DBpedia entity, TWOLAR demonstrates a clear advantage over InPars. This highlights the robustness of TWOLAR's topic-agnostic approach and its applicability in a broad range of scenarios.

When reranking the top-1000 documents (Table 4), our model did not perform as well as when reranking the top-100 documents. However, the difference with the best performing model on BEIR 18 is minor (54.4 vs 54.5) and TWOLAR-xl outperforms every competitor in 5 out of 18 tasks. A possible explanation for this result lies in our model's training setup. Since our method optimizes for reranking a subset of 30 documents, it seems plausible that it can easily scale up to 100 documents, less easily to 1000.

Table 5. We perform ablation studies on the eight smallest datasets in BEIR benchmark. The reported scores are nDCG@10. In comparing our scoring strategy against the RankT5 scoring strategy, the statistical tests yielded a p-value of $p = 0.268$. *COV: TREC-COVID, SCI: SciFact, NFC: NFCorpus, TOU: Tóuche-2020, DBP: DBPedia, ROB: Robust04, SIG: Signal-1M, NEW: TREC-NEWS.

	COV	SCI	NFC	TOU	DBP	ROB	SIG	NEW	avg
Score Strategy	effectiveness of score strategy - 19K train samples								
Difference	74.0	67.9	**31.9**	**35.7**	**38.8**	**47.4**	32.5	**43.7**	**46.5**
RankT5	**74.1**	**69.2**	31.5	32.2	36.2	47.0	**34.1**	41.2	45.7
# documents	effectiveness of amount of documents - 19K train samples								
30	**74.0**	67.9	31.9	**35.7**	**38.8**	47.4	**32.5**	**43.7**	**46.5**
20	73.0	**69.8**	**32.5**	31.7	37.9	**47.5**	31.9	40.8	46.3
10	72.3	65.6	29.6	28.4	34.2	43.2	30.4	37.9	42.7
Not used source	effectiveness of first source of supervision - ~14K train samples								
- BM25	72.7	70.3	31.8	31.8	37.7	47.0	31.9	41.5	45.6
- SPLADE	73.9	**70.9**	**33.6**	32.7	**38.6**	**48.8**	32.4	42.5	**46.3**
- DRAGON	**74.0**	67.7	32.9	**33.9**	37.6	47.7	**33.1**	43.2	46.2
- monoT5	73.9	69.4	31.8	33.0	36.3	46.6	32.5	**43.6**	45.9
Type of query	effectiveness of type of query - 9.5K train samples								
Mixed	**75.5**	67.3	30.4	**34.0**	37.2	**46.2**	**31.8**	41.6	**45.5**
Sentence	67.2	**67.9**	**31.4**	32.7	32.2	44.8	31.7	39.7	43.4
docT5query	74.6	59.4	31.2	33.4	**37.8**	44.9	28.1	**44.0**	44.2

5.1 Ablation Study

We conducted an extensive ablation study in order to validate our design choices. Due to computational constraints, these experiments were performed using the smaller `flan-t5-small` checkpoint, with 77M parameters. Furthermore, we evaluated the models on reranking the top-100 documents retrieved by BM25 from a subset of 8 smallest datasets from the BEIR benchmark, including TREC-COVID, SciFact, NFCorpus, Tóuche-2020, DBPedia, Robust04, Signal-1M, and TREC-NEWS.

The results are summarized in Table 5.

Scoring Strategy Effectiveness. Our scoring strategy achieved an average nDCG@10 of 46.5, showing superior performance compared to 45.7 for the RankT5 scoring approach, which indicates the importance of properly exploiting the knowledge from the language modeling head of PLMs. We did not make a direct comparison with the softmax method used in monoT5, due to the inherent differences in the pipeline structure: while our method and the RankT5 method allow for direct finetuning of the model to rank documents, the monoT5 approach operates on a fundamentally different mechanism, making a direct comparative analysis less feasible and potentially misleading in evaluating the distinct methodologies.

Documents per Training Samples. We trained models with varying numbers of documents per training sample. Our results suggest a clear advantage in using more than 10 documents per sample. The trade-off between 20 and 30 is less clear, with nDCG@10 scores of 46.3 and 46.5 respectively, suggesting diminishing returns beyond 20 documents for the top-100 reranking task.

Effectiveness of First Source of Supervision. We conducted four individual experiments by excluding each source of supervision (BM25, SPLADE, DRAGON, monoT5) from the training set and training the model on the residual data. This allowed us to evaluate the individual contribution of each retrieval strategy to the overall performance of the model.

Our results demonstrate that BM25, even being a traditional bag-of-words method, still plays a critical role in the model's performance. This result may be also due to the fact that, following standard practice, during the test the top-100 documents have been retrieved using BM25 itself. Conversely, when SPLADE and DRAGON were excluded during training, the performance drop was not substantial, which suggests that the main contribution comes from blending lexical and semantic models rather than including multiple and possibly equivalent semantic models.

Impact of Query Type. We also trained models exclusively on cropped sentences, docT5query generated queries, and a mixed subset of both types. The model trained only with docT5query generated queries, which are formulated as natural language questions, had a higher average performance than the model trained only on cropped sentences. This suggests that training with grammatically correct questions is more important.

Interestingly, for datasets where the queries were predominantly formed as 'what' or 'how' questions, such as TREC-COVID, the model trained on docT5query queries delivered a strongly superior performance. Conversely, the model trained with cropped sentences performed better in specific datasets where the queries are not expressed as a question. For example, the queries in SciFact are expert-written claims, aiming to find evidence in annotated abstracts. Here, the model trained with cropped sentences achieved an nDCG@10 score of 67.9, significantly outperforming the model trained with docT5query queries, which scored 59.4.

When we trained the model on a mixed subset comprising an equal proportion of both query types, it exhibited the best overall performance. This highlights the benefit of a diverse training regimen incorporating natural questions (docT5query) and sentences cropped directly from documents.

In summary, these results of the ablation study underscore the value of our proposed scoring strategy, the importance of incorporating sufficient documents per training sample, the significant contribution of BM25 as a supervision source, and the advantages of a mixed query approach.

6 Conclusion

The paradigm shift, enabled by LLMs, suggests that traditional methods relying heavily on handcrafted labeled data might no longer be the most effective or efficient approach for certain machine learning tasks. Indeed, as LLMs continue to showcase their prowess, there is a promising realization that they can be harnessed to provide the needed supervision, reducing the need for manual data labeling. However, tasks that demand efficiency, such as information retrieval, often cannot deploy LLMs directly due to their substantial computational overhead. In such scenarios, distillation enables the retention of the LLM's capabilities in a more computationally amenable format.

In this paper, we presented a novel two-step LLM-augmented distillation approach for passage reranking. Our method capitalizes on the strengths of LLMs to enable computationally efficient information retrieval systems, with performance comparable or even superior to that of state-of-the-art baselines and a reduction in size by several orders of magnitude. Our experiments, conducted across various benchmarks, demonstrate robustness and generality of our approach across domains. An ablation offers further insight about the crucial elements of our architectural design.

Looking forward, TWOLAR offers promising avenues for scalability. In the future, we plan to further our experimentation by substituting the 3B model with an 11B version, expanding the number of queries, increasing the sources of supervision, or even refining the quality of the LLM used for distillation, for example by experimenting with more powerful generative language models.

References

1. Baktash, J.A., Dawodi, M.: Gpt-4: a review on advancements and opportunities in natural language processing (2023)
2. Bonifacio, L., Abonizio, H., Fadaee, M., Nogueira, R.: InPars: unsupervised dataset generation for information retrieval. In: Proceedings of the 45th International ACM SIGIR Conference on Research and Development in Information Retrieval, SIGIR 2022, pp. 2387–2392. Association for Computing Machinery, New York, NY, USA (2022). https://doi.org/10.1145/3477495.3531863
3. Brown, T.B., et al.: Language models are few-shot learners. In: Proceedings of the 34th International Conference on Neural Information Processing Systems. NIPS 2020. Curran Associates Inc., Red Hook, NY, USA (2020)
4. Burges, C., et al.: Learning to rank using gradient descent. In: Proceedings of the 22nd International Conference on Machine Learning, ICML 2005, pp. 89–96. Association for Computing Machinery, New York, NY, USA (2005). https://doi.org/10.1145/1102351.1102363
5. Chung, H.W., et al.: Scaling instruction-finetuned language models (2022)
6. Craswell, N., Mitra, B., Yilmaz, E., Campos, D.: Overview of the TREC 2020 deep learning track (2021)
7. Craswell, N., Mitra, B., Yilmaz, E., Campos, D., Voorhees, E.M.: Overview of the TREC 2019 deep learning track (2020)
8. Devlin, J., Chang, M.W., Lee, K., Toutanova, K.: BERT: pre-training of deep bidirectional transformers for language understanding. In: Proceedings of the 2019 Conference of the North American Chapter of the Association for Computational Linguistics: Human Language Technologies, vol. 1 (Long and Short Papers), pp. 4171–4186. Association for Computational Linguistics, Minneapolis, Minnesota, June 2019. https://doi.org/10.18653/v1/N19-1423, https://aclanthology.org/N19-1423
9. Fan, Y., et al.: Pre-training methods in information retrieval. Found. Trends® Inf. Retrieval **16**(3), 178–317 (2022)
10. Formal, T., Lassance, C., Piwowarski, B., Clinchant, S.: SPLADE v2: sparse lexical and expansion model for information retrieval. arXiv preprint arXiv:2109.10086 (2021)
11. Formal, T., Lassance, C., Piwowarski, B., Clinchant, S.: From distillation to hard negative sampling: making sparse neural IR models more effective. In: Proceedings of the 45th International ACM SIGIR Conference on Research and Development in Information Retrieval, pp. 2353–2359 (2022)
12. Formal, T., Piwowarski, B., Clinchant, S.: SPLADE: sparse lexical and expansion model for first stage ranking. In: Proceedings of the 44th International ACM SIGIR Conference on Research and Development in Information Retrieval, SIGIR 2021, pp. 2288–2292. Association for Computing Machinery, New York, NY, USA (2021). https://doi.org/10.1145/3404835.3463098
13. Guo, J., Fan, Y., Ai, Q., Croft, W.B.: A deep relevance matching model for ad-hoc retrieval. In: Proceedings of the 25th ACM International on Conference on Information and Knowledge Management. ACM, October 2016. https://doi.org/10.1145/2983323.2983769
14. Guo, J., et al.: A deep look into neural ranking models for information retrieval. Inf. Process. Manag. **57**(6), 102067 (2020)
15. He, P., Gao, J., Chen, W.: DeBERTaV 3: improving DeBERTa using ELECTRA-style pre-training with gradient-disentangled embedding sharing. arXiv preprint arXiv:2111.09543 (2021)

16. Jeronymo, V., et al.: InPars-v2: large language models as efficient dataset generators for information retrieval (2023). https://doi.org/10.48550/ARXIV.2301.01820, https://arxiv.org/abs/2301.01820
17. Karpukhin, V., et al.: Dense passage retrieval for open-domain question answering. In: Proceedings of the 2020 Conference on Empirical Methods in Natural Language Processing (EMNLP), pp. 6769–6781. Association for Computational Linguistics, Online, November 2020. https://doi.org/10.18653/v1/2020.emnlp-main.550, https://aclanthology.org/2020.emnlp-main.550
18. Kwiatkowski, T., et al.: Natural questions: a benchmark for question answering research. Trans. Assoc. Comput. Linguist. **7**, 452–466 (2019). https://doi.org/10.1162/tacl_a_00276, https://aclanthology.org/Q19-1026
19. Li, H.: Learning to rank for information retrieval and natural language processing, second edition. Synth. Lect. Hum. Lang. Technol. **7**, 1–123 (2015). https://doi.org/10.2200/S00607ED2V01Y201410HLT026
20. Lin, J., Nogueira, R., Yates, A.: Pretrained Transformers for Text Ranking: BERT and Beyond. Springer, New York (2022). https://doi.org/10.1007/978-3-031-02181-7
21. Lin, S.C., et al.: How to train your dragon: diverse augmentation towards generalizable dense retrieval. arXiv preprint arXiv:2302.07452 (2023)
22. Liu, T.Y.: Learning to rank for information retrieval. Found. Trends Inf. Retr. **3**(3), 225–331 (2009). https://doi.org/10.1561/1500000016
23. Loshchilov, I., Hutter, F.: Decoupled weight decay regularization. In: International Conference on Learning Representations (2019). https://openreview.net/forum?id=Bkg6RiCqY7
24. Ma, X., Zhang, X., Pradeep, R., Lin, J.: Zero-shot listwise document reranking with a large language model. arXiv preprint arXiv:2305.02156 (2023)
25. Mitra, B., Craswell, N.: Neural models for information retrieval. arXiv preprint arXiv:1705.01509 (2017)
26. Nguyen, T., et al.: MS MARCO: a human-generated MAchine reading COmprehension dataset (2017). https://openreview.net/forum?id=Hk1iOLcle
27. Nogueira, R., Cho, K.: Passage re-ranking with BERT. arXiv preprint arXiv:1901.04085 (2019)
28. Nogueira, R., Jiang, Z., Pradeep, R., Lin, J.: Document ranking with a pretrained sequence-to-sequence model. In: Findings of the Association for Computational Linguistics: EMNLP 2020, pp. 708–718. Association for Computational Linguistics, Online, November 2020. https://doi.org/10.18653/v1/2020.findings-emnlp.63, https://aclanthology.org/2020.findings-emnlp.63
29. Nogueira, R., Lin, J., Epistemic, A.: From doc2query to docTTTTTquery. Online Preprint **6**, 2 (2019)
30. Nogueira, R., Yang, W., Cho, K., Lin, J.: Multi-stage document ranking with BERT. arXiv preprint arXiv:1910.14424 (2019)
31. Qin, Z., et al.: Large language models are effective text rankers with pairwise ranking prompting. arXiv preprint arXiv:2306.17563 (2023)
32. Radford, A., et al.: Language models are unsupervised multitask learners. OpenAI Blog **1**(8), 9 (2019)
33. Raffel, C., et al.: Exploring the limits of transfer learning with a unified text-to-text transformer. J. Mach. Learn. Res. **21**(1), 5485–5551 (2020)
34. Robertson, S.E., Walker, S., Jones, S., Hancock-Beaulieu, M.M., Gatford, M., et al.: Okapi at TREC-3. NIST Special Publ. SP **109**, 109 (1995)
35. Rosa, G.M., et al.: No parameter left behind: how distillation and model size affect zero-shot retrieval. arXiv preprint arXiv:2206.02873 (2022)

36. Sun, W., Yan, L., Ma, X., Ren, P., Yin, D., Ren, Z.: Is ChatGPT good at search? Investigating large language models as re-ranking agent. arXiv preprint arXiv:2304.09542 (2023)
37. Thakur, N., Reimers, N., Rücklé, A., Srivastava, A., Gurevych, I.: BEIR: a heterogeneous benchmark for zero-shot evaluation of information retrieval models. In: Thirty-fifth Conference on Neural Information Processing Systems Datasets and Benchmarks Track (Round 2) (2021). https://openreview.net/forum?id=wCu6T5xFjeJ
38. VanBuskirk, A.: GPT-3.5 turbo vs GPT-4: what's the difference? March 2023. https://blog.wordbot.io/ai-artificial-intelligence/gpt-3-5-turbo-vs-gpt-4-whats-the-difference/
39. Wang, B., Komatsuzaki, A.: GPT-J-6B: A 6 Billion Parameter Autoregressive Language Model, May 2021. https://github.com/kingoflolz/mesh-transformer-jax
40. Yang, Z., et al.: HotpotQA: a dataset for diverse, explainable multi-hop question answering. arXiv preprint arXiv:1809.09600 (2018)
41. Zhai, C., et al.: Statistical language models for information retrieval a critical review. Found. Trends® Inf. Retr. **2**(3), 137–213 (2008)
42. Zhao, W.X., Liu, J., Ren, R., Wen, J.R.: Dense text retrieval based on pretrained language models: a survey. arXiv preprint arXiv:2211.14876 (2022)
43. Zhu, Y., et al.: Large language models for information retrieval: a survey. arXiv preprint arXiv:2308.07107 (2023)
44. Zhuang, H., et al.: RankT5: fine-tuning T5 for text ranking with ranking losses. In: Proceedings of the 46th International ACM SIGIR Conference on Research and Development in Information Retrieval, pp. 2308–2313 (2023)

Author Index

N. Goharian et al. (Eds.): ECIR 2024, LNCS 14608, pp. 487–488, 2024.
https://doi.org/10.1007/978-3-031-56027-9

Printed in the United States
by Baker & Taylor Publisher Services